Primate Anti-Predator Strategies

DEVELOPMENTS IN PRIMATOLOGY: PROGRESS AND PROSPECTS

Series Editor: Russell H. Tuttle, University of Chicago, Chicago, Illinois

This peer-reviewed book series melds the facts of organic diversity with the continuity of the evolutionary process. The volumes in this series exemplify the diversity of theoretical perspectives and methodological approaches currently employed by primatologists and physical anthropologists. Specific coverage includes: primate behavior in natural habitats and captive settings; primate ecology and conservation; functional morphology and developmental biology of primates; primate systematics; genetic and phenotypic differences among living primates; and paleoprimatology.

ANTHROPOID ORIGINS: NEW VISIONS
Edited by Callum F. Ross and Richard F. Kay

MODERN MORPHOMETRICS IN PHYSICAL ANTHROPLOGY
Edited by Dennis E. Slice

BEHAVIORAL FLEXIBILITY IN PRIMATES: CAUSES AND CONSEQUENCES
By Clara B. Jones

NURSERY REARING OF NONHUMAN PRIMATES IN THE 21ST CENTURY
Edited by Gene P. Sackett, Gerald C. Ruppenthal and Kate Elias

NEW PERSPECTIVES IN THE STUDY OF MESOAMERICAN PRIMATES: DISTRIBUTION, ECOLOGY, BEHAVIOR, AND CONSERVATION
Edited by Paul Garber, Alejandro Estrada, Mary Pavelka and LeAndra Luecke

HUMAN ORIGINS AND ENVIRONMENTAL BACKGROUNDS
Edited by Hidemi Ishida, Martin Pickford, Naomichi Ogihara and Masato Nakatsukasa

PRIMATE BIOGEOGRAPHY
Edited by Shawn M. Lehman and John Fleagle

REPRODUCTION AND FITNESS IN BABOONS: BEHAVIORAL, ECOLOGICAL, AND LIFE HISTORY PERSPECTIVES
Edited By Larissa Swedell and Steven R. Leigh

RINGAILED LEMUR BIOLOGY: *LEMUR CATTA* IN MADAGASCAR
Edited by Alison Jolly, Robert W. Sussman, Naoki Koyama and Hantanirina Rasamimanana

PRIMATE ORIGINS: ADAPTATIONS AND EVOLUTION
Edited by Matthew J. Ravosa and Marian Dagosto

LEMURS: ECOLOGY AND ADAPTATION
Edited by Lisa Gould and Michelle L. Sauther

PRIMATE ANTI-PREDATOR STRATEGIES
Edited by Sharon L. Gursky and K.A.I. Nekaris

Primate Anti-Predator Strategies

Edited by

Sharon L. Gursky
Texas A&M University
College Station, TX, USA

and

K.A.I. Nekaris
Oxford Brookes University
Oxford, UK

 Springer

Sharon L. Gursky
Department of Anthropology
Texas A&M University,
College Station, TX 77843
USA
gursky@tamu.edu

K.A.I. Nekaris
School of Social Sciences and Law
Department of Anthropology
Oxford Brookes University
Oxford OX3 0PB
UK
anekaris@brookes.ac.uk

Library of Congress Control Number: 2006926879

ISBN-10: 0-387-34807-7   e-ISBN-10: 0-387-34810-7
ISBN-13: 978-0387-34807-0   e-ISBN-13: 978-0-387-34810-0

Printed on acid-free paper.

9 8 7 6 5 4 3 2 1

springer.com

S.L. Gursky dedicates this volume to the memory of her grandfather, Herman Gursky.

K.A.I. Nekaris dedicates this volume to S.K. Bearder.

Preface

The impact of predation on the morphology, behavior, and ecology of animals has long been recognized by the primatologist community (Altmann, 1956; Burtt, 1981; Curio, 1976; Hamilton, 1971; Kruuk, 1972). Recent thorough reviews of adaptations of birds and mammals to predation have emphasized the complex role that predation threat has played in modifying proximate behaviors such as habitat choice to avoid predator detection, degree and type of vigilance, and group size and defense, as well as ultimate factors including the evolution of warning systems, coloration, and locomotor patterns (Thompson et al., 1980; Sih, 1987; Lima & Dill, 1990; Curio, 1993; Caro, 2005).

From the late 1960s, primatologists have adopted similar techniques to analyze the impacts of predation on the social systems of monkeys and apes (Crook & Gartlan, 1966; Eisenberg et al., 1972; Goss-Custard et al., 1972; Clutton-Brock, 1974; van Schaik & van Hooff, 1983). The fact that actual predation was witnessed but rarely fueled a debate regarding whether predation or food acquisition played a more important role in primate evolution (Wrangham, 1980; van Schaik, 1983; Anderson, 1986; Janson, 1987; Wrangham, 1987; Rodman, 1988; Janson, 1998). More recent studies are more subtle in their design, and have worked from a hypothetical framework that an animal's being eaten is more costly than its missing a meal; they have thus attempted to quantify how animals perceive and act upon predation risk rather than the act of predation itself (Cords, 1990; Boesch, 1994; Isbell, 1994; Cowlishaw, 1997; Hill & Dunbar, 1998; Cowlishaw, 1998). Other investigations have been founded on how primate cognitive abilities and complex social learning aid them in avoiding predators (Seyfarth et al., 1980; Bshary and Noë , 1997; Zuberbühler et al. , 1999; Zuberbühler, 2000; Shultz et al., 2004). These, and a multitude of other studies, are beginning to elucidate our understanding of the impact of predation on primate evolution. Or are they?

We have conducted research on nocturnal primates for more than ten years. Immersed as we have been in the literature of nocturnal primatology we recognize a spectrum of diversity amongst the nocturnal primates in their social organization, cognitive behavior, and ecology (Charles-Dominique, 1978; Bearder, 1999; Müller and Thalmann, 2000). Our studies on tarsiers and lorises showed that these species were highly social and that resource distribution was not sufficient to explain why they defied the supposed "stricture" of being solitary (Gursky, 2005a; Nekaris, 2006). Furthermore, our animals defied another supposed "rule" — namely, that all nocturnal primates should avoid predators by crypsis (Charles-Dominique, 1977). Even recent reviews of primate social organization and predation theory included one-sentence write-offs, excluding nocturnal primates from discussions of primate social evolution on the basis that crypsis is their only mechanism of predator avoidance (Kappeler, 1997; Stanford, 2002).

An analysis of the mammalian literature shows this type of generalization to be crude at best. Small mammals are known to have extraordinarily high rates of predation, and a plethora of studies of rodents, insectivores, and lagomorphs, among others, have shown that predation is a viable and powerful ecological force (Lima & Dill, 1990; Caro, 2005). Furthermore, although researchers have long considered it critical to include prosimian studies in a general theoretical framework concerning the evolution of the order Primates (Charles-Dominique & Martin, 1970; Cartmill, 1972; Oxnard et al., 1990), a pervading view contends that prosimians are too far removed from humans for the former's behavior to shed any light on the patterns of behavior seen in anthropoids (Kappeler & van Schaik, 2002; Stanford, 2002).

Such notions are perhaps fueled by a paucity of predation research on prosimians in general. This lack of literature may relate to the fact that the study of nocturnal primates is still in the descriptive rather than the theoretical phase; with so many species still being described, data collection on endangered species may begin with recording basic parameters of the diet and home range of these animals (Bearder, 1999). Furthermore, any study of nocturnal and cathemeral primates that goes beyond collecting radio-tracking fixes has proved to be a challenge; much more difficult has been the actual observation of predation events (Sterling et al., 2000). However, an excellent review by Goodman et al. demonstrates the dramatic effect predation can have on lemurs, and it remains the most highly quoted resource on lemur predation, despite that it was published in 1993. Studies of referential signaling aid in dispelling the view that prosimians are primitive and not worthy of comparison with monkeys and apes (Oda, 1998; Fichtel & Kappeler, 2002). A handful of studies further reveal that prosimians are not always cryptic and may engage in social displays toward predators (Sauther, 1989; Schülke, 2001; Bearder et al., 2002; Gursky, 2005b).

In addition to the above cited works, our colleagues regaled us with tales of lemurs, bushbabies, and lorises that demonstrate how these animals employ numerous tactics against predators beyond crypsis. Their observations showed that strategies of nocturnal and cathemeral primates they studied were not unlike anti-predator strategies exhibited by the better-studied diurnal primates — conclusions that contradict the popular view (Stanford, 2002). The anecdotal nature of many of these observations, however, suggested that an outlet was needed to report them; thus the idea for this volume was formed.

The original goal of this volume was to synthesize current research on the anti-predator behavior of nocturnal and cathemeral primates. We quickly realized, however, that although we could, in this volume, emphasize these less-studied species, we would fall into the same trap as previous researchers if we did not consider primates as an order. Thus, the seventeen chapters in this volume consider anti-predator strategies exhibited across primates including: crypsis, alarm calling (referential or otherwise), mobbing behavior, production of toxins, group cohesion, behavioral modification due to environmental factors (habitat choice, sleeping site choice, visibility, moonlight), and vigilance, among others. This volume is organized into three sections: predation theory, anti-predator strategies of nocturnal and cathemeral primates, and anti-predator strategies of diurnal primates. Although we have divided it in this manner, we hope the reader can see the common theoretical and behavioral threads that unite these primate studies as emphasized here.

The two chapters of Section One bring together an immense volume of literature and observations on two important areas of primate predation studies. Zuberbühler fuels a discussion on the effect of predation on primate cognitive evolution with examples from long-term research by himself and colleagues at the Taï forest in West Africa. The studies at the site benefited from complementary observations of the predators themselves — a component often lacking in primate fieldwork. Zuberbühler's comprehensive experience of this ecological system leads him to the controversial conclusion that at Taï, predation does not drive traditionally recognized traits such as group size, body size, life history etc. Rather, it has selected for the evolution of sophisticated cognitive processes, including semantic predator-specific calls, amongst the sympatric primates. This straying from typical predation theory is also emphasized by Hart in her biogeographical analysis of primate predation. Hart's comprehensive dataset of inferred and observed instances of predation on primates allowed her to search for regional patterns to predation. Although primates in some regions (the Neotropics and Madagascar) seem to be more heavily preyed upon than in others (Africa and Asia), Hart found the overall scarcity of data a limiting factor in interpreting them. She did uncover, however, that primates of all body sizes, activity cycles, and ecological niches as determined by strata were preyed upon. This study reminds us that the range of primates from small-bodied nocturnal primates to large-bodied apes cannot be removed from our consideration of predation theory.

Section Two on cathemeral and nocturnal primates contains two major areas: reviews from long-term studies of multiple species, and specific field studies of one or more species. Dollar et al., in line with Zuberbühler, contribute the first research project directly aimed at analyzing the foraging strategies of the largest Malagasy predator — the fossa. Their research shows, contradictorily to the findings of Hart, that although the fossa is capable of taking many species, the taxon most likely to fall victim to it is *Lepilemur*, possibly due to that primate's predictable pattern of sleeping in tree holes, slow locomotion, and solitary lifestyle. Indeed, in her contribution Nash sheds further light on factors that might influence the desirability of *Lepilemur* as a prey item. *Lepilemur* at Nash's study site did not significantly moderate its behavior in relation to different quantities of moonlight, as did many nocturnal mammals. Although they did reduce their time in the highest part of the canopy during moonlit times, perhaps as a device to avoid aerial predators, in general, Nash proposes, their nutrient-poor diet does not allow for much behavioral flexibility.

Karpanty & Wright, Scheumann et al., and Colquhoun present informative reviews on lemur predation that link to other theoretical perspectives in this volume. Karpanty & Wright synthesize an enormous dataset on lemur predation collected over nineteen years in Ranomafana National Park. *Ad libitum* observations, combined with playback experiments, analyses of predator scats, and systematic fieldwork, aided in formulating a picture of the impact of predation on the rainforest primate community. Although some lemurs relied on the traditional pattern of crypsis, others were highly vocal. Furthermore, they found that both activity pattern and body size *did* have an effect on predation, again in contradiction to Hart. Scheumann et al. review the scanty body of studies on lemur predation and examine their own long-term studies in northwestern Madagascar. They find that, as in the Taï forest, predation clearly has had an impact on cognitive evolution in lemurs, with numerous lemurs using predator-specific referential signals that appear to be socially learned. Body size not only seems to relate to predation risk, but also seems to influence what types of predator strategies lemurs use to combat potential predators. Colquhoun reviews the anti-predator strategies of cathemeral primates, which, due to their potential for activity in the day or night, may need a defense system against a greater array of predators. He suggests that in these small, group-living primates crypsis may play a role; however, cathemerality is not fully understood amongst most of the larger-bodied taxa. For the better-studied *Eulemur*, all of which are sexually dichromatic, he puts forth the interesting hypothesis that this coloration may be a form of polymorphic strategy to counter apostatic predation. All three of these studies are excellent illustrations for revaluating the "crypsis only" view.

Ultimate strategies for avoiding predation are explored by Crompton & Sellers Hagey et al., and Nekaris et al. Crompton & Sellers consider the function of the unique locomotor pattern exhibited by many nocturnal primates: vertical clinging and leaping. By showing that galagos, tarsiers, and some lemurs are capable of leaping far beyond their average distance, they suggest that, from an energetic perspective, the most likely selective factor influencing this ability would be confounding, avoiding, or escaping from a predator. Hagey et al. present long-awaited data on the function and composition of the brachial gland of *Nycticebus*. Their study confirms that small levels of toxic compounds are indeed present in the gland, possibly having evolved as a complex chemical signal to conspecifics and as a toxin for immobilizing prey. The ecological ramifications of chemical communication are presented by Nekaris et al. Novel field data on West African pottos and Sri Lankan slender lorises are compared to revaluate the role that crypsis supposedly plays in the anti-predator strategies of these primates. Although the authors report that lorises engaged in noisy displays and were faster and more vocal than the relatively cryptic pottos, they also describe the prevalence of olfactory communication in all taxa. Comparing their work with the mammalian literature in general, they stress the olfactory capabilities of predators and warn against dismissing olfactory communication as cryptic.

Bearder & Gursky both present data from long-term field studies on strategies that nocturnal primates employ to cope with danger. Both papers reinforce elements presented by Scheumann et al. and by Zuberbühler in that both find evidence for referential signaling. In his paper on calling patterns in two species of galagos, Bearder not only describes, for the first time, the large array of alarm calls emitted by these species, but also contextualizes them. Not only are the calls acoustically distinct, based on the level of fear or arousal of the emitter, they also vary in speed and intensity. Although calls may not have predator-specific contexts, they were situation-specific and were uttered according to the level of danger. Similarly, Gursky discovered that tarsiers incorporate a wide range of tactics to cope with potential predators. Systematic presentation of avian and terrestrial predator models allowed for detailed observations on how spectral tarsiers react in the presence of a potential predator. Not surprisingly, although some predators elicited cryptic responses from them, tarsiers also vocalized, banded together and mobbed predators, again contradicting the view of the cryptic prosimians.

The studies of diurnal primates have been grouped together in Section Three, but certain themes continue throughout these studies. Long-term studies of *Lemur catta* allow for a review by Gould & Sauther supplemented with novel data. These data emphasize that ring-tailed lemurs, like other well-studied diurnal primates, form larger groups and increase vigilance in areas of vulnerability and during vulnerable times of the year (during and after weaning). Referential signaling is also evident, reinforcing the postulation of Scheumann et al. that this cognitive system is characteristic of lemurs. In line with Dollar et al., Gil da Costa approached the predator and prey relationship between harpy eagles and howler monkeys. In a unique situation whereby harpy eagles were released into an area from which they had been extirpated, Gil da Costa was able to analyze both the eagles' and the monkeys' tactics. Whereas monkeys immediately adopted strategies such as group repositioning and vigilance, the eagles too adopted their own mechanisms, learning situations where stealth or attack would improve their capture rate. This study shows how quickly primates and prey can adapt in only one generation.

Long-term field studies on estimating predation risk are presented by Enstam and by Hill & Weingrill. Enstam reviews the impact on predation risk on cercopithecines and illustrates the importance of the study of multiple aspects of habitat structure rather than ecosystem type alone in order to estimate risk. She demonstrates through her own studies of vervet and patas monkeys that even with a highly flexible suite of anti-predator strategies, these primates still can suffer high predation pressure. We have stressed throughout this review that studies of predation on nocturnal primates are in their infancy. In their chapter Hill & Weingrill reiterate this point for diurnal primates and provide elegant guidelines for the measurement of predation risk in terrestrial environments. By focusing on their work with chacma baboons, they show how baboons respond behaviorally to habitat-specific levels of predation risk, even in environments where predators are scarce. These results suggest a deeper ultimate impact of predation on the primate behavior.

The volume is concluded with a review by Treves & Palmqvist. When we were organizing this volume, our colleagues asked us if we would include humans. The scope of understanding the development of *Homo* as a predator would require a volume in its own right. However, Treves & Palmqvist attempt to reconstruct the interactions of hominins prior to *Homo ergaster*, particularly with respect to them as prey to mammalian carnivores. By reconstructing the hunting habits and the diets of paleocarnivores, Treves & Palmqvist suggest that strong group cohesion, vigilance, and last but not least, extreme crypsis, would have characterized early hominin anti-predator strategies.

Many of the contributors to this volume stress how the study of predation, in whichever form it takes, is still at an early stage. A number of the authors outline areas of further study or present compelling hypotheses worthy of additional testing. Although this book focuses mainly on prosimians, we hope that the unifying themes running through all the essays will aid the reader in considering predation theory in a broader light. These studies show that species, regardless of their activity rhythm, body size, or brain size, do not engage in uniform or predictable strategies. At the very least we hope that this volume will dispel a myth as well as encourage a new spectrum of research on primate anti-predator strategies.

References

Altmann, S.A. (1956). Avian mobbing behavior and predator recognition. *Condor*, 58: 241–253.

Anderson, C.M. (1986). Predation and primate evolution. *Primates*, 27: 15–39.

Bearder, S.K. (1999). Physical and social diversity among nocturnal primates: A new view based on long term research. *Primates*, 40(1): 267–282.

Bearder, S.K., Nekaris, K.A.I., and Buzzell, C.A. (2002). Dangers in the night: Are some nocturnal primates afraid of the dark? In E. Miller (Ed.), *Eat or be eaten: Predator sensitive foraging in primates* (pp. 21–40). Cambridge: Cambridge Univ. Press.

Boesch, C. (1994). Chimpanzee-red colobus monkeys: A predator-prey system. *Animal Behaviour*, 47: 1135–1148.

Bshary, R., and Noë, R. (1997). Anti-predation behaviour of red colobus monkeys in the presence of chimpanzees. *Behavioral Ecology and Sociobiology*, 41: 321–333.

Burtt, E.H. (1981). The adaptiveness of animal colors. *BioScience*, 31: 721–729.

Caro, T. (2005). *Antipredator defenses in birds and mammals*. Chicago: Univ. of Chicago Press.

Cartmill, M. (1972). Arboreal adaptations and the origin of the order Primates. In R. Tuttle (Ed.), *The functional and evolutionary biology of the primates* (pp.97–122). Chicago: Aldine Atherton.

Charles-Dominique, P. (1977). *Ecology and behaviour of nocturnal primates*. London: Duckworth Press.

Charles-Dominique, P. (1978). Solitary and gregarious prosimians: Evolution of social structures in primates. In D.J. Chivers and K.A. Joysey (Eds.), *Recent advances in primatology, Vol.3* (pp. 139–149). London: Academic Press, London.

Charles-Dominique, P. and Martin, R.D. (1970). Evolution of lorises and lemurs. *Nature*, 227: 257–260.

Clutton-Brock, T.H. (1974). Primate social organization and ecology. *Nature*, 250: 539–542.

Cords, M. (1990). Vigilance and mixed-species association of some East African forest monkeys. *Behavioral ecology and sociobiology*, 26: 297–300.

Cowlishaw, G.C. (1997). Trade-offs between foraging and predation risk determine habitat use in a desert baboon population. *Animal Behaviour*, 53: 667–686.

Cowlishaw, G.C. (1998). The role of vigilance in the survival and reproductive strategies of desert baboons. *Behaviour*, 135: 431–452.

Crook, J.H., and Gartlan, J.S. (1966). Evolution of primate societies. *Nature*, 210: 1200–1203.

Curio, E. (1993). Proximate and developmental aspects of anti-predator behavior. *Advanced study of behavior*, 22: 135–238.

Curio, E. (1976). *The ethology of predation*. New York: Springer-Verlag.

Eisenberg, J.E., Mukenhirn, N.A., and Rudran, R. (1972). The relation between ecology and social structure in primates. *Science*, 176: 863–874.

Fichtel, C., and Kappeler, P.M. (2002). Anti-predator behavior of group-living Malagasy primates: Mixed evidence of a referential alarm call system. *Behavioural Ecology and Sociobiology*, 51: 262–275.

Goodman, S.M, O'Connor, S., and Langrand, O. (1993). A review of predation on lemurs: Implications for the evolution of social behavior in small, nocturnal primates. In P.M. Kappeler and J.U. Ganzhorn (Eds.), *Lemur social systems and their ecological basis* (pp. 51–66). New York: Plenum Press.

Goss-Custard, J.D., Dunbar, R.I.M., and Aldrich-Blake, F.P.G. (1972). Survival, mating, and rearing strategies in the evolution of primate social structures. *Folia Primatologica*, 17: 1–19.

Gursky, S. (2005a). Associations between adult spectral tarsiers. *American Journal of Physical Anthropology*, 128(1): 74–83.

Gursky, S. (2005b). Predator mobbing in spectral tarsiers. *International Journal of Primatology*, 26: 207–221.

Hamilton, W.D. (1971). Geometry for the selfish herd. *Journal of Theoretical Biology*, 31: 295–311.

Hill, R.A. and Dunbar, R.I.M. (1998). An evaluation of the roles of predation risk and predation rate as selective pressures on primate grouping behaviour. *Behaviour*, 135: 411–430.

Isbell, L.A. (1994). Predation on primates: Ecological patterns and evolutionary consequences. *Evolutionary Anthropology*, 3: 61–71.

Janson, C.H. (1987). Food competition in brown capuchin monkeys (*Cebus apella*): Quantitative effects of group size and tree productivity. *Behaviour*, 105: 53–75.

Janson, C.H. (1998). Testing the predation hypothesis for vertebrate sociality: Prospects and pitfalls. *Behaviour*, 135: 289–410.

Lima, S.L., and Dill, L.M. (1990). Behavioral decisions made under the risk of predation: A review and prospectus. *Canadian Journal of Zoology*, 68: 619–640.

Kappeler, P.M. (1997). Determinants of primate social organisation: Comparative evidence and new insights from Malagasy lemurs. *Biological Reviews of the Cambridge Philosophical Society*, 72: 111–151.

Kappeler, P.M., and van Schaik, C.P. (2002). Evolution of primate social systems. *International Journal of Primatology*, 23(4): 707–740.

Kruuk, H. (1972). *The spotted hyena: A study of predation and social behavior*. Chicago: Univ. of Chicago Press.

Müller, A.E., and Thalmann, U. (2000). Origin and evolution of primate social organisa-tion: A reconstruction. *Biological Reviews of the Cambridge Philosophical Society*, 75: 405–435.

Nekaris, K.A.I. (2006). The social lives of Mysore slender lorises. *American Journal of Primatology*, 68: 1–12.

Oda, R. (1998). The response of Verreaux's sifakas to anti-predator alarm calls given by sympatric ring-tailed lemurs. *Folia Primatologica*, 69: 357–360.

Oxnard, C.E., Crompton, R.H., and Lieberman, S.S. (1990). *Animal lifestyles and anatomies*. Seattle: Univ. of Washington Press.

Rodman, P.S. (1988). Resources and group sizes of primates. In C.N. Slobodchikoss (Ed.), *The ecology of social behavior* (pp. 83–108). New York: Academic Press.

Sauther, M.L. (1989). Anti-predator behavior in troops of free-ranging *Lemur catta* at Beza Mahafaly Special Reserve, Madagascar. *International Journal of Primatology*, 10: 595–606.

van Schaik, C.P. (1983). Why are diurnal primates living in groups? *Behaviour*, 87: 120–144.

van Schaik, C.P., and van Hooff, J.A.R.A.M. (1983). On the ultimate causes of primates social systems. *Behaviour*, 85: 91–117.

Schülke, O. (2001). Social anti-predator behaviour in a nocturnal lemur. *Folia Primatolog-ica*, 72: 332–334.

Sih, A. (1987). Predator and prey lifestyles: An evolutionary and ecological overview. In W.C. Kerfoot and A. Sih (Eds.), *Predation: Direct and indirect impacts on aquatic com-munities* (pp. 203–224). Hanover, NH: University Press of New England.

Seyfarth, R.M., Cheney, D.L., and Marler, P. (1980). Monkey responses to three different alarm calls: Evidence of predator classification and semantic communication. *Science*, 210: 801–803.

Shultz, S., Noë, R., McGraw, W.S., and Dunbar, R.I.M. (2004). A community-level eval-uation of the impact of prey behavioural and ecological characteristics on predator diet composition. *Proceedings of the Royal Society London. Series B*, 271: 725–732.

Stanford, C. (2002). Avoiding predators: Expectations and evidence in primate anti-predator behavior. *International Journal of Primatology*, 23(4): 741–757.

Sterling, E.J., Nguyen N., and Fashing, P. (2000). Spatial patterning in nocturnal prosimi-ans: A review of methods and relevance to studies of sociality. *American Journal of Primatology*, 51: 3–19.

Thompson, S.D., MacMillen, R.E., Burke, E.M., and Taylore, C.R. (1980). The energetic cost of bipedal hopping in small mammals. *Nature*, 287: 223–224.

Wrangham, R.W. (1980). An ecological model of female-bonded primate groups. *Behav-iour*, 75: 262–300.

Zuberbühler, K. (2000). Referential labeling in Diana monkeys. *Animal Behaviour*, 59: 917–927.

Zuberbühler, K., Jenny, D., and Bshary, R. (1999). The predator deterrence function of primate alarm calls. *Ethology*, 105: 477–490.

Acknowledgments

The contributions and comments of many people were critical to the development of this edited volume and we would like to acknowledge their pivotal role. We would like to thank the many anonymous reviewers for the time and effort they put into critiquing the papers presented in this volume; Andrea Macalusco, the editor at Springer-Verlag; and Dr. Russell Tuttle, for allowing this volume to be a part of the Developments in Primatology series.

S.L. Gursky thanks Texas A&M University Anthropology Department Head David Carlson for providing her with time off from teaching responsibilities so she could organize this volume. K.A.I. Nekaris was supported in part by a Promising Researcher Fellowship from Oxford Brookes University during the production of this volume.

Contents

Preface   vii

Acknowledgments   xv

List of Contributors   xxi

Part 1 Predation Theory

---

1 Predation and Primate Cognitive Evolution
Klaus Zuberbühler   3

2 Predation on Primates: A Biogeographical Analysis
D. Hart   27

Part 2 Anti-Predator Strategies of Nocturnal Primates

**3 Primates and Other Prey in the Seasonally Variable
Diet of *Cryptoprocta* ferox in the Dry Deciduous Forest
of Western Madagascar**
Luke Dollar, Jörg U. Ganzhorn, and Steven M. Goodman   63

**4 Predation on Lemurs in the Rainforest of Madagascar by Multiple
Predator Species: Observations and Experiments**
Sarah M. Karpanty and Patricia C. Wright   77

5 Predation, Communication, and Cognition in Lemurs
*Marina Scheumann, Andriatahiana Rabesandratana,
and Elke Zimmermann*   100

6 A Consideration of Leaping Locomotion as a Means of Predator Avoidance in Prosimian Primates
Robin Huw Crompton and William Irvin Sellers   127

7 Anti-Predator Strategies of Cathemeral Primates: Dealing with Predators of the Day and the Night
Ian C. Colquhoun   146

8 Moonlight and Behavior in Nocturnal and Cathemeral Primates, Especially *Lepilemur leucopus*: Illuminating Possible Anti-Predator Efforts
Leanne T. Nash   173

9 A Comparison of Calling Patterns in Two Nocturnal Primates, *Otolemur crassicaudatus* and *Galago moholi* as a Guide to Predation Risk
Simon K. Bearder   206

10 Predator Defense by Slender Lorises and Pottos
K. Anne-Isola Nekaris, Elizabeth R. Pimley, and Kelly M. Ablard   222

11 The Response of Spectral Tarsiers Toward Avian and Terrestrial Predators
Sharon L. Gursky   241

12 Talking Defensively, a Dual Use for the Brachial Gland Exudate of Slow and Pygmy Lorises
Lee R. Hagey, Bryan G. Fry, and Helena Fitch-Snyder   253

Part 3 Anti-Predator Strategies of Non-Nocturnal Primates

13 Anti-Predator Strategies in a Diurnal Prosimian, the Ring-Tailed Lemur (*Lemur catta*), at the Beza Mahafaly Special Reserve, Madagascar
Lisa Gould and Michelle L. Sauther   275

14 Howler Monkeys and Harpy Eagles: A Communication Arms Race
Ricardo Gil-da-Costa   289

15 Effects of Habitat Structure on Perceived Risk of Predation and Anti-Predator Behavior of Vervet (*Cercopithecus aethiops*) and Patas (*Erythrocebus patas*) Monkeys
Karin L. Enstam   308

16 Predation Risk and Habitat Use in Chacma Baboons (*Papio hamadryas ursinus*)
Russell A. Hill and Tony Weingrill   339

17 Reconstructing Hominin Interactions with Mammalian Carnivores (6.0–1.8 Ma)

Adrian Treves and Paul Palmqvist 355

Index 383

List of Contributors

Kelly Ablard
Oxford Brookes University
Nocturnal Primate Research Group
School of Social Science and Law
Department of Anthropology
Oxford OX3 0BP
England
UK

Simon Bearder
Oxford Brookes University
Nocturnal Primate Research Group
School of Social Science and Law
Department of Anthropology
Oxford OX3 0BP
England
UK

Ian C. Colquhoun
Department of Anthropology
University of Western Ontario
London, Ontario
CANADA N6A 5C2

Robin Huw Crompton
University of Liverpool
Sherrington Buildings, Ashton Street
L69 3GE
England
UK

Luke Dollar
Duke University
Nicholas School of the Environment
and Earth Sciences
Department of Ecology
P.O. Box 90329
Durham, NC 27708
USA

Karin L. Enstam
Department of Anthropology
Sonoma State University
1801 East Cotati Avenue
Rohnert Park, CA 94928
USA

Bryan G. Fry
School of Medicine
University of Melbourne
Parkville, Victoria
3010 Australia

Jörg Ganzhorn
Abt. Tierökologie und Naturschutz
Biozentrum Grindel
Martin-Luther-King Platz 3
20146 Hamburg
Germany

Ricardo Gil da Costa
Laboratory of Brain and Cognition
National Institute of Health
10 Center Dr MSC 1366,
Bldg 10, Room 4C103
Bethesda, MD 20892 USA

Lisa Gould
Dept. of Anthropology
University of Victoria
Victoria, BC V8W-3P5
Canada

Sharon Gursky
Texas A&M University
Department of Anthropology,
MS 4352 College Station,
TX 77843
USA

Lee R. Hagey
Center for Reproduction of Endangered
Species (CRES)
Zoological Society of San Diego
P.O. Box 551 San Diego, CA 92112
USA

Donna Hart
Pierre Laclede Honors College &
Department of Anthropology
University of Missouri–St. Louis
One University Boulevard
St. ouis, MO 63121
USA

Russell Hill
Evolutionary Anthropology
Research Group
Department of Anthropology
Durham University
43 Old Elvet Durham DH1
3HN England UK

Sarah Karpanty
Department of Fisheries
and Wildlife Science
Virginia Polytechnic Institute
and State University Blacksburg,
VA 24061-0321
USA

Leanne Nash
Professor of Anthropology
School of Human Evolution and
Social Change (SHESC) Arizona State
University Box 872402 Tempe, AZ
85287-2402 USA

K. Anne-Isola Nekaris
Oxford Brookes University Nocturnal
Primate Research Group School of
Social Science and Law Department
of Anthropology Oxford OX3 0BP
England UK

Paul Palmqvist
Departamento de Geología y Ecología
(Área de Paleontología)
Facultad de Ciencias Universidad de
Málaga 29071 Málaga Spain

Elizabeth Pimley
University of Cambridge
Nocturnal Primate Research Group
Madingley Cambridge CB3 8AA
England UK

**Andriatahiana Zatovonirina
Rabesandratana**
Département de Biologie Animale,
Université d'Antananarivo,
BP 906, Antananarivo 101,
Madagascar

Michelle Sauther
University of Colorado at Boulder
Department of Anthropology
233 UCB Boulder, CO 80309-0233
USA

Marina Scheumann
Institut fuer Zoologie Tierärztliche
Hochschule Hannover
Buenteweg 17 D- 30559 Hannover
Germany

William Irvin Sellers
Department of Human Sciences
Loughborough University
Loughborough LE11 3TU
England UK

Helena Fitch-Snyder
Zoological Society of San Diego
P.O. Box 120551
San Diego, California, 92112-0551
USA

Adrian Treves
Conservation International Center
for Applied Biodiversity Science
Madison, WI 53705
USA

Tony Weingrill
Anthropological Institute and Museum
University of Zurich
CH-8057 Zurich
Switzerland

Patricia C. Wright
SUNY-Stony Brook
Institute for the Conservation
Tropical Environments
Stony Brook, NY 11794 USA

Elke Zimmermann
Institut für Zoologie Tieraerztliche
Hochschule Hannover
Bünteweg 17 D-30559 Hannover
Germany

Klaus Züberbühler
University of St. Andrews
School of Psychology
St. Andrews Fife KY16
9JP Scotland UK

Part 1
Predation Theory

1
Predation and Primate Cognitive Evolution

Klaus Zuberbühler

Introduction

Predation is a major cause of mortality in non-human primates (Cheney et al., 2004), but its impact as a selective force on primate evolution is not well understood. Predation has long been thought to affect traits such as body size, group size, group composition, and ecological niche, as well as the traits of vigilance and vocal and reproductive behaviour (Anderson, 1986). The general assumption is that if a trait has evolved as an adaptation to predation, then there should be a negative relationship between the expression of the trait and the individual's vulnerability to predation. First, if large body size is an adaptation to predation (Isbell, 1994), then larger primates should be underrepresented in a predator's prey spectrum compared to smaller ones. Second, if individuals living in large groups are less susceptible to a certain kind of predation, leopard predation, for example, due to enhanced levels of predator vigilance, then individuals of larger groups should be underrepresented in the prey spectrum (van Schaik, 1983; Cords, 1990). Third, if multi-male groups are an adaptation to predation, for example due to the possibility of cooperative defence (Stanford, 1998), then species living in multi-male groups should be underrepresented in a predator's prey spectrum compared to single-male groups. Fourth, if females shorten their inter-birth intervals to increase their lifetime reproductive success to compensate for higher levels of predation (Hill & Dunbar, 1998), then individuals with short inter-birth intervals should be over-represented in the prey spectrum. Finally, if living in higher forest strata is an adaptive response to predation (Enstam & Isbell, 2004), then individuals that are generally more exposed to the ground should be over-represented in the prey spectrum of ground predators.

Although intuitively convincing, strong empirical data for these ideas are often lacking. Another problem is that in most primate habitats the predator fauna is severely understudied, often rendering statements about possible selective pressures a matter of speculation. Moreover, there are theoretical reasons to remain cautious about some of the proposed relationships. Predation is not a homogeneous evolutionary force, and predators differ considerably in their hunting behaviour and the selective pressure they exert on a primate population (Struhsaker, 1969; Treves, 1999). It is also the case that the current selection pressure exerted by a predator may not be representative for those of the evolutionary past. Finally, individual predators may be forced into particular ecological niches due to competition with other non-predatory species, and many of the traits mentioned above could be the direct product to these habitat-related factors.

The predator-prey system in the Taï Forest, Ivory Coast, is an ideal system for addressing some of these questions. A body of recent work conducted on the hunting behaviour of the four main predators on the Taï monkeys has enabled direct assessments of their impact as a force of natural selection. This chapter reviews some of the major findings, and it makes two basic claims. First, a number of behavioural and morphological traits that are classically interpreted as adaptations to predation fail to explain the differences in vulnerability in the Taï monkeys. At the same time, experimental studies have shown that the predation context reliably triggers sophisticated and flexible vocal behaviour in monkeys in a previously undescribed way. This has lead to the hypothesis that the main legacy of predation was not upon the morphology and social behaviour of these monkeys, but upon the evolution of communicative and cognitive abilities necessary to avoid predation. The most important impact of predation on primate evolution, therefore, may have been a cognitive one.

Methods

Habitat

The Taï forest is one of the largest blocks of intact tropical forest in West Africa, originally stretching from Ghana to Sierra Leone. The Taï National Park consists of a protected area of 330,000 ha covered with dense evergreen ombrophilous forest vegetation, habitat to 47 species of large mammals and a largely intact predator fauna. UNESCO declared the Taï Forest a World Heritage Site in 1982 (http://whc.unesco.org). The land is state property located in the southwest Ivory Coast between the Cavally and Sassandra rivers. Rainfall ranges from 1,700 mm to 2,200 mm, reaching peaks in June and September/October, which are followed by a marked dry season from December to February. Humidity is constantly 85% or higher, and temperatures range from 24 to 27°C.

The following eight monkey species can regularly be observed in the park: Diana monkeys (*Cercopithecus diana*), Campbell's monkeys (*C. campbelli*), lesser spot-nosed monkeys (*C. petaurista*), putty-nosed monkeys (*C. nictitans*, northern parts of the forest only), red colobus monkeys (*Procolobus badius*), black-and-white (or King) colobus monkeys (*Colobus polykomos*), olive colobus monkeys (*Procolobus verus*), and sooty mangabeys (*Cercocebus atys*). Table 1.1 provides an overview of some of the most important species differences.

TABLE 1.1. Population density, group size, body weight, strata use, number of males per group, birth rate, and usage of the lower forest strata for the Taï primates (Data from Zuberbühler & Jenny, 2002).

Species	Density (ind/km²)	Body Size (kg)	Group Size (ø N)	Males (ø N)	Annual Reproduction Rate	Habitat (% on ground)
Cercopithecus diana	48.2	3.9	20.2	1	0.62	6.1
C. campbelli	24.4	2.7	10.8	1	0.63	36.8
C. petaurista	29.3	2.9	17.5	1	0.52	9.9
C. nictitans	2.1	4.2	10.5	1	0.50	0.7
Procolobus badius	123.8	8.2	52.9	10.1	0.42	0.4
Colobus polykomos	35.5	8.3	15.4	1.42	0.59	1.3
Procolobus verus	17.3	4.2	6.7	1.43	0.61	13.2
Cercocebus atys	11.9	6.2	69.7	9.0	0.40	88.9
Pan troglodytes	2.6	47.5	61.1	6.7	0.23	85.0

Polyspecific Associations

Primate mixed-species associations are common in forest habitats throughout Africa; only very few species, such as the DeBrazza monkeys (*C. neglectus*), avoid them (Gautier-Hion & Gautier, 1974; Waser, 1982). In the Taï forest, mixed-species associations tend to be individualized, that is, the same groups form these associations over many consecutive years, and groups often use similar home ranges. There is good evidence that mixed-species behaviour has evolved because of its merit as an anti-predator strategy (Wachter et al., 1997; Noë & Bshary, 1997). For example, in the presence of Diana monkeys, red colobus use lower strata more often, are less vigilant, and forage in canopy parts that are more exposed to the forest floor than when Diana monkeys are absent, suggesting that associations provide protection against ground predators (Bshary & Noë, 1997a, b). Mixed-species associations appear to be beneficial in the face of raptor predation as well. Red colobus and Diana monkeys are less vigilant and use exposed locations more often when in the presence of their partner species than when either are alone (Bshary & Noë, 1997a, b).

A number of anti-predation benefits from polyspecific associations are probably a direct consequence of the increased number of individuals, rather than of species complementing each other in their anti-predator skills. For example, individuals living in large groups run a smaller risk of being singled out by a predator than individuals living in small groups (Krebs & Davis, 1993). Moreover, it is likely that chances of detecting a stalking or approaching predator is a function of group size (Treves, 2000). Finally, the adult males of several monkey species have been observed to approach and attack eagles, suggesting that mixed-species groups consisting of various males may have a dissuasive effect on some predators (e.g., Eckardt and Zuberbühler, 2004).

Why do the different monkey species not simply increase their own conspecific group sizes? One likely explanation is that the relationship between feeding competition and group size is stronger in monospecific than in polyspecific groups. At the same time, the relationship between anti-predation benefits and group size appears to be similar in both cases. In the Taï forest, the different monkey species reveal unique food and habitat preferences, leading them to only exploit a sub-segment of the available resources. Niche separation of this kind automatically decreases interspecies competition and appears to make coexistence of closely related species possible (Wachter et al., 1997; McGraw, 1998, 2000; Korstjens, 2001; Wolters & Zuberbühler, 2003). These observations may also explain the substantial primate biomass in the Taï forest, which can reach densities of more than two groups per square kilometre (Zuberbühler & Jenny, 2002).

There are some striking exceptions to this general pattern. For example, the Diana monkey–putty-nosed monkey association shows a remarkable overlap in food preferences and habitat use, yet the two species do not avoid each other but form associations throughout much of the year. However, during periods of food shortage Diana monkeys have been observed to become increasingly aggressive towards putty-nosed monkeys, and association rates consequently plummet to very low levels during these times (Eckardt and Zuberbühler, 2004). It appears that the mixed-species associations follow the logic of economy, determined by the balance of anti-predator benefits and costs of feeding competition.

Primate Predators

The Taï monkeys are hunted by chimpanzees (*Pan troglodytes*), crowned eagles (*Stephanoaetus coronatus*), leopards (*Panthera pardus*), and human poachers. Snakes and other reptiles are common throughout the forest but they seem to be irrelevant as monkey predators, except in areas along the few big forest rivers, where large pythons and crocodiles can be observed. The following section describes some key characteristics of the four main monkey predators: human poachers, chimpanzees, crown eagles, and leopards.

Human Poachers

Illegal hunting by humans is responsible for significant predation pressure on the primate population. Bush meat is widely consumed throughout West Africa; lax law enforcement and understaffing of forest police combine to make poaching a lucrative business (Martin, 1991). Poaching activity is continuous throughout the park both during day and at night. At night, animals are blinded with flashlights and subsequently killed with shotguns. Used batteries and shotgun shells litter the forest floor and poacher camps can be found regularly throughout the forest. Hunting with slings and snares is more common in the border zones of the park.

The impact of human poaching is tremendous. A recent study by Refisch & Kone (2005) estimated the annual mortality of monkeys due to human hunting in and around the Taï Forest to be at least 50,000 individuals, an estimated mass weight of 250,000 kg total. As a result primate densities in many peripheral areas of the park are already substantially reduced. Some species, such as red colobus monkeys, are already locally extinct, forcing poachers to enter the more central areas of the park. Professional hunters build and inhabit temporary camps in the park for several days and employ carriers to transport the carcasses out of the park borders to supply the various local restaurants and bush meat markets.

Archaeological evidence suggests that anthropogenic effects have been present in this area at least since the late Holocene (Mercader et al., 2003), suggesting that human predation has been present for considerable time. The human population density in the area of the Taï forest increased massively in the 20th century, suggesting that the immense hunting pressure currently exerted by humans is a recent phenomenon. Clearly, primates did not have the time to evolve efficient anti-predator behaviour to cope with this level of human predation, although research has shown that some species can employ general cognitive abilities to improve their protection from human predation and other sudden changes in the predator fauna (Bshary, 2001; Gil-da-Costa et al., 2003).

Chimpanzee Hunting

Taï chimpanzees use their home ranges in a clumped way, with small central core areas visited preferentially. There are no major shifts in home range use over consecutive months. Home range sizes of Taï chimpanzees tend to be larger than those of other African communities (Herbinger et al., 2001). Decreases in home range size is related to decreases in the number of males in the group, not overall group size or food availability (Lehmann & Boesch, 2003). The average density of the Taï chimpanzee population throughout the park was estimated at 1.84 individuals/km^2, suggesting a total population of 7,500 individuals for the Taï Forest (Marchesi et al., 1995; Herbinger et al., 2001).

Taï chimpanzees regularly hunt monkeys, which the chimpanzees probably locate by acoustic cues. They decide to hunt apparently in the absence of monkey groups, then they search for a suitable target group (Boesch & Boesch, 1989). Once a group is located, a small hunting party, usually consisting of adult males, climbs into the trees near the group to single out and capture an individual in the high canopy. In Taï, hunting is mainly focused on red or black-and-white colobus monkeys, while the other monkey species are rarely caught. However, due to the monkeys' tendency to associate in mixed-species associations all species are equally exposed to significant hunting pressure, and all take immediate evasive action in the presence of chimpanzees. There are seasonal variations in hunting activity (Boesch & Boesch, 1989). Hunts are particularly common from September through November, following a period of low food availability from June to August when chimpanzee groups are dispersed. Even though there are more than 50 different red colobus groups in an average chimpanzee home range, some monkey groups are likely to suffer attacks several times per year, especially those that live in the core area of a chimpanzee group. Thus, the hunting pressure exerted by this predator apparently varies both locally and seasonally.

In Taï, successful hunting depends on coordination with other hunters. If a hunt is successful, the meat is typically shared amongst the hunters and sometimes with other group members (Boesch, 2002). Males begin hunting at about age 10, and performance improves at a very slow rate, suggesting that hunting skills are not easily learned (Boesch, 2002). Wild chimpanzees have a life expectancy at birth of less than 15 years, which is considerably lower than that of modern human hunter-gatherers (Hill et al., 2001). It is also noteworthy that in other chimpanzee study sites across Africa, high rates of cooperative hunting for monkeys is not a particularly prominent feature (Uehara, 1997; Reynolds, 2005). It remains a distinct possibility, therefore, that the high predation pressure described for Taï is a transient cultural rather than a biological feature of chimpanzees. As with humans, it may thus be that the Taï monkeys did not have the time to evolve specialised anti-predator behaviour to cope with chimpanzee predation in an efficient way, but that they relied on their more general cognitive abilities to avoid predation. For example, the authors have never been able to identify an acoustically distinct alarm call for chimpanzees or humans in any monkey species, even though such calls exist for other predators such as leopards or crowned eagles (Zuberbühler et al., 1997).

Crowned Eagles

The African crowned eagle (*Stephanoaetus coronatus*) is the primary aerial predator for primates throughout sub-Saharan forests. Breeding pairs defend the areas surrounding their nests, which they build in high emergent trees with open flight paths to facilitate transport of prey (Malan & Shultz, 2002). The eagle density in the Taï forest is estimated to be 0.4 individual per km^2, suggesting a total population of about 1,500 individuals in the park (Shultz, 2002). Breeding is seasonal, with one or two chicks fledging in March, followed by a prolonged period of provisioning of the surviving one (Brown, 1982; Shultz & Thomsett, 2007).

Crowned eagles rely on surprise to hunt successfully. Pairs of eagles sweeping through the canopy to attack monkey groups have been observed, possibly older juveniles following a parent (Gautier-Hion & Tutin, 1988; Shultz & Thomsett, 2007). In Taï, however, their preferred hunting strategy is to sit and wait in the high canopy until a sudden surprise attack on an unsuspecting prey individual becomes possible (Shultz & Thomsett, 2007). Eagles have been observed to track monkey groups and to fly around them to position themselves in front of the approaching monkey group to drop down onto unwary individuals. Interestingly, during these observations monkeys showed no signs of being aware of the eagle's presence, suggesting that vigilance behaviour is not very effective in avoiding eagles (Shultz, 2001; Shultz & Thomsett, 2007). An interesting implication is that travelling in a straight line, for example, to reach an anticipated food tree, may not be the most adaptive way of foraging for primates because it would allow sit-and-wait predators to predict group movement. A study on the sooty mangabeys at Taï has found that although the monkeys know which trees carry fruits their approach path to these trees often deviates substantially from a straight line (Janmaat et al., 2006a & b). Whether or not this is an adaptation to eagle hunting behaviour is currently not known.

Predation rates by Taï eagles are not uniformly distributed: a radio tracking study has shown that hunting activity is concentrated to the core areas of the eagles' home ranges (Shultz & Noë, 2002), suggesting that predation pressure varies not only by time of year but also with location. By and large eagles are opportunistic predators, taking prey roughly according to abundance, although there are some interesting deviations (see below).

Leopards

As the largest of carnivore predators, leopards are an important force of natural selection. Leopards traditionally have been studied in the African savannah (Bailey, 1993), but data are now also available for the forest leopards in Taï (Jenny, 1996; Jenny & Zuberbühler, 2005). For this project, four adult leopards were captured, fitted with radio collars, and subsequently monitored for about two years (Table 1.2; Fig. 1.1). Two of the study animals were monitored systematically both by triangulation from treetops and by directly following them through the forest. Results showed that both animals were more active during the day than at night, with relative peaks at dawn and dusk, a pattern that differed from those of savannah leopards (Bailey, 1993; Fig. 1.2). At night, activity patterns consisted of either complete inactivity or of travelling over large distances. Daytime activity showed a more evenly distributed pattern, and inactive periods were always less than five hours (Jenny & Zuberbühler, 2005). The adult male covered a home

TABLE 1.2. Information on the four radio-tracked leopards (Jenny, 1996).

Individual	Capture Date	Sex	Age (years)	Weight (kg)
Cosmos	5 Feb 93	Male	3–5	56
Adele	16 Aug 93	Female	3–5	34
Cora	16 Jun 94	Female	2–3	32
Arthur*	11 Oct 94	Male	3–4	49

*Dind et al. (1996)

FIGURE 1.1. Left: Adult male "Cosmos" passing a photo-trap; Right: adult female "Adele" is fitted with a radio collar (Photos: D. Jenny)

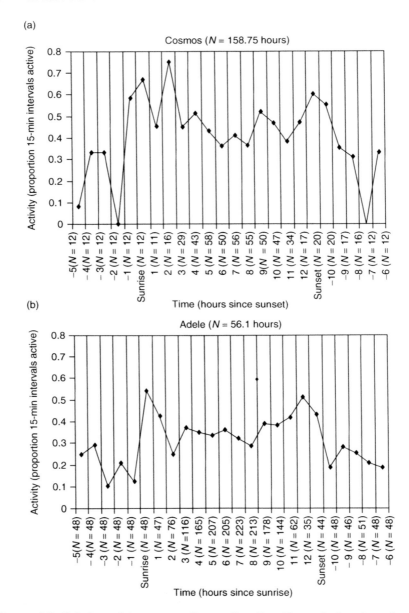

FIGURE 1.2. Relative activity patterns of two radio-collared leopards plotted as a function of onset of sunrise and sunset (Reprinted Jenny, D. & Zuberbühler, K., "Hunting behaviour in West African forest leopards," *African Journal of Ecology*, 43, 197–200, 2005, with permission from Blackwell Publishing)

range 85.6 km^2; the adult female had a total home range size of less than 28.5 km^2 (Jenny, 1996). The overall density of leopards in the Taï forest was estimated to be 0.1/km^2, suggesting a total population of about 400 individuals in the park (Jenny, 1996). This relatively high density, compared to other African forests, might be a direct consequence of the large population of forest duikers (*Cephalophus spp.*), frequent prey items of Taï leopards (see Table 1.3). The lowest monthly activity rates were observed during the rainy period (particularly in October), perhaps because heavy rainfall increased hunting success. During heavy rains, it could be more difficult for prey to detect an approaching or hiding leopard, a rationale that may also explain the increased hunting activity of chimpanzees during this time of the year.

Observations during direct follows of radio-tracked individuals revealed that individuals often hid in dense thickets. When in close vincinity of a resting leopard, the observer encountered significantly more monkey groups than when sitting alone at ten randomly chosen observation points throughout the study area, indicating that leopards selectively chose hiding spots close to monkey groups (Jenny, 1996; Jenny & Zuberbühler, 2005).

Forest leopards hide and attack by surprise, presumably from the lower branches of a tree. All eight species of monkeys occasionally come to the ground to forage or play (McGraw, 1998). Studies of leopard feces in the Taï forest allow some estimate of predation rates inflicted by leopards (Hoppe-Dominik, 1984; Zuberbühler & Jenny, 2002). Of the roughly 140 mammal species that have been

TABLE 1.3. Prey spectrum of Taï leopards.

Scientific Name	Common Name	Zuberbühler & Jenny (2002)	Hoppe-Dominik (1984)
Procolobus badius	Red colobus	21	8
Colobus polykomos	Black-white colobus	16	5
Procolobus verus	Olive colobus	1	0
Cercopithecus diana	Diana monkey	5	17
Cercopithecus petaurista	White-nosed monkey	1	5
Cercopithecus campbelli	Campbell's monkey	3	4
Cercopithecus nictitans	Putty-nosed monkey	0	0
Cercocebus atys	Sooty mangabey	6	9
Cercopithecidae	Unknown monkeys	10	3
Pan troglodytes	Chimpanzee	1	0
Perodicticus potto	Potto	0	1
Primates Total		**64**	**61**
Cephalophus spp. total	Duikers	82	82
Manis spp.	Pangolins	43	10
Sciuridae (undet.)	Squirrels	8	9
Panthera pardus	Leopards	6	6
Other mammals	Other mammals	18	62
Mammalia (undet.)	Unknown mammals	6	26
Non-Primates Total		**163**	**195**
Aves Total		**2**	**2**

described for the Taï forest, a wide variety has been found in leopard feces, most of them are mammals weighing less than 10 kg. Monkeys and duikers make up the largest proportion (Table 1.3). Leopards are known scavengers so a small proportion of leopard prey remains may have come from animals that died of other causes (Hart et al., 1996).

Savannah leopards are typically described as opportunistic predators, hunting prey in proportion to abundance. However, at least one adult female monitored in Taï exhibited clear preferences (Fig. 1.3). This female consumed duikers and monkeys significantly more often, and pangolins less often than other leopards (Zuberbühler & Jenny, 2002). Also, after hearing drumming or screaming from a nearby chimpanzee party, she started moving in the opposite direction, while approach was never recorded (Zuberbühler & Jenny, 2002). It may be that such individual hunting preferences are only temporary, and that the female changed them again after some time. Nevertheless, the fact that the resident leopards can develop hunting preferences for particular prey species is evolutionarily relevant. Since leopards are territorial (Jenny, 1996), a particular monkey group is likely to interact with the same few resident leopards over many years. It may thus be of additional importance to actively dissuade leopards from hunting, not only to secure one's own survival and those of close genetic relatives, but also to avoid preference formation. The subsequent sections will show how the Taï primates have evolved specialised and highly conspicuous anti-predator behaviour in response to leopards.

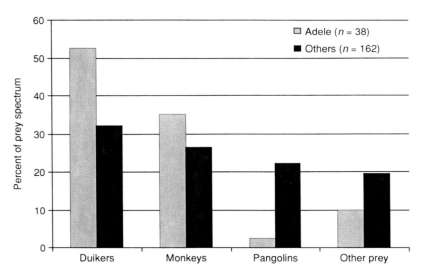

FIGURE 1.3. Prey selectivity of the focal animal Adele compared to other leopards in the Taï forest (Reprinted Jenny, D. & Zuberbühler, K., "Hunting behaviour in West African forest leopards," *African Journal of Ecology*, 43, 197–200, 2005, with permission from Blackwell Publishing)

Results

Adaptations to Predation

Predator-prey interactions lead to evolutionary "arms races," although the nature of co-adaptations will depend on various factors. The following section focuses on the possible effects of predation on primate evolution by the four predators just discussed, especially morphological, social, and cognitive adaptations.

The Effects of Human Predation

Red colobus monkeys are usually the first species to disappear in areas of high poaching (Refisch, 2001), suggesting that large multi-male groups, arboreal live, and large body size are ineffective measures in the face of human predation. All monkeys respond to approaching humans or playback of human speech with no or very few alarm calls, following their response by silent flight and prolonged cryptic behaviour (Zuberbühler et al., 1997; Zuberbühler, 2000). In Taï, poachers use deceptive tactics to localize and attract individuals, mainly by imitating the presence of a crowned eagle or a leopard, predators to which monkeys normally react with high calling rates and approach (Zuberbühler et al., 1997). Experimental work has shown that monkey groups frequently exposed to poachers are less likely to respond to these imitations than monkeys living in more protected areas. This work has demonstrated that adaptive discrimination can be acquired within the lifespan of individual monkeys using general learning abilities (Bshary, 2001).

The Effects of Chimpanzee Predation

Predatory chimpanzees have a bias towards the heavier arboreal monkeys, suggesting that large body size and arboreal live are ineffective deterrents against chimpanzee predation. Once a monkey group is located and a hunt is initiated by a group of chimpanzees, Taï monkeys no longer have very effective anti-predation responses. This is because chimpanzees can reach individuals in the high canopy and their multi-male hunting parties make escape difficult, especially for larger and less agile colobines. Not surprisingly, the presence or vocalisations of chimpanzees reliably elicits prolonged cryptic behaviour in all monkey species, sometimes lasting for several hours (Zuberbühler et al., 1999a). The pattern found in Taï is not necessarily representative for other parts of Africa, however. The Sonso chimpanzees of Budongo forest, for example, only hunt occasionally, and they have been observed to avoid black-and-white colobus monkeys, possibly because of their highly aggressive behaviour towards chimpanzees (Reynolds, 2005).

Playback experiments with red colobus monkeys have shown that, when hearing chimpanzee vocalisations nearby, individuals hide higher up the trees in positions where exposure to the forest floor is minimal and they become silent, often at close range (Bshary & Noë, 1997a, b). When a chimpanzee group is still at some distance, however, the monkeys move away silently through the canopy or they seek the presence of their Diana monkey group, if it is nearby, even if they have to move towards the chimpanzees. Interestingly, chimpanzees tend to refrain from hunting red colobus–Diana monkey groups, probably because Diana monkeys are excellent sentinels for predators approaching over the forest floor (Bshary & Noë, 1997a, b). Bshary (2007) found that in only about 5% of cases did a chimpanzee group approach a Diana–red colobus group if they heard Diana monkeys first. However, if the chimpanzees heard the red colobus first, then approach was much more common—almost 30% of occasions (Bshary & Noë, 1997a; Bshary, 2007). As mentioned previously, the situation in East Africa is somewhat different, suggesting that the Taï chimpanzees may not be representative for the species (Boesch, 1994).

The Effects of Eagle Predation

Crown living species, such as the Diana and red colobus monkeys, are underrepresented in the prey spectrum of crowned eagles, while the large, more terrestrial sooty mangabeys are strongly overrepresented (Shultz et al., 2004). The fact that mangabeys travel as large groups on the ground apparently makes them particularly easy targets for perched eagles. Mangabeys are the largest monkey species in the Taï forest and their large multi-male groups can surpass one hundred individuals, showing that neither group size nor body size of Taï primates was significantly related to crowned eagle prey preference. Crowned eagle predation has also been studied at other sites (e.g., Skorupa, 1989). In the Kibale forest, monkeys were also the predominant prey of crowned eagles, although hunting activity was apparently biased towards monkeys of an intermediate size, such as juvenile red colobus and adult guenons (Struhsaker & Leakey, 1990).

When detecting an eagle Taï monkeys responded with producing high rates of alarm calls. The continuous canopy of the Taï forest provides some protection from attacks and individuals often responded to eagle presence with rapid flight responses towards the middle of the tree or into thick vegetation. In several species, particularly in Diana monkeys, putty-nosed monkeys, black-and-white colobus and red colobus monkeys, males have been observed to attack a perched eagle, which then flies away and presumable leaves the area (Zuberbühler et al., 1997; Eckardt and Zuberbühler, 2004; Bshary & Noë, 1997a, b). Similar observations have been made in other parts of Africa (Gautier-Hion & Tutin, 1988).

The Effects of Leopard Predation

For leopards, the relationship between vulnerability to predation and morphological and behavioural traits has been analyzed (Zuberbühler & Jenny, 2002). Contrary to predictions, leopard predation rates were positively related to body size because the larger monkey species were preyed upon more often than smaller ones, even if population was controlled for overall density. Similarly, the relationships between leopard predation rate and monkey group size on the one hand and the number of adult males per group on the other hand were positive. Predation rates by leopards were unrelated to the reproductive rate of adult females and to a species' use of the lower forest strata, again suggesting that these traits are not very effective measures against leopard predation.

When they detect a leopard, monkeys react by giving myriad alarm calls and sometimes approaching the predator in the lower canopy. This has a striking effect on the leopard's hunting behavior: Radio-tracking data have shown that leopards typically give up hiding positions to move on and find another group. Zuberbühler et al. (1999a) reported 18 cases in which a group of monkeys detected a hiding leopard: The leopard's spent significantly less time hiding underneath a monkey group after detection the leopard usually abandoned its hiding spot within a few minutes to move onto another area (Fig. 1.4). The monkeys' strategy of signalling detection and making further hunting attempts futile is an adaptive response to leopard hunting behavior.

To further investigate the detection signalling hypothesis, we simulated predator presence by playing back typical vocalisations of two major ground predators of Taï monkeys—leopards and chimpanzees—from a concealed speaker (Zuberbühler et al., 1999a). Various monkey groups throughout a large 100-km² study area were tested this way, but never more than once on each stimulus type. Once a group was located, usually by the sound of its members' vocalisations, the speaker was hidden about 50 m from it and a trial conducted if no monkey had detected the observer or part of the equipment and no predator alarm calls had occurred for at least 30 minutes. The focal group's vocal response was recorded.

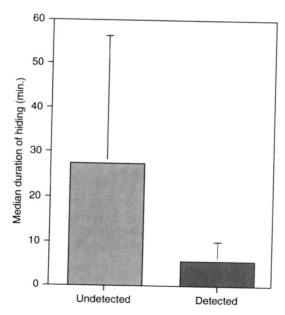

FIGURE 1.4. Median duration of hiding behaviour of the focal animal before and after detection by a group of monkeys (Reprinted from Zuberbühler, K., et al. "The predator deterrence function of primate alarm calls," *Ethology*, 105, 477–490, 1999, with permission from Blackwell Publishing)

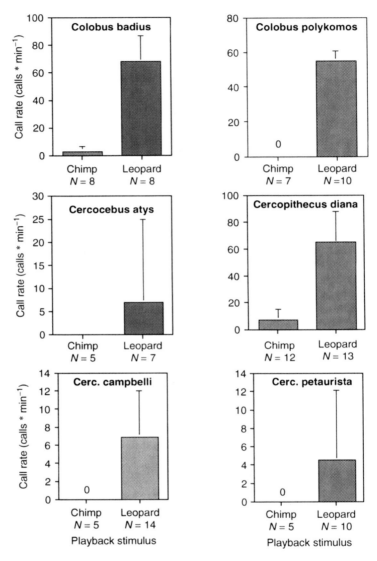

FIGURE 1.5. Alarm call behaviour of six Taï monkeys in response to chimpanzee pant hoots and leopard growls (Reprinted from Zuberbühler, K., et al. "The predator deterrence function of primate alarm calls," *Ethology*, 105, 477–490, 1999, with permission from Blackwell Publishing)

All tested monkey species gave significantly higher rates of alarm calls to playbacks of leopard growls than to playbacks of chimpanzee pant hoots (Fig. 1.5). Groups occasionally approached the speaker after hearing playback stimuli, but only during playback of leopard growls and never after playback of chimpanzee pant hoots, which typically caused flight away from the speaker. In sum, data support the hypothesis that monkey alarm calls to leopards have a predator deterrence function because leopards, in contrast to chimpanzees, elicited conspicuously high alarm call rates, which drove the leopards away.

Predation and Primate Cognitive Evolution

Behavioral and cognitive flexibility appear to be the crucial traits in dealing with Taï predators as compared to traits such as large body size, group size, number of males, inter-birth interval, and so on. A good deal of this flexibility becomes apparent in the monkeys' alarm call behavior. Diana monkeys, Campbell's monkeys, putty-nosed monkeys, and possibly most or all other primates have evolved acoustically distinct and predator-specific alarm calls that function, amongst other ways, to warn each other about the presence of specific types of predators (Zuberbühler et al., 1997; Zuberbühler, 2001; Eckardt and Zuberbühler, 2004; Arnold & Zuberbühler, 2006a & b; Wright, 1998).

Playback experiments have shown that individuals respond to the recordings of different alarm calls as if the corresponding predator were present. The alarm calls from other monkey species are just as effective in this respect as the calls of conspecific individuals (Zuberbühler, 2000b & c; Eckardt and Zuberbühler, 2004). Other work has shown that reactions to alarm calls are not simple responses to the acoustic features of the calls. Instead, monkeys associate the alarm calls of other individuals to the presence of a particular corresponding predator, rather than simply responding to the acoustic features of the calls (Zuberbühler et al., 1999a; Zuberbühler, 2000a & b), demonstrating a level of processing that goes beyond simple stimulus-response arithmetic.

Diana monkeys, and possibly other Taï monkeys, possess relatively detailed knowledge of their main predators' behavior. For example, chimpanzees produce various types of screams, which are given both in social situations (social screams) (Slocombe & Zuberbühler, 2005) and in response to leopards (SOS screams) (Goodall, 1986). Playback experiments have demonstrated that Diana monkeys are able to distinguish between the various types of chimpanzee screams, even though the acoustic differences are only very subtle (Zuberbühler, 2000b). Diana monkey groups whose home ranges were in the core area of a chimpanzee community responded with cryptic behavior to playbacks of chimpanzee agonistic screams and with their own leopard alarm calls when hearing playbacks of chimpanzee SOS-screams (indicating the presence of a leopard). In contrast, Diana monkey groups living in the peripheral areas of a chimpanzee group were more likely to respond cryptically to both types of screams (Zuberbühler, 2000b). Direct encounters between chimpanzees and leopards are probably quite rare events, suggesting that in core areas Diana monkey groups had more learning opportunities for forming these associations. These observations further stress the importance of cognitive abilities in dealing appropriately with these predators. Griffin and Galef (2005) have recently argued that predation has favoured the evolution of a specialised learning apparatus, which also accepts arrangements in which the conditioned stimulus follows, rather than precedes, the unconditioned stimulus.

Although monkeys are able to attend to the meaning encoded by other individuals' alarm calls, this is not always possible, particularly if species do not encode information about the predator type when producing alarm calls. For example, crested Guinea fowls (*Guttera pulcheri*) forage in large groups on the forest floor and, when chased, produce conspicuously loud alarm calls that can be heard over long distances. These birds are not hunted by chimpanzees but may be attacked by leopards and humans (Zuberbühler & Jenny, 2002). Interestingly, the default response of Diana monkeys to Guinea fowl alarm calls is to behave as if a leopard were present. However, this kind of behavioral pattern is not the result of a rigid link between one particular acoustic structure and a behavioral response. Instead, monkeys appear to take into account the pragmatic information obtained from the environment before selecting a response (Zuberbühler, 2000a). Experiments have shown that if Guinea fowl alarm calls are caused by the presence of a human, then the Diana monkeys remain mostly silent to the birds' alarm calls. Cryptic behavior is the typical response of wild Diana monkeys to humans, suggesting that the monkeys assume the presence of humans when responding to the alarm calls. However, if the birds' alarm calls are given in response to a leopard, then the monkeys' response to the alarm calls is strong, that is, as if a leopard were present, suggesting that the monkeys' response is determined by the most likely cause of the birds' alarm calls. A number of control experiments were conducted to rule out simpler mechanisms. Since the same bird alarm calls were used in both the leopard and human situation, monkeys could not have been responding to some subtle acoustic cues unnoticed by the researcher.

Finally, some monkeys are able to alter the meaning of their alarm calls by constructing simple call combinations using existing elements of their vocal repertoire (Zuberbühler, 2002). As mentioned before, Campbell's males give acoustically distinct alarm calls to leopards and eagles and Diana monkeys respond to these calls with their own corresponding alarm calls (Zuberbühler, 2000). However, in less dangerous situations, Campbell's males often emit a pair of low, resounding 'boom' calls before their alarm calls. Playbacks of boom-introduced Campbell's eagle or leopard alarm calls no longer elicited alarm calls in Diana monkeys, indicating that the 'booms' have affected the semantic specificity of the subsequent alarm calls. Diana monkeys themselves do not produce booms and combining Campbell's booms with Diana monkey alarm calls had no effect, indicating that they were only meaningful in conjunction with Campbell's alarm calls.

Another surprising finding emerged from a field study on putty-nosed monkey alarm calls (Arnold & Zuberbühler, 2006a & b). In the Nigerian subspecies, males produced two types of alarm calls to leopards and eagles, but individual calls were given as parts of long sequences that often involved both alarm call types. However, playback experiments demonstrated that call production was not random, but that leopards and eagles elicited structurally unique call sequences. It is quite likely, therefore, that receivers are able extract semantic information from the structural features of a call sequence, as opposed to the individual calls.

Discussion

There is a large literature that suggests that a species' traits, including its morphology, group size, and so on, are a consequence of migration patterns that resulted from global climate changes. African rainforests have undergone dramatic changes in size in the relatively recent evolutionary past. Changing Pleistocene climate led to the compartmentalisation of the once continuous Upper Guinea forest, of which the Taï forest is part. During the dry and cold periods the forest contracted, forcing inhabitants into increasingly restricted refuges. The Taï forest sits in between two main West African refuges, one located in the border region of Sierra Leone and Liberia, the other one at the border of the Ivory Coast with Ghana (Booth, 1958a,b; Hamilton, 1988; Oates, 1988). When warmer, moister climates lead to forest expansion, the primate populations isolated in these refuges diverged and radiated outward to colonize new areas, including the Taï forest. However, it is likely that the newly emerging forests were limited in the number of available niches to fill in, resulting in inter-species competition (Fleagle & Reed, 1996; Tutin & White, 1999; Struhsaker, 1999; Reed & Bidner, 2004). The Taï primate fauna is fairly representative of most primate communities in other parts of Africa. These communities typically consist of a several arboreal frugivores, 2–3 arboreal folivores, terrestrial cryptic foragers, and some nocturnal prosimian species, suggesting that the number and types of available niches a tropical forest can offer is limited and roughly the same throughout the tropical forest belt (Reed & Bidner, 2004; Fleagle & Reed, 1996, 1999; Chapman et al., 1999). Phylogenetic history and the constraints of a rainforest habitat, in other words, may explain the current trait differences in the Taï monkeys much better than predation alone.

Polyspecific Associations

Rather than tolerating more conspecific group members, forest primates appear to prefer living with members of other species. In some cases they may even be forced to do so because their own rigid social system prevents them from increasing group size. What keeps these mixed-species groups together is difficult to understand, particularly if two species have similar feeding requirements. A number of observations are consistent with the idea that monkeys trade their services in predation defence for increased tolerance by a more dominant partner at the feeding site, a prerequisite of a biological market game (Noë et al., 2001). An important feature of a biological market is that there is an element of choice and that the partners can adjust the benefits they offer each other. Some species, such as the Diana monkeys and putty-nosed monkeys, are astonishingly tolerant towards each other, despite a high degree of feeding competition (Eckardt and Zuberbühler, 2004). In this case, biological market theory predicts that interspecies tolerance should be determined by differences in predation pressure and food availability, as well as level of competition between the two partner species. The mechanisms governing mixed-species groups are thus fundamentally different from those determining monospecific groups. In polyspecific groups individuals have a large degree of control over how many partner individuals they want to associate with simply by refusing to form a group with other species. Individuals can thus change their effective group size on an almost ad hoc basis, allowing rapid adjustments to changes in the environment. Successful living in poly-specific groups thus requires a cognitive apparatus that can deal with the signals of the partner species, especially its alarm call and anti-predator behavior. The studies reviewed in this chapter have illustrated this capacity extensively. Predation, in other words, has lead to the formation of polyspecific associations, whose value is multiplied by adequate semantic abilities. The forest may thus have been the breeding ground for advanced communicative abilities.

Interaction Effects

It is unclear how the various behavioral and morphological traits interact with one another in their efficiency in predation aversion. For instance, the habitat may put a limit on how much biomass it can support, leading to a trade-off between body size and group size. However, that is not what is observed in Taï. Sooty mangabeys and red colobus are amongst the heaviest Taï primates but they also form the largest groups, while the smaller guenons and olive colobus monkeys live in substantially smaller groups. It may also be that different predators put opposing selection pressures on a particular trait. For instance, small body size may be advantageous to escape pursuit hunters, such as chimpanzees, but it may provide a substantial disadvantage when interacting with crowned eagles. Although this is interesting, the finding was that all predators consistently preferred the larger monkeys, and that multi-male groups seemed to provide little or no protection for other group members.

It would be desirable to analyze within species differences. For example, it would be useful to ask, "Are individual red colobus monkeys living in larger groups attacked less often by chimpanzees, eagles, or leopards than individuals living in smaller groups?" Hill & Lee (1998) have found some evidence for this sort of relation, but no such data are available for the Taï monkeys. Nevertheless, differences among the various Taï monkey species with respect to vulnerability to predators need to be explained. If predation did play a crucial role in the monkeys' morphological and behavioral evolution, then the outcome of this process was not very simple or straightforward. There seem to be two options: (a) predation was the main factor in the evolution of these traits, but its effects were so intricate and complex that it is impossible to see any clear effects in the species' current traits; (b) predation played some role during evolution as part of a more general adaptation process, which was mainly determined by interspecies competition and the colonization history of the habitat. The latter scenario seems to be the more likely one.

Evolution of Cognitive Abilities

The idea that the principal evolutionary effect of predation was on primate cognitive evolution is obviously a controversial one, and it will be nothing short of a challenge to think of studies capable of providing conclusive evidence. The generally accepted view is that predation selected for sociality (e.g., van Schaik, 1983) and that living in groups then resulted in selective pressure that lead to advanced cognitive abilities: the 'social' or 'Machiavellian' intelligence hypothesis (Humphrey, 1976). What is proposed in this chapter is a more direct route, one in which predation directly affected primate cognitive evolution, independent of social structure. It is noteworthy that some of the monkey species described in this chapter, particularly the Diana monkeys, live in very primitive mammalian social systems with small groups consisting of one adult male and several philopatric adult females and their offspring. Their social behavior is decisively unremarkable, and classic indicators of social complexity, such as differentiated grooming and social relationships or complex triadic interactions, are not normally observed in these animal (e.g., Buzzard & Eckardt, 2007).

What exactly are 'advanced' communicative abilities? A useful way of addressing this problem is to invoke the notion of 'flexibility' (Tomasello & Call, 1997). In the vocal domain this relates to the flexible, context- and audience-dependent meaningful use of vocalisations, rather than a reflex-like direct response to some sorts of stimuli. The various examples discussed in this chapter suggest that much of the monkey behavior in the predation context is of the former kind.

Conclusions

The Taï forest primate fauna is the product of a series of evolutionary events. It is not completely clear for how long each species has existed at Taï, nor is the exact order of their arrival known. Just like in other African forests, the different monkey species in Taï occupy specific niches, as defined by body size, diet, locomotion, or activity patterns, which suggests that there is a deterministic element to the composition and structure of the primate community at Taï and elsewhere in Africa (Fleagle et al., 1999). The various species were probably forced to adapt to one of the few available ecological niches and this process may better explain the interspecies differences in their traits than predation alone. Predation may have been an important factor in this process, but it fails to account for many of the observed patterns: There was no evidence that the four predators drove the evolution of traits, such as group size, body size, multi-male grouping pattern, inter-birth interval, stratum use, and so on, in the predicted way.

Instead, the most striking adaptations displayed by the monkeys are highly predator-specific behavioral strategies, apparently designed to interfere with a predation event at various levels. Several field experiments have demonstrated that some of these behavioral responses are based on relatively sophisticated cognitive processes. Monkeys use their alarm call behavior not only to warn each other about the type of predator present but also to interfere with some of the predators' hunting techniques. They are attuned to responding to the alarm calls of other individuals, which are interpreted by flexibly taking into account a variety of additional information.

Acknowledgments. Much of the field research reviewed in this chapter was funded by grants of the U.S. and Swiss National Science Foundations, the University of Zurich, the Leakey Foundation, the National Geographic Society, and the Max-Planck Society. Further funding came from the British Academy, the Royal Society, the Biotechnology and Biological Sciences Research Council, the European Science Foundation ('Origins of Man, Language, and Languages') and a grant by the European Commission (FP6 Pathfinder 'Origins of referential communication').

References

Anderson, C.M. (1986). Predation and primate evolution. *Primates*, 27: 15–39.

Arnold, K., and Zuberbühler, K. (2006a). The alarm calling adult male putty-nosed monkeys, Cercopithecus nictitans martini. *Animal Behaviour*, 72.

Arnold, K., and Zuberbühler, K. (2006b). Semantic combinations in primate calls. *Nature*, 441: 303.

Bailey, T.N. (1993). *The African leopard*. New York: Columbia Univ. Press.

Boesch, C. (1994). Chimpanzee-red colobus monkeys: A predator-prey system. *Animal Behaviour*, 47, 1135–1148.

Boesch, C. (1994). Hunting strategies of Gombe and Taï chimpanzees. In R. Wrangham, W. McGrew, F. de Waal, and P. Heltne (Eds.), *Chimpanzee Cultures* (pp. 77–92). Cambridge: Harvard University Press.

Boesch, C. (2002). Cooperative hunting roles among Taï chimpanzees. *Human Nature*, 13: 27–46.

Boesch, C., and Boesch, H. (1989). Hunting behavior of wild chimpanzees in the Taï National Park. *Am. J. Phys. Anthr.*, 78: 547–573.

Boesch, C., and Boesch-Achermann, H. (2000). *The chimpanzees of the Taï forest. Behavioural ecology and evolution*. Oxford: Oxford Univ. Press.

Brown, L. (1982). The prey of the crowned eagle *Stephanoaetus coronatus* in Central Kenya. *Scopus*, 6: 91–94.

Bshary, R. (2001). Diana monkeys, *Cercopithecus diana*, adjust their anti-predator response behaviour to human hunting strategies. *Behav. Ecol. Sociobiol.*, 50: 251–256.

Bshary, R. (2007). Interactions between red colobus monkeys and chimpanzees. In W.S. McGraw, K. Zuberbühler & R. Noë (Eds.), *Monkeys of the Taï forest: An African primate community* Cambridge: Cambridge Univ. Press.

Bshary, R., and Noë, R. (1997). Anti-predation behaviour of red colobus monkeys in the presence of chimpanzees. *Behavioral Ecology and Sociobiology*, 41: 321–333.

Bshary, R., and Noë, R. (1997). Red colobus and Diana monkeys provide mutual protection against predators. *Animal Behaviour*, 54: 1461–1474.

Buzzard, P., & Eckardt, W. (2007). Social systems of the Taï guenons (*Cercopithecus spp.*). In R. McGraw, K. Zuberbühler & R. Noë (Eds.), *Monkeys of the Taï forest: An African monkey community.* Cambridge: Cambridge Univ. Press.

Cheney, D.L., Seyfarth, R.M., Fischer, J., Beehner, J., Bergman, T., Johnson, S.E., Kitchen, D.M., Palombit, R.A., Rendall, D., and Silk, J.B. (2004). Factors affecting reproduction and mortality among baboons in the Okavango Delta, Botswana. *International Journal of Primatology,* 25: 401–428.

Cheney, D.L., and Wrangham, R.W. (1987). Predation. In B. Smuts, et al. (Eds.), *Primate societies* (pp. 227–239). Chicago: Univ. of Chicago Press.

Cords, M. (1990). Vigilance and mixed-species association of some East African forest monkeys. *Behavioral Ecology and Sociobiology,* 26: 297–300.

Dind, F., Jenny, D., and Boesch, C. (1996). Écologie et prédation du léopard en foret de Taï (Côte d'Ivoire). In J.F. Graf, M. Zinsstag, and J. Zinsstag (Eds.), *Centre Suisse de Recherches Scientifiques en Côte d'Ivoire (CSRS). Rapport d'activité 1992–1994* (pp. 22–26). Abidjan: Centre Suisse de Recherches Scientifiques en Côte d'Ivoire.

Eckardt, W., and Zuberbühler, K. (2004). Cooperation and competition in two forest monkeys. *Behavioral Ecology,* 15: 400–411.

Enstam, K.L., and Isbell, L.A. (2004). Microhabitat preference and vertical use of space by patas monkeys (*Erythrocebus patas*) in relation to predation risk and habitat structure. *Folia Primatologica,* 75: 70–84.

Gautier-Hion, A., and Gautier, J.P. (1974). Les associations polyspecifiques de Cercopitheques du Plateau de M'Passa (Gabon). *Folia primatol.,* 22: 134–177.

Gautier-Hion, A., and Tutin, C.E.G. (1988). Simultaneous attack by adult males of a polyspecific troop of monkeys against a crowned hawk eagle. *Folia Primatol.,* 51: 149–151.

Gil-da-Costa, R., Palleroni, A., Hauser, M.D., Touchton, J., and Kelley, J.P. (2003). Rapid acquisition of an alarm response by a neotropical primate to a newly introduced avian predator. *Proceedings of the Royal Society of London Series B—Biological Sciences,* 270: 605–610.

Goodall, J. (1986). *The chimpanzees of Gombe: Patterns of behavior.* Cambridge: Harvard Univ. Press.

Hart, J.A., Katembo, M., and Punga, K. (1996). Diet, prey selection and ecological relations of leopard and golden cat in the Ituri forest, Zaire. *African Journal of Ecology,* 34: 364–379.

Herbinger, I., Boesch, C., and Rothe, H. (2001). Territory characteristics among three neighbouring chimpanzee communities in the Taï National Park, Côte d'Ivoire. *Int. J. Primatol.,* 22: 143–167.

Hill, K., Boesch, C., Pusey, A., Williams, J., and Wrangham, R. (2001). Mortality rates among wild chimpanzees. *J. Hum. Evol.,* 40: 437–450.

Hill, R.A., and Dunbar, R.I.M. (1998). An evaluation of the roles of predation rate and predation risk as selective pressures on primate grouping behaviour. *Behaviour,* 135: 411–430.

Hill, R.A., and Lee, P.C. (1998). Predation risk as an influence on group size in cercopithecoid primates: Implications for social structure. *J. of Zoology,* 245: 447–456.

Hoppe-Dominik, B. (1984). Etude du spectre des proies de la panthere, *Panthera pardus,* dans le Parc National de Taï en Côte d'Ivoire. *Mammalia,* 48: 477–487.

Humphrey, N.K. (1976). The social function of intellect. In P.P.G. Bateson and R.A. Hinde (Eds.), *Growing points in ethology* (pp. 303–317). Cambridge, UK: Cambridge Univ. Press.

Isbell, L.A., Cheney, D.L., and Seyfarth, R.M. (1993). Are immigrant vervet monkeys, *Cercopithecus aethiops*, at greater risk of mortality than residents? *Animal Behaviour*, 45: 729–734.

Janmaat, K.R.L., Byrne, R.W., and Zuberbühler, K. (2006a). Evidence for a spatial memory of fruiting states of rainforest trees in wild mangabeys. *Animal Behaviour*.

Janmaat, K.R.L., Byrne, R.W., and Zuberbühler, K. (2006b). Primates take weather into account when searching for fruits. *Curr. Biol.*, 16: 1232–1237.

Jenny, D. (1996). Spatial organization of leopards (*Panthera pardus*) in Taï National Park, Ivory Coast: Is rain forest habitat a tropical haven? *J. Zool.*, 240: 427–440.

Jenny, D., and Zuberbühler, K. (2005). Hunting behaviour in West African forest leopards. *African J. of Ecology*, 43: 197–200.

Korstjens, A.H. (2001). *The mob, the secret sorority, and the phantoms: An analysis of the socio-ecological strategies of the three colobines of Taï*. Unpublished Ph.D., Utrecht.

Krebs, J.R., and Davis, N.B. (1993). Predator versus prey: Evolutionary arms races. In J.R. Krebs and N.B. Davies (Eds.), *An introduction to behavioral ecology* (pp. 77–101). Oxford: Blackwell Scientific Publications.

Lehmann, J., and Boesch, C. (2003). Social influences on ranging patterns among chimpanzees (*Pan troglodytes verus*) in the Taï National Park, Côte d'Ivoire. *Behavioral Ecology*, 14: 642–649.

Malan, G., and Shultz, S. (2002). Nest-site selection of the crowned hawk-eagle in the forests of KwaZulu-Natal, South Africa, and Taï, Ivory Coast. *Journal of Raptor Research*, 36: 300–308.

Marchesi, P., Marchesi, N., Fruth, B., and Boesch, C. (1995). Census and distribution of chimpanzees in Côte -Divoire. *Primates*, 36: 591–607.

Martin, C. (1991). *The rainforests of West Africa: Ecology, threats, conservation*. Basel: Birkhäuser.

McGraw, W.S. (1998). Comparative locomotion and habitat use of six monkeys in the Taï Forest, Ivory Coast. *Am. J. Phys. Anthrop.*, 105: 493–510.

McGraw, W.S. (2000). Positional behavior of *Cercopithecus petaurista*. *Inter. Jour. of Primatology*, 21: 157–182.

Mercader, J., Panger, M., Scott-Cummings, L., and Boesch, C. (2003). Late Holocene archaeological remains from chimpanzee and human sites in the rainforest of Côte d'Ivoire. *American Journal of Physical Anthropology*, 150–151.

Noë, R., and Bshary, R. (1997). The formation of red colobus–Diana monkey associations under predation pressure from chimpanzees. *Proceedings of the Royal Society of London Series B—Biological Sciences*, 264: 253–259.

Refisch, J. (2001). Einfluss der Wilderei auf Affen und die sekundären Effekte auf die Vegetation im Taï Nationalpark, Elfenbeinküste. In: *Bayreuther Forum Ökologie* (pp. 1–189). Bayreuth: Bayreuth.

Refisch, J., and Kone, I. (2005). Impact of commercial hunting on monkey populations in the Taï region, Côte d'Ivoire. *Biotropica*, 37: 136–144.

Reynolds, V. (2005). *The chimpanzees of the Budongo forest—Ecology, behaviour, and conservation*. Oxford: Oxford Univ. Press.

Shultz, S. (2001). Notes on interactions between monkeys and African crowned eagles in Taï National Park, Ivory Coast. *Folia primatol.*, 72: 248–250.

Shultz, S. (2002). Population density, breeding chronology and diet of crowned eagles (*Stephanoaetus coronatus*) in Taï National Park, Ivory Coast. *Ibis*, 144: 135–138.

Shultz, S., Faurie, C., and Noë, R. (2003). Behavioural responses of Diana monkeys to male long-distance calls: Changes in ranging, association patterns and activity. *Behavioral Ecology and Sociobiology*, 53: 238–245.

Shultz, S., and Noë, R. (2002). The consequences of crowned eagle central-place foraging on predation risk in monkeys. *Proceedings of the Royal Society of London Series B—Biological Sciences*, 269: 1797–1802.

Shultz, S., Noë, R., McGraw, W.S., and Dunbar, R.I.M. (2004). A community-level evaluation of the impact of prey behavioural and ecological characteristics on predator diet composition. *Proceedings of the Royal Society of London Series B—Biological Sciences*, 271: 725–732.

Shultz, S., and Thomsett, S. (2007). Interactions between African crowned eagles and their primate prey community. In W.S. McGraw, K. Zuberbühler and R. Noë (Eds.), *Monkeys of the Taï forest: An African monkey community*. Cambridge: Cambridge Univ. Press.

Skorupa, J.P. (1989). Crowned eagles (*Stephanoaetus coronatus*) in rainforest: Observations on breeding chronology and diet at a nest in Uganda. *Ibis*, 131: 294–298.

Slocombe, K.E., and Zuberbühler, K. (2005). Agonistic screams in wild chimpanzees (*Pan troglodytes schweinfurthii*) vary as a function of social role. *J. Comp. Psych.*, 119: 67–77.

Stanford, C.B. (1998). Predation and male bonds in primate societies. *Behaviour*, 135: 513–533.

Struhsaker, T.T., and Leakey, M. (1990). Prey selectivity by crowned hawk-eagles on monkeys in the Kibale Forest, Uganda. *Behav. Ecol. Sociobiol.*, 26: 435–444.

Tomasello, M., and Call, J. (1997). *Primate cognition*. New York: Oxford Univ. Press.

Treves, A. (1999). Has predation shaped the social systems of arboreal primates? *International Journal of Primatology*, 20: 35–67.

Treves, A. (2000). Theory and method in studies of vigilance and aggregation. *Animal Behaviour*, 60: 711–722.

Uehara, S. (1997). Predation on mammals by the chimpanzee (*Pan troglodytes*). *Primates*, 38: 193–214.

van Schaik, C.P. (1983). Why are diurnal primates living in groups? *Behaviour*, 87: 120–144.

Wachter, B., Schabel, M., and Noë, R. (1997). Diet overlap and polyspecific associations of red colobus and Diana monkeys in the Taï Park, Ivory Coast. *Ethology*, 103: 514–526.

Waser, P.M. (1982). Primate polyspecific associations: Do they occur by chance? *Animal Behaviour*, 30: 1–8.

Wolters, S., and Zuberbühler, K. (2003). Mixed-species associations of Diana and Campbell's monkeys: The costs and benefits of a forest phenomenon. *Behaviour*, 140: 371–385.

Wright, P. (1998). Impact of predation risk on the behaviour of *Propithecus diadema edwardsi* in the rainforest of Madagascar. *Behaviour*, 135: 483–512.

Zuberbühler, K. (2000). Causal cognition in a non-human primate: Field playback experiments with Diana monkeys. *Cognition*, 76: 195–207.

Zuberbühler, K. (2000). Causal knowledge of predators' behaviour in wild Diana monkeys. *Animal Behaviour*, 59: 209–220.

Zuberbühler, K. (2000). Interspecific semantic communication in two forest monkeys. *Proc. R. Soc. London B*, 267: 713–718.

Zuberbühler, K. (2000). Referential labelling in Diana monkeys. *Animal Behaviour*, 59: 917–927.

Zuberbühler, K. (2001). Predator-specific alarm calls in Campbell's guenons. *Behav. Ecol. Sociobiol.*, 50: 414–422.

Zuberbühler, K. (2002). A syntactic rule in forest monkey communication. *Animal Behaviour*, 63: 293–299.

Zuberbühler, K., Cheney, D.L., and Seyfarth, R.M. (1999). Conceptual semantics in a nonhuman primate. *Journal of Comparative Psychology*, 113: 33–42.

Zuberbühler, K., and Jenny, D. (2002). Leopard predation and primate evolution. *Journal of Human Evolution*, 43: 873–886.

Zuberbühler, K., Jenny, D., and Bshary, R. (1999). The predator deterrence function of primate alarm calls. *Ethology*, 105: 477–490.

Zuberbühler, K., Noë, R., and Seyfarth, R.M. (1997). Diana monkey long-distance calls: Messages for conspecifics and predators. *Animal Behaviour*, 53: 589–604.

2
Predation on Primates: A Biogeographical Analysis

D. Hart

Introduction

Measuring the magnitude of predation has been deemed an important task in clarifying aspects of primate ecology (Terborgh & Janson, 1986). This goal is in keeping with a general theoretical shift noted by Sih et al. (1985) toward acknowledgment that predation often has a greater impact than resource competition on individual animals through behavior and life history; on prey populations through size and stability; and on ecosystems through diversity and relative abundance patterns. Biogeography, as a comparative observational science dealing with spatial and temporal scales too large for experimentation, seeks patterns of biodiversity upon which theories may be formulated (Brown & Lomolino, 1996). Primate predation studies benefit from a biogeographical approach when primates and their predators are assessed from the standpoint of four major regions: Africa, Madagascar, Asia, and the Neotropics. Since predation is thought to have affected morphological, ecological, and behavioral traits in primates (Hart, 2000; Zuberbühler & Jenny, 2002), a comparison of the four regions may facilitate identification of broad biogeographic patterns that are associated with predation.

Erroneous assumptions concerning predation as a demographic variable find their way into published comments. One commonplace, but erroneous, assumption is that "mortality due to predation appears to be negligible" (Dunbar, 1988, p.53). Opinions have ranged from a belief that the role of predation in primate evolution is minimal (Raemakers & Chivers, 1980; Wrangham, 1980; Cheney & Wrangham, 1987) to theories that predation is a powerful force in shaping social patterns (van Schaik, 1983; Terborgh & Janson, 1986; Dunbar, 1988; Hart, 2000; Hart & Sussman, 2005). Recent reviews and studies on topics such as the vulnerability of baboons to predation (Cowlishaw, 1994), ecological patterns of predation on primates (Isbell, 1994a), the status of predation research (Boinski & Chapman, 1995), predation rate versus predation risk (Hill & Dunbar, 1998), and the influence of predation on arboreal primates (Treves, 1999) have expanded theoretical discussion of this topic. Nonetheless, because observation of predation in the primate literature is often anecdotal rather than quantitative, there has been a tendency to underestimate the pervasive influence predation has on the behavior and ecology of primates (Caine, 1990).

Large-scale patterns in predation have been discussed in broad theoretical terms but never assessed using quantified data. Moreover, there have been few attempts to recruit research carried out on various predators as an aid to understanding the impact of predation on primates. Predator-prey relationships are best studied from the perspective of the predator (Washburn et al., 1965; Cheney & Wrangham, 1987; Isbell, 1994; Boinski & Chapman, 1995; Mitani et al., 2001). Observation of only one group of one species (the typical parameters of primate research) provides limited data and often skews the perception of predation, whereas fieldwork on predatory species gives an ecosystemic view of several trophic levels. The home range of a solitary predator usually overlaps numerous prey groups and species; while the predator hunts on a daily basis, it may only occasionally attack the primate group under study. Primatologists have rarely viewed their subjects as prey, and the inclusion of predators into the realm of primate ecology has not been common. To that end, I conducted a meta-analysis of predation on primates that can serve as a basis for objective review of this topic.

Methods

Meta-analysis is the branch of statistics wherein data from various sources are combined (Halvorsen, 1986). Because the broad overview of data collected in my study is a first attempt to quantify the entire spectrum of predation on primates, a descriptive numerical summary is needed to deal with the data in manageable form (Sokal & Rohlf, 1981; Mansfield, 1986). Since it was not possible to collect data from all research sites in a random sampling, I collected raw data and used descriptive statistics throughout to summarize these data. Frequency distributions are used for comparison of variables, and summaries based on percentages are employed to interpret specific issues.

Primate data were categorized as follows: (1) by geographic region: Africa, Madagascar, Asia, and the Neotropics; (2) by body size divided into two categories: because body weights of primates extend along a continuum from 60 g to 169.5 kg, I selected a reasonable arbitrary weight division of under and over 2 kg to separate small-bodied from large-bodied primates; (3) by stratum generally occupied: arboreal or terrestrial; and (4) by daily activity cycle: diurnal or nocturnal (a decision to limit the activity cycles to these two divisions was based on the realization that more precise divisions, such as cathemeral, would constitute very small fractions of the data set).

For each primate prey, the equivalent data on its predator were also collated as follows: (1) by broad predator categories: felids, raptors, canids and hyaenids, small carnivores (which included the vivverid, herpestid, and mustelid families), reptiles, or unidentified (if the predator left a dead or dying primate but was not itself observed); (2) by geographic region: Africa, Madagascar, Asia, and the

Neotropics; (3) by weight in kilograms (predators ranged from *Vanga curvirostris*, weighing 72 g, to *Crocodylus palustris*, weighing 227 kg); (4) by stratum occupied: aerial, arboreal, terrestrial, or aquatic; and (5) by daily activity cycle: diurnal or nocturnal.

Data were drawn from both published and unpublished sources (viz., the scientific literature and my own questionnaires) based on the fieldwork of primate researchers, ornithologists, herpetologists, and mammalogists. Data were derived from observed predation events and studies of predation that have produced quantitative results. The latter are heavily dependent on predator research and offer information on the entire spectrum of prey in the diet of many of the 174 primate predator species identified by Hart (2000). Along with other food items, primate remains—ranging from the smallest (*Microcebus*) to the largest (*Gorilla*)—have been found in predator scats, pellets, nests, and dens.

One hundred and seventy-four predator species identified in this meta-analysis were divided into five broad categories: *felids* (21 species of wild cats), *raptors* (82 species of hawks, eagles, owls, and other predatory birds), *canids* and *hyaenids* (10 species of wild dogs and jackals and 3 species of hyenas), *small carnivores* (22 species of civets, genets, mongoose, the fossa, and a tropical weasel, among others), and *reptiles* (36 species of snakes, crocodilians, and monitor lizards). Ecology rather than taxonomy was emphasized in the predator categories; for instance, taxonomically the hyenas are more closely related to felids than canids, but the predation strategies of dogs and hyenas (i.e., pack hunting and coursing after prey) justify combining the two carnivore families.

While all categories of primate mortality are pertinent and deserve further research, the meta-analysis described in this chapter was deliberately limited to the relationship between primates and the groups of carnivorous animals that are predatory by definition. Neither an analysis of human predation on non-human primates nor predation by primates on other primates was attempted. There is a large body of literature detailing human exploitation of primates (see Mittermeier, 1987; Mittermeier & Cheney, 1987; Peres, 1990; Alvard & Kaplan, 1991; Alvard, 1994; Oates, 1994, 1996; McRae, 1997; Redmond, 1998; McNeil, 1999). Less is known about the effects of non-human primate predation on other primates. Chimpanzees, orangutans, baboons, blue monkeys (*Cercopithecus mitis*), capuchin monkeys (*Cebus spp.*), red-fronted brown lemurs (*Eulemur fulvus rufus*), and dwarf lemurs (*Microcebus coquereli*) have been observed hunting and eating smaller primates (see Hart, 2000). A few instances of primates preying on other primates are relatively well studied, particularly chimpanzee predation (Uehara et al., 1992; Stanford et al., 1994; Stanford, 1995; Stanford & Wrangham, 1998). At Gombe National Park in Tanzania, chimpanzee predation on red colobus (*Procolobus badius*) is extensive, alleged to account for "an annual harvest of from 16.8 to 32.9% of the red colobus population, depending on the number of male chimpanzees and the precise size of the red colobus population in a given year" (Stanford et al., 1994, p. 221).

Results

A total of 3,592 primate mortalities and unsuccessful predation attempts were identified. This establishes a baseline for understanding the biogeographical patterns of predation on primates. General patterns will be examined prior to discussing the four regions separately.

Figure 2.1 is an overall representation of 3,592 instances of predation cited in questionnaires and literature, classified by geographic region and predator category. Table 2.1 separates the predation incidents into unsuccessful attacks ($n = 679$, 18.9%), successful predations ($n = 2,229$, 62.1%), and suspected predations ($n = 684$, 19.0%). (See (Hart, 2000) for data sources and a discussion of the number of reported predation events as a function of the number of sources from which they were collected.) Felids and raptors accounted for the most predations on primates (34.6%, $n = 1,243$ and 40.7%, $n = 1,461$, respectively), followed by unidentified predators (9.0%, $n = 323$), canids and hyaenids (7.0%, $n = 253$), reptiles (5.4%, $n = 194$), and small carnivores (3.3%, $n = 118$).

Table 2.1 requires explanation lest the reader equate the number of predation events listed with the number of identified primate predator species. There is no direct cause and effect relationship between these two variables because the number of predation events is not random but, rather, the outcome of studies directed at specific primates or predators. Thus, the data on unsuccessful attacks, successful predations, and suspected predations by felids, raptors, canids and hyaenids, small carnivores, and reptiles are representative of those primates or predator species that

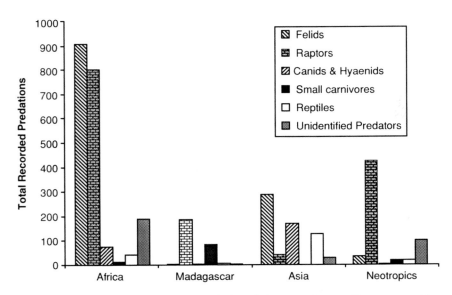

FIGURE 2.1. Overall magnitude of recorded predation on primates (Data source: Hart, 2000)

TABLE 2.1. Summary of recorded predations from questionnaires and literature (Data source: Hart, 2000).

Region and Predator	Unsuccessful Attacks	Successful Attacks	Suspected Predation	Number of Identified Predator Species
AFRICA				
Felid	66	725	123	7
Raptor	199	573	36	22
Canid & hyaenid	26	40	10	7
Small carnivore	4	9	0	9
Reptile	4	36	3	12
Unidentified	5	37	149	–
Total Africa	**304**	**1420**	**321**	**57**
MADAGASCAR				
Felid	0	3	1	–
Raptor	18	158	10	17
Canid & hyaenid	0	3	0	–
Small carnivore	5	63	17	7
Reptile	0	6	0	5
Unidentified	0	0	2	–
Total Madagascar	**23**	**233**	**30**	**29**
ASIA				
Felid	8	254	27	8
Raptor	13	26	4	15
Canid & hyaenid	13	58	100	6
Small carnivore	0	0	0	3
Reptile	2	41	83	7
Unidentified	0	10	19	–
Total Asia	**36**	**389**	**233**	**39**
NEOTROPICS				
Felid	10	20	6	6
Raptor	263	146	15	30
Canid & hyaenid	1	1	1	1
Small carnivore	15	3	2	4
Reptile	11	7	1	13
Unidentified	16	10	75	–
Total Neotropics	**316**	**187**	**100**	**54**
TOTAL	**679**	**2229**	**684**	

have been studied. On the other hand, the number of predator species associated with primate predation derives from both anecdotal and quantitative observations.

After making an initial assessment to gauge the magnitude of recorded predation in the four geographic regions, I eliminated data on suspected predations from further analysis. I based this decision on a simple rationale that there was an inherent margin of error built into the "suspected" classification. Even with the most conservative approach to judging suspected predation, it would be problematic to combine these data with those gathered from eyewitness observations and results from controlled studies. At this point in the meta-analysis I also combined the data from the remaining two classes—unsuccessful attacks and successful predations—since these categories were empirical in nature.

After graphing the magnitude of recorded predation, the next stage of data analysis explores primate predation separately in each geographical region and, further, attempts to isolate the variables that determine which groupings of primate species are preyed upon. In all four regions I examine the possible combinations of primate body size, stratum occupied, and activity cycle to see whether there are primates that are exempt from predation. Data indicate that none of the characteristics examined protects primates from predators. Although the exact rates of predation are often unknown, it is apparent from these data that primates are preyed upon if they are small or large, nocturnal or diurnal, arboreal or terrestrial.

Africa

African felids and raptors together accounted for the highest frequencies of primate predation, 53.7%, $n = 1,563$ (Figure 2.2). That more than half of all reported predation events can be attributed to felids and raptors in one region is most likely an artifact of the greater quantity of questionnaire returns and scientific articles based on field research in Africa than in other regions. Leopards are opportunistic ambush hunters that are a key predator of primates, particularly in African tropical rainforests (Boesch, 1991, 1992; Zuberbühler & Jenny, 2002). Two major studies in the Taï forest calculated relatively similar percentages of primates in leopard diets; Zuberbühler & Jenny, (2002) estimated that 27.9% ($n = 64$) of the leopard diet consisted of primates; Hoppe-Dominik (1984) estimated 24.2% ($n = 61$). Outside of rainforest habitat, leopards are also major predators of primates. During a study of vervet monkeys (*Cercopithecus aethiops*) in Amboseli National Park, Kenya, Isbell (1990) lost 45.0% ($n = 23$) of her study population to leopards in one year.

Other African wild cats also prey on primates. In the Mahale mountains of Tanzania, for example, lions (*Panthera leo*) appear to be major predators of primates (Tsukahara, 1993). Until recently, predation as a mortality factor for Mahale chimps was assumed to be negligible. However, this assumption has been challenged by evidence of chimpanzee hair, bones, and teeth in 4 out of 11 samples of lion feces collected over a four-month period. As another example, two lionesses near Mana Pools in Zimbabwe were known to favor baboons as prey. Over a six-year period, a safari guide in the area observed the lionesses killing six baboons (T. Williamson, pers. comm.).

African raptors have numerous failed predation attempts on primates ($n = 199$). This figure is borne out by observations of frequent unsuccessful attacks on Diana monkeys (*Cercopithecus diana*) by crowned hawk-eagles (*Stephanoaetus coronatus*, Figure 2.3) (Zuberbühler et al., 1997). The crowned hawk-eagle is one of the largest of the African eagles and is immensely powerful (Steyn, 1973; Williams & Arlott, 1980; Brown et al., 1982). Its thick tarsi, robust toes, and long talons enable it to kill large prey; with an average adult weight of 3.6 kg, the eagle routinely subdues animals four to five times its own size (Brown, 1971; Steyn, 1973, 1983; Brown et al., 1982; Tarboton, 1989). The crowned hawk-eagle

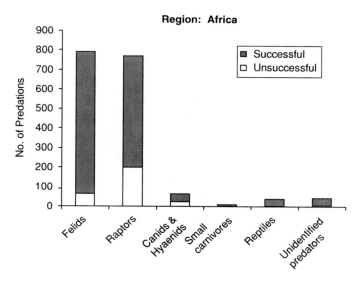

FIGURE 2.2. Comparison of successful and unsuccessful attacks by six categories of predators in Africa (Data source: Hart, 2000)

may be a primate specialist. Studies of this raptor in the Kibale forest of Uganda found high percentages of primates in eagle diets; Skorupa (1989) noted that 87.9% ($n = 29$) of eagle prey were monkeys, and Struhsaker & Leakey (1990) estimated this figure to be 83.7% ($n = 41$). Mitani et al., (2001) determined that primates composed the vast majority of crowned hawk-eagle prey items (82%, $n = 74$) in the Ngogo study site in Kibale during a 37-month study. At another research site in the Kiwengoma Forest Reserve, Tanzania, the skeletal remains found in one crowned hawk-eagle nest were "90% dominated by blue monkey" (Msuya, 1993, p.120). The geographic range of this raptor is extensive throughout the tropical belt of Africa. New research is finding that crowned hawk eagles exert much the same predation pressure on monkeys across different parts of their range (cf. Mitani et al., 2001; Shultz, 2001, 2002).

There are no arboreal-nocturnal primates weighing more than 2 kg in Africa, and there are no terrestrial-diurnal primates weighing less than 2 kg. (Of course, no terrestrial-nocturnal primates exist of any weight in any region.) Predation was recorded in the remaining four ecological categories identified in Figure 2.4. The single data point representing small, arboreal-diurnal primates refers to predation on the talapoin monkey (*Miopithecus talapoin*), the only African primate species in this category. The remaining three groups are dominated by guenons, mangabeys, and colobus in the arboreal-diurnal, over-2-kg category; arboreal-nocturnal primates under 2 kg refer to galagos and lorisids; terrestrial-diurnal primates over 2 kg include apes and baboons.

There are some interesting patterns that can be inferred from Figure 2.4. More terrestrial primate genera ($n = 7$) have evolved in Africa than other regions, and

FIGURE 2.3. Forest-hunting raptors, such as the African crowned hawk-eagle, are the major and most competent predators on primates (Steve Bird/Birdseekers Tours)

Africa is the only region in which there are more terrestrial-diurnal than arboreal-diurnal genera. Some of the information contained in Figure 2.4 likely represents an artifact of the numerous studies carried out on terrestrial primate species weighing over 2 kg, particularly baboons and chimpanzees. But it is difficult to say whether the 806 predations recorded in this category might also reflect an abundance of terrestrial primates, or might even point to a striking difference between arboreal and terrestrial primates as far as vulnerability to predators.

Madagascar

Corresponding information for Madagascar (Figure 2.5) shows an emphasis on raptor and small carnivore predation. Madagascar is the only region in which

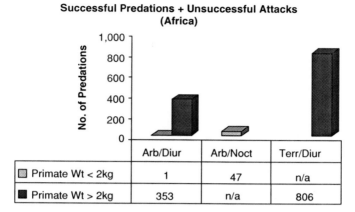

FIGURE 2.4. Comparison of recorded predation on African primates weighing <2 kg and >2 kg in three ecological groups; n/a denotes no primate species exist in that category (Data source: Hart, 2000)

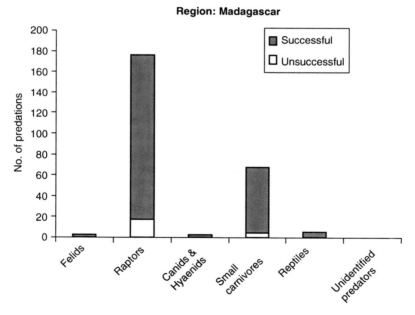

FIGURE 2.5. Comparison of successful and unsuccessful attacks by six categories of predators in Madagascar (Data source: Hart, 2000)

small carnivores (specifically, the fossa, *Cryptoprocta ferox*) are important as primate predators. Indeed, more than half of the predation data for all four regions included in the small carnivore category of Table 2.1 refer to the fossa. This is easily attributed to the fossa's unique status on the island of Madagascar. No wild cats are indigenous to the island, and the fossa (a viverrid that weighs 20 kg and resembles a small North American puma) occupies the ecological niche of the island's absent felids. (The few instances of felid predation shown in Figure 2.5 are due to feral cats.) Some studies reveal that small carnivores, such as the fossa, may not target any particular age or sex of primate prey (Wright et al., 1997, 1998). Wright et al. (1998) described fossa as "equal opportunity" predators; deaths due to fossa predation in three groups of Milne-Edward's sifakas (*Propithecus diadema edwardsi*) were spread over all age and sex classes.

The fossa is the only species of small carnivore that has been the subject of repeated studies that have the objective of understanding the ecological relationship between a predator and its primate prey (Rasolonandrasana, 1994; Rasoloarison et al., 1995; Goodman et al., 1997; Wright et al., 1997). It is interesting to speculate that many of the small, fast-moving, arboreal carnivores may have the same capacity as fossa to inflict heavy predation on arboreal primates. At least six other species of these small carnivores prey on Madagascar primate fauna; they are Indian civet (*Viverricula indica*), Malagasy civet (*Fossa fossana*), narrow-striped mongoose (*Mungotictis decemlineata*), ring-tailed mongoose (*Galidia elegans*), Malagasy brown-tailed mongoose (*Salanoia concolor*), and broad-striped mongoose (*Galidictis spp.*). Small carnivores may be important predators on primates in other regions also, but no quantitative information exists on diets of African, Asian, or Neotropical small carnivores that have been identified as primate predators.

Malagasy prosimians (Figure 2.6) occupy five of the ecological groupings identified here. Arboreal-diurnal primates weighing less than 2 kg are represented only by bamboo lemurs (*Hapalemur spp.*); those over 2 kg include *Propithecus, Indri, Varecia,* and *Eulemur.* (For the purpose of comparison, cathemeral species,

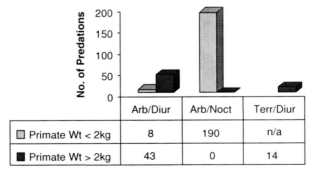

FIGURE 2.6. Comparison of recorded predation on Malagasy primates weighing <2 kg and >2 kg in three ecological groups; n/a denotes no primate species exist in that category; 0 denotes no predation events were reported (Data source: Hart, 2000)

such as *Eulemur*, were analyzed with arboreal-diurnal species.) The category of arboreal-nocturnal primates weighing less than 2 kg is occupied by the *Cheirogaleidae*. The terrestrial-diurnal, over-2-kg category is filled by the ring-tailed lemur (*Lemur catta*). The aye-aye (*Daubentonia madagascariensis*) is the only primate in Madagascar that is arboreal-nocturnal and weighs more than 2 kg. Except for *Daubentonia*, predation has been recorded for all other families of Malagasy primates.

Nocturnal raptors (the Malagasy owls) and the diurnal Madagascar harrier hawk (*Polyboroides radiatus*) are frequent predators on prosimians (Goodman et al., 1991; Goodman & Langrand, 1993; Goodman et al., 1993a, 1993b; Karpanty & Goodman, 1999; Brockman, 2003). The increasing number of studies that document Malagasy raptor diets has served to reveal the extent to which primates incur predation. Diurnal raptors, such as the Madagascar harrier hawk, are major predators of Verreaux's sifaka (*Propithecus verreauxi*), even though the primates are two to three times the size of the raptor (Karpanty & Goodman, 1999). Henst's goshawks (*Accipiter henstii*) weigh only 1.2 kg but successfully prey on large-bodied, arboreal-diurnal species as well as small-bodied, nocturnal primates (Goodman et al., 1998; Karpanty, 2003).

There is conspicuously high predation on small, arboreal-nocturnal primates in Madagascar. This may reflect the fact that Madagascar is the only region in which more nocturnal than diurnal primate genera have evolved. Ornithological research has made it apparent that small nocturnal primates on Madagascar constitute a prey base for many species of endemic owls, for example, *Tyto soumagnei*, *Otus rutilus*, and *Asio madagascariensis*, along with the Malagasy subspecies of barn owl (*Tyto alba affinis*) (Goodman et al., 1991; Goodman & Langrand, 1993; Goodman et al., 1993a, 1993b).

Asia

Leopards and tigers (*Panthera tigris*) incur a substantial impact on Asian primates. A good example comes from research in the Periyar Tiger Reserve, South India, where 81.4% (*n* = 79) of the leopard diet from September 1991–September 1994 consisted of Nilgiri langur (*Trachypithecus johnii*) (Srivastava et al., 1996). In Meru-Betiri Reserve, Indonesia, langurs and macaques were the predominant food of the leopard (56.9%, *n* = 33) in a study carried out by Seidensticker and Suyono (1980). Perhaps less intuitive than the leopard's reliance on primate prey is the tiger's penchant for primates. Tigers are usually assumed to take only very large ungulate prey. Nevertheless, Hanuman langurs (*Presbytis entellus*) are frequent prey of tigers in the forest of Ranthambhore, India, where the monkeys are often captured when moving between trees (Thapar, 1986). Schaller (1967) calculated that langurs made up 7.0% (*n* = 21) of the tiger diet in Kanha Park, India; Sunquist (1981) studied the composition of tiger diets in Chitawan Park, Nepal, finding that 5.7% (*n* = 7) consisted of langurs. Two recent studies carried out in Bangladesh and India indicate that rhesus macaques (*Macaca mulatta*) and langurs were the third highest components in tiger diets (Reza et al., 2001; Sankar & Johnsingh, 2002).

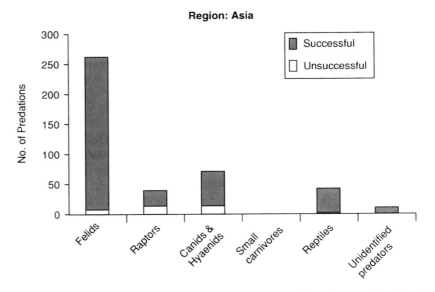

FIGURE 2.7. Comparison of successful and unsuccessful attacks by six categories of predators in Asia (Data source: Hart, 2000)

With regard to currently available data on primate predators, canids and hyaenids are not heavily represented in any region. Nevertheless, Asian canids—the golden jackal (*Canis aureus*) and the dhole (*Cuon alpinus*)—figure prominently as predators (Johnsingh, 1980; Newton, 1985; Stanford, 1989; D'Cunha, 1996; see Figure 2.7). Several Asian canids not previously considered primate predators have been identified in recent years. N. Itoigawa (pers. comm.) related that he has received anecdotal reports concerning red fox (*Vulpes vulpes*) and raccoon dog (*Nyctereutes procyonoides*) predation on Japanese macaques *(Macaca fuscata)*.

Wolves (*Canis lupus*) still exist in Saudi Arabia and other parts of Southwest Asia and are known to be quintessential opportunists throughout their nearly global range. Remains of *Papio hamadryas* were found in wolf scats in the Arabian Peninsula (Biquand et al., 1994). The decline in large Asian carnivores has been dramatic over the last several decades, but in the early 1970s wolves and Asian black bears (*Selenarctos thibetanus*) in Nepal were alleged to prey on Hanuman langurs (Bishop, 1975).

Asia is also notable for a relatively high incidence of reptile predation on primates. There are more reptile predations ($n = 43$) in Asia than in other geographic regions, although Africa has nearly as many ($n = 40$). When "suspected" reptile predations are added to successful and unsuccessful categories (refer to Table 2.1), the Asian figure ($n = 126$) is nearly three times higher than the figure for Africa

($n = 43$), over six times higher than that of the Neotropics ($n = 19$), and twenty-one times greater than the number for Madagascar ($n = 6$).

The first quantitative study of large tropical snake diets was published less than a decade ago (Shine et al., 1998). Specimens of *Python reticulatus* (an Asian snake in which females routinely reach a length of 7 m) were examined for stomach contents within the context of commercial exploitation for the skin trade. Although large ungulate prey were more easily identified in the hindgut than smaller primate species, Shine et al. (1998) calculated that 3.4% ($n = 14$) of the identifiable remains of food in the python alimentary tracts consisted of macaques and langurs. Pythons are also known to consume small, nocturnal Asian primates (Wiens & Zitzmann, 1999). During a study of slow loris (*Nycticebus coucang*) in Indonesia, weak signals from a radio-collared focal animal were traced to dense ferns on the forest floor. When these signals continued over a three-day period from such an unlikely location for an arboreal primate, researchers investigated and found a reticulated python. The signals were being emitted from the interior of the python, which had swallowed the loris.

Compared to other regions, the level of primate predation by raptors in Asia is low. Probably correlated with this minimal level is the fact that fewer raptor species have been identified as primate predators in Asia than other regions. Another reason may be a lack of field studies on South and Southeast Asian raptors. (Other than the Philippine eagle, *Pithecophaga jeffery*, I found no literature on the diets of Asian raptor species known to prey on primates.) If a similar body of field research becomes available for Asian raptors, as now exists for African birds of prey, this picture may change.

Asian primates (Figure 2.8) occupy only three of the ecological groups identified here: arboreal-diurnal primates over 2 kg in weight (*Pongo, Presbytis, Trachypithecus, Nasalis*, and others), terrestrial-diurnal primates over 2 kg

Successful Predations + Unsuccessful Attacks (Asia)

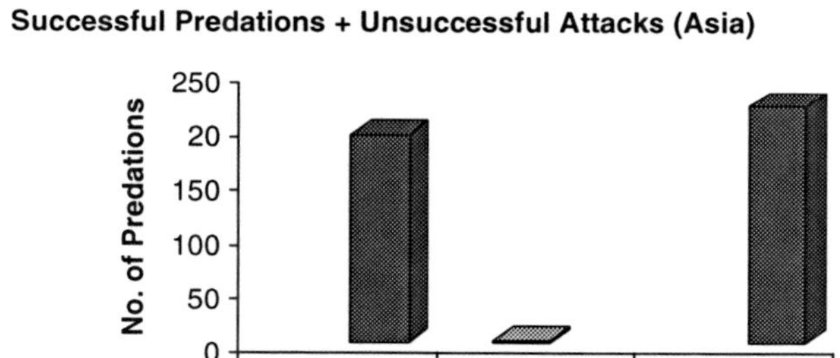

	Arb/Diur	Arb/Noct	Terr/Diur
Primate Wt < 2kg	n/a	2	n/a
Primate Wt > 2kg	193	n/a	219

FIGURE 2.8. Comparison of recorded predation on Asian primates weighing <2 kg and >2 kg in three ecological groups; n/a denotes no primate species exist in that category (Data source: Hart, 2000)

(*Macaca*), and arboreal-nocturnal primates under 2 kg (*Tarsius, Nycticebus, Loris*). The large diurnal species are preyed on by leopards, tigers, dholes, jackals, crocodiles, and snakes, but until recently there were so few studies on small, nocturnal Asian primates (Rasmussen, 1997) that only two incidents were available for examination at the time of this meta-analysis. There are three genera of prosimians in Asia, half the number found in Africa and Madagascar, but the current surge in field research on nocturnal Asian primates has greatly expanded knowledge about predation on these species (see Wiens and Zitzmann, 1999, 2003; Gursky, 2002, 2003, 2005; Lakshmi and Mohan, 2002; Nekaris, 2003; Nekaris and Jayewardene, 2004).

The Neotropics

Figure 2.9 represents an overview of primate predation in the Neotropics. The paucity of felid predation is readily apparent despite the fact that two large cat species: jaguar (*Panthera onca*) and puma (*Felis concolor*), and four small felids: ocelot (*F. pardalis*), jaguarundi (*F. yagouroundi*), margay (*F. wiedii*), and oncilla (*F. tigrina*), have been identified as primate predators.

A variety of small hawk and falcon species inhabit Central and South American forests. Neotropical raptor species are twice as numerous as Old World species mainly because of the ubiquitous small forest falcons of the genus *Micrastur*. Thiollay (1985) describes the hunting techniques of small rainforest hawks and falcons as a combination of active and inactive behaviors; sitting motionless

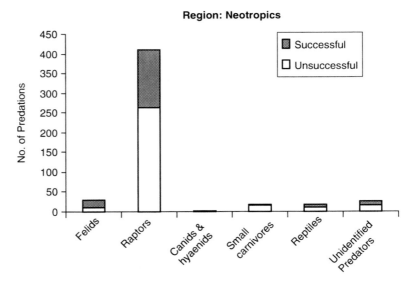

FIGURE 2.9. Comparison of successful and unsuccessful attacks by six categories of predators in the Neotropics (Data source: Hart, 2000)

FIGURE 2.10. The harpy eagle is the premier raptor of the Neotropics (Used by permission of R.W. Sussman)

and inconspicuous, they intersperse inactivity with occasional swift, soundless flights from tree to tree. Some species, such as the collared forest falcon (*M. semitorquatus*), pursue active hunting. This raptor actually runs along branches in pursuit of prey (Thiollay, 1985). Many of the predation attempts by Neotropical hawks, falcons, and toucans are unsuccessful, but this does not deter frequent attacks on callitrichids and very young squirrel monkeys (Terborgh, 1983; Boinski, 1987; Goldizen, 1987; Mitchell et al., 1991).

The harpy eagle (*Harpia harpyja*, Figure 2.10) is one of the largest and strongest raptors in the world (Brown & Amadon, 1989). This species exhibits the same short, broad wings and relatively long, graduated tail as the crowned hawk-eagle of

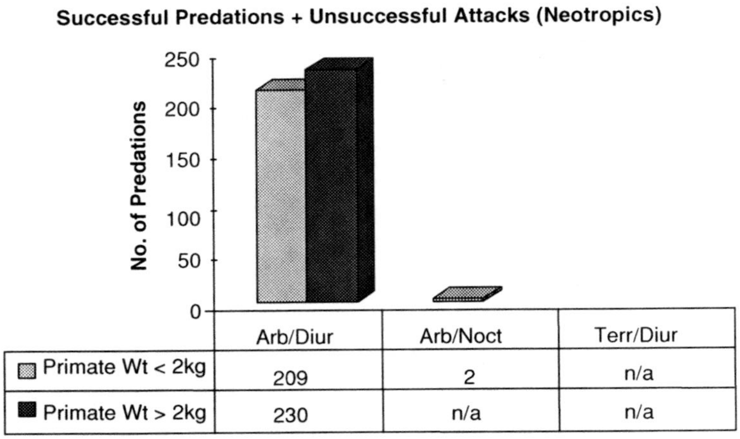

Successful Predations + Unsuccessful Attacks (Neotropics)

	Arb/Diur	Arb/Noct	Terr/Diur
▨ Primate Wt < 2kg	209	2	n/a
■ Primate Wt > 2kg	230	n/a	n/a

FIGURE 2.11. Comparison of recorded predation on Neotropical primates weighing <2 kg and >2 kg in three ecological groups; n/a denotes no primate species exist in that category (Data source: Hart, 2000)

Africa. Ecological equivalents, the two raptors have garnered similar reputations as premier predators on monkeys (Izor, 1985).

There are two features of the Neotropical primate component not found in other regions (Figure 2.11): it lacks a terrestrial species and it has only a single nocturnal genus. Considerable predation is recorded for small and large arboreal-diurnal Neotropical primates, i.e., the callitrichids and the cebids. The only New World primate that is arboreal and nocturnal is *Aotus,* the owl monkey, for which a small number of predations by owls has been recorded (Wright, 1985; Brooks, 1996). There are no Neotropical primates inhabiting other ecological divisions identified here.

Estimated Predation Rates

Estimated predation rate (EPR), the percentage of a primate population killed annually by predators, provides a valuable insight into the effect predation has on a primate group. Additionally, EPR calculations measure the effect of predator mortality on all components of the population, including the reproductively active portion. This is an important caveat since estimated rates of predation on immature primates (infant and juvenile age classes) may be higher in comparison to adults. Janson and van Schaik (1993) compared immature versus mature primates and estimated the predation rate was 3–17 times higher for immature individuals than for adults in species of cercopithecines and 3–6 times higher in cebids.

Figure 2.12 displays mean estimated predation rates for four regions. Madagascar has the highest mean EPR (8.9 %, $n = 6$), and Asia has the lowest (3.0 %, $n = 19$). Mean EPRs for Africa and the Neotropics are 5.6% ($n = 57$) and 6.7% ($n = 14$), respectively. Estimated predation rates ranged from zero to

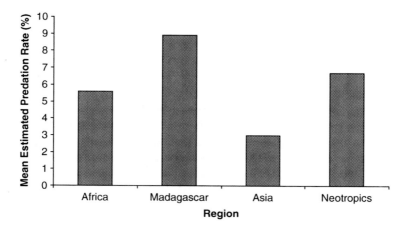

FIGURE 2.12. Comparison of mean estimated predation rates by region (Data sources: Hart, 2000; Mitani et al., 2001; Shultz, 2003)

TABLE 2.2. Estimated predation rates for primate weight and ecological groupings (Data sources: Hart, 2000; Mitani et al., 2001; Shultz, 2003).

Primates < 2 kg Arboreal Diurnal ($n = 7$)	Primates < 2 kg Arboreal Nocturnal ($n = 4$)	Primates > 2 kg Arboreal Diurnal ($n = 38$)	Primates > 2 kg Terrestrial Diurnal ($n = 44$)
Mean EPR 7.0%	Mean EPR 15.8%	Mean EPR 5.4%	Mean EPR 4.4%
Range 1.0–15.0%	Range 8.6–25.0%	Range 0–18.0%	Range 0–15.0%

25.0% in this sample. The inclusion of a zero predation rate is due to calculations by questionnaire respondents who lost no study animals to predators over a number of years. (Of course, it is possible that aberrant conditions existed at these study sites, such as the eradication of predators in the area or human disturbance causing predators to disperse.) The highest rate in the sample is 25.0% of a *Microcebus* population lost to predation each year (Goodman et al., 1993c). This EPR is based on predation by two genera of owls and does not include additive predation by diurnal raptors, snakes, or small carnivores. The high reproductive potential of *Microcebus* counteracts what would seem to be an intolerable level of predation (Goodman et al., 1993c; Hill & Dunbar, 1998). Unlike most primate species, some Malagasy prosimians (including *Microcebus*) produce an average of two infants twice per year (Martin, 1972). The mouse lemur is able to sustain a predation rate of 25.0% because, for a primate, it has a very high reproductive potential (Goodman et al., 1993c).

In Table 2.2 estimated predation rates for primates are summarized from the perspective of ecological groupings used in this chapter. The highest predation rate was incurred by small, arboreal-nocturnal primates. This may be partially reflective of the 25.0% EPR calculated for mouse lemurs; in addition, the sample sizes are very small for two of the categories ($n = 4, n = 7$) and relatively large for the other two ($n = 38, n = 44$).

Frequency of Occurrence of Primates in Predator Diets

Frequency of occurrence is defined as the number of individual prey animals of one taxon relative to all prey eaten (Rabinowitz & Nottingham, 1986). Expressed as a percentage of all food intake by a predator, frequencies of occurrence can be estimated using various methods. (Table 2.3 lists these techniques along with the number of studies in the data set that used each sampling method.) Boshoff et al. (1994) give an excellent explanation of how frequencies of occurrence provide a good approximation of the composition and species richness of prey; any biases can be assumed to be common to all samples, so comparison between samples is valid.

It should be noted that frequencies of occurrence of primates in predator diets are based on conservative estimates. These methods usually result in underestimates since biases against finding the remains of young, small, or nocturnally active prey are exacerbated by several processes (Rice, 1986; Thapar, 1986). Primates are often underrepresented when frequencies of occurrence are calculated from direct observation of kills or examination of prey carcasses due to the rapidity with which small carcasses are consumed by large carnivores (Schaller, 1972; Eloff, 1973; Floyd et al., 1978; Bothma & Le Riche, 1986). Furthermore, the chance that skeletal remains pass through the digestive tract of a carnivore in recognizable form is greater for large prey animals than for smaller ones (Muckenhirn, 1972). Even when the largest primates fall prey to a carnivore, the remains disappear rapidly in tropical climates. All traces of a western lowland gorilla killed by a leopard in Gabon were nearly gone three or four days after death due to consumption by the primary predator, scavengers, and insects (Tutin & Benirschke, 1991). A similar amount of time was noted for the disappearance of a chimpanzee carcass after leopard predation in the Taï forest, Côte d'Ivoire (Boesch, 1991). Fecal samples from predators are also difficult to collect in tropical forests because they may be destroyed within hours by dung beetles and trigonid bees; only those containing large amounts of fur or those placed in sunny areas survive a few days (Emmons, 1987).

The most commonly used methods (fecal sampling, pellet/regurgitation sampling, analysis of nest or den remains, and analysis of prey carcasses) provide

TABLE 2.3. Frequency of occurrence sampling methods (Data source: Hart, 2000).

Type of Sampling Method	Number of Studies
Stomach contents	3
Fecal sampling	33
Pellets and regurgitations	8
Nest and den remains	38
Analysis of prey carcasses	9
Direct observation of kills	5

information on food ingested over an extended period of time and are non-invasive, unlike analysis of stomach contents, which involves dissection of the predator. Direct observation of kills has the advantage of providing indisputable confirmation of predation rather than scavenging, but it requires both perseverance and luck. As a sampling method it yields more limited information since only one meal at a time can be identified. Another drawback to direct observation is that prey are often alerted to predators or made more vigilant when human observers are present (Isbell & Young, 1993). Observing the kill of a secretive, nocturnal predator, such as the leopard, is particularly problematic. Despite nearly half of the vervet population under study falling victim to leopards during one year at Amboseli National Park, Kenya, no monkeys were killed within sight of researchers (Isbell, 1990). The sampling of feces, regurgitations, nest or den remains, and prey carcasses provides an estimate of the minimum number of preyed-upon individuals of one taxon, and it requires a tedious cleaning and reconstruction process (Figure 2.13). Nest and den remains yield excellent data for compilation of predator diets since several nesting cycles result in large build-ups of prey bones within and below raptor nests (Sanders et al., 2003; Shultz et al., 2004). The larger the collection of nest and den remains the greater the accuracy of dietary content.

The percentage of a predator's diet composed of primates ranged widely in the data set described here. At the upper end of a continuum, nest remains of forest-hunting African crowned hawk-eagles identified 80–90% of their diet as primates

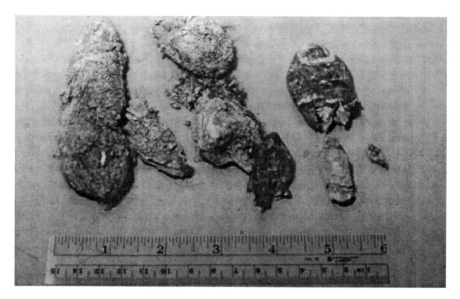

FIGURE 2.13. Leopard scat containing two gorilla hind digits was found by researchers in the Central African Republic; one intact toe has been removed from the fecal matter and is clearly visible on the right (Used by permission of Michael Fay)

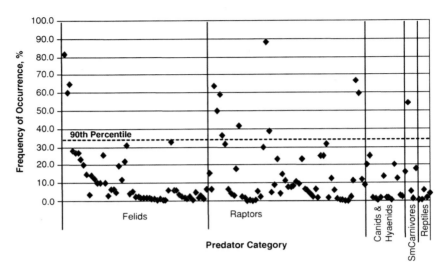

FIGURE 2.14. Frequency of occurrence of primates in predator diets. Each point represents data from a field study on a species of felid, raptor, canid and hyaenid, small carnivore, or reptile plotted as a percentage of primates found in the diet of a single predator. Median values: felids, 5.2%, $n = 53$; raptors, 7.6%, $n = 59$; canids and hyaenids, 2.0%, $n = 13$; small carnivores, 10.6%, $n = 4$; reptiles, 3.0%, $n = 6$ (Data source: Hart, 2000)

of various species (Skorupa, 1989; Struhsaker & Leakey, 1990; Msuya, 1993; Mitani et al., 2001). At the lower end a study of Verreaux's eagles (*Aquila verreauxii*), an African savanna raptor, estimated only 0.05% ($n = 27$) of the diet was composed of primates (Boshoff et al., 1991). Frequencies of occurrence were available from 96 studies on 35 species of predators. More data were collated on felid and raptor diets containing primates than for other predators (Figure 2.14). Extensive research is available analyzing the total range of hyena and wild canid prey, mostly large savanna ungulate species (Estes & Goddard, 1967; Henschel & Tilson, 1988; Johnsingh, 1980, 1983; Kruuk, 1970, 1972; Kruuk & Turner, 1967; Mills & Biggs, 1989), so it is plausible to compare them with felids and raptors in Figure 2.14. Reptile and small carnivore species inhabiting the same geographic ranges as primates have not been the focus of many studies intended to generate information on diet composition (reptiles $n = 5$, small carnivores $n = 4$). Taking this into consideration, however, it is still apparent that felids and raptors are major predator groups where the killing of primates is concerned. Only felids, raptors, and one small carnivore, the fossa, have frequencies of occurrence that fall above the 90th percentile.

In Figures 2.15–2.18 means were determined for the percentage of primates in the diets of different predator groups by first averaging each separate species' frequency of occurrence percentages and then calculating the mean for all species within each predator group. These means are presented separately for Africa, Madagascar, Asia, and the Neotropics to facilitate comparison across regions.

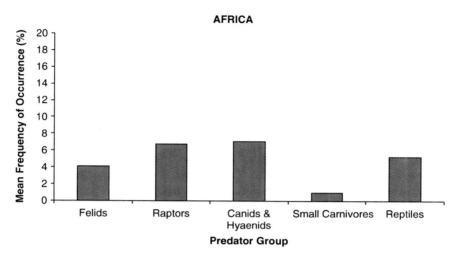

FIGURE 2.15. Five groups of African predators are compared by frequency of occurrence of primates in their diets. Number of identified primate predators in Africa: felids, $n = 7$; raptors, $n = 22$; canids and hyaenids, $n = 7$; small carnivores, $n = 9$; reptiles, $n = 12$ (Data sources: Hart, 2000; Mitani et al., 2001; Shultz, 2002; Zuberbühler & Jenny, 2002)

In Africa (Figure 2.15) there is a relatively narrow range of percentages of primate prey among the five predator groups; the highest mean component of primates occurs in canid and hyaenid diets (7.1%, $n = 2$ species), and the lowest occurs in small carnivore diets (1.0%, $n = 1$ species). Within the narrow range, raptor diets averaged 6.8% primate prey ($n = 7$ species), reptiles 5.3% ($n = 2$ species), and felids 4.1% ($n = 4$ species).

Frequency of occurrence of primates in Malagasy predator diets (Figure 2.16) reveals an emphasis on raptor and small carnivores. Mean raptor frequency of occurrence was 17.2% ($n = 6$ species), and mean small carnivore frequency was 25.1% ($n = 1$ species). Seventeen raptor species have been identified as primate predators in this region (58.6% of the total predator component). This is the highest ratio of raptor to total predator numbers in any region. The highest estimated predation rates in any region are also due to Malagasy birds of prey.

The frequency with which primates appear in the diets of Asian predators (Figure 2.17) is similar to Africa except that felids have a much higher mean frequency of primates in their diets (15.0%, $n = 2$ species). Raptors averaged 4.4% primate prey ($n = 1$ species), canids and hyaenids 4.0% ($n = 3$ species), and reptiles 4.1% ($n = 2$ species). No frequency of occurrence data were available for small carnivores in Asia.

Neotropical raptors have the highest mean percentage of primates in their diets (36.6%, $n = 2$ species) of any predator group in any region. All other predator consumption of primates in the Neotropics is negligible by comparison. Figure 2.18 also presents an apparent association between the number of identified Neotropical raptor species that prey on primates ($n = 30$) and these high

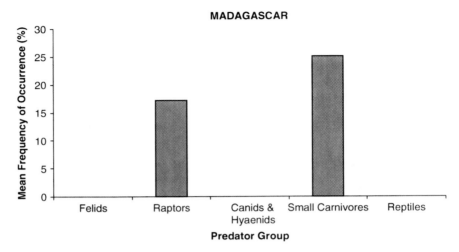

FIGURE 2.16. Five groups of Malagasy predators are compared by frequency of occurrence of primates in their diets. Number of identified primate predators in Madagascar: felids, $n = 0$; raptors, $n = 17$; canids and hyaenids, $n = 0$; small carnivores, $n = 7$; reptiles, $n = 5$ (Data sources: Karpanty & Goodman, 1999; Hart, 2000; Thorstrom & La Marca, 2000)

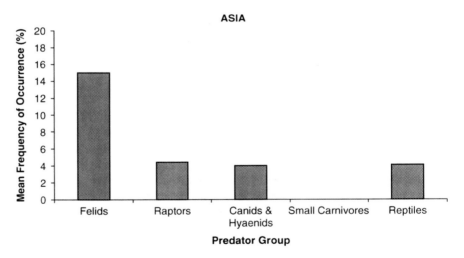

FIGURE 2.17. Five groups of Asian predators are compared by frequency of occurrence of primates in their diets. Number of identified primate predators in Asia: felids, $n = 8$; raptors, $n = 15$; canids and hyaenids, $n = 6$; small carnivores, $n = 3$; reptiles, $n = 7$ (Data sources: Hart, 2000; Reza et al., 2001; Sankar & Johnsingh, 2002; Uhde & Sommer, 2002)

frequencies of occurrence. The mean primate component in raptor diets in the Neotropics is more than twice as high as this figure in Madagascar, more than four times higher than Africa's, and more than eight times higher than the figure in

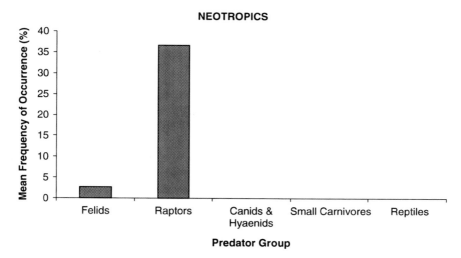

FIGURE 2.18. Five groups of Neotropical predators are compared by frequency of occurrence of primates in their diets. Number of identified primate predators in the Neotropics: felids, $n = 6$; raptors, $n = 30$; canids and hyaenids, $n = 1$; small carnivores, $n = 4$; reptiles, $n = 13$ (Data source: Hart, 2000)

Asia. Therefore, while there are more Neotropical raptor species, they also prey on many more primates than raptors of other regions.

Discussion

Biogeographical associations and insights have emerged from this meta-analysis despite the necessary reliance on preliminary and non-random data. What are the biogeographical patterns that account for links between primate regions and certain types of predation? With some exceptions, there appears to be a possibility of two primate predation patterns based on frequency of occurrence data. One pattern is apparent in Africa and Asia and consists of modest levels of predation spread among many predator taxa. It may be that shared predators (the leopard, lion, cheetah, striped hyena, and several species of canids) in combination with shared primate taxa (Catarrhini) enhance the perceived similarity between the two regions.

The other pattern found in Madagascar and the Neotropics consists of heavy predation by a narrower range of predators. High levels of raptor predation define a common link between Madagascar and the Neotropics. The four highest mean frequencies of primates in individual predator species diets were raptors indigenous to Madagascar and the Neotropics—Henst's goshawks, Madagascar long-eared owls (*Asio madagascariensis*), harpy eagles, and Guiana crested eagles (*Morphnus guianensis*). There is a complete absence of wild felids, canids, or hyaenids in Madagascar. While this pattern is not paralleled in the Neotropics, especially concerning wild cat species, there are no hyaenids, and only one wild canid predator—the coyote (*Canis latrans*)—is suspected to be a primate predator in Central America. Nevertheless, it would be presumptuous to infer indelible patterns from the analysis in this chapter, due to the many limiting factors. Perhaps the most limiting factor is that extensive research efforts are made on certain species of primate predators while other identified predators remain known only through anecdotal reports. This lack of random data collection skews the picture of primate predation to an unknown degree.

Primates are "generalist" prey in the sense that, as a taxon, they range in size from 60 g to 169.5 kg, they inhabit geographic ranges throughout the tropics, subtropics, and a few temperate forests, they range from completely arboreal to wholly terrestrial, and they include both nocturnal and diurnal species. Their successful radiation into many ecological niches carried with it the potential to interact with many predators. The 174 primate predators identified in Hart (2000) include many opportunistic feeders. While there are key primate predators among these species, there are no examples of predators with a rigidly narrow food base that forces them to prey only on primates.

Co-evolution between predators and their primate prey is most visible from the behavioral and morphological adaptations in primates that are traceable to specific predators (Terborgh, 1983). For example, primate polyspecific associations are limited to geographic regions inhabited by monkey-eating raptors (e.g., harpy eagles of the Neotropics and crowned hawk-eagles of Central and West Africa), which are predators that provide a strong incentive for aggregation (Gautier-Hion et al., 1983; Terborgh, 1990a). Terborgh (1983) discussed the relationship between body size and methods of escape from raptor predation among Neotropical primates at Cocha Cashu, Peru. He identified three distinct strategies adopted by primates: crypsis, group living, and escape from predators through an increase in size. The smallest primates (tamarins and marmosets) spend many hours per day in safe hiding places; medium-sized *Cebus* and *Saimiri* seek protection in groups. The remaining evolutionary adaptation in Terborgh's model, that of size increase, applies to adults of the largest Neotropical species, i.e., *Ateles, Brachyteles, Lagothrix,* and *Alouatta*. These primates often rest in conspicuous exposed perches in the canopy, from which they scan for harpy and Guiana crested eagles. Although the two raptors are known as capable predators of the largest Neotropical primates, they do not pass up primates of any size. Harpy eagles prey most frequently on *Cebus* (Voous, 1969); Guiana crested eagles even prey on infant tamarins (Vasquez & Heymann, 2001).

Consistently high predation rates on primates may indicate long-term predator-prey relationships. Many years of recording leopard predation on vervets at Amboseli have produced an estimated predation rate of 11.0–15.0% (Cheney & Wrangham, 1987; Isbell, 1990). Owl predation on mouse lemurs (Goodman et al., 1993b) was estimated to be 25.0% annually. However, a "high" estimated predation rate is not the only, or necessarily most important, criterion for determining the levels at which certain predators may kill primates for food. The estimated predation rates for crowned hawk-eagle exploitation of red colobus, black and white colobus, mangabey, and blue monkeys in the Kibale forest range from 0.3–3.0%, depending on the species of primate, but the frequency of occurrence of primates in the diet of the eagle pairs under study was 83.7% (Struhsaker & Leakey, 1990). Thus, frequency of occurrence of primates in the diet of a predator may be a more precise measure of the predator-prey relationship than EPR since the latter can be calculated as the collective effect from many predators in an ecosystem. Frequencies of occurrence, on the other hand, present a clear connection between the predator and its prey.

Primates have been observed to be secondary prey in some geographic locations and primary prey for the same predator species in another (Brown, 1966; Seidensticker & Suyono, 1980). Differences may exist in levels of predation on primates due to richness of other fauna or because other prey species have been eliminated by natural or human-induced causes. Seidensticker (1983, 1985, 1991) examined field studies containing reliable data in order to identify the environmental correlates in which primates account for a major portion of African and Asian leopard diets. He credits primate body size and availability of ungulate prey as key factors: If there were abundant ungulate species in the 20–50-kg range, leopards ate few primates; if ungulates in this size class were present but at low density, leopards had intermediate numbers of primates (i.e., <30%) in their diet; if this size class of ungulate was missing from the faunal composition, leopards had high proportions of primates in their diet. In four cases this pattern is substantiated: (1) Seidensticker & Suyono (1980) discovered that *Trachypithecus cristata* and *Macaca fascicularis* were the predominant food of tiger, leopard, and dhole in Meru-Betiri Reserve, Indonesia, because small ungulates have been extirpated by humans. Primates in the reserve are the substitute for a range of other prey normally available to large Asian carnivores. (2) In the Periyar Tiger Reserve, India, where Nilgiri langur account for 81.4% of the leopard diet, there is an absence of large ungulate species such as chital (*Axis axis*), hog deer (*A. porcinus*), and swamp deer (*Cervus duvanceli*). Ungulates weighing 20–50 kg are also not available to leopards; the Nilgiri tahr (*Hemitragus hylocrius*) exists only in isolated pockets, and sambar (*Cervus unicolor*) is the major prey item in the diet of a competing predator, the pack-hunting dhole (Srivastava et al., 1996). (3) At another site in India, Eravikulam National Park, where small ungulates, such as Nilgiri tahr, sambar, and barking deer (*Muntiacus muntjak*) were common, the remains of these animals occurred in 94.0% of tiger droppings collected for analysis and in 77.0% of leopard droppings. Remains of Nilgiri langurs appeared in no tiger droppings and in 27.0% of leopard droppings (Rice, 1986). In addition, all leopard sightings occurred within the home ranges of tahr, and leopards were seen hunting tahr in 36.0% of the sightings (Rice, 1986). (4) Niokolo-Koba National Park in Senegal does not contain dense concentrations of ungulates, and Guinea baboons (*Papio papio*) are the commonest large herbivore (Byrne, 1982). A high risk of predation from healthy populations of both diurnal African hunting dogs (*Lycaon pictus*) and nocturnal predators, such as leopards, lions, and spotted hyenas (*Crocuta crocuta*), was inferred from baboon behavior and social structure, specifically through frequent alarm vocalizations, extreme wariness of open spaces, and unusually high numbers in baboon troops. In addition, a paucity of secure sleeping sites may increase the likelihood that considerable predation on baboons occurs. Baboons are "likely to be more important in the diet of all large predators than would be the case in East Africa" (Byrne, 1982, p. 308).

Studies of geographically variable interactions have been credited with furthering an understanding of how evolution affects predator-prey systems (Abrams, 2000). Before true comparisons can further our understanding of the evolutionary ecology of primate predation, however, it will be necessary to study many more predator species throughout the four regions in which primates exist. When more of this critical information is forthcoming, the biogeographic emphasis can then shift from the search for mere associations to that of statistical correlations that may exist between predation and primate ecology, morphology, and behavior. That said, the four regional analyses, in which all possible combinations of primate body size, stratum occupied, and activity cycle were examined for any ecological groups that might be exempt from predation (see Figures 2.4, 2.6, 2.8, and 2.11), indicate the extent and all-encompassing character of predation on primates. There were no variable combinations of body size, stratum, activity cycle, or geographic region that protected primates from predators. Even without knowledge of the exact rates of predation, it is safe to hypothesize that primates are preyed upon no matter what size they are or what ecological variables they exhibit.

Acknowledgments. I wish to express my appreciation to two anonymous reviewers for their thorough and thoughtful suggestions to improve this chapter, as well as Pamela Ashmore, Robert W. Sussman, and Mary Willis for helpful comments. Thanks to Susanne Shultz for supplying predation rates by crowned hawk-eagles and leopards and to Scott McGraw for facilitating the transfer of that information. Thanks also to Michael Fay and Robert W. Sussman for the use of photographs. Gratitude is extended to Sharon Gursky and Anna Nekaris, editors of *Primate Anti-Predator Strategies*, for the opportunity to contribute this chapter. Acknowledgments to 279 questionnaire respondents can be found in Hart (2000).

References

Abrams, P. (2000). The evolution of predator-prey interactions: Theory and evidence. *Ann. Rev. Ecol. Syst.*, 31: 79–105.

Alvard, M. (1994). The sustainability of primate hunting in the Neotropics: Data from native communities. *Am. J. Phys. Anthr.*, (Suppl.18): 49.

Alvard, M., and Kaplan, H. (1991). Procurement technology and prey mortality among indigenous Neotropical hunters. In M. Steiner (Ed.), *Human predators and prey mortality* (pp. 79–104). Boulder, Colorado: Westview Press.

Biquand, S., Urios, V., Boug, A., Vila, C., Castroviejo, J., and Nader, I. (1994). Fishes as diet of a wolf (*Canis lupus arabs*) in Saudi Arabia. *Mammalia*, 58(3): 492–494.

Bishop, N. (1975). Social behavior of langur monkeys (*Presbytis entellus*) in a high altitude environment. Doctoral dissertation. University of California, Berkeley, California.

Boesch, C. (1991). The effects of leopard predation on grouping patterns in forest chimpanzees. *Behav.*, 117(3–4): 220–242.

Boesch, C. (1992). Predation by leopards on chimpanzees and its impact on social grouping. *Bulletin of the Chicago Academy of Science*, 15(1): 5.

Boinski, S. (1987). Birth synchrony in squirrel monkeys: A strategy to reduce neonatal predation. *Behavioral Ecology and Sociobiology*, 21(6): 393–400.

Boinski, S., and Chapman, C. (1995). Predation on primates: Where are we and what's next? *Evol. Anthropol.*, 4: 1–3.

Boshoff, A., Palmer, N., Avery, G., Davies, R., and Jarvis, M. (1991). Biogeographical and topographical variation in the prey of the black eagle in the Cape Province, South Africa. *Ostrich*, 62: 59–72.

Boshoff, A., Palmer, N., Vernon, C., and Avery, G. (1994). Comparison of the diet of crowned eagles in the savanna and forest biomes of southeastern South Africa. *S. Afr. J. Wildl. Res.*, 24: 26–31.

Bothma, J. du, and Le Riche, E. (1986). Prey preference and hunting efficiency of the Kalahari desert leopard. In S. Miller and D. Everett. (Eds.), *Cats of the world: Biology, conservation and management* (pp. 389–414). Washington, DC: National Wildlife Federation.

Brockman, D. (2003). *Polyboroides radiatus* predation attempts on *Propithecus verreauxi*. *Folia Primatol.*, 74: 71–74.

Brooks, D. (1996). Notes from the Paraguayan Chaco on the night monkey (*Aotus azarae*). *Neotropical Primates*, 4(1): 15–19.

Brown, J., and Lomolino, M. (1996). *Biogeography*. Sunderland, MA: Sinauer Associates.

Brown, L. (1966). Observations on some Kenya eagles. *Ibis*, 108: 531–572.

Brown, L. (1971). The relations of the crowned eagle *Stephanoaetus coronatus* and some of its prey animals. *Ibis*, 113: 240–243.

Brown, L., and Amadon, D. (1989). *Eagles, hawks and falcons of the world*. Secaucus, NJ: Wellfleet.

Brown, L., Urban, E., and Newman, K. (1982). *The birds of Africa*, Vol 1. London: Academic Press, Inc.

Byrne, R. (1982). Distance vocalisations of Guinea baboons (*Papio papio*) in Senegal: An analysis of function. *Behav.*, 78: 283–312.

Caine, N. (1990). Unrecognized anti-predator behavior can bias observational data. *Animal Behav.*, 39(1): 195–197.

Cheney, D., and Wrangham, R. (1987). Predation. In B. Smuts, D. Cheney, R. Seyfarth, R. Wrangham, and T. Struhsaker (Eds.), *Primate societies* (pp. 227–239). Chicago: Univ. of Chicago Press.

Cowlishaw, G. (1994). Vulnerability to predation in baboon populations. *Behav.*, 131(3–4): 293–304.

D'Cunha, E. (1996). Jackal (*Canis aureus*) hunting common langur (*Presbytis entellus*) in Kanha National Park. *J. Bombay Nat. Hist. Soc.* 93(2): 285–286.

Dunbar, R. (1988). *Primate social systems*. Ithaca, NY: Comstock Publications.

Eloff, F. (1973). Lion predation in the Kalahari Gemsbok National Park. *J. South Afr. Wildl. Manage. Assoc.*, 3: 59–63.

Emmons, L. (1987). Comparative feeding ecology of felids in a Neotropical rainforest. *Behavioral Ecology and Sociobiology*, 20: 271–283.

Estes, R., and Goddard, J. (1967). Prey selection and hunting behavior of the African wild dog. *J. Wildlife Manage.*, 31: 52–70.

Floyd, T., Mech, L., and Jordan, P. (1978). Relating wolf scat content to prey consumed. *J. Wildlife Manage.*, 42: 528–532.

Gautier-Hion, A., Quris, R., and Gautier, J. (1983). Monospecific vs. polyspecific life: A comparative study of foraging and anti-predatory tactics in a community of *Cercopithecus* monkeys. *Behavioral Ecology and Sociobiology*, 12(4): 325–335.

Goldizen, A. (1987). Tamarins and marmosets: Communal care of offspring. In B. Smuts, D. Cheney, R. Seyfarth, R. Wrangham, and T. Struhsaker (Eds.), *Primate societies* (pp. 34–43). Chicago: Univ. of Chicago Press.

Goodman, S. and Langrand, O. (1993). Food habits of the barn owl, *Tyto alba*, and the Madagascar long-eared owl, *Asio madagascariensis*, on Madagascar: Adaptation to a changing environment. In R. Trevor Wilson (Ed.), *The proceedings of the 8th Pan-African Ornithological Congress: Birds and the environment* (pp. 118–120). Tervuren, Belgium: Sciences Zoologiques.

Goodman, S., Creighton, G., and Raxworthy, C. (1991). The food habits of the Madagascar long-eared owl, *Asio madagascariensis*, in southeastern Madagascar. *Bonn. Zool. Beitr.*, 42(1): 21–26.

Goodman, S., de Roland, L., and Thorstrom, R. (1998). Predation on the eastern woolly lemurs (*Avahi laniger*) and other vertebrates by Henst's goshawk (*Accipiter henstii*). *Lemur News* (3): 14–15.

Goodman, S., Langrand, O., and Rasolonandrasana, B. (1997). The food habits of *Cryptoprocta ferox* in the high mountain zone of the Andringitra Massif, Madagascar. *Mammalia*, 61(2): 185–192.

Goodman, S., Langrand, O., and Raxworthy, C. (1993a). Food habits of the Madagascar long-eared owl, *Asio madagascariensis*, in two habitats in southern Madagascar. *Ostrich*, 64: 79–85.

Goodman, S., Langrand, O., and Raxworthy, C. (1993b). The food habits of the barn owl (*Tyto alba*) at three sites on Madagascar. *Ostrich*, 64: 160–171.

Goodman, S., O'Connor, S., and Langrand, L. (1993c). A review of predation on lemurs: Implications for the evolution of social behavior in small, nocturnal primates. In P. Kappeler and J. Ganzhorn (Eds.), *Lemur social systems and their ecological basis* (pp. 51–66). New York: Plenum Press.

Gursky, S. (2002). Predation on a wild spectral tarsier (*Tarsius spectrum*) by a snake. *Folia Primatol.*, 73(1): 60–62.

Gursky, S. (2003). Predation experiments on infant spectral tarsiers (*Tarsius spectrum*). *Folia Primatol.*, 74(5–6): 272–284.

Gursky, S. (2005). Predator mobbing in *Tarsius spectrum*. *Int. J. Prim.*, 26(1): 207–221.

Hart, D. (2000). Primates as prey: Ecological, morphological, and behavioral relationships between primate species and their predators. Doctoral dissertation. Washington University, St. Louis, Missouri.

Hart, D., and Sussman, R. (2005). *Man the hunted:Primates, predators, and human evolution*. New York: Westview/Perseus.

Henschel, J., and Tilson, R. (1988). How much does a spotted hyaena eat? Perspective from the Namib Desert. *Afr. J. Ecol.*, 26: 247–255.

Hill, R., and Dunbar, R. (1998). An evaluation of the roles of predation rate and predation risk as selective pressures on primate grouping behaviour. *Behav.*, 35(4): 411–430.

Hoppe-Dominik, B. (1984). Prey frequency of the leopard (*Panthera pardus*) in the Taï National Park of the Ivory Coast. *Mammalia*, 48(4): 477–488.

Isbell, L. (1990). Sudden short-term increase in mortality of vervet monkeys (*Cercopithecus aethiops*) due to leopard predation in Amboseli National Park, Kenya. *Am. J. Primatol.*, 21(1): 41–52.

Isbell, L. (1994). Predation on primates: Ecological patterns and evolutionary consequences. *Evol. Anthropol.*, 3: 61–71.

Isbell, L., and Young, T. (1993). Human presence reduces predation in a free-ranging vervet monkey population in Kenya. *Animal Behav.*, 45: 1233–1235.

Izor, R. (1985). Sloths and other mammalian prey of the harpy eagle. In G. Montgomery (Ed.), *The evolution and ecology of armadillos, sloths, and vermilinguas* (pp. 343–346). Washington, DC: Smithsonian Institution Press.

Janson, C., and van Schaik, C. (1993). Ecological risk aversion in juvenile primates: Slow and steady wins the race. In M. Pereira and L. Fairbanks (Eds.), *Juvenile primates: Life history, development, and behavior* (pp. 57–74). New York: Oxford University Press.

Johnsingh, A. (1980). Ecology and behavior of the dhole or Indian wild dog, *Cuon alpinus* Pallas 1811, with special reference to predator-prey relations at Bandipur. Doctoral dissertation. Madurai Univ., Tamil Nadu, India.

Johnsingh, A. (1983). Large mammalian prey-predators in Bandipur. *J. Bombay. Nat. Hist. Soc.*, 80(1): 1–57.

Karpanty, S. (2003). Rates of predation by diurnal raptors on the lemur community of Ranomafana National Park, Madagascar [Abstract]. *Am. J. Phys. Anthr.*, (Suppl.36): 126.

Karpanty, S., and Goodman, S. (1999). Diet of the Madagascar harrier hawk, *Polyboroides radiatus*, in southeastern Madagascar. *Journal of Raptor Research*, 33: 313–316.

Kruuk, H. (1970). Interactions between populations of spotted hyaenas (*Crocuta crocuta*, Erxleben) and their prey species. In A. Watson (Ed.), *Animal populations in relation to their food resources* (pp. 359–374). Oxford: Blackwell Scientific.

Kruuk, H. (1972). *The spotted hyena: A study of predation and social behavior*. Chicago: Univ. of Chicago Press.

Kruuk, H., and Turner, M. (1967). Comparative notes on predation by lion, leopard, cheetah, and wild dog in the Serengeti area, East Africa. *Mammalia*, 31(1): 1–27.

Lakshmi, B., and Mohan, B. (2002). Behavioural ecology, distribution and status of *Loris tardigradus* (slender loris) in Andhra Pradesh. *Journal of Nature Conservation* 14(1): 27–31.

Mansfield, E. (1986). *Basic statistics with applications*. New York: W.W. Norton.

Martin, R. (1972). A preliminary field study of the lesser mouse lemur (*Microcebus murinus* J. Miller 1777). *Z. Tierpsychol.*, 9: 43–89.

McNeil, D. (1999). The great ape massacre. *The New York Times Magazine*. May 9, 1999: pp. 54–57.

McRae, M. (1997). Road kill in Cameroon. *Natural Hist.*, 06: 36–47.

Mills, M., and Biggs, H. (1993). Prey apportionment and related ecological relationships between large carnivores in Kruger National Park. In N. Dunstone and M. Gorman, (Eds.), *Mammals as predators* (pp. 253–268). Oxford: Clarendon Press.

Mitani, J., Sanders, W., Lwanga, J., and Windfelder, T. (2001). Predatory behavior of crowned hawk-eagles (*Stephanoaetus coronatus*) in Kibale National Park, Uganda. *Behavioral Ecology and Sociobiology*, 49: 187–195.

Mitchell, C., Boinski, S., and van Schaik, C. (1991). Competitive regimes and female bonding in two species of squirrel monkeys (*Saimiri oerstedi* and *S. sciureus*). *Behavioral Ecology and Sociobiology*, 28: 55–60.

Mittermeier, R. (1987). Effects of hunting on rain forest primates. In C. Marsh and R. Mittermeier (Eds.), *Primate conservation in the tropical rain forest* (pp. 109–146). New York: Alan R. Liss.

Mittermeier, R., and Cheney, D. (1987). Conservation of primates and their habitats. In B. Smuts, D. Cheney, R. Seyfarth, R. Wrangham, and T. Struhsaker (Eds.), *Primate societies* (pp. 477–496). Chicago: Univ. of Chicago Press.

Msuya, C. (1993). Feeding habits of crowned eagles, *Stephanoaetus coronatus*, in Kiwengoma Forest Reserve, Matumbi Hills, Tanzania. *Musee Royal de L'Afrique Centrale Annales Sciences Zoologiques*, 268: 118–120.

Muckenhirn, N. (1972). Leaf-eaters and their predators in Ceylon: Ecological roles of gray langurs, *Presbytis entellus*, and leopards. Doctoral dissertation. University of Maryland, College Park, Maryland.

Nekaris, K. (2003). Observations of mating, birthing and parental behaviour in three subspecies of slender loris (*Loris tardigradus* and *Loris lydekkerianus*) in India and Sri Lanka. *Folia Primatol.*, 74(5–6): 312–336.

Nekaris, K., and Jayewardene, J. (2004). Survey of the slender loris (Primates, Lorisidae Gray, 1821: *Loris tardigradus* Linnaeus, 1758 and *Loris lydekkerianus* Cabrera, 1908) in Sri Lanka. *J. Zool.*, 262(4): 327–338.

Newton, P. (1985). A note on golden jackals (*Canis aureus*) and their relationship with langurs (*Presbytis entellus*) in Kanha Tiger Reserve. *J. Bombay. Nat. Hist. Soc.*, 82: 633–636.

Oates, J. (1994). Africa's primates in 1992: Conservation issues and options. *Am. J. Primatol.*, 34: 61–71.

Oates, J. (1996). Habitat alteration, hunting and the conservation of folivorous primates in African forests. *Aust. J. Ecol.*, 21: 1–9.

Peres, C. (1990). Effects of hunting on western Amazonian primate communities. *Biol. Conser.*, 54: 47–59.

Rabinowitz, A., and Nottingham, B. (1986). Ecology and behaviour of the jaguar (*Panthera onca*) in Belize, Central America. *J. Zool. (Lond.)*, A210: 145–159.

Raemaekers, J., and Chivers, D. (1980). Socioecology of Malayan forest primates. In D. Chivers (Ed.), *Malayan forest primates: Ten years' study in tropical rain forest* (pp. 279–331). New York: Plenum Press.

Rasmussen, D. (1997). African and Asian prosimian field studies. In F. Spencer (Ed.), *History of physical anthropology*, Vol. 1 (pp. 23–26). New York: Garland Publishers.

Rasoloarison, R., Rasolonandrasana, B., Ganzhorn, J., and Goodman, S. (1995). Predation on vertebrates in the Kirindy forest, western Madagascar. *Ecotropica*, 1: 59–65.

Rasolonandrasana, B. (1994). Contribution a l'etude de l'alimentation de *Cryptoprocta ferox* Bennet (1833) dans son milieu naturel. Doctoral dissertation. Universite d'Antananarivo, Antananarivo, Madagascar.

Redmond, I. (1998). *The African bushmeat trade—A recipe for extinction*. UK: Ape Alliance.

Reza, A., Feeroz, M., and Islam, M. (2001). Food habits of the Bengal tiger (*Panthera tigris tigris*) in the Sundarbans. *Bangladesh J. Zool.*, 29(2): 173–179.

Rice, C. (1986). Observations on predators and prey at Eravikulam National Park, Kerala, India. *J. Bombay. Nat. Hist. Soc.*, 83(2): 283–305.

Sanders, W., Trapani, J., and Mitani, J. (2003). Taphonomic aspects of crowned hawk-eagle predation on monkeys. *J. Hum. Evol.*, 44: 87–105.

Sankar, K., and Johnsingh, A. (2002). Food habits of tiger (*Panthera tigris*) and leopard (*Panthera pardus*) in Sariska Tiger Reserve, Rajasthan, India, as shown by scat analysis. *Mammalia*, 66(2): 285–289.

Schaller, G. (1967). *The deer and the tiger*. Chicago: Univ. of Chicago Press.

Schaller, G. (1972). *The Serengeti lion: A study of predator-prey relations*. Chicago: Univ. of Chicago Press.

Seidensticker, J. (1983). Predation by *Panthera* cats and measures of human influence in habitats of South Asian monkeys. *Int. J. Prim.*, 4(3): 323–327.

Seidensticker, J. (1985). Primates as prey of *Panthera* cats in South Asian habitats. Paper given at the seventh annual meeting of the American Society of Primatologists, State University of New York at Buffalo, Niagara Falls, New York, June 1–4, 1985.

Seidensticker, J. (1991). Leopards. In J. Seidensticker and S. Lumpkin (Eds.), *Great cats: Majestic creatures of the wild* (pp. 106–115). Emmaus, PA: Rodale Press.

Seidensticker, J., and Suyono, I. (1980). *The Javan tiger and the Meru-Betiri Reserve: A plan for management*. Gland, Switzerland: International Union for Conservation of Nature and Natural Resources.

Shine, R., Harlow, P., Keogh, J., and Boeadi. (1998). The influence of sex and body size on food habits of a giant tropical snake, *Python reticulatus*. *Functional Ecology*, 12(2): 248–258.

Shultz, S. (2001). Notes on interactions between monkeys and African crowned eagles in Taï National Park, Ivory Coast. *Folia Primtatol.*, 72(4): 248–250.

Shultz, S. (2002). Population density, breeding chronology and diet of crowned eagles, *Stephanoaetus coronatus* in Taï National Park, Ivory Coast. *Ibis*, 144(1): 133–138.

Shultz, S. (2003). Of monkeys and eagles: Predator-prey relationships in the Taï National Park, Cote d'Ivoire. Doctoral thesis. University of Liverpool, Liverpool, UK.

Shultz, S., Noë, R., McGraw, W., and Dunbar, R. (2004). A community-level evaluation of the impact of prey behavioral and ecological characteristics on predator diet composition. *Proc. R. Soc. London B*, 271: 725–732.

Sih, A., Crowley, P., McPeek, M., Petranka, J., and Strohmeier, K. (1985). Predation, competition, and prey communities: A review of field experiments. *Ann. Rev. Ecol. Syst.* 16: 269–311.

Skorupa, J. (1989). Crowned eagles, *Stephanoaetus coronatus*, in rainforest: Observations on breeding chronology and diet at a nest in Uganda. *Ibis*, 131: 294–298.

Sokal, R., and Rohlf, J. (1981). *Biometry: The principles and practice of statistics in biological research* (2nd ed.). New York: W.H. Freeman.

Srivastava, K., Bhardwaj, A., Abraham, C., and Zacharias, V. (1996). Food habits of mammalian predators in Periyar Tiger Reserve, South India. *Indian For.*, 122(10): 877–883.

Stanford, C. (1989). Predation on capped langurs (*Presbytis pileata*) by cooperatively hunting jackals (*Canis aureus*). *Am. J. Primatol.*, 19: 53–56.

Stanford, C. (1995). The influence of chimpanzee predation on group size and anti-predator behaviour in red colobus monkeys. *Animal Behav.*, 49(3): 577–587.

Stanford, C., and Wrangham, R. (1998). *Chimpanzee and red colobus: The ecology of predator and prey*. Cambridge: Harvard Univ. Press.

Stanford, C., Wallis, J., Matama, H., and Goodall, J. (1994). Patterns of predation by chimpanzee on red colobus monkeys in Gombe National Park, 1982–1991. *Am. J. Phys. Anthr.*, 94(2): 213–228.

Steyn, P. (1973). *Eagle days: A study of African eagles at the nest*. Johannesburg: Purnell.

Steyn, P. (1983). *Birds of prey of southern Africa*. Dover: Tanager.

Struhsaker, T., and Leakey, M. (1990). Prey selectivity by crowned hawk-eagles on monkeys in the Kibale forest, Uganda. *Behavioral Ecology and Sociobiology*, 26(6): 435–443.

Sunquist, M. (1981). The social organization of tigers (*Panthera tigris*) in Royal Chitwan National Park, Nepal. *Smithsonian Contributions to Zoology*, 336: 1–98.

Tarboton, W. (1989). *African birds of prey*. Ithaca, NY: Cornell Univ. Press.

Taylor Halvorsen, K. (1986). Combining results from independent investigations: Meta-analysis in medical research. In J. Bailar and F. Mosteller (Eds.), *Medical Uses of Statistics* (pp. 392–416). Waltham: MA: New England Journal of Medicine Books.

Terborgh, J. (1983). *Five New World primates: A study in comparative ecology*. Princeton, NJ: Princeton Univ. Press.

Terborgh, J. (1990). Mixed flocks and polyspecific associations: cost and benefits of mixed groups to birds and monkeys. *Am. J. Primatol.*, 21(2): 87–100.

Terborgh, J., and Janson, C. (1986). The socioecology of primate groups. *Ann. Rev. Ecol. Syst.* 17: 111–135.

Thapar, V. (1986). *Tiger: Portrait of a predator*. New York: Facts on File Publications.

Thiollay, J. (1985). Species diversity and comparative ecology of rainforest falconiforms on three continents. In I. Newton and R. Chancellor (Eds.), *Conservation studies on raptors*. Cambridge: ICBP Technical Publication No. 5.

Thorstrom, R., and La Marca, G. (2000). Nesting biology and behavior of the Madagascar harrier-hawk (*Polyboroides radiatus*) in northeastern Madagascar. *J. Raptor Res.*, 34(2): 120–125.

Treves, A. (1999). Has predation shaped the social systems of arboreal primates? *Int. J. Prim.*, 20(1): 35–67.

Tsukahara, T. (1993). Lions eat chimpanzees: The first evidence of predation by lions on wild chimpanzees. *Am. J. Primatol.*, 29(1): 1–11.

Tutin, C., and Benirschke, K. (1991). Possible osteomyelitis of skull causes death of a wild lowland gorilla in the Lopé Reserve, Gabon. *J. Med. Prim.*, 20: 357–360.

Uehara, S., Nishida, T., Hamai, M., Hasegawa, T., Hayaki, H., Huffman, M., Kawanaka, K., Kobayashi, S., Mitani, J., Takahata, U., Takasaki, H., and Tsukahara, T. (1992). Characteristics of predation by the chimpanzees in the Mahale Mountains National Park, Tanzania. In T. Nishida, W. McGrew, P. Marler, M. Pickford, and F. de Waal (Eds.), *Topics in primatology, human origins*, Vol. 1 (pp. 143–158). Tokyo: Univ. of Tokyo Press.

Uhde, N., and Sommer, V. (2002). Antipredatory behavior in gibbons (*Hylobates lar*, Khao Yai/Thailand). In L. Miller (Ed.), *Eat or be eaten: Predator sensitive foraging among primates* (pp. 268–291). Cambridge: Cambridge Univ. Press.

van Schaik, C. (1983). Why are diurnal primates living in groups? *Behav.*, 87(1–2): 120–143.

Vasquez, M., and Heymann, E. (2001). Crested eagle (*Morphnus guianensis*) predation on infant tamarins (*Saguinus mystax* and *Saguinus fuscicollis*, Callitrichinae). *Folia Primatol.*, 72: 301–303.

Voous, K. (1969). Predation potential in birds of prey from Surinam. *Ardea T. Ned.*, 57: 117–148.

Washburn, S., Jay, P., and Lancaster, J. (1965). Field studies of Old World monkeys and apes. *Science*, 150: 1541–1547.

Wiens, F., and Zitzmann, A. (1999). Predation on a wild slow loris (*Nycticebus coucang*) by a reticulated python (*Python reticulatus*). *Folia Primatol.*, 70(6): 362–364.

Wiens, F., and Zitzmann, A. (2003). Social structure of the solitary slow loris *Nycticebus coucang* (Lorisidae). *J. Zool.*, 261(1): 35–46.

Williams, J., and Arlott, N. (1980). *The Collins field guide to the birds of east Africa.* Lexington: Stephen Greene.

Wrangham, R. (1980). An ecological model of female-bonded primate groups. *Behav.*, 75: 262–300.

Wright, P. (1985). The costs and benefits of nocturnality of *Aotus trivirgatus* (the night monkey). Doctoral dissertation. The City University of New York, New York.

Wright, P., Heckscher, S., and Dunham, A. (1997). Predation on Milne-Edwards' sifaka (*Propithecus diadema edwardsi*) by the fossa (*Cryptoprocta ferox*) in the rain forest of southeastern Madagascar. *Folia Primatol.*, 68(1): 34–43.

Wright, P., Heckscher, S., and Dunham, A. (1998). Predation of rain forest prosimians in Ranomafana National Park, Madagascar [Abstract]. *Folia Primatol.*, 69(Suppl.1): 401.

Zuberbühler, K., and Jenny, D. (2002). Leopard predation and primate evolution. *J. Hum. Evol.*, 43: 873–886.

Zuberbühler, K., Noë, R., and Seyfarth, R. (1997). Diana monkey long-distance calls: Messages for conspecifics and predators. *Animal Behav.*, 53(3): 589–604.

Part 2
Anti-Predator Strategies of Nocturnal Primates

3
Primates and Other Prey in the Seasonally Variable Diet of *Cryptoprocta* ferox in the Dry Deciduous Forest of Western Madagascar

Luke Dollar, Jörg U. Ganzhorn, and Steven M. Goodman

Introduction

The puma-like *Cryptoprocta ferox* is the largest living Carnivora on Madagascar (Goodman et al., 2003). *Cryptoprocta* has been a taxonomic enigma until recently (cf. Veron & Catzeflis, 1993; Veron, 1995), showing numerous convergent morphological characters with members of the Felidae. Some of these attributes, such as semi-retractable claws used in both climbing and hunting, contributed to the long-running uncertainty as to the phylogenetic relationships of this animal. Recent molecular studies indicate that *Cryptoprocta* is part of a radiation of Carnivora endemic to Madagascar, which unites all of the native species on the island into a single clade (Yoder et al., 2003), now recognized as the endemic family Eupleridae (Wozencraft, in press). On the basis of molecular data this radiation of Carnivora is slightly younger than that of lemurs, but the two groups have co-existed on Madagascar for something on the order of 20 million years (Yoder et al., 2003). Until the Holocene a second member of *Cryptoprocta* occurred on the island that was notably larger than the living species (Goodman et al., 2004).

Though many aspects of its behavior and ecology remain unstudied or unpublished, *Cryptoprocta* (Fig. 3.1) is a formidable predator, equally agile on the ground or in trees (Laborde, 1986a,b). Over the past few years, studies involving natural populations of *Cryptoprocta* in different biomes of Madagascar have been conducted to examine certain aspects of the behavioral ecology and reproduction of the predator (Hawkins, 1998; Dollar, 1999; Rahajanirina, 2003; Hawkins & Racey, 2005), as well as aspects of its diet (Rasolonandrasana, 1994; Rasoloarison, 1995; Goodman, 1997; Rasamison, 1997; Goodman, 2003). The different food habit studies were based on limited seasonal samples of relatively small collections of fecal material. However, they clearly indicated that the diet of this animal

FIGURE 3.1. *Cryptoprocta ferox*, camera trapped in Ankarafantsika National Park, Madagascar

is remarkably plastic, with considerable regional and seasonal variation in the types of prey it consumes. Despite this plasticity, *C. ferox* has been considered an important predator of lemurs, one that can largely reduce or even eliminate local populations. It has been referred to as a "lemur specialist" (Wright et al., 1997).

As long-standing debates over the role of predators in shaping primate life history continue (van Schaik & van Hooff, 1983; van Schaik & Kappeler, 1996), empirical data on relative rates of predation are needed to assess and test various hypotheses. In this paper, we provide empirical data on the prey of *Cryptoprocta*, including an assessment of dietary composition relative to primate abundance and certain aspects of their life history traits. Here we expand on earlier studies and analyze a large collection of scats from *Cryptoprocta* collected in both wet and dry seasons at two sites in Ankarafantsika National Park, northwest Madagascar.

Study Sites and Methods

Study Site

Scat samples of *Cryptoprocta ferox*, from two different sites in a single western Madagascar protected area, are analyzed in this paper. The two sites (Ampijoroa Forestry Station and Lake Tsimalato) are in the Ankarafantsika National Park (Fig. 3.2), which was originally named as an Integrated Natural Reserve and Forestry Reserve.

The Ankarafantsika reserve complex is one of Madagascar's largest remaining tracts of dry deciduous forest and contains approximately 200,000 ha, of

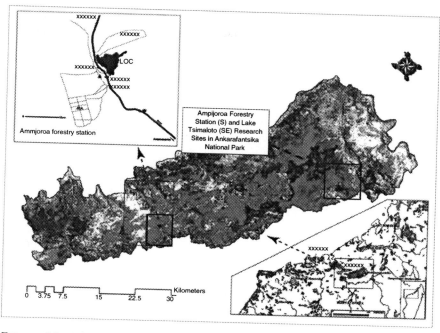

FIGURE 3.2. Ankarafantsika National Park, Madagascar, with the Ampijoroa and Lake Tsimaloto research sites highlighted. Ampijoroa is in the central-southern portion and Lake Tsimaloto is at the southeastern end of the park. Ankarokaroka, less than 5 km southwest of the Ampijoroa Forestry Station, is the site closest to Ampijoroa for which complete lemur surveys are available. Figure and maps adapted from Liu (2005) & Bradt (2002)

which the Ankarafantsika National Park covers 65,520 ha and, within it, the Ampijoroa Forestry Station holds about 20,000 ha (Nicoll & Langrand, 1989; Alonso & Hannah, 2002). The study areas were in the Lake Tsimaloto region, in the far southeastern portion of the park (110–370 m above sea level) and in the Ampijoroa area (80–370 m above sea level) in the south central portion. Both of these sites rest on sandy soils and contain notable floristic diversity (Rajoelison et al., 2002).

The Ankarafantsika region experiences considerable seasonal fluctuations in meteorological conditions, particularly rainfall. The dry season is normally between May and September, a period when there is often no recorded monthly precipitation. This is in contrast to the rainy season, occurring between October and April, when much of the yearly average of approximately 1500 mm of rain falls (Rahajanirina, 2003). This marked seasonality in weather conditions may have important implications for population cycling in the local biological community.

Scat Samples and Analysis

Scat samples were collected opportunistically along trails or from live animal traps. Scat from *Cryptoprocta* is easily distinguishable from that of other local

Carnivora by its size and shape. Specimens were placed in plastic bags, labeled, dated, and stored (due to the climate, the samples were usually desiccated when collected). In cases when a scat sample contained remains of chicken (*Gallus*), clearly associated with the trap bait, this sample was excluded from the analysis.

Scats were collected in natural forest formations, in anthropogenic savannah, and the ecotone between these two formations at Ampijoroa during the months of June to November 2000, March 2001, and June to December 2001, particularly in the vicinities of Jardin Botanique A and Jardin Botanique B. Samples from the second site, Lake Tsimaloto, were collected in June 2001.

In the laboratory, individual fecal samples were frozen for about one week and then submerged in soapy water or an alcohol solution for several days. Thereafter, each sample was broken down into small pieces, sifted, and bone, tooth, and various forms of epidermal fragments were removed. The osteological material was identified by SMG to the most precise taxonomic level feasible, using an extensive comparative collection of Malagasy mammals at the Université d'Antananarivo. Scales, feathers, or fur where used in identification in cases were samples contained unidentifiable bone or no bone at all.

Paired osteological elements of any taxon recovered from the scats were separated and the largest number of elements from either the left or right side was considered to be the minimum number of individuals. When possible, the identified prey items were placed in age categories. For birds and some mammal post-cranial remains, this was based on bone ossification: unossified or partially ossified = sub-adults; fully ossified = adults. When tooth-bearing mammalian remains were present, individuals with partially non-erupted permanent teeth = sub-adults or juveniles; fully erupted permanent dentitions = adults. For the purposes of this analysis, we have combined all age classes into overall numbers of individuals as single prey species. Samples from the months of March, October, and November were assigned to the wet season and those from the months of June to September were assigned to the dry season.

The average body mass of prey animals was taken from the literature in the case of mammals (Goodman et al., 2003) and birds (Ravokatra et al., 2003), while for reptiles and amphibians these were based on the weights of liquid-preserved specimens of a given taxon held in the Université d'Antananarivo collection. Moderate-sized insects were estimated to have a body mass of 2 g and large insects a mass of 5 g.

Statistical Analyses

When scats contained more than one prey type, the minimum number of individuals as defined above was identified per taxon and each was considered as a datum point. Sample size differs between certain analyses. For example, the contents of scats that contained unidentifiable bone remains or scales, feathers or hair and could not be identified to species or genus were used exclusively in higher taxon level comparisons.

To include as many scats as possible in the statistical analyses, samples were pooled when feasible. This included combining samples from the same site collected during the same month during different years. In Ampijoroa, only one sample was collected during the month of May, which contained a *Setifer setosus* and was combined with the samples collected in June. Undated samples were not included in these analyses. The distribution of body mass differed significantly from a normal distribution and could not be transformed to allow for the application of parametric analyses. Therefore non-parametric analyses were used for all comparisons. Significance levels are two-tailed. Tests were run with the help of SPSS (1999).

Results

In total, 220 and 67 *Cryptoprocta ferox* scats were analyzed from Ampijoroa and Lack Tsimaloto, respectively. Identifiable contents were isolated from 185 and 59 scats of the two sites. These provided the bases for the present analyses. Further, samples that were not from *Cryptoprocta*, but from other species of locally occurring Carnivora, most notably *Felis silvestris* or *Viverricula indica*, both introduced to Madagascar, were excluded from the analyses.

For Ampijoroa, the percentages of mammals in the prey are uncorrelated with the percentage of birds and reptiles (Spearman rank correlation: $r_s = -0.07$ and $r_s = -0.38$, respectively, $p > 0.40, n = 7$). The percentage of birds in the prey tends to be negatively correlated with the percentage of reptiles ($r_s = -0.72$, $p = 0.07, n = 7$).

The analysis of taxonomic differences in prey composition from Ampijoroa between months was restricted to mammals, birds, and reptiles due to small sample sizes of the other types of prey (Tables 3.1, 3.2). The month of March was also removed due to small sample size ($n = 3$). On a monthly basis there was no significant difference in the taxonomic composition of prey at the level of class ($x^2 = 8.88, \text{df} = 10, p = 0.54; n = 242$). Further, at this taxonomic level prey composition also did not differ between the wet season (months of March, October, and November) and the dry season (months of June to September) ($x^2 = 1.49, \text{df} = 2, p = 0.48; n = 245$). However, the proportion of primates between seasons was different ($x^2 = 5.32, \text{df} = 1, p = 0.02$; test based on real numbers, not on proportions).

At the Class level, prey composition varied significantly between the two sites during the dry season (the single frog, one egg from Ampijoroa and one egg from Tsimaloto were removed to eliminate cells with expected counts less than 2: $x^2 = 13.02, \text{df} = 3, p = 0.005$; Table 3.3). Insects and birds were under represented and mammals over represented in the samples from Lake Tsimaloto.

Temporal and Geographic Variation in Body Mass of Prey Items

In the samples collected at Ampijoroa, average live body mass of prey animals differed significantly between months (Kruskal Wallis Test: $x^2 = 18.84$,

TABLE 3.1. Taxonomic composition of prey. For Ampijoroa data listed for the month of June include one scat collected in May containing a single *Setifer setosus*.

Site	Month		Insecta	Mammalia	Amphibia	Aves	Reptilia	Total
Ampijoroa	March	Count		2		1		3
		%		66.70%		33.30%		100.00%
	June	Count	7	22		8	7	44
		%	15.90%	50.00%		18.20%	15.90%	100.00%
	July	Count	3	43		16	5	67
		%	4.50%	64.20%		23.90%	7.50%	100.00%
	Aug	Count	1	14		4		19
		%	5.30%	73.70%		21.10%		100.00%
	Sept	Count		11		4	1	16
		%		68.80%		25.00%	6.30%	100.00%
	Oct	Count	4	54	1	21	12	92
		%	4.30%	58.70%	1.10%	21.70%	13.00%	100.00%
	Nov	Count		15		2	4	21
		%		71.40%		9.50%	19.00%	100.00%
	Total	Count	15	161	1	56	29	262
		%	5.70%	61.50%	0.40%	21.00%	11.10%	100.00%
Tsimaloto	June	Count	2	46		4	9	61
		%	3.30%	75.40%		6.60%	14.80%	100.00%
	July	Count		2			1	3
		%		66.70%			33.30%	100.00%
	Total	Count	2	48		4	10	64
		%	3.10%	75.00%		6.30%	15.60%	100.00%
Ankarafantsika	All		17	209	1	60	39	326
			5.21%	64.11%	0.31%	18.40%	11.96%	100.00%

df $= 6$, p $= 0.004$; $n = 245$). The monthly medians differed unsystematically between months (Table 3.4). Median body mass of all samples available for the dry season was 120 g (quartiles: 45–980 g; $n = 245$) and 80 g for the wet season (quartiles 15–500 g, $n = 105$). Prey items had higher body mass during the dry season than during the wet season (Mann-Whitney U test: $z = 2.02$, $p = 0.04$). This difference in prey body mass is due to the large proportion of chicken (16 individuals) in the diet of *Cryptoprocta* at Ampijoroa during the dry season. Chicken have been classified as prey with 800–1599 g body mass. If they are removed from the calculation, the distribution of prey body mass remains fairly constant between seasons.

Median dry season prey body mass for the pooled sample of Lake Tsimaloto was 900 g (quartiles: 50–2500 g; $n = 58$). Thus, in general, prey animals had higher body mass at Tsimaloto than at Ampijoroa during the dry season (Mann Whitney U test: $z = 2.61$, $p = 0.009$; Table 3.3.

Focusing on median body mass of prey items might be somewhat misleading as none of the body mass distributions at the three sites is unimodal (Fig. 3.3).

TABLE 3.2. Taxonomic composition of *Cryptoprocta* prey at two different sites. To provide detailed information on prey composition at Ampijoroa, the samples were subdivided into scats collected during the dry and wet seasons. These two seasons did not differ significantly in prey composition and this distinction was not considered in the statistical analysis for differences between sites. All vertebrate prey, whether sub-adult or adult, are grouped into the overall species count shown here, unless specifically stated otherwise. Taxa marked with an asterisk were introduced to the island.

		Ampijoroa wet season	Ampijoroa dry season	Tsimaloto dry season	
Mammals					**Total**
Lipotyphla	*Microgale brevicaudata*	2			
	Setifer setosus	3	2	2	7
	Tenrec ecaudatus	2			2
	Suncus murinus	1	1		2
Rodentia	*Eliurus myoxinus*		2		2
	Eliurus spp.	1			1
	Macrotarsomys spp.	1			1
	M. bastardi		2	1	3
	M. ingens		4	1	5
	Rattus rattus	20	11	4	35
	Unidentified rodent	2		1	3
Primates	*Avahi occidentalis*	1	1	1	3
	Cheirogaleus medius	7	15		22
	Eulemur fulvus			8	8
	E. mongoz	1	2	1	4
	Lepilemur edwardsi	19	29	10	58
	Microcebus spp.	4	12	3	19
	Propithecus verreauxi	1		10	11
	Unidentified lemur fur	1	2	2	5
	Unidentified mammal fur	5	2	3	10
Birds	*Coua coquereli*	1	2		3
	Coua cristata	1	1		2
	Dicrurus forficatus			1	1
	Gallus gallus	2	16	1	19
	Gallus gallus (sub-adult)	2			2
	Margaroperdix madagascariensis		1		1
	Mirafa hova	2	2		4
	Turnix nigricollis	6	4		10
	Unknown bird	5	2		7
	Small bird	4	2		6
	Medium bird	3	1	1	5
	Large bird (> 150 g)		1		1
	Galliform egg			1	1
	Egg	1			1

TABLE 3.2. (Continued).

		Ampijoroa wet season	Ampijoroa dry season	Tsimaloto dry season	
Insects	Coleoptera	2	1	1	4
	Orthoptera	2	10	1	13
Amphibians	Large frog	1			1
Reptiles	Gecko			1	1
	Uroplatus spp.		1	1	2
	Medium-sized chameleon	2			2
	Small lizard		1		1
	Large snake	3	1	1	5
	Medium-size snake	5	5	5	15
	Small snake	3	5	2	10
Total		**116**	**141**	**63**	**318**

TABLE 3.3. Prey composition of scats (by minimum number of individuals) from *cryptoprocta ferox* in ankarafantsika national park, at higher taxonomic levels (without eggs). values are counts and percentages per site and season.

Site	Season	Taxon					Total
		Insects	Mammals	Birds	Reptiles	Amphibians	
Ampijoroa	Wet	4	71	26	13	1	115
		3.40%	61.74%	22.61%	11.30%	0.87	
Ampijoroa	Dry	11	85	32	13		141
		7.80%	60.28%	22.70%	9.22%	0%	
Tsimaloto	Dry	2	48	3	10		63
		3.17%	76.19%	4.76%	15.87%	0%	

TABLE 3.4. Prey biomass (in g) per month in Ampijoroa and Tsimaloto in relation to season; values are medians (top number), quartiles (at center), and sample size (bottom number). For March, the actual values for both samples are given.

Site	Dry season					Wet season	
	June	July	August	September	Oct	Nov	March
Ampijoroa	63	120	900	245	61	120	980 /
	15 / 820	50 / 980	64 / 1010	65 / 980	15 / 408	50 / 835	3500
	40	67	18	15	84	19	2
Tsimaloto	980	62					
	50 / 2500	50 / 385					
	55	3					

Discussion

Cryptoprocta as a Predator on Primates in Ankarafantsika

Based on these data, primates account for a substantial proportion of the prey in the diet of *Cryptoprocta*. This predator takes all primate species known to occur in

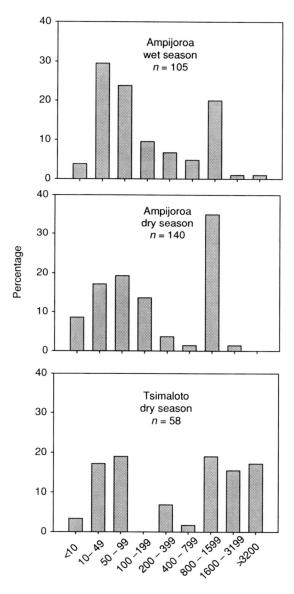

FIGURE 3.3. Frequency distribution of prey items of different biomass (in grams) at different sites in Ankarafantsika National Park during different seasons

Ankarafantsika (Table 3.2). During the wet season primates accounted for 28.5% of prey animals taken by *C.ferox* and during the dry season 41.8% and 52.4% at Ampijoroa and Tsimaloto, respectively (Fig. 3.4). However, this Carnivora also

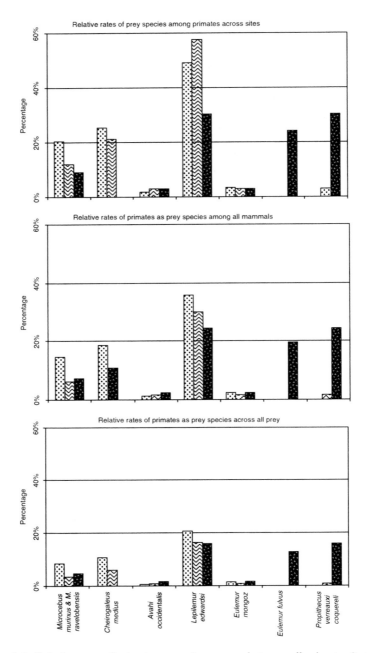

FIGURE 3.4. Relative rates of primates occurring as prey between all primates (top), relative to all mammals (center), and relative to all prey items and all taxa (bottom). Columns with black dots on white background represent Ampijoroa in the dry season, wavy lines are Ampijoroa in the wet season, and black background with white dots represent data from Lake Tsimaloto

takes a wide variety of non-primate prey and previous assertions that the diet of *Cryptoprocta* is remarkably plastic in its prey choice are supported. *Cryptoprocta* is not a lemur specialist per se, as on average less than 50% of the prey consumed is primates.

In Tsimaloto, the larger lemur species (*Eulemur* spp. and *Propithecus verreauxi*) are over represented in the diet compared to their relative density in this part of Ankarafantsika forest, while *Microcebus spp.* and *Cheirogaleus medius* are under represented (Table 3.5). *Lepilemur edwardsi* and *Avahi occidentalis* were taken in close proportion to their relative densities. Small sample size prohibits statistical analyses in these comparisons. *Microcebus spp.* are also under represented in the prey of *Cryptoprocta* at Ampijoroa. In contrast, the situation in Tsimaloto shows that *Avahi*, *Eulemur* spp., and *P. verreauxi* are also under represented, while *L. edwardsi* is grossly over represented accounting for more than 50% of all primates eaten by *Cryptoprocta*. *Lepilemur* is one of the moderate-sized local primates, being similar in body mass to *Avahi* and among the prey least taken. *Microcebus* is the smallest lemur species and is under represented in the diet of *Cryptoprocta* at both study sites, suggesting discrimination against small body mass or something particular about the life history traits of members of this genus. Selection criteria with respect to the largest species (*Eulemur* spp. and *P. verreauxi*) differ between sites. While a number of the anecdotes of primates falling prey to *Cryptoprocta* focus on members of these larger genera (e.g., Overdorff et al., 1996; Wright et al., 1997) these should not be extrapolated as a rule for an otherwise little-studied predator. Prey body mass may not play a key role in the rate of occurrence of primates in the diet of *Cryptoprocta*.

TABLE 3.5. Primate species taken as prey of *Cryptoprocta* in Ankarafantsika National Park, including notes on their life history and density. Body mass from Goodman et al., 2003). Lemur density estimates are for Ankarokaroka (for comparison with Ampijoroa) and Tsimaloto (from Schmid & Rasoloarison, 2002). Estimates are mean number of individuals per km transect.

Species	Activity pattern	Average body mass	Foraging party size	Ankarokaroka ind./km	Tsimaloto ind./km
Microcebus murinus & *M. ravelobensis*	nocturnal	≈ 60	solitary	400	297
Cheirogaleus medius	nocturnal	120	solitary/ pairs	227	110
Avahi occidentalis	nocturnal	816	small group	193	30
Lepilemur edwardsi	nocturnal	980	solitary	97	240
Eulemur mongoz	cathemeral	1650	small group	23	0
E. fulvus fulvus	cathemeral	2600	large group	33	50
Propithecus verreauxi coquereli	diurnal	3500	small–large groups	33	23

Direct comparison between two nocturnal lemur species, *A. occidentalis* and *L. edwardsi*, both of similar body mass, home ranges, and diets, but different predation pressure, suggest that other life-history traits contribute to the differences in risks taken by *Cryptoprocta*. *Lepilemur* is a solitary forager and spends the day in tree holes or in other partially exposed arboreal settings, and, in contrast, *Avahi* forages in family groups and rests in the open (Thalmann, 2001). Thus, solitary foraging and the utilization of tree holes for shelter might increase risks of predation. However, while this interpretation is appealing it is contradicted by varying predation rates on other nocturnal species that are solitary foragers and rest in tree holes (*Microcebus, Cheirogaleus*), which have low predation pressure from *Cryptoprocta*.

These smaller primate species such as *Microcebus* and *Cheirogaleus* also experience notably high predation pressure from a variety of diurnal and nocturnal avian predators (Goodman, 2003). Certain species of owls and hawks hunting beneath the canopy take a greater percentage of smaller primate prey and rarely feed on animals the size of *Lepilemur* or larger. Other types of predators (e.g., snakes, smaller Carnivora) also take lemurs. Further studies on specific attributes of predators and prey are needed to understand aspects of various predation and anti-predation strategies at the level of the ecosystem.

Another comparison of the dietary regime of *Cryptoprocta*, across portions of its range and in a variety of biomes (Goodman, 2003), suggests the possibility for regional differences in the prey taken by this animal. Within certain humid forest ecosystems these animals make up the largest percentage of the biomass obtained by this predator. However, above tree line on the Andringitra Massif, ground-dwelling animals make up a large percentage of *Cryptoprocta*'s prey biomass; they even feed upon aquatic animals such as frogs and crabs (Goodman et al., 1997). This wide niche breadth of prey types underlines the remarkable adaptability of Madagascar's top predator and that it should not be considered in particular as a "lemur specialist," although predation pressure on these animals has certainly played an important role in the evolution of their behavior.

Acknowledgments. We would like to thank Ken Glander and Julie Pomerantz for reviewing earlier drafts of this manuscript. Shaun Dunn, Haley Hougton, Ted Gilliland, Jodie LaPoint, Erica Strand, and several Earthwatch teams collected many of the scats used in these analyses. The work was funded in part by Conservation International, the Center for Field Research, and the Earthwatch Institute. Leon Pierrot Rahajanirina provided invaluable logistical assistance. We thank the Association Nationale pour le Gestion des Aires Protégées (ANGAP) and the Direction des Eaux et Forêts for permission to conduct this work. We are grateful to Dr. Daniel Rakotondravony and Dr. Olga Ramilijaona, Département de Biologie Animale, Université d'Antananarivo, for access to comparative osteology collections. This is Duke Primate Center publication number *xx*.

References

Albignac, R. (1973). Faune de Madagascar. *Mammifères Carnivores*, No. 36. Paris: ORSTOM/CNRS.

Alonso, L.E., and Hannah, L. (2002). Introduction to the Réserve Naturelle Intégrale and Réserve Forestière d'Ankarafantsika and to the rapid assessment program. In L.E. Alonso, T.S. Schulenburg, S. Radilofe, and O. Missa (Eds.), *Une évaluation biologique de la Réserve Naturelle Intégrale d'Ankarafantsika, Madagascar* (pp. 28–32). RAP Bulletin of Biological Assessment No. 23. Washington, DC: Conservation International.

Dollar, L. (1999). Preliminary report on the status, activity cycle, and ranging of *Cryptoprocta ferox* in the Malagasy rainforest, implications for conservation. *Small Carnivore Conservation*, 20: 7–10.

Goodman, S.M. (2003). Predation on lemurs. In S.M. Goodman and J.P. Benstead (Eds.), *The natural history of Madagascar* (pp. 1221–1228). Chicago: Univ. of Chicago Press.

Goodman, S.M., Ganzhorn, J.U., and Rakotondravony, D. (2003). Introduction to the mammals. In S.M. Goodman and J.P. Benstead (Eds.), *The natural history of Madagascar* (pp. 1159–1186). Chicago: Univ. of Chicago Press.

Goodman, S.M., Langrand, O., and Rasolonandrasana, B.P.N. (1997). The food habits of *Cryptoprocta ferox* in the high mountain zone of the Andringitra Massif, Madagascar (Carnivora, Viverridae). *Mammalia*, 61: 185–192.

Goodman, S.M., Rasoloarison, R., and Ganzhorn, J.U. (2004). On the specific identification of subfossil *Cryptoprocta* (Mammalia, Carnivora) from Madagascar. *Zoosystema*, 26: 129–143.

Hawkins, C.E. (1998). The behaviour and ecology of the fossa, *Cryptoprocta ferox* (Carnivora: Viverridae) in a dry deciduous forest in western Madagascar. Doctoral thesis. University of Aberdeen.

Hawkins, C.E., and Racey, P.A. (2005). Low population density of a tropical forest carnivore, *Cryptoprocta ferox*: Implications for protected area management. *Oryx*, 39: 35–43.

Laborde, C. (1986a). Description de la locomotion arboricole de *Cryptoprocta ferox* (Carnivore Viverridé Malgache). *Mammalia*, 50: 369–378.

Laborde, C. (1986b). Caractères d'adaptation des membres au mode de vie arboricole chez *Cryptoprocta ferox* par comparaison avec d'autres Carnivores Viverridés. *Annales des Sciences Naturelles, Zoologie*, Paris 13: 25–39.

Nicoll, M.E., and Langrand, O. (1989). *Madagascar: Revue de la conservation et des aires protégées*, Gland, Switzerland: WWF.

Rahajanirina, L.P. (2003). Contribution à l'étude biologique, écologique et éthologique de *Cryptoprocta ferox* (Benett 1883) dans la région du lac Tsimaloto, Parc National d'Ankarafantsika, Madagascar. Mémoire de DEA-Biologie Animale, Université d'Antananarivo.

Rajoelison, G., Raharimalala, J., Rahajasoa, G., Andriambelo, L.H., Rabevohitra, R., and Razafindrianilana, N. (2002). Evaluation de la diversité floristique dans la Réserve d'Ankarafantsika. In L.E. Alonso, T.S. Schulenberg, S. Radilofe, and O. Missa (Eds.), *Une évaluation biologique de la Réserve Naturelle Intégrale d'Ankarafantsika, Madagascar* (pp. 37–62). RAP Bulletin of Biological Assessment No. 23. Washington, DC: Conservation International.

Rasamison, A.A. (1997). Contribution à l'étude biologique, écologique et éthologique de *Cryptoprocta ferox* (Bennett, 1833) dans la forêt de Kirindy à Madagascar. D.E.A., Faculté des Sciences, Université d'Antananarivo.

Rasoloarison, R.M., Rasolonandrasana, B.P.N., Ganzhorn, J.U., and Goodman, S.M. (1995). Predation on vertebrates in the Kirindy forest, western Madagascar. *Ecotropica*, 1: 59–65.

Rasolonandrasana, B.P.N. (1994). Contribution à l'étude de l'alimentation de *Cryptoprocta ferox* Bennett (1833) dans son milieu naturel. Mémoire de D.E.A., Service de Paléontologie, Université d'Antananarivo.

Ravokatra, M., Wilmé, L., and Goodman, S.M. (2003). Bird weights. In S.M.Goodman and J.P. Benstead (Eds.), *The natural history of Madagascar* (pp. 1059–1063). Chicago: Univ. of Chicago Press, Chicago.

Schmid, J., and Rasoloarison, R.M. (2002). Lemurs of the Réserve Naturelle Intégrale d'Ankarafantsika, Madagascar. In L.E. Alonso, T.S. Schulenberg, S. Radilofe, and O. Missa (Eds.), *Une évaluation biologique de la Réserve Naturelle Intégrale d'Ankarafantsika, Madagascar* (pp. 73–82). RAP Bulletin of Biological Assessment No. 23. Washington, DC: Conservation International.

SPSS (1999). *SPSS base 9.0 user's guide*, SPSS Inc., Chicago.

Thalmann, U. (2001). Food resource characteristics in two nocturnal lemurs with different social behavior: *Avahi occidentalis* and *Lepilemur edwardsi*. *International Journal of Primatology*, 22: 287–324.

van Schaik, C.E., and van Hooff, J.A.R.A.M. (1983). On the ultimate causes of primate social systems. *Behaviour*, 85: 91–117.

van Schaik, C.P., and Kappeler, P.M. (1996). The social systems of gregarious lemurs: Lack of convergence with anthropoids due to evolutionary disequilibrium? *Ethology*, 102: 915–41.

Veron, G. (1995). La position systématique de *Cryptoprocta ferox* (Carnivora). Analyse cladistique des caractères morphologiques de carnivores Aeluroidea actuels et fossiles. *Mammalia*, 59: 551–582.

Veron, G., and Catzeflis, F.M. (1993). Phylogenetic relationships of the endemic Malagasy carnivore *Cryptoprocta ferox* (Aeluroidea): DNA/DNA hybridization experiments. *Journal of Mammalian Evolution*, 1: 169–185.

Wright, P.C., Heckscher, S.K., and Dunham, A.E. (1997). Predation on Milne-Edwards's sikafa (*Propithecus diadema edwardsi*) by the fossa (*Cryptoprocta ferox*) in the rain forest of southeastern Madagascar. *Folia Primatologica*, 68: 34–43.

Yoder, A.D., Burns, M.M., Zehr, S., Delefosse, T., Veron, G., Goodman, S.M., and Flynn, J.J. (2003). Single origin of Malagasy Carnivora from an African ancestor. *Nature*, 421: 734–737.

Wozencraft, W.C. (In press). Order Carnivora. In D.E. Wilson and D.M. Reeder (Eds.), *Mammal species of the World: A taxonomic and geographic reference* (3rd ed.). Baltimore: Johns Hopkins Univ. Press.

4
Predation on Lemurs in the Rainforest of Madagascar by Multiple Predator Species: Observations and Experiments

Sarah M. Karpanty and Patricia C. Wright

Introduction

Predation by raptors, snakes, and carnivores is a constant risk for most wild primates (Cheney & Seyfarth, 1981; Anderson, 1986; Cheney & Wrangham, 1987; Janson & van Schaik, 1993; Cowlishaw, 1994; Isbell, 1994; Hill & Dunbar, 1998; Treves, 1999; Bearder et al., 2002; Gursky, 2002a, b; Shultz & Noë, 2002). In Madagascar, the problem may be especially severe since prosimians are the largest, most abundant and conspicuous mammals in the forest (Wright, 1998). Lemur behavior may be strongly influenced in its avoiding predation by stealthy predators, such as Henst's goshawk (*Accipiter henstii*), the fossa (*Cryptoprocta ferox*), or the Madagascar boa constrictor (*Boa manditra*) (Sauther, 1989; Goodman et al., 1993a; Gould, 1996; Wright, 1998; Karpanty & Goodman, 1999; Karpanty & Grella, 2001; Fichtel & Kappeler, 2002; Goodman, 2004). Most studies of predator and prey concentrate on one taxon of predator, such as hawks or leopards (Isbell, 1990; Peres, 1990; Struhsaker & Leakey, 1990, Boesch, 1991; Shultz, 2001, 2002), while the forest reality is that an animal avoids several distinct predators simultaneously. This is certainly true in Madagascar, where day-hunting hawks and eagles hunt both sleeping nocturnal and active diurnal lemurs, and fossas and boas hunt day and night (Wright, 1998; Karpanty, 2006). Therefore, ability to develop foraging and resting strategy for risk avoidance might be a major factor in primate sociality (Janson & van Schaik, 1993; Janson & Goldsmith, 1995; Stanford, 1995).

Predation on primates is a factor governing patterns in species' social assembly, travel, resting tactics, and community composition (van Schaik, 1983; van Schaik & van Hooff, 1983; Janson, 1992; Isbell, 1994; Wright, 1998). It has been suggested that over evolutionary time predators may impact a change in the primate nocturnal or diurnal activity cycle (Wright, 1989; van Schaik & Kappeler, 1993; Wright, 1994; van Schaik & Kappeler, 1996). However, few authors have considered the real life complexity inherent in the avoidance of simultaneous

TABLE 4.1. Lemur species in Ranomafana National Park. The five nocturnal species are marked by asterisk. The two cathemeral species are marked with a C.

Species	Body Mass (g)	Biomass (kg/km^2)
Avahi laniger	900	20
Propithecus edwardsi	5,800	125
Cheirogaleus major*	320	18
Microcebus rufous*	45	4
Daubentonia madagascariensis*	3,500	?
Lepilemur seallii*	970	1
Hapalemur griseus	900	20
Hapalemur aureus	1,800	9.6
Prolemur simus	2,800	12
Eulemur fulvus rufus C	2,100	66
Eulemur rubriventer C	2,100	48
Varecia variegata variegate	3,500	4

predation by multiple species of predators with distinct hunting strategies (Lima & Dill, 1990; Sih et al., 1998; Wright, 1998).

Extensive fieldwork on the lemurs of Madagascar has shown that lemurs, once thought to have few predators, actually have multiple species of predators (van Schaik & Kappeler, 1996; Wright, 1998, 1999; Goodman, 2004; Karpanty, 2006). Little is known about the effects of multiple predators on lemur social and foraging behavior. It is possible that a lemur's response to one predator may bring a greater risk from another predator (e.g., risk enhancement or reduction, Sih et al., 1998). Wright (1998) outlined possible behaviors that would protect or decrease risk to lemurs from raptors and carnivores.

Twelve sympatric lemur species live in Ranomafana National Park (RNP) (Table 4.1). Of the five carnivore species observed at Ranomafana (Table 4.2), only two have been observed to prey on lemurs. The fossa is the largest extant carnivore in Madagascar and is found in forested areas in both the western dry and eastern rainforests. An agile mammalian predator in the trees, with retractile claws, strong mandible, and formidable canines, the fossa is able to kill prey nearly its own size (Wright et al., 1997). Fossas captured at RNP weighed 8.5 kg (adult male) and 6.5 kg (sub-adult male), and radio-collared fossas traveled 2–5 km per day (Dollar et al., 1997; Dollar, 1999). The ring-tailed mongoose (*Galidia elegans*) is a small (700 g) carnivore found in most forested areas throughout Madagascar (Garbutt, 1999). This diurnal carnivore eats birds, beetles, fruits, and small-bodied lemurs (Wright & Martin, 1995).

All four diurnal raptor species (Table 4.2) are large enough to take lemurs. The Madagascar harrier-hawk was observed to take lemur remains to its nest in gallery forest near spiny desert (Karpanty & Goodman, 1999). In contrast, in a study in the rainforest of Masoala peninsula of nest remains of *Buteo brachypterus*— Madagascar buzzard— no lemurs were found (Berkelman, 1994; Watson & Lewis, 1994). In the dry forests of Madagascar, owls eat small lemurs (Goodman et al., 1993a,b,c). There are no reports of the contents of owl pellet remains in the rainforest of Ranomafana.

TABLE 4.2A. Sympatric birds of prey, Ranomafana National Park. Raptors marked by an asterisk are known to eat lemurs in this region.

Species Name	Common Name
Buteo brachypterus	Madagascar buzzard
Polyboroides radiatus*	Madagascar Harrier-hawk *
Accipiter henstii*	Henst's goshawk*
Eutriorchis astur	Madagascar Serpent-eagle
Tyto alba	Madagascar barn owl
Asio madagascariensis	Madagascar long-eared owl

TABLE 4.2B. Sympatric viverrid carnivores in Ranomafana National Park. Viverrids marked by an asterisk are known to eat lemurs in this region.

Species Name	Common Name
Euplores goudoti	Falanouc
Galidia elegans*	Ring-tailed mongoose*
Fossa fossa	Fanalouc
Galidictus fasciata	Broad-striped mongoose
Cryptoprocta ferox*	Fossa*

Lemurs have several possible evolutionary strategies to avoid these predators, including (1) lowering susceptibility to predation via group defense, increased vigilance, or dilution of risk (Hamilton, 1971; Pulliam & Caraco, 1984; Janson, 1992); and (2) increasing crypsis and hiding (Vine, 1973; Janson, 1986, 1992; Cowlishaw, 1994; Terborgh & Wright, 1998). It has also been hypothesized that since predation rates vary with prey activity cycle, nocturnality may protect primates against diurnal raptor predation (Wright, 1989, 1994). Understanding the impact of predation on the evolution of lemur behavior and social systems as hypothesized above requires understanding the interactions of lemurs with all of their many predators.

As a first step in understanding the complex relationships between multiple predators and multiple lemur prey, we review and update information on direct observations of predator attacks on lemurs in the rainforest of Ranomafana National Park, and conduct an experiment to better understand how the lemurs react to and avoid multiple predators. By using audio playbacks, we compare the responses of three species of lemurs to experimental exposure to aerial and terrestrial predator vocalizations, and we examine whether differences in lemur responses to different predators are correlated with observed predation rates for these lemur species.

Methods

Study Site

Ranomafana National Park (RNP), established in 1991, contains 43,500 ha of continuous rainforest in southeastern Madagascar and is situated at 21°16' S latitude

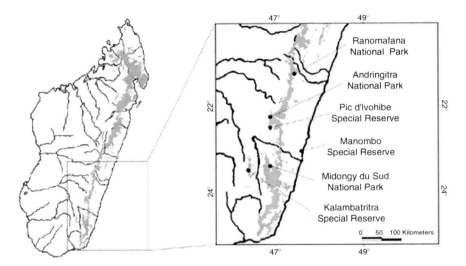

FIGURE 4.1. Map of Madagascar with study site, Ranomafana National Park, noted in the context of other protected areas in this region (Irwin & Arrigo-Nelson, pers. comm.)

and 47°20' E longitude (Wright, 1992; Wright & Andriamihaja, 2004). The park is 25 km from Fianarantsoa and 60 km from the Indian Ocean (Fig. 4.1). Elevations range from 500 to 1500 m, and annual rainfall ranges from 1600 to 3600 mm (RNP records). Most of the rainfall occurs during the months of December to March. Temperatures range from 4–12°C (June to September) to 30–32°C (December-February). The park contains moist evergreen forest and the canopy height range is 18–25 m. The study groups of lemurs were located in the 5 km^2 Talatakely study site (TTS) which was selectively logged by hand in the period 1986–1989, and the Vatoharanana study site (VATO), 5 km south, which is undisturbed by humans. Human impact on predation rates has been minimal as there has been a non-hunting tradition in the last 50 years (Wright, 1997).

The faunal diversity in RNP is high for Madagascar (Wright, 1992), with 116 species of birds including six species of raptors, five species of viverrid, and twelve species of primates (Table 4.1). Total biomass of primates at this site was approximately 330 kg/km^2, comparable to *terra firme* forests in Central Amazon and Lope Reserve in Gabon, but roughly half the primate biomass of the alluvial floodplain forest of Manu, Peru or Kirindy dry forest in western Madagascar (Terborgh, 1983; Oates et al., 1990; Peres, 1993; White, 1994; Ganzhorn & Kappeler, 1996; Wright, 1998).

Review of Reported Kills

Although predation is difficult to quantify, we are lucky that at RNP, where 13 dissertations and 15 masters theses (DEA) have been completed on the behavior and ecology of lemurs, incidental observations have been recorded and accumulated over time (Wright and Andriamihaja, 2004). We began by reviewing existing information, including reports from researchers and research technicians and data from field notebooks, with the objective of ascertaining all the known acts of predation on lemurs. We especially reviewed the data books from the long-term continuous behavioral study of *Propithecus edwardsi* (Wright, 1995; Pochron et al., 2004). In this study, predation events were scored as "kills" when the predator was near the corpse or when there were signs of predation (i.e., discarded entrails, or teeth or talon marks on bones) (Wright et al., 1997; Wright, 1998). Animals abruptly missing from a group and never seen again were scored as "possible kills."

Observations at Raptor Nest Sites

During four raptor nesting seasons (August–January) between 1999 and 2002, 11 nests of *B. brachypterus* were observed for a total of 1,204 hrs with 204 observed prey deliveries; 7 nests of *A. henstii* were observed for a total of 1,703 hrs with 284 observed prey deliveries; and 7 nests of *P. radiatus* were observed for a total of 1,007 hrs with 186 observed prey deliveries (Fig. 4.2). Nest observations included behavioral sampling of a nest through continuous recording of prey deliveries, feeding behavior and instantaneous sampling every 5 min for nest attendance by the adult male and female. Focal nest observations on all three raptor species were conducted from sunrise to sunset with each nest being observed one to two days per week throughout the four-year study from a distance of at least 150 m, to minimize nest disturbance (Karpanty, 2005, 2006).

Experiments

Playback experiments were conducted on five previously habituated groups (n = 15 groups total) of *Eulemur fulvus rufus*, *Hapalemur griseus griseus*, and *P. edwardsi* in the Talatakely and Vatoharanana trail systems of RNP. The three diurnal lemur species were chosen as they represent a range in body size, group size, and anti-predator tactics. Individuals in most groups were collared to allow the researcher individual recognition. The design of this experiment was modeled after Zuberbühler et al. (1999) and Hauser & Wrangham (1990).

Vocalizations used included the fossa, Henst's goshawk, Madagascar harrier-hawk, Madagascar buzzard, Madagascar serpent eagle (*Eutriorchis astur*), and the greater vasa parrot (*Coracopsis vasa*), the latter as a control. To avoid pseudoreplication, a collection of four different vocalization tapes was made for predator and control species, with each tape containing a different individual from RNP. Raptor calls were recorded from birds near their nest sites. Calls of *E. astur* were provided by the Peregrine Fund and were recorded from two nesting birds in the Masoala Peninsula of northeastern Madagascar. The tapes were then merged for each species so that each playback consisted of calls from different individuals of the same species. Two tapes of the common vocalizations of the fossa were provided by Deutsches Primatenzentrum, recorded from one individual from Zoo

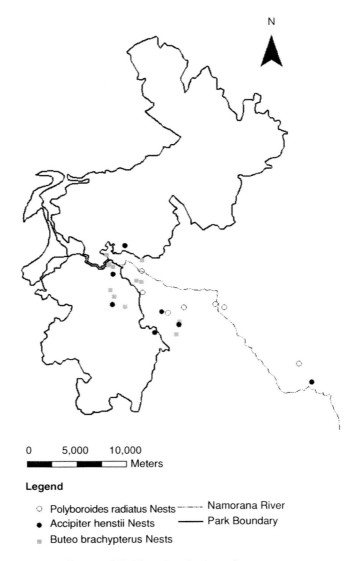

FIGURE 4.2. Map of study site and raptor nests

Duisburg in Germany, and by Animal Sound Archives (Tierstimmenarchiv), from a collection of calls from three individuals. Vocalizations were broadcast using a SONY WMD6C with Nagra DSM speakers. Sound level was set to mimic natural intensity (85–105 dB SPL) and was calibrated using a Radio Shack sound level meter planed one meter from the speaker.

All subjects and groups were tested only once with each of the six stimuli in a randomized order in either September to December of 2001 or the same period in 2002. The playback trial was only conducted if (1) no lemur had detected the observer as a predator risk (e.g., they were engaged in normal activity) and (2) no predator alarms had occurred within thirty minutes. Statistical independence was maintained within species by sampling different groups and by using groups from both the Talatakely and Vatoharanana trail systems of RNP, which are separated by approximately 5 km of contiguous rainforest. It was assumed that the natural predation risk and predator experience were constant between the groups of the same species. Human presence was minimized and experiments separated by at least seven days so lemur subjects would not habituate to predator calls. Playback stimuli in the rainforest generally can only be detected up to 300 m, so other groups and species should not have been affected by the playback.

When a lemur was located, its location was marked on a map and behavioral observations were conducted on two adult focal individuals (one male, one female) chosen at random from the group. Twelve 5-minute focal samples on each male and female individual were collected before the playback (pre-playback time period), the playback occurred during the 13th focal sample, and 12 more focal samples were collected after the playback (post-playback period), giving a total of 2 hr 5 min of sampling per playback experiment. While the animals were in an observable location during the 13th focal sample and engaging in normal activities, the speaker was hidden 50 m away from the groups, and an observer conducted the playback. Immediately before the playback, the speaker was raised with a stick to 4 m above the ground to control for speaker-induced downward vigilance. Four observers stayed within viewing distance of the group and continued conducting the focal sample and documenting the response of the group to the playback. Two observers were responsible for writing the data and checking the observations of the primary observers and two for continuously watching the lemurs and verbally reporting the data. The focal group and individuals' responses included vigilance type and duration, height, activity, vocal alarm responses, nearest neighbor distance and individual. Vigilance types were defined to be fixed stares either greater than 3 sec duration in an upward direction; greater than 3 sec downward; or greater than 3 sec in a horizontal direction (from the lemur's point of view) with cessation of other activity. Height was classified as low (<5 m), medium (5–15 m) and high (>16 m). Activity classes included feeding, traveling, grooming, resting, sleeping, and playing. Any aerial predator alarms, terrestrial predator alarms, general predator alarms, contact calls, or lost calls were also recorded. Finally, ad libitum notes were made when animals dropped or ascended in the canopy or approached the speakers. Vigilance, height, and activity behaviors were recorded through focal individual sampling and continuous recording in 5-minute intervals. Proximity data and trail locations were recorded at the end of each 5-mininterval by instantaneous recording.

Analysis of Experimental Data

Data were summarized from each focal sample to give the percentage of each 5-min sample that a focal animal spent vigilant (summing upward, downward and horizontal vigilance; fixed stares >3 sec duration), active (summing feeding, playing, traveling, and grooming), and low (<5 m high). The effects of time in relation to the playback sex, playback type, lemur species, and all possible interactions on percent time vigilant, active, and low were tested by ANOVA and adjusted for multiple comparisons while controlling for variation across groups. The effect of time in relation to playback was coded as the pre-playback period (behavioral samples 1–12), the short-term post-playback period (0–15 min after playback, samples 14–16), and the longer-term post-playback period (16–60 min after the playback, samples 17–25).

Three types of a priori contrasts were conducted on the data set: (1) contrasts of responses over time: pre-playback versus longer-term post-playback (significance indicative of slow reaction to the playback and a delayed reaction) and pre-playback and longer-term playback responses versus the short-term response (significance indicative of a quick reaction to the playback and a quick decay of the reaction; (2) contrasts of playback type by risk or predator category: Control versus All Predators, Control vs. Aerial Predators, Control vs. Ground Predators, Aerial vs. Ground Predators, *A. henstii* vs. other aerial predators; and (3) contrasts of lemur species effects (*Eulemur* vs. *Hapalemur* vs. *Propithecus*).

Results

Observations of Predation on Lemurs from Reported Kills and Scat

Long-term studies have resulted in observations of corpses immediately following fossa kills for four species of lemurs, *Eulemur rubriventer, Varecia variegata, Avahi laniger,* and *Propithecus edwardsi* (Andrea Baden, pers. comm.; Overdorff & Strait, 1995; Overdorff et al., 1999; Stacey Tecot, pers. comm.; Wright et al., 1997; Wright, 1998). Including data from behavioral ecology studies between 1986 and 2005, we observed both actual kills and possible kills in four groups of *P. edwardsi* that we followed year round (Table 4.3). During the 19-year study of the 87-member *P. edwardsi* community (four groups), a maximum of 19 and minimum of 9 individuals were killed and eaten by the fossa. The fossa ate all age, sex classes (see Table 4.3) with the minimum toll: 1 adult male, 3 adult females, 1 three-year old female, 1 one-year old female, and 2 infants that died with their mothers (Wright, 1995, 1998, unpubl. data). The data show that predation on *Propithecus* is seasonal, and all verified fossa kills occurred in May–September, the cold, dry season and the season when infants are 1–3 months old. *Propithecus* has been observed giving a ZZUSS! call at ground predators including the fossa (Wright, 1998).

Additionally, five species of lemurs have been identified from fossa scats found at Ranomafana National Park, including two diurnal (*P. edwardsi* and *Hapalemur simus*), two cathemeral (*E. rubriventer* and *E. fulvus rufus*) (Wright et al., 1997), and one nocturnal (*Microcebus rufus*) (Goodman, 2004, Table 4.4).

TABLE 4.3. *Cryptoprocta ferox* kills (corpse observed) and suspected kills of *Propithecus edwardsi* at Talatakely Trail System (TTS) in Ranomafana National Park during a 19-year continuous study of two groups (1986–2005), as well as two additional groups since 1993 (Group III) and 1996 (Group IV).

C. *ferox* kills	Suspected C. *ferox* kills	Yr/month	Group
adult male (RR)		1990/Jul	I
adult female (BY)		1994/Aug	III
2 month infant (BYI)		1994/Aug	III
1 year old female (PYI)		1994/Sep	
3 yr old female (BB)		1994/Sep	I
3 year old male (PS)		2003/May	I
adult female (Radio Silver)		2005/May	II
	adult male (I)	1987/Jan	I
	one yr old male (GGI)	1987/Oct	II
	adult female (RG)	1989/Feb	II
	6 mo old (GGI)	1989/Dec	II
	adult female (Y)	1993/Sep	I
	3 mo infant (YI)	1993/Sep	I
	adult female (GG)	1993/Jun	II
	6 mo old (BYI)	1993/Jan	III
	one yr old (GGI)	1992/Jun	II
	2 yr old female (TSI)	2000/Jul	I
	2 yr old female (BI)	2000/Jul	III
	1 yr old male (BGI)	2000/May	IV
	adult female (TS)	2001/Jun	I

TABLE 4.4. Lemur remains identified in *Cryptoprocta ferox* scats within RNP. Each asterisk represents a separate scat. These scats were found by Luke Dollar (Wright et al., 1997; Goodman, 2004), Summer Arrigo-Nelson, pers. comm., and Deborah Overdorff, pers. comm.

Species	Common Name	MNI
Propithecus edwardsi	Milne Edwards' sifaka	***
Hapalemur griseus	Lesser bamboo lemur	***
Avahi laniger	Eastern woolly lemur	*
Cheirogaleus major	Greater dwarf lemur	*
Microcebus rufus	Brown mouse lemur	*
Eulemur fulvus rufus	Red-fronted brown lemur	***
Eulemur rubriventer	Red-bellied lemur	**
Hapalemur simus	Greater bamboo lemur	*
Varecia variegate	Black and white ruffed lemur	*

All *P. edwardsi* group members give the aerial predator call, a very loud, low-pitched series of roars and barks, which can continue for 5–15 minutes (Wright, 1998). In observations we made during follows, we saw sifakas react to raptor sightings, or to group members alarm-barking in response to raptors by looking up, alarm barking, and dropping to lower levels of the forest. During the 19 years of sifaka follows, the four species of raptors observed to elicit alarm barking were *B. brachypterus*, *P. radiatus*, *A. henstii*, and *E. astur*. Other large birds such as the crested ibis, vasa parrot, or blue coua occasionally received an alarm bark. Only one attack on *P. edwardsi* by raptors was observed. The hawk, talons extended (*A. henstii*, although it happened too fast for positive identification), swooped at a mother with infant during July.

During the long-term study of *H. simus* (Tan, 1999), C. Tan and P. Wright observed an *A. henstii* attempt an attack on an eight-month old infant. The group of nine individuals gave an alarm call, dropped to the forest floor and hid for over two hours. The infant (3/4 the size of the mother) leaped into his mother's arms and remained ventrally cradled for over an hour, low in the understory.

Sightings of the serpent eagle (*E. astur*) are rare, but L. Rasabo reports a serpent eagle eating an adult *A. laniger* (Wright, 1998). A nest of this eagle was not found for this study, and no further observations of kills of lemurs by the serpent eagle have been made.

Raptor Nest Site Observations

The remains from seven species of lemur were observed taken to the nests of *A. henstii* and *P. radiatus* for ingestion by chicks and parents during October–December 1999–2002. Three of these lemur species are nocturnal, two are diurnal and two are cathemeral (active equally in day and night hours). No lemurs were observed to be delivered to nests of *B. brachypterus* during this same time period. Predation rates on these lemurs were calculated by taking the percentage of the lemur population killed per year by each raptor predator or by a combination of the two hawk species (Figs. 4.3, 4.4). The highest predation rate was on the 1-kg primates, *Hapalemur* (diurnal) and *Avahi* (nocturnal). The 2-kg *E. rubriventer* and *E. fulvus rufus* (cathemeral) were also eaten at a high rate.

The diet of *A. henstii* (Table 4.5) comprises at least 26 different prey species, including three nocturnal, two diurnal, and two cathemeral lemur species. The largest component of the prey profile in terms of individuals is avian prey (59%); however, lemurs are second in terms of percent individuals (23%). In terms of percent of total biomass, the trends hold the same with avian prey accounting for 70.51% of all biomass delivered to the nest, primates 28.43% of all biomass, and reptiles 0.32% of all biomass. The diet of *A. henstii* is highly variable, ranging from endangered species such as *Varecia variegata* (black and white ruffed lemur) and *Lophotibis cristata* (crested ibis)—found only in old-growth forest—to domestic chickens and rats. Both *A. henstii* and *P. radiatus* delivered lemurs to the nest only during the nestling and fledgling stages of the nesting cycle. During this period, *A. henstii* individuals delivered a lemur to the nest every 21 hrs, or 0.047 lemur per hr of observation. Extrapolating this prey delivery rate to the incubation period, we would have expected to see at least 15 lemur prey deliveries during our observations at the nest during incubation, instead of the zero we did observe.

The diet of *P. radiatus* (Table 4.6) is composed of at least 24 different prey species, including 3 nocturnal and 1 diurnal species of lemur. The diet of this generalist predator ranges from prey relying on high quality forest (*Accipiter madagascariensis*) to prey associated only with human disturbance (*Rattus rattus*). The largest percentage of prey deliveries is avian (27%), followed by reptiles

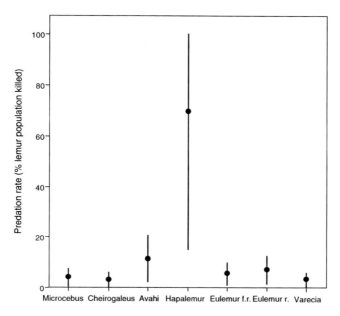

FIGURE 4.3. Predation rates on lemurs by *Accipiter henstii* and *Polyboroides radiatus* as calculated from direct nest observations in this study. Predation rates equal percentage of the lemur population killed per year by a combination of the two hawk species. The filled black dot is the median of the possible projected rates

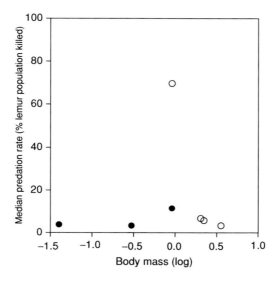

FIGURE 4.4. Predation rates on lemurs by raptors with the weights of the lemurs log transformed. The filled black dot is the median of the possible projected rates

TABLE 4.5. Primate prey of *Accipiter henstii* at RNP from direct nest observations 1999–2002. Data are combined from observations at 7 nest sites. MNI is the minimum number of individuals. Additional prey includes primarily birds with a few amphibians, reptiles, tenrecs and rodents (see Karpanty, 2005, for a complete list of prey taxa).

Species	Common Name	MNI	% Total Individuals	% Total Biomass
Hapalemur griseus	Lesser bamboo lemur	28	9.86	9.96
Avahi laniger	Eastern woolly lemur	13	4.58	7.37
Cheirogaleus major	Greater dwarf lemur	10	3.52	2.41
Microcebus rufus	Brown mouse lemur	8	2.82	2.56
Eulemur fulvus rufus	Red-fronted brown lemur	5	1.76	5.59
Eulemur rubriventer	Red-bellied lemur	1	0.35	1.01
Varecia variegata	Black and white ruffed lemur	1	0.35	1.83

TABLE 4.6. Primate prey of *Polyboroides radiatus* at RNP from direct nest observations 1999–2002. Data are combined from observations of 7 nests. MNI is the minimum number of individuals. Additional prey include mainly reptiles and birds with a few frogs, bats, rodents, tenrecs (see Karpanty, 2005, for a complete list of taxa).

Species	Common Name	MNI	% Total Individuals	% Total Biomass
Microcebus rufus	Brown mouse lemur	14	7.53	3.73
Avahi laniger	Eastern wooly lemur	4	2.15	18.86
Cheirogaleus major	Greater dwarf lemur	3	1.61	6.02
Hapalemur griseus	Lesser bamboo lemur	3	1.61	8.88

(18%) and primates (13%). In terms of percent of total biomass, the trend is reversed, with the most important taxa being primates (37.49%), followed by reptiles (24.65%) and birds (19.05%). All deliveries of lemurs occurred during the nestling and fledgling stages of the nesting cycle. During this study, a lemur was delivered to the nest every 31 hrs (0.0327 lemur/ hour of observation). If the delivery rate were to be the same during incubation, we would expect to have seen at least 6 lemurs (instead of zero) delivered to the nest during the hours that *Polyboroides* nests were observed.

Playback Experiments

Intra-species responses to playback experiments

Eulemur fulvus rufus

After the playbacks of predator vocalizations, *Eulemur* individuals generally exhibited a cryptic anti-predator strategy by increasing vigilance, moving to higher portions of the canopy, and decreasing activity levels in the hour after the playback. Changes were most marked in the last 45 min of the experiments, indicating that *Eulemur* have a delayed, but long-term, cryptic response to predator vocalization exposures.

Changes in vigilance were generally short term for *Eulemur*, with overall levels of vigilance highest in the first 15 min after the playback and lowest in the last 45 min, when the lemurs were quiet and cryptic (*Percent Time Vigilant*: Pre-Playback: 33.1%; Short-term Post-Playback: 35.1%; Long-term Post-Playback: 18.5%; $F_{2,119} = 29.63$, $p < 0.001$). As another indication of the cryptic response, the activity levels of *Eulemur* did decrease during the last 45 min of the playback experiments (*Percent Time Active*: Pre-Playback: 45.7%; Short-term Post-Playback: 43.9%; Long-term Post-Playback: 32.4%; $F_{2,119} = 8.20$, $p < 0.001$). *Eulemur* individuals did move up in the canopy during the last 45 minutes of the experiments (*Percent Time Low in Canopy*: Pre-Playback: 3.7%; Short-term Post-Playback: 14.1%; Long-term Post-Playback: 11.9%; $F_{2,119} = 6.81$, $p < 0.01$).

Hapalemur griseus griseus

Hapalemur individuals exhibited an even greater cryptic response than *Eulemur* by decreasing vigilance and activity levels for the entire one hour after the predator playbacks. (*Percent Time Vigilant*: Pre-Playback: 22.9%; Short-term Post-Playback: 17.5%; Long-term Post-Playback: 16.4%; $F_{2,119} = 3.32$, $p < 0.05$. *Percent Time Active*: Pre-Playback: 52.9%; Short-term Post-Playback: 23.9%; Long-term Post-Playback: 38.4%; $F_{2,119} = 18.29$, $p < 0.0001$). Instead of moving up in the canopy to hide, *Hapalemur* generally moved lower and were significantly lower in the canopy following playbacks of the aerial predators. (After Aerial Predators: 40.6 % Time Low) than the terrestrial predator (After Fossa: 31.1% Time Low; $F_{1,119} = 8.67$, $p < 0.01$). *Hapalemur* individuals decreased their activity levels in response to all predator playbacks versus the control. (*Percent Time Active*: After Predator Playbacks: 29.9%; After Control Playbacks: 41.2%; $F_{1,119} = 13.42$, $p < 0.001$).

Propithecus edwardsi

In contrast to the smaller lemur species, *Propithecus* individuals altered vigilance only following playbacks of the most important aerial predator in this rainforest system, *A. henstii*. (*Percent Time Vigilant*: After *A. henstii*: 31.9%; After Other Predators: 25.7%; After Control Playback: 22.3%; $F_{1,119} = 6.47$, $p < 0.01$.) There were no overall effects of the playbacks on vigilance or height choice for this species. The only general effect observed was that *Propithecus* exhibited higher activity levels following the control playbacks than those of the aerial predators. (*Percent Time Active*: After Control Playback: 63.1%; After Aerial Predators: 48.0%; $F_{1,119} = 15.72$, $p < 0.001$.)

Inter-species responses to playback experiments

The effect of species identity on the behavioral responses of lemurs to the predator playback experiments was tested along with the effects of time since playback, playback type, and all possible interactions.

Vigilance

Both *Eulemur* (24.7% more vigilant) and *Propithecus* (25.9%) were significantly more vigilant than *Hapalemur* (17.9%; $F_{2,395} = 20.61$, $p < 0.001$), but there were no differences between *Eulemur* and *Propithecus*.

Activity

For all species, activity decreased significantly from the pre-playback through the hour after the playbacks. *Propithecus* spent significantly more time active (51.9%) than either *Eulemur* (37.4%) or *Hapalemur* (38.4%; $F_{2,395} = 8.32$, $p < 0.001$), but the latter two species did not differ significantly from each other. *Hapalemur* individuals were significantly less active after the playbacks than both of the other lemur species.

Height

Hapalemur spent significantly more time at a low height, under 5 m in canopy, (36.6%) than either *Eulemur* (10.7%) or *Propithecus* (18.2%; $F_{2,395} = 58.68$, $p < 0.001$), while *Propithecus* spent significantly more time low than *Eulemur*.

Playback type

There were no consistent responses across the three lemur species with regard to vigilance levels to the playbacks. For all species, the percent of time spent active was significantly greater following playbacks of the control than of the other aerial predators ($F_{2,395} = 11.62$, $p < 0.001$). Further, for all three lemur species, the percent of time spent low was greater after playbacks of *A. henstii* than of the other aerial predators ($F_{2,395} = 5.21$, $p < 0.05$).

Discussion

Predation Rates and Lemur Anti-predator Tactics

Long-term data on raptor predation on lemurs in Ranomafana, including this study on raptor nest prey, show that no individuals of *Propithecus* were observed in the diets of *A. henstii* or *P. radiatus*, or any other raptor. In comparison, *E. fulvus rufus* experienced a minimum predation rate of 1.25% of the population killed per year by *Accipiter* and *Polyboroides*, while *H. griseus* experienced a predation rate of at least 15.12% per year by these two raptor species (Figs. 4.3, 4.4). The cryptic habits of *Eulemur* and *Hapalemur* documented in the experimental part of this study did not appear to be totally effective in protecting these lemurs from raptor predation.

Lemur Behavioral Responses to a Multiple-Predator Community

In general, the results of these experiments are in agreement with previous research using playbacks of predator vocalizations in showing that the initial responses by lemurs are often predator specific (Macedonia & Polak, 1989; Macedonia, 1990; Zuberbühler et al., 1999; Fichtel & Kappeler, 2002). All species of lemurs recognized aerial vs. terrestrial predators and all three lemur species became more vigilant after the playbacks of calls. As reported in Karpanty & Grella (2001) and Wright (1998) and observed in these experiments, the initial response of each of these lemurs is most frequently to search the sky, drop in the canopy, and alarm or flee from the source of the vocalization when the playback is of a raptor predator. When the playback is of the fossa, the lemurs more frequently ascend in the canopy, increase downward vigilance, and give a general excitement alarm.

The data on vigilance, height, and activity choice from this experiment indicate that after the initial alarm and flight reaction, *Eulemur* and *Hapalemur* switch behavior to employ a cryptic anti-predator strategy, while *Propithecus* individuals respond very specifically only to predators that pose a serious threat (*C. ferox* and *A. henstii*). Studies of nests of the diurnal raptors reveal that *A. henstii* kills more lemurs than other raptors in this system (this study, Karpanty, 2006). *Propithecus* increased their vigilance more significantly to playbacks of *Accipiter* than to other raptors.

These findings that lemurs may alter vigilance, height, and activity after the initial alarm response according to general predator type and specific level of risk provide new information on the anti-predator strategies of diurnal and cathemeral lemurs. The contrast in the general cryptic strategy of *Hapalemur* and *Eulemur* versus the predator-specific strategy of *Propithecus* may have important implications in this multiple-predator community. Lima (1992) and Matsuda et al. (1993, 1994, 1996) demonstrate that predator-specific, anti-predator behaviors, such as those exhibited by *P. edwardsi*, may lead to greater predation rates than what would be expected if one simply extrapolated the predation rates of single predators alone.

Lemur Social Aggregations and Risk of Predation

Primates may join in larger groups to reduce risk from predators (Hamilton, 1971; Alexander, 1974; van Schaik, 1983; Terborgh & Janson, 1986; Janson, 1992).

Compared to many primates on other continents, group size in lemurs is small, ranging from monogamous groups of 3–6 and polygynous groups of 3–25 (Wright, 1999). In this study of a community of lemurs of different social group sizes, we have begun to understand the nuances of variability in group size as a predator deterrent or protection by comparing the effects of predation on two same-sized lemurs, *E. rubriventer* (monogamous groups of 3–5) and *E. fulvus rufus* (polygynous groups of 5–18, Overdorff, 1996). We would predict that larger group size would be a more successful strategy due to dilution effect, as well as having "more eyes and ears" for an early warning alert. Both lemur species are taken by the goshawk and the fossa. The prediction based on socio-ecological theory would be that the species with smaller group size would be preyed upon more. However, our data suggest that five times as many *E. f. rufus* were eaten by Henst's goshawk as *E. rubriventer*. When the predation rate is calculated to equal the percentage of the lemur population killed per year, this difference evens out. The sample size is not large and should be taken with caution, but there is a suggestion that commonness rather than group size may be a factor in predator choice. There may be a slight advantage to large group size in the "dilution effect," but there is also the possibility that larger, noisier, groups may attract predators.

Body Mass and Risk of Predation

The absence of adult *Propithecus* from the diets of raptors may be accounted for by *Propithecus'* large body mass: three times the body mass of *Eulemur* and six times the body mass of *Hapalemur*. However, infants only reach the body mass of *Hapalemur* at six months, and of *Eulemur* after a year, and yet these vulnerable infants are not preyed on by the raptors; perhaps because of the high levels of vigilance of the adult *Propithecus*. This strategy makes sense in light of the life history of the lemurs. *Propithecus* females give birth only once every two years, on average (Wright, 1995; Pochron et al., 2004), and each offspring is perhaps more valuable than infants of species that reproduce every year, such as *Eulemur* and *Hapalemur* (Overdorff et al., 1999, Tan, 1999). Vigilance may be well worth the foraging cost to *Propithecus*.

Both goshawks and harrier hawks preferred primates that were 1 kg in body weight. There are three lemur species—*Hapalemur griseus*, *Avahi laniger*, and *Lepilemur seallii*—with this adult body weight at RNP, but *Lepilemur* was not observed eaten by the raptors, perhaps because of their rarity or their habit of sleeping deep in tree holes during the day (Porter, 1998). *Avahi* was the preferred choice by the harrier hawk and *Hapalemur* was the preferred choice of the goshawk. When predation rate was calculated, taking into account percentage of lemur population killed by raptors, *H. griseus* had the highest predation rate at three and a half times the rate of *A. laniger*. Body mass is equal in *Avahi* and *Hapalemur*, so the higher rate of *Hapalemur* cannot be accounted for by body mass alone. This difference in observed predation rate might be related to differences in anti-predator strategies, group size, activity patterns, or habitat use patterns.

Species Rarity and Hibernation

A factor in predation rate that is obvious, but not often discussed as a strategy, is patchiness or rarity in the environment. In this community there are two endangered (*P. edwardsi* and *V. variegata*) and two critically endangered species (*H. aureus* and *P. simus*) (IUCN, 2005) lemurs in the RNP community. Three of these species (*Varecia, Hapalemur, Prolemur*) have populations that are extremely patchy in all forests (Arrigo-Nelson & Wright, 2004; Balko & Underwood, 2005; Irwin et al., 2005). These lemurs were rarely eaten by raptors, partially because they were difficult to find. Making oneself "scarce" may be a strategy in the case of two lemur species that are commonly eaten, *Microcebus rufus* and *Cheirogaleus major*. For many weeks or months of the year, these species go into torpor, and because they are not active they are not easily found by aerial and terrestrial predators. However, during periods of torpor, these lemurs may be more vulnerable to snake predation (Wright & Martin, 1995). The avoidance of predation by certain species of predators may be another advantage of torpor.

Birth Synchrony of Lemurs

One strategy to help alleviate the risk of predation is the synchronization of births, which results in a "dilution" effect (Boinski, 1989). Same-size lemurs do synchronize their birth seasons (Wright, 1999; Wright et al., 2005). More research would be needed to determine if birth synchrony is successful against predation.

Activity Cycle and Risk of Predation

During the 1990s a lively scientific controversy arose regarding lemur evolution and raptor predation. With the discovery of two extinct genera of large eagles in the sub-fossil record, Goodman (1994a,b) and Goodman & Rakotozafy (1995) suggested that present-day raptor alarm calls by lemurs could be remnants of behaviors evolved to avoid the giant extinct raptors. A series of papers expanded on this idea to suggest that diurnal lemurs had only recently become day-active, after day-active giant raptors went extinct (van Schaik & Kappeler, 1993, 1996; Kappeler & Heyman, 1996). This change would have been quite recent since sub-fossil lemur bones are dated 500–20,000 yrs BP (Simons, 1997). Field evidence on present-day raptor predation on lemurs was sparse (Goodman et al., 1993a,b,c), but lemur studies cast doubt on this theory of lemur evolution (summarized in Wright, 1999).

Recent evidence, including the results from this paper, confirms that in present-day Madagascar, nocturnal, diurnal, and cathemeral lemurs are vulnerable to both raptors and carnivores (Table 4.7). We show that in the Ranomafana rainforest, diurnal raptors such as *A. henstii* and *P. radiatus* eat almost equal numbers of nocturnal, cathemeral, and diurnal prey. In addition, *C. ferox*, active in all hours of the day and night, eats nocturnal and diurnal lemurs (Wright, 1998; Dollar, 1999; this study). However, the most eaten prey of the raptors by a factor of 10 was a diurnal primate (*H. griseus*), and the second most popular menu item was *A. laniger* of equivalent body mass and social system, but with a nocturnal lifestyle. Therefore, the advantage of being nocturnal to avoid diurnal raptors, as seen in

TABLE 4.7. Activity cycle of lemurs and predation evidence in Ranomafana National Park, ranked high, medium or low. Population density (P) is ranked common (C) or patchy and rare (R). Common species of possible size are preferred by all predators. Note that no data have been collected for owls at this site.

Activity Cycle	Species	P	Accipiter	Polyboroides radiatus	Cryptoprocta ferox	Galidia elegans
Diurnal	Propithecus edwardsi	C	–	–	medium	–
Diurnal	Varecia variegata	R	low	–	low	–
Diurnal	Prolemur simus	R	–	–	low	–
Diurnal	Hapalemur aureus	R	–	–	–	–
Diurnal	Hapalemur griseus	C	high	medium	–	–
Cathemeral	Eulemur rubriventer	C	medium	–	low	–
Cathemeral	Eulemur fulvus rufus	C	medium	–	low	–
Nocturnal	Daubentonia	R	–	–	–	–
Nocturnal	Lepilemur seallii	R	–	–	–	–
Nocturnal	Avahi laniger	C	medium	medium	low	–
Nocturnal	Cheirogaleus major	C	medium	medium	–	low
Nocturnal	Microcebus rufus	C	medium	high	low	low

the South American owl monkey (Wright, 1989, 1994), may be important to the rainforest primates of Madagascar as well, but will not offer complete protection from predation by diurnal predators. Future research is needed to determine if this is true of other sites or at all times of year as the raptor predation data reported here were collected during the raptor nesting season alone.

Acknowledgments. We acknowledge the support and assistance of ANGAP, the University of Antananarivo, the University of Fianarantsoa, MICET, and the Ministry of the Environment. The SUNY Stony Brook Research Foundation and ICTE are thanked for their support. We give special thanks to our Centre ValBio research assistants: the late Georges Rakotonirina, Raymond Ratsimbazafy, Remi Rakotosoa, Georges LaDaDa, Ralasoa Laurent, Ramarojoana Perle Mamisoa, Sabo, the late Rafanomezantsoa Sedrarimanona Angelo, Rakotomalala Miandrisoa Jeannot, Randrianantenaina Johnny, Samuel, Randrianalenarina Mamy, Razafy Mahatratra Maurice, Fahatelo Gista, Rakotonirina Telo, Randrianarivo Jeannot, Rakotonjatovo Justin, Filybert, Velotsara Jean Baptiste, and Djaizandry Bertrin. The Wenner Gren Foundation for Anthropological Research (PCW), the John D. and Catherine T. MacArthur Foundation (PCW), the

Earthwatch Institute (PCW), the Douroucouli Foundation (PCW), National Science Foundation USA (SK), Environmental Protection Agency STAR Fellowship Program (SK), Wildlife Conservation Society (SK), National Geographic Society Committee for Research and Exploration (SK), The Allstate Foundation (SK), Margot Marsh Biodiversity Fund (SK), and the International Osprey Foundation are thanked for assistance with funding. Fossa vocalizations were graciously provided by Claudia Fichtel and Karl Heinz Frommolt, and serpent-eagle calls by The Peregrine Fund. We thank Mitch Irwin and Summer Arrigo-Nelson for the map provided in Figure 4.1. Jukka Jernvall and Jean Claude Razafimahaimodison are thanked for helpful comments. The editors of this volume, Sharon Gursky and Anna Nekaris, are thanked for their diligence, expertise and patience.

References

Alexander, R. (1974). The evolution of social behavior. *Annu. Rev. Ecol. Syst.*, 5: 325–383.

Anderson, C.M. (1986). Predation and primate evolution. *Primates*, 27: 15–39.

Arrigo-Nelson, S.J., and Wright, P.C. (2004). Census and survey results from Ranomafana National Park: New evidence for the effects of habitat preference and disturbance on the Genus *Hapalemur. Folia Primatol.*, 75(5): 331–334 .

Balko, E., and Underwood, B. (2005) Effects of forest structure and composition on food availability for *Varecia variegata* at Ranomafana National Park, Madagascar. *Amer. J. Primatol.*, 66: 45–70.

Bearder S.K., Nekaris, K.A.I., and Buzzell, A. (2002). Dangers in the night: Are some nocturnal primates afraid of the dark? In L.E. Miller (Ed.), *Eat or be eaten: Predator sensitive foraging among primates* (pp. 21–43). Cambridge: Cambridge Univ. Press.

Berkelman, J. (1994). The ecology of the Madagascar buzzard *Buteo brachypterus*. In B. Meyburg and R. Chancellor (Eds.), *Raptor conservation today* (pp. 255–256). London: The Working Group on Birds of Prey and Owls.

Boesch, C. (1991). The effects of leopard predation on grouping patterns in forest chimpanzees. *Behaviour*, 117: 220–242.

Boinski, S. (1989). Birth synchrony in squirrel monkeys (*Saimiri oerstedi*): A strategy to reduce neonatal predation. *Behav. Ecol. Sociobiol.*, 21: 393–400.

Cheney, D.L., and Seyfarth, R. (1981). Selective forces affecting the predator alarm calls of vervet monkeys. *Behaviour*, 76: 25–61.

Cheney, D.L., and Wrangham, R.W. (1987). Predation. In B.B. Smuts, D.L. Cheney, R.M. Seyfarth, R.W. Wrangham , and T.T. Struhsaker (Eds), *Primate societies* (pp. 227–239). Chicago: Univ. of Chicago Press.

Cowlishaw, G. (1994). Vulnerability to predation in baboon populations. *Behaviour*, 131: 293–304.

Dollar, L., Forward, Z., and Wright, P.C. (1997). First study of *Cryptoprocta ferox* in the rainforests of Madagascar. *Am. J. Phys. Anthro.*, (Suppl.24): 102–103.

Dollar, L. (1999). Preliminary report on the status, activity cycle, and ranging of *Cryptoprocta ferox* in the Malagasy rainforest, with implications for conservation. *IUCN Small Carnivore Newsletter*: 3–6.

Fichtel, C., and Kappeler, P.M. (2002). Anti-predator behavior of group-living Malagasy primates: Mixed evidence for a referential alarm call system. *Behav. Ecol. Sociobiology*, 51: 262–275.

Ganzhorn, J.U., and Kappeler, P.M. (1996). Lemurs of the Kirindy forest. In J.U. Ganzhorn and J.P. Sorg (Eds.), *Ecology and economy of a tropical dry forest in Madagascar. Primate Report*, 46: 257–274.

Garbutt, N. (1999). *The mammals of Madagascar*. New Haven: Yale Univ. Press.

Goodman, S.M. (1994a). The enigma of anti-predator behavior in lemurs: Evidence of a large extinct eagle on Madagascar. *Int. J. Primatol.*, 15: 129–134.

Goodman, S.M. (1994b). Description of a new species of subfossil eagle from Madagascar: Stephanoatus (Aves: Falconiformes) from the deposits of Ampasambazimba. *Proc. Biol. Soc. Wash.*, 107: 421–426.

Goodman, S.M. (2004). Predation on lemurs. In S.M. Goodman and J. Bedstead (Eds.), *The natural history of Madagascar* (pp. 1221–1228). Chicago: Univ. of Chicago Press.

Goodman, S.M., O'Connor, S., and Langrand, O. (1993a). A review of predation on lemurs: Implications for the evolution of social behavior in small, nocturnal primates. In P.M. Kappeler and J.U. Ganzhorn (Eds.), *Lemur social systems and their ecological basis* (pp. 51–63). New York: Plenum Press.

Goodman, S.M., Langrand, O., and Raxworthy, C.J. (1993b). The food habits of the barn owl *Tyto alba* at three sites on Madagascar. *Ostrich*, 64: 160–171.

Goodman, S.M., Langrand, O., and Raxworthy, C.J. (1993c). The food habits of the Madagascar long eared owl *Asio madagascariensis* in two habitats in southern Madagascar. *Ostrich*, 64: 160–171.

Goodman, S.M., and Rakotozafy, L.M. (1995). Evidence for the existence of two species of *Aquila* on Madagascar during the Quaternary. *Geobios*, 28: 241–246.

Gould, L. (1996). Vigilance behavior during the birth and lactation season in naturally occurring ring-tailed lemurs (*Lemur catta*) at the Beza-Mahafaly Reserve, Madagascar. *Int. J. Primatol.*, 17: 331–347.

Gursky, S. (2002a). Determinants of gregariousness in the spectral tarsier (Prosimian: *Tarsius spectrum*). *J. Zool. Lond.*, 256: 401–410.

Gursky, S. (2002b). The behavioral ecology of the spectral tarsier, *Tarsius spectrum. Evol. Anthropol.*, 11: 226–234.

Hamilton, W. (1971). Geometry for the selfish herd. *J. Theor. Biol.*, 32: 295–311.

Hauser, M.D., and Wrangham, R.W. (1990). Recognition of predator and competitor calls in nonhuman primates and birds: A preliminary report. *Ethology*, 86: 116–130.

Hill, R.A., and Dunbar, R.I.M. (1998). An evaluation of the roles of predation rate and predation risk as selective pressures on primate grouping behaviour. *Behaviour*, 135: 411–430.

Irwin, M.T., Johnson, S.E., and Wright, P.C. (2005). The state of lemur conservation in south-eastern Madagascar: Population and habitat assessments for diurnal and cathemeral lemurs using surveys, satellite imagery and GIS. *Oryx*, 39 (2).

Isbell, L. (1990). Sudden short-term increase in mortality of vervet monkeys (*Cercopithecus aethiops*) due to predation in Amboseli National Park, Kenya. *Am. J. Primatol.*, 21: 41–52.

Isbell, L.A. (1994). Predation on primates: ecological patterns and evolutionary consequences. *Evol. Anthropol.*, 61–74.

IUCN (2005). The lemur working group report. *Global Mammal Assessment*. Switzerland: IUCN.

Janson, C.H. (1992). Evolutionary ecology of primate social structure. In E.A. Smith (Ed.), *Evolutionary ecology and human behavior* (pp. 95–130). New York: Walter de Gruyter, Inc.

Janson, C.H. (1998). Testing the predation hypothesis for vertebrate sociality: Prospects and pitfalls. *Behaviour*, 135: 389–410.

Janson, C.H. and Goldsmith, M.L. (1995). Predicting group size in primates: Foraging costs and predation risks *Behav. Ecol.*, 6: 326–336.

Janson, C.H., and van Schaik, C. (1993). Ecological risk aversion in juvenile primates: Slow and steady wins the race. In M.E. Pereira and L.A. Fairbanks (Eds.), *Juvenile primates life history, development and behavior* (pp. 57–74). Oxford: Oxford Univ. Press.

Kappeler, P.M., and Heyman, E.W. (1996). Non-convergence in the evolution of primate life history and socioecology. *Biol. J. Linn. Soc.*, 59: 297–326.

Karpanty, S.M. (2005). Behavioral and ecological interactions of raptors and lemurs in southeastern Madagascar: A multiple-predator approach. Doctoral dissertation. Stony Brook, New York.

Karpanty, S.M. (2006). Direct and indirect impacts of raptor predation on lemurs in southeastern Madagascar. *Int. J. Primatol.*, 27(1): 239–261.

Karpanty, S.M., and Goodman S.M. (1999). Diet of the Madagascar harrier-hawk, *Polyboroides radiatus*, in southeastern Madagascar. *J. Raptor Res.*, 33: 313–316.

Karpanty, S.M., and Grella, R. (2001). Lemur responses to diurnal raptor calls in Ranomafana National Park, Madagascar. *Folia Primatol.*, 72: 100–103.

Lima, S.L. (1992). Life in a multi-predator environment: some considerations for anti-predatory vigilance. *Ann. Zool. Fennici.*, 29: 217–226.

Lima, S.L., and Dill, L.M. (1990). Behavioral decisions made under the risk of predation: A review and prospectus. *Can. J. Zool.*, 68: 619–640.

Macedonia, J.M. (1990). What is communicated in the antipredator calls of lemurs: Evidence from playback experiments with ringtailed and ruffed lemurs. *Ethology*, 86: 177–190.

Macedonia, J.M., and Polak, J.F. (1989). Visual assessment of avian threat in semi-captive ringtailed lemurs (*Lemur catta*). *Behaviour*, 111: 291–304.

Matsuda, H., Abrams, P.A., and Hori, M. (1993). The effect of anti-predator behavior on exploitive competition and mutualism between predators. *OIKOS*, 68: 549–559.

Matsuda, H., Hori, M., and Abrams, P.A. (1994). Effects of predator-specific defence on community complexity. *Evolutionary Ecology*, 8: 628–638.

Matsuda, H., Hori, M., and Abrams, P.A. (1996). Effects of predator-specific defence on biodiversity and community complexity in two-trophic-level communities. *Evolutionary Ecology*, 10: 13–28.

Oates, J.F., Whitesides, G.H., Davies, A.F., Waterman, P.G., Green, S.M., Dasilva, G.L., and Mole, S. (1990). Determinants of variation in tropical forest primate biomass: New evidence from West Africa. *Ecology*, 71: 328–343.

Overdorff, D.J. (1996). Ecological correlates to social structure in two lemur species in Madagascar. *Am.J. Phys. Anthro.*, 100: 487–506.

Overdorff, D.J., and Strait, S.G. (1995). Life history and predation in *Eulemur rubriventer* in Madagascar. *Am. J. Phys. Anthro.*, 20: 164–165.

Overdorff, D.J., Merenlender, A.M., Talata, P., Telo, A., and Forward Z.A. (1999). Life history of *Eulemur fulvus rufus* from 1988–1998 in southeastern Madagascar. *Am. J. Phys. Anthropol.*, 108: 295–310.

Peres, C.A. (1990). A harpy eagle successfully captures an adult male red howler monkey. *Wilson Bull.*, 102: 560–561.

Peres, C.A. (1993). Structure and spatial organization of an Amazonian *terre firme* primate community. *J. Trop. Ecol.*, 9: 259–276.

Pochron, S.T., Tucker, T., and Wright, P.C. (2004). Demography, life history, and social structure in *Propithecus diadema edwardsi* from 1986–2000 in Ranomafana National Park, Madagascar. *Amer. J. Phys. Anthro.*, 125(1): 61–72.

Porter, L. (1998). Influences on the distribution of *Lepilemur microdon* in Ranomafana National Park, Madagascar. *Folia Primatol.*, 69: 172–176.

Pulliam, H.R., and Caraco, T. (1984). Living in groups: Is there an optimal group size? In J.R. Krebs and N.B. Davies (Eds), *Behavioral ecology: An evolutionary approach* (pp. 122–147). Sunderland: Sinauer.

Sauther, M.L. (1989). Antipredator behavior in troops of free-ranging *Lemur catta* at Beza Mahafaly Special Reserve, Madagascar. *Int. J. Primatol.*, 10: 595–606.

Shultz, S. (2001). Interactions between crowned eagles and monkeys in Taï National Park, Ivory Coast. *Folia Primatol.*, 72: 248–250.

Shultz, S. (2002). Crowned eagle population density, diet, and behavior in the Taï National Park, Ivory Coast. *Ibis*, 144: 135–138.

Shultz, S., and Noë, R. (2002). The consequences of crowned eagle central-place foraging on predation risk in monkeys. *Proc. Roy. Soc. Lond.*, 269: 1797–1802.

Sih, A., Englund G., and Wooster, D. (1998). Emergent impacts of multiple predators on prey. *Trends Ecol. and Evol.*, 13(9): 350–355.

Simons, E.L. (1997). Lemurs: Old and new. In S.M. Goodman and B.D. Patterson (Eds.), *Natural change and human impact in Madagascar* (pp. 142–166). Washington, DC: Smithsonian Institution Press.

Stanford, C.B. (1995). The influence of chimpanzee predation on group size and anti-predator behaviour in red colobus monkeys. *Anim. Behav.*, 49: 577–587.

Struhsaker, T., and Leakey, M. (1990). Prey selectivity by crowned hawk-eagles on monkeys in the Kibale Forest, Uganda. *Behav. Ecol. Sociobiol.*, 26: 435–443.

Tan, C.L. (1999). Group composition, home range size and diet of three sympatric bamboo lemurs (genus *Hapalemur*) in Ranomafana National Park, Madagascar. *Int. J. Primatol.*, 20: 547–566.

Terborgh, J. (1983). *Five New World primates*. Princeton, NJ: Princeton Univ. Press.

Terborgh, J., and Janson, C.H. (1986). The socioecology of primate groups. *Ann. Rev. Ecol. System.*, 17: 111–135.

Treves, A. (1999). Has predation shaped the social systems of arboreal primates? *Int. J. Primatol.*, 20: 35–67.

van Schaik, C.P. (1983). Why are diurnal primates living in groups? *Behaviour*, 87: 120–144.

van Schaik, C.P., and van Hooff, J.A.R.A.M. (1983). On the ultimate causes of primate social systems. *Behaviour*, 85: 91–117.

van Schaik, C.P., and Kappeler, P.M. (1993). Life history, activity period and lemur social systems. In P.M. Kappeler and J.U. Ganzhorn (Eds.), *Lemur social systems and their ecological basis* (pp. 241–260). New York: Plenum Press.

van Schaik, C.P., and Kappeler, P.M. (1996). The social systems of gregarious lemurs: Lack of convergence with anthropoids due to evolutionary disequilibrium? *Ethology*, 102: 915–941.

Vine, I. (1973). Detection of prey flocks by predators. *J. Theor. Biol.*, 40: 207–210.

Watson, R., and Lewis, R. (1994). Raptor studies in Madagascar's rain forest. In B.U. Meyberg and R. Chancellor (Eds.), *Raptor conservation today* (pp. 283–290). London: World Working Group on Birds of Prey and Owls.

White, L.J.T. (1994). Biomass of rain forest mammals in the Lope Reserve, Gabon. *J. An. Ecol.*, 63: 499–512.

Wright, P.C. (1989). The nocturnal primate niche in the new world. *J. Hum. Evol.*, 18: 635–658.

Wright, P.C. (1992). Primate ecology, rainforest conservation, and economic development: Building a national park in Madagascar. *Evol. Anthropol.*, 1: 25–33.

Wright, P.C. (1994). Behavior and ecology of the owl monkey in the wild. In J.F. Baer, R.E. Weller, and I. Kakoma (Eds.), *Aotus: The owl monkey* (pp. 97–112). San Diego: Academic Press.

Wright, P.C. (1995). Demography and life history of free-ranging *Propithecus diadema edwardsi* in Ranomafana National Park, Madagascar. *Int. J. Primatol.*, 16: 835–853.

Wright, P.C. (1997). The future of biodiversity in Madagascar A view from Ranomafana National Park. In S.M. Goodman and B.D. Patterson (Eds.), *Natural change and human impact in Madagascar* (pp. 381–405). Washington, DC: Smithsonian Institution Press.

Wright, P.C. (1998). Impact of predation risk on the behaviour of *Propithecus diadema edwardsi* in the rain forest of Madagascar. *Behaviour*, 135: 483–512.

Wright, P.C. (1999). Lemur traits and Madagascar ecology: Coping with an island environment. *Yearbook Phys. Anthropology*, 42: 31–72.

Wright, P.C., and Martin, L.B. (1995). Predation, pollination and torpor in two nocturnal primates: *Cheirogaleus major* and *Microcebus rufus* in the rain forest of Madagascar. In L. Alterman, M.K. Izard, and G. Doyle (Eds.), *Creatures of the dark: The nocturnal prosimians* (pp. 45–60). New York: Plenum Press.

Wright, P.C., Heckscher, S.K., and Dunham, A.E. (1997). Predation on Milne-Edwards' sifaka (*Propithecus diadema edwardsi*) by the fossa (*Cryptoprocta ferox*) in the rain forest of southeastern Madagascar. *Folia Primatol.*, 97: 34–43.

Wright, P.C., and Andriamihaja, B. (2004). Making a rain forest national park work in Madagascar: Ranomafana National Park and its long-term research commitment. In J. Terborgh, C. van Schaik, L. Davenport, and M. Rao (Eds.), *Making parks work: Strategies for preserving tropical nature* (pp. 112–136). Covelo, CA: Island Press.

Wright, P.C. Razafindratsita V., Pochron, S.T. and Jernvall, J.J. (2005). The key to Madagascar frugivores. In J.L. Dew and H. Boubli (Eds.), *Tropical fruits and frugivores: Strong interactors* (pp. 1–18). New York: Springer Press.

Zuberbühler, K., Jenny, D., and Bshary R. (1999). The predator deterrence function of primate alarm calls. *Ethology*, 105: 477–490.

5
Predation, Communication, and Cognition in Lemurs

Marina Scheumann, Andriatahiana Rabesandratana, and Elke Zimmermann

Introduction

Predation represents an important selective force shaping the evolution of primate behavior. Primates confronted with predators have evolved various strategies to minimize the probability of being eaten. Predation risk and hunting styles of predators should have selected for communicative and cognitive abilities linked to socioecology and life history. As studies on several socially cohesive mammals indicate, the study of anti-predator behavior represents an important tool for gaining insight into cognition, e.g., to understand how animals classify objects and events in the world around them (e.g., marmots: Blumstein, 1999; vervet monkeys: Seyfarth et al., 1980; Diana monkeys: Zuberbühler, 2000; suricates: Manser et al., 2002).

Malagasy lemurs belong to the most ancient extant primate radiation (Yoder, 2003). They show the largest variation in body sizes, activity, feeding patterns, locomotion styles, and sociality patterns among the strepsirrhine and provide, therefore, important models to explore the origin and evolution of primate behavior. Previously, Goodman et al. (1993) stated that predation pressure on lemurs was highly underestimated. Recent data supported that this pressure is comparable to, and in some cases even higher than, that of primates on other continents (Goodman, 2003). We therefore expected that lemurs would not only show crypsis to avoid predators, but would adapt to their predatory world by evolving distinct anti-predator strategies similar to those of anthropoid primates.

In this review we will have two major goals: We will estimate predation risk of lemurs based on current data on the number of predator species. By relating this information to the variation of life history and ecology in lemurs we will explore whether these traits are shaped by predation. Furthermore, we will summarize data on predation-related behavior of lemurs, including our own data on nocturnal lemurs, to investigate the general hypothesis that predation risk, perception abilities, and hunting styles of predators explain the variation of anti-predator strategies and associated communicative and cognitive abilities.

100

Methods

We have reviewed the literature on predation and anti-predation strategies in lemurs from 1940 to 2005 using PrimateLit (http://primatelit.library.wis.edu.). We also included unpublished predator-related information from the Ankarafantsika National Park in northwestern Madagascar obtained by our group and those of unpublished diploma and Ph.D. theses. In total, we included 49 references, 24 for diurnal, 10 for cathemeral, and 29 for nocturnal lemurs.

Predation in primates is difficult to assess (Goodman et al., 1993; Goodman, 2003), in particular for nocturnal species. One possible approach toward estimating predation pressure for different lemur genera is to take as a rough indirect estimate the number of predator species to which lemurs are exposed. In this study we have used the number of predator species as an index to estimate predation risk in a respective lemur genus (see also Anderson, 1986). We summarized information of predator species for each lemur genus and displayed it in Table 5.1. We distinguished three different predation risk classes by taking the highest reported number of predator species ($N = 13$) and dividing it by three. The following classes were then set up: *low risk* (0 to 3 predator species), *medium risk* (4 to 8 predator species) and *high risk* class (9 to 13 predator species). To relate the information on predation risk to life history traits and ecology of lemurs, we extracted data on activity, body mass, number of predators, foraging group size, maximum female reproductive output per year (Mueller & Thalmann, 2000; Goodman et al., 2003; Zimmermann & Radespiel, in press). We calculated the mean for the following traits per genus across the number of those species for which this information was accessible: body mass, foraging group size, and female reproductive output per year (Table 5.2). We related the number of predator species to life history traits and ecology using Spearman rank correlation. We calculated a regression model (curve estimation procedure) according to SPSS 13.0 to explore the relation between predation risk and the particular trait being considered when values for the latter were normally distributed. We compared predation risk between nocturnal and diurnal/cathemeral lemurs using the Mann-Whitney U test for two independent samples.

Results

Predation and Its Relation to Variation in Life History and Ecology of Lemurs

More than twenty different predator species (ten raptors, six carnivores, five reptiles, as well as two lemur species) are reported to prey on lemurs (Goodman, 2003). Three lemur genera belong to the low predation risk class (*Daubentonia, Varecia, Indri*), eight to the medium predation risk class (*Mirza, Phaner,*

TABLE 5.1. Overview of known lemur predators based on Goodman (2003), [1]Schülke (2001), [2]Burney (2002), [3]Fietz & Dausmann (2003), [4]Dollar et al. (Chapter 3), [5]Karpanty & Wright (Chapter 4) and [6]Gould & Sauther (Chapter 13). Lemurs and predators are arranged according to body mass (data of lemurs and carnivores from Goodman et al., 2003; of birds from Langrand, 1990, and of reptiles from Glaw & Vences, 1996).

Predator	Primate Species Preyed Upon												
	Microcebus	Cheirogaleus	Mirza	Phaner	Lepilemur	Avahi	Hapalemur	Eulemur	Daubentonia	Lemur	Varecia	Propithecus	Indri
BIRDS													
Vanga curvirostris	+												
Tyto soumagnei	+												
Tyto alba	+	+[3]	+		+								
Asio madagascariensis	+	+			+	+	+						
Aviceda madagascariensis				+									
Buteo brachypterus	+	+	+	+	+							+	
Accipiter henstii	+	+[5]				+	+[5]	+			+[5]	+[5]	
Accipiter francesii						+							
Eutriorchis astur						+[5]				+			
Polyboroides radiatus	+	+	+		+		+[5]			+		+	
CARNIVORES													
Mungotictis decemlineata	+	+	+		+								
Galidia elegans	+	+			+								
Viverricula indica*		+[3]								+			
Felis sylvestris*										+			
Cryptoprocta ferox	+	+	+	+	+	+[4]	+	+	+		+	+	
Canis lupus*	+									+[6]			
REPTILES													
Ithycyphus miniatus	+												
Sanzinia madagascariensis		+		+[1]			+						
Acrantophis madagascariensis								+				+[2]	
Acrantophis dumerilii								+					
Crocodylus niloticus													
PRIMATES													
Mirza coquereli	+												
Eulemur fulvus										+			
TOTAL NUMBER OF PREDATORS	13	10	5	4	7	5	5	4	1	6	2	5	0

* introduced predators

TABLE 5.2. Predation risk, life history traits and ecology of lemurs (according to [1]Goodman et al. (2003), [2]Zimmermann & Radespiel (in press), [3]Müller & Thalmann (2000); mean value of trait across all species of a genus was calculated.

Genus	Family	Activity[1]	Body Mass[1] (g)	Number of Predators[1]	Foraging Group Size[1]	Maximum Female Reproductive Output/Year[2]	Social System[3]
Microcebus (n = 8)	Cheirogaleidae	n	53	13	1	2	dispersed MM
Cheirogaleus (n = 2)	Cheirogaleidae	n	281	10	1	2	dispersed monogamy
Mirza (n = 1)	Cheirogaleidae	n	294	5	1	2	dispersed MM
Phaner (n = 1)	Cheirogaleidae	n	327	4	2	?	dispersed monogamy
Lepilemur (n = 4)	Lepilemuridae	n	740	7	1	1	dispersed monogamy
Avahi (n = 2)	Indriidae	n	1012	5	3.25	1	gregarious monogamy
Hapalemur (n = 3)	Lemuridae	c	1644	5	4.83	1.09	gregarious MM gregarious monogamy
Eulemur (n = 5)	Lemuridae	c	2081	4	5.70	0.75	gregarious MM gregarious monogamy
Daubentonia (n = 1)	Daubentoniidae	n	2516	1	1	0.6	dispersed MM
Lemur (n = 1)	Lemuridae	d	2678	6	16.9	0.86	gregarious MM
Varecia (n = 1)	Lemuridae	d	3548	2	11.0	2	gregarious MM
Propithecus (n = 3)	Indridae	d	4196	5	5.47	1	gregarious MM
Indri (n = 1)	Indridae	d	6480	0	3.10	1	gregarious monogamy

N = number of species; n = nocturnal;
c = cathemeral; d = diurnal; MM MF = multimale-multifemale system

Lepilemur, Avahi, Hapalemur, Eulemur, Lemur, Propithecus) and two genera belong to the high predation risk class (*Microcebus, Cheirogaleus*) (Table 5.2).

Life history traits are supposed to be an adaptation to predation in anthropoid primates (Isbell, 1994; Janson, 2003), therefore we explored to what extent this is also true for lemurs by relating particular life history and ecological traits to predation risks.

It is predicted that predation shapes group size in anthropoid primates (e.g., Van Schaik, 1983). Individuals living in large groups are assumed to be less threatened by predation because of safety-in-number effects and/or improved predation detection (Alcock, 1997). Lemurs living in cohesive groups that forage together should consequently be eaten by fewer numbers of predators than those foraging in pairs or solitarily. If grouping pattern at a sleeping site protects against predators, as assumed by various studies (e.g., Radespiel et al., 1998, 2003), lemur species forming sleeping groups should be exposed to a lesser predation risk than those sleeping solitarily. The first part of this hypothesis is not supported by our data. Thus, foraging group size and predation risk are neither correlated for nocturnal (Spearman correlation: $r = -0.360$, $N = 7$, $P = 0.428$) nor for diurnal/cathemeral genera (Spearman correlation: $r = 0.464$, $N = 6$, $P = 0.354$) nor for the whole lemur sample (Spearman correlation: $r = -0.278$, $N = 13$, $p = 0.357$). The second part of the hypothesis is hard to investigate since up until this time, only two genera were described in which individuals of both sexes sleep solitarily (*Mirza, Daubentonia*).

A further hypothesis established for anthropoid primates predicts that high female reproductive output per year is an adaptation to predation (e.g., Hill & Dunbar, 1998). According to this hypothesis, genera with a high number of offspring per year should be exposed to a higher number of predator species than those with a lower reproductive rate. Reproductive rate and predation risk are, however, not significantly related neither in the whole lemur sample (Spearman correlation: $r = 0.408$, $N = 12$, $p = 0.188$) nor in nocturnal (Spearman correlation: $r = 0.705$, $N = 6$, $p = 0.188$) or diurnal/cathemeral lemurs (Spearman correlation: -0.279, $N = 6$, $p = 0.592$).

Others discuss that predation selects for activity mode (e.g., Clutton-Brock & Harvey, 1980; Terborgh & Janson, 1986; Bearder et al., 2002). Nocturnal activity is assumed to be a response to high predation pressure during the day. According to this hypothesis, lemurs foraging during the night should be exposed to a lower predation risk than those foraging during the day. Instead, nocturnal lemurs seem to suffer a similar predation risk to diurnal/cathemeral lemurs (Mann-Whitney $U = 12, 5$, $N = 13$, $p = 0.218$).

Body size (or body mass) is often assumed to be an adaptation to predation (e.g., Isbell, 1994), in so far as larger species are less vulnerable than smaller ones. This hypothesis is supported by data in lemurs. Variation of body mass is indeed significantly related to predation risk and can be best explained by a logarithmic model ($r^2 = 0.639$, df $= 11$, $p = 0.001$, Figure 5.1). Accordingly, the small-bodied mouse lemurs (*Microcebus ssp.*) were found to be eaten by the highest number of predator species, from smaller to larger ones and from nocturnal to

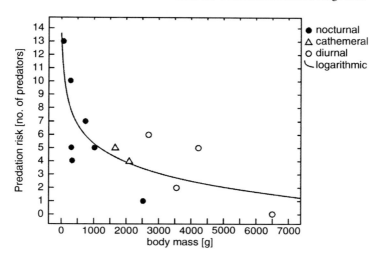

FIGURE 5.1. Relationship between mean adult body mass and predation risk in 13 lemur genera

crepuscular to diurnal ones, including aerial and terrestrial predators, whereas the largest extant lemur, the indri (*Indri indri*), does not seem to have a single extant predator (except human poachers, who represent evolutionarily new predators that cause an increasing threat to all extant lemurs).

These findings imply that predation risk and body mass are closely linked to each other in lemurs. Consequently they should act as important selective forces shaping the evolution of sensory and brain mechanisms and related antipredator behaviors minimizing the risk to be eaten.

Strategies and Alarm Call Systems

Current theory suggests that different hunting styles of predators shape escape behaviors as well as communication and cognitive abilities of anthropoid primates (e.g., Seyfarth & Cheney, 2003). "Alarm calls" were thereby defined as calls given by the prey when they encountered a predator. Calls may not only inform about the presence of a predator, but may also encode information about the urgency of escape (urgency-based alarm call system sensu: Owings & Hennessy, 1984) and the type of predator (functionally referential alarm call system sensu: Seyfarth & Cheney, 2003). A researcher needs to perform audio playback experiments to make the differentiation between an urgency-based and a functionally referential alarm system.

Lemurs face different risks of predation related to body mass. Here, we will transfer the theory outlined for anthropoid primates for the first time to lemurs to explore whether different perception abilities and hunting styles of evolutionarily old lemur predators have selected for particular predation recognition and signaling systems in the lemur's brain. How lemurs express and recognize fear evoked by predators and how they will categorize their predatory world we will assess by their predator-advertisement and -avoidance behavior. Based on the previously described lemur predators, we expect four major anti-predator strategies:

1. *Terrestrial snake anti-predator strategy*: Snakes locate prey by olfactory cues and boas additionally by infrared detection of body heat (e.g., Neuweiler, 2003; Safer & Grace, 2004). They are usually sit-and-wait hunters, but may also actively search for their prey on the ground, in dense vegetation or in nests or tree holes. Snakes cannot hear, do not seem to see well, and are not able to move too fast, especially when temperatures are low, but they can climb well on bushes and trees. A prey living in a dispersed or cohesive social system with kin, mate, or social partners nearby, will bear almost no cost, but may benefit (with regard to fitness), if snake detection will induce acoustic and visual snake advertisement, e.g., pointing to the snake while circling around and steering at the snake from a safe distance while giving alarm calls. Not only group members at visual distance but also visually separated ones profit from receiving alerts and from searching for the sender and the alerting stimulus. A snake confronted by several mammals moving around will most probably get distracted and retreat.

2. *Terrestrial carnivore anti-predator strategy*: Carnivores such as viverrids locate prey by olfaction, audition, and vision (e.g., Neuweiler, 2003) and are therefore not easy to avoid. Viverrids hunt either by surprise attacks from hidden places, by pursuing their prey on the ground as well as through bushes and trees, or by grabbing it out of vegetation, nest, or tree hole (Goodman, 2003). Different anti-predator responses may be evolutionarily beneficial, depending on body size of prey and its actual location. During foraging, small-bodied prey should be expected to retreat as cryptically as possible into dense strata of the forest without any calling, whereas larger-bodied prey should flee to cover and advertise predator detection by loud mobbing calls (Curio, 1993), in this way recruiting mobbing conspecifics. A carnivore predator mobbed by loud calls of a number of mobile animals will most likely give up and retreat (e.g., Curio, 1993; Zuberbühler & Jenny, 2002). During resting at the sleeping site, e.g., when being grabbed out of a hole, a surprised prey should produce a loud and noisy threat display to distract the predator (e.g., Owings & Morton, 1998).

3. *Aerial anti-predator strategy*: Diurnal raptors, e.g., hawks, buzzards, or eagles, locate prey primarily by vision while flying around and scanning their territories from the sky and by surprise attack, whereas nocturnal raptors such as owls, for example, locate their prey primarily by audition (e.g., Konishi, 1973; Gaffney & Hodos, 2003; Neuweiler, 2003) while sitting motionless on perches followed by almost noiseless surprise attacks. Foraging and sleeping in dense vegetation or sleeping in shelters such as nests or tree holes should provide prey with the best protection against both diurnal and nocturnal raptors. Besides, detection of a flying raptor should induce alarm calling accompanied by a sudden flight-to-cover reaction in the detecting animal (e.g., flight to denser vegetation, nest, hole), sky scanning, and similar reactions in nearby conspecifics; detection of

a perched raptor should evoke a mobbing response as described for viverrids. Diurnal and nocturnal raptors are constrained in their hearing capabilities to frequencies below 10 kHz (Fay, 1988), which provides small-bodied lemurs with the possibility of exploiting a range above 10 kHz, for less costly predator advertisement.

4. *Panic cry anti-predator strategy*: If a predator has already seized its prey, the captured prey should use panic, distress calls, or screams as a last-ditch effort to manage the predator that holds it. These calls may startle the predator, bring on mobbing, or attract a larger predator to compete for it (e.g., Driver & Humphries, 1969; Hogstedt, 1983; Owing & Morton, 1998).

A fifth anti-predator strategy, described for a variety of birds and mammals (e.g., Zuberbühler, 2003; Rainey et al., 2004), is not directly related to perception abilities and hunting styles of predators, but may depend on cognitive abilities of lemurs:

5. *Semantic predator recognition strategy*: Individuals living in dispersed or cohesive groups should benefit if they relate predator alarm calls of sympatric species and calls produced by the predator itself to the same predator category as their own conspecific predator alarm calls irrespective of their acoustic structure. We will explore our hypotheses with regard to the expected anti-predator strategies by reviewing our current knowledge on the behavior of lemurs in the predation context.

Nocturnal Lemurs

Nocturnal lemurs consist of genera with low to high predation risk (Table 5.1 and 5.2). All studied genera forage solitarily during the night (other than *Avahi* (cohesive pairs), Table 5.2) and sleep either solitary (*Mirza coquereli, Daubentonia madagascariensis*) or form sleeping groups of stable composition during the day (e.g., *Cheirogaleus medius, Lepilemur edwardsi, Microcebus murinus*, and *M. ravelobensis, Phaner furcifer*). Anecdotal information suggests that, other than *Cheirogaleus*, all nocturnal lemur genera produce calls in the presence of predators (e.g., Petter & Charles-Dominique, 1979; Stanger, 1995; Zimmermann 1995; Rakotoarison et al., 1996; Table 5.3). In most cases, however, predator-prey interactions were not specified. We will summarize in the following account studies in which this information was documented.

In four nocturnal species direct snake–lemur interactions were seen in the natural environment. Schmelting (2000) observed a confrontation of a gray mouse lemur male (*Microcebus murinus*) with the Madagascar boa (*Sanzinia madagascariensis)* in the dry deciduous forest of Ankarafantsika in northwestern Madagascar. The mouse lemur detecting the snake jumped around it, and approached and retreated from it to a safe distance producing whistle calls (calls with Fo above 10 kHz, see Zimmermann et al., 2000). Other mouse lemurs in the vicinity are attracted to the sender, themselves producing whistle calls. Whistle calls are not exclusively produced in the predation context but also during various

TABLE 5.3. Alarm calls of different lemur species.

Species	Calls in Response to Predators	Panic Call	Calls in Response To			Response Toward Heterospecific Alarm Calls	References
			Snake	Raptor	Carnivore		
Microcebus murinus	grunt, whistle		whistle[m]			*Microcebus ravelobensis*	Petter & Charles-Dominique, 1979; Stanger, 1995; Zimmermann,1995; Schmelting, 2000; Zietemann, 2000; Braune et al., 2005
Microcebus ravelobensis	grunt, whistle						Zietemann, 2000; Braune et al., 2005
Microcebus rufus	grunt, whistle						Zietemann, 2000
Microcebus myoxinus	grunt, whistle						Zietemann, 2000
Cheirogaleus major	grunt	distress call					Petter & Charles-Dominique, 1979
Cheirogaleus medius	grunt						Petter & Charles-Dominique, 1979; Stanger, 1995; Gilbert, 2001; this paper
Mirza coquereli	ptiak/zek, croak, grunt		zek[m]				Petter & Charles-Dominique, 1979; Pages, 1980; Stanger, 1995; Schülke, 2001; this paper
Phaner furcifer	ki, kiu, kea		kiu[m]			*Mirza coquereli*	Petter & Charles-Dominique, 1979; Charles-Dominique & Petter, 1980; Stange. 1995; Schülke, 2001
Lepilemur leucopus	hein, hee						Petter & Charles-Dominique, 1979
Lepilemur edwardsi	oooai, oui oui oui, grunt, bark, shrill, chatter	scream			bark, shrill, chatter		Petter & Charles-Dominique, 1979; Rasoloharijaona, 2001; this paper
Lepilemur ruficaudatus	"boako-boako", oooai,	distress call					Schülke & Ostner, 2001
Lepilemur septentrionalis	oui oui oui						Petter & Charles-Dominique. 1979
Avahi spec.	grunt, "ava hy"						Petter & Charles-Dominique, 1979
Hapalemur aureus	whimpering			whimpering			Goodmann et al., 1993
Hapalemur griseus	grunt	muffled alarm call					Petter & Charles-Dominique, 1979; Rakotondravony et al. 1998
Hapalemur sinus	roar						Petter & Charles-Dominique, 1979
Eulemur coronatus	grunt, high pitched call						Petter & Charles-Dominique, 1979
Eulemur fulvus	woofs, huvvs, (grunts), chutter			chutter[r]	woofs[u], huvvs[u]	*Propithecus v. verreauxi*	Sussmann, 1975; Paillette & Petter, 1978; Petter & Charles-Dominique, 1979; Fichtel & Kappeler, 2002; Fichtel, 2004

Species					References
Eulemur macaco	huff-grunts, rasp, scream whistle, (crouiii), alarm hack	scream, rasp, whistle, alarm hack	huff-grunts[m], rasp[m]		Petter & Charles-Dominique, 1979; Colquhoun, 1993
Eulemur mongoz					Petter & Charles-Dominique, 1979
Eulemur rubiventer	grunt, hon grunt, high pitched call				Petter & Charles-Dominique, 1979
Daubentonia madagascariensis	ron-tsi/sneeze, grunt				Petter & Charles-Dominique, 1979;
Lemur catta	gulp, click, closed-mouth, click, open-mouth, click, chirp, shriek, rasp, yap	gulp, shriek[r], rasp[r], chirp	gulp, click, closed-mouth click, open-mouth click, yap[r]	*Propithecus v. verreauxi*	Stanger & Macedonia, 1994; this paper Sussmann, 1975; Sauther, 1989; Macedonia, 1993, 1990; Pereira & Macedonia, 1991; Macedonia & Evans, 1993; Oda & Masataka, 1996; this paper
Varecia variegata	abrupt roars with wails, growls, growl snorts, pulsed squaks	abrupt roars with wails[u/m]	growls, growl snorts, pulsed squaks[u/m]		Macedonia, 1990, 1993; Pereira et al., 1988
Propithecus diadema	roar, "simpona"				Petter & Charles-Dominique, 1979
Propithecus verreauxi	growl, alarm barks, roar	growl[u], roar[r]	growl[u]	*Lemur catta, Eulemur fulvus*	Sussmann, 1975; Richard, 1978; Oda, 1998; Fichtel & Kappeler, 2002; Brockman, 2003; Fichtel, 2004
Indri indri	bark, honk	alarm	barks		Pollock, 1975; Petter & Charles-Dominique, 1979

m = mobbing call; u = urgency-based alarm call; r = referential alarm call social interactions (Stanger, 1995; Zimmermann, 1995; Zimmermann et al., 1995; Zietemann, 2000; Braune et al., 2005). Their high variability in acoustic structure provides the potential for predator specificity. In a sympatric association of fork-marked lemur (*Phaner furcifer*) and Coquerel's dwarf lemur (*Mirza coquereli*) in the dry deciduous forest Kirindy of western Madagascar, Schülke (2001) observed another direct snake–lemur interaction, which provided the first evidence for semantic predator recognition in nocturnal lemurs. The sub-adult male of a dispersed foraging group of fork-marked lemurs detected a snake and gave "kiu" calls while circling around the snake, staring at it, and moving toward and away from it at a safe distance. This vocal reaction induced not only a sudden attraction of and mobbing by other group members, but also the attraction of a sympatrically foraging Coquerel's dwarf lemur, which, after detecting the snake, also started circling around it while giving "zek" calls. There was no indication of mobbing behavior in *Cheirogaleus medius*, but (Fietz & Dausmann, 2003) observed a female fat-tailed dwarf lemur defending its offspring by attacking a snake (*Madagascarophis colobrinus*) next to its sleeping hole. Induced snake–lemur confrontations were investigated in the laboratory. Here, predator-naïve pairs of two nocturnal species–the gray mouse lemur (*Microcebus murinus*; $N = 4$) and the brown mouse lemur (*Microcebus rufus*; $N = 2$)–were visually exposed either to a living python or to a python dummy in front of the enclosure (Zimmermann et al., 2000). In contrast to the situation in the natural environment, there was no significant difference in locomotion and vocal activity before and after stimulus presentation (Bunte, 1998). No information on snake anti-predator strategies exists so far for *Allocebus, Avahi, Lepilemur, Cheirogaleus*, and *Daubentonia*.

To date, direct viverrid interactions with nocturnal lemurs in nature were reported in two genera. Rabesandratana et al. (2005) observed a fossa (*Cryptoprocta ferox*) chasing a Milne-Edwards' sportive lemur through the forest in the Ankarafantsika National Park in northwestern Madagascar. The sportive lemur fled by jumping rapidly from tree to tree into the vicinity of its sleeping site emitting loud bark call sequences (Figure 5.2). As the fossa had almost gripped the lemur, the latter gave much louder shrill and chatter calls. It seemed as if an increase in arousal was encoded in an increasing noisiness and an increasing calling rate, as well as in a change of call types. Sportive lemurs in the vicinity got attracted to the interaction while emitting loud bark calls. The fossa finally gave up and retreated. Schülke (pers. communication) observed an interaction between Phaner and a fossa at the Kirindy research station in central western Madagascar. The lemur sat high in the canopy and produced kiu calls while the fossa was walking over the ground.

Up until now, two direct confrontations between nocturnal lemurs and raptors have been described from the field. Schülke & Ostner (2001) observed a Madagascar harrier-hawk (*Polyboroides radiatus*) attacking a red-tailed sportive lemur (*Lepilemur ruficaudatus*) at its sleeping site in the Kirindy forest in western Madagascar. The Madagascar harrier-hawk seized the lemur with its bill and pulled it out from its hole while the lemur emitted loud distress calls (= panic call). Gilbert & Tingay (2001) saw a Madagascar harrier-hawk preying

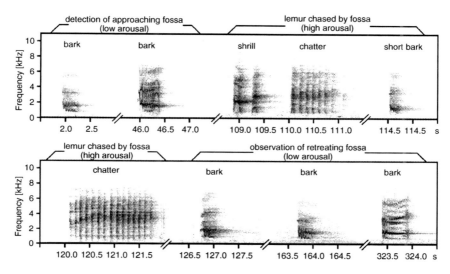

FIGURE 5.2. Sonogram of an alarm calling sequence of a Milne-Edwards' sportive lemur (*Lepilemur edwardsi*) given during a fossa encounter. Differences in call structure and call repetition rate appear to reflect different levels of arousal

on a fat-tailed dwarf lemur (*Cheirogaleus medius*) in the Tsimembo forest in western Madagascar. The lemur emitted a shrill incessant squeaking sound (= panic call) while it was being picked up by the raptor. A playback experiment (Karpanty & Grella, 2001) performed in the Ranomafana National Park with two nocturnal genera, the weasel sportive lemur *(Lepilemur mustelinus)* and the eastern woolly lemur *(Avahi laniger)*, gave first evidence on how nocturnal lemurs reacted toward sounds of sympatric diurnal raptors (Madagascar serpent-eagle (*Eutriorchis astur*), Henst's goshawk (*Accipiter henstii*), and Madagascar harrier-hawk (*Polyboroides radiatus*). The sportive lemur ($N = 1$) did not respond to the playback of a Madagascar serpent-eagle at all, whereas it scanned the sky after playbacks of the Henst's goshawk and the Madagascar harrier-hawk. One group of woolly lemurs looked toward the loudspeaker irrespective of the raptor species diffused whereas the other group reacted only toward the Henst's goshawk. Induced raptor–lemur confrontations were studied in the laboratory. Predator-naïve pairs of gray mouse lemurs (*Microcebus murinus*; $N = 4$) and brown mouse lemurs (*Microcebus rufus*; $N = 2$) were exposed to either a moving barn owl silhouette or a perched barn owl dummy in front of the enclosure (Zimmermann et al., 2000). As in the induced snake–lemur confrontation experiment, no significant difference in vocal activity before and after stimulus presentation was found. *M. murinus* showed, however, a significantly higher locomotion rate afterward (Bunte, 1998). In a confrontation experiment 24 gray mouse lemurs (*Microcebus murinus*) were placed next to a cage containing the predator *Mirza coquereli*, a non-predatory rodent (*Eliurus myoxinus*) or an empty cage (Rakotonirainy, Schülke & Kappeler unpublished data). In response to the predator 17 of 24

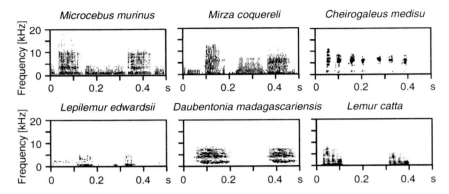

FIGURE 5.3. Sonograms of noisy grunt calls of different lemur genera given as an anti-predator response

mouse lemurs produce vocalisations whereas only 6 of 24 produce vocalisation in response to the non-predatory rodent and none of them to the empty cage.

Observations on how *Allocebus, Cheirogaleus, Microcebus, Mirza, Lepilemur*, and *Daubentonia* reacted against being captured by a human experimenter from a covered sleeping site were also made (see Table 5.3, Zimmermann unpublished data). In all genera, a reflexive lunge toward the disturbing stimulus was observed, accompanied by loud and noisy grunts. These calls show a similar call structure across different lemur genera (Figure 5.3) and may be effective in inducing a startle reflex and escape behaviors in predators. For example, even a human experimenter, who knows that a tiny mouse lemur cannot really hurt him, will show sudden recoil as a reaction toward these calls. Larger-bodied nocturnal lemurs also produce screams (= panic calls) when they are captured and seized by a human experimenter. Thus, for example, some individuals of the Milne-Edward's sportive lemur, which we captured for radio-collaring, produced sequences of these calls under these circumstances of most likely extreme fear (Rasoloharijaona, 2001). Most interesting, these screams sometimes attracted conspecifics from the vicinity. They circled around us giving alarm calls (unpublished data, Rasoloharijaona & Zimmermann) and seemed to mob us, similar to what they did during lemur–snake interactions. Despite the fact that sportive lemurs forage solitarily during the night, it seems as if they are included in a social network of dispersed pairs (families) by long distance vocal communication.

Altogether these findings provide first evidence for the evolution of the expected anti-predator strategies and for semantic predator recognition in nocturnal lemurs. Furthermore, they suggest a possible influence of learning on predator recognition. However, missing are quantitative and experimental studies that address to what extent our hypothesis is supported and whether the emitted calls refer to particular predator categories. There is a strong need for further studies.

Cathemeral and Diurnal Lemurs

Cathemeral and diurnal lemurs face a medium to low predation risk (Table 5.1). All genera forage and sleep either in cohesive pairs or family groups or in cohesive multimale-multifemale groups (Table 5.2). Alarm calls are known from all of them (Table 5.3). In the following discussion we summarize the information of researches in which predator-prey interactions were specified.

Three snake–lemur interactions were reported from the natural environment. Burney (2002) studied Coquerel's sifakas (*Propithecus verreauxi coquereli*) at Anjohibe in northwestern Madagascar. He described that a sifaka group reacted with a roar chorus while observing a boa (*Acrantophis madagariensis*) strangling an adult group member. Another boa-lemur interaction with *Sanzinia madagascariensis* was observed in black lemurs (*Eulemur macaco*) at Ambato Massif (Colquhoun, 1993). They produced mobbing calls and showed mobbing behavior. An eastern lesser bamboo lemur (*Hapalemur griseus griseus*) seized by a boa of the same species at the littoral forest of the Forestiére de Tampolo Station in eastern Madagascar emitted panic calls (Rakotondravony, 1998).

Direct lemur–viverrid interactions in nature were reported only rarely. Sussman (1975) described that a red-fronted brown lemur (*Eulemur fulvus rufus*) reacted toward ground predators similar to their reaction toward humans by producing grunts and wagging their tails. Black lemurs (*Eulemur macaco*) at Ambato Massif responded to dogs and endemic viverrids by producing huff-grunts and by tail wagging (Colquhoun, 1993). These calls may integrate into rasping loud calls while mobbing.

To date, three field observations describe direct lemur interactions with raptors. Colquhoun (1993) observed that black lemurs (*Eulemur macaco*) produced alarm hacks and rasping loud calls in response to raptors (*Accipiter madagascariensis*, *Buteo brachypterus*, and *Milyus migrans*) circling at the sky. If lemurs detected a Madagascar harrier-hawk (*Polyboroides radiatus*) there, they emitted alarm hacks and rasping loud calls with a sharply ascending and descending scream whistle while they climbed down and searched for cover in the inner trunk of the tree. In Beza Mahafaly Special Reserve, southwest Madagascar, a harrier-hawk attacking from the air a group of *Propithecus verreauxi* immediately elicited roars and climbing into the dense canopy (Brockmann, 2003). In the same forest, Sauther (1989) observed that troops of ring-tailed lemurs (*Lemur catta*), travelling on the ground, responded to the presence of Madagascar harrier hawk (*Polyboroides radiatus*) and Madagascar buzzard (*Buteo brachtypterus*) by approaching the tree and produced chirp and moaning vocalisations which could escalate into shriek vocalisations. In contrast, in the presence of Black kite (*Milvus migrans*) they produced no vocalisations and moved silently into the bush. A playback experiment (Karpanty & Grella, 2001) performed in the Ranomafana National Park with four cathemeral and two diurnal lemur species provided some insight into how these species reacted toward sounds of sympatric diurnal raptors Madagascar serpent-eagle (*Eutriorchis astur*), Henst's goshawk (*Accipiter henstii*), and Madagascar harrier-hawk (*Polyboroides radiatus*). In contrast to the tested nocturnal lemurs, all of these species responded to the playbacks of all three diurnal raptors. The eastern lesser bamboo lemur (*Hapalemur griseus griseus*; $N = 3$) and the golden bamboo lemur (*Hapalemur aureu*; $N = 2$) gave alarm calls in response to the calls of Madagascar serpent-eagle and Henst's goshawk while dropping into the canopy, whereas they looked toward the sound source in response to the Madagascar harrier-hawk (*Polyboroides radiatus*). The red-fronted brown lemur (*Eulemur fulvus rufus*; $N = 4$) and the red-bellied lemur (*Eulemur rubriventer*; $N = 2$) produced alarm calls to all three diurnal raptors but fled more often in response to Madagascar serpent-eagle and Henst's goshawk. The diurnal red ruffed lemur (*Varecia variegata rubra*; $N = 2$) emitted aerial alarm calls in response to the Madagascar serpent and Henst's goshawk but dropped into the canopy only in response to Henst's goshawk. The Milne-Edwards' sifaka (*Propithecus diadema edwardsi*; $N = 2$) showed a stronger response to *A. henstii* by producing alarm calls and fleeing from the sound source. In summary, the lemurs in this study showed significantly stronger responses to playbacks of Henst's goshawk than to the two other raptor species. In an experimental study with semi-free-living ring-tailed lemurs at the Duke Primate Center, the visual assessment of avian threat was investigated (Macedonia & Polak, 1989). Five different moving silhouettes were used as visual stimuli: (1) a naturalistic silhouette of a hawk (*Buteo jamaicensis*), (2) a stylized hawk, (3) a stylized goose, (4) a diamond, and (5) a square control. Besides the hawk silhouette, all were presented in two different sizes, large and small. Individuals responded with a higher calling rate (rasps and shrieks) to the large naturalistic hawk silhouette than to the large stylized goose, diamond, or square silhouette. Furthermore they responded with a higher calling rate and longer calls to the large stylized hawk silhouette than to the small one. Furthermore, the large naturalistic and stylized hawk silhouettes led to significantly longer calls than the large goose and the large square, but not the large diamond silhouette. Regarding the small silhouettes it was found that individuals produced significantly longer calls in response to small hawk and goose silhouettes compared to the small square but not to the diamond silhouette. In summary, it seems that a stylized hawk shape was perceived as equally threatening as a realistic hawk shape. Likewise, the shapes of the stylized hawk were perceived as more threatening than the goose and the square shape but not more than the diamond shape. This suggests that features of size and proportions of silhouettes could trigger visual avian predator recognition.

Quantitative playback experiments studying the responses of cathemeral and diurnal genera toward ground and aerial predators to illuminate how they are perceives and categorized are available for only four different species of four genera, one of which is cathemeral and the three others, diurnal. All of these lemurs belong to the medium predation risk class, are relatively large-bodied (between 2 kg and 4 kg), but differ in their degree of arboreality and in the used habitat. The cathemeral red-fronted brown lemurs and the diurnal Verreaux's sifakas as well as the black and white ruffed lemurs are primarily arboreal. Whereas the two former species live in the dry deciduous forest, the latter is distributed in the evergreen rainforest. The diurnal ring-tailed lemur is semi-terrestrial and occurs in dry deciduous forests. Major results are outlined as follows.

Red-Fronted Brown Lemurs (*Eulemur fulvus rufus*)

Red-fronted brown lemurs were investigated at the Kirindy research station in central western Madagascar (Fichtel & Kappeler, 2002; $N = 8$–9). They emitted three different call types in the context of predation, tentatively described according to their acoustic structure as "chutters," "woofs," and "huvvs." Chutters were given to raptors (*Polyboroides radiatus*) circling at the sky, woofs and huvvs to ground predators (*Cryptoprocta ferox, Canis familaris*). Woofs were not given exclusively in the predation context, but in other social contexts such as group encounters, also. Chutters were only produced in response to aerial predators. An acoustic analysis of alarm calls given to the different predator species was not performed. In playback experiments, woofs were diffused as the terrestrial alarm call and chutters as the aerial alarm call. The subjects responded to the terrestrial alarm call by woofs and to the aerial alarm call by chutters and woofs. After playbacks of calls of the aerial predator and of the aerial alarm call, lemurs looked up more often and climbed down, whereas they looked down and climbed up more often in response to terrestrial alarm calls and terrestrial predator calls.

Red-fronted brown lemurs respond to heterospecific alarm calls of Verreaux's sifakas. They emitted woofs (general alarm call) in response to sifakas' aerial and terrestrial alarm calls, but showed an appropriate escape strategy with regard to the type of alarm call (Fichtel, 2004; $N = 8$).

The authors conclude that red-fronted brown lemurs recognize their conspecific aerial alarm calls semantically, whereas they produce and recognize ground alarm calls based on the urgency of a response. Furthermore, they showed heterospecific alarm call recognition.

Verreaux's Sifakas (*Propithecus v. verreauxi*)

Verreauxi sifakas were observed at the same study site as red-fronted brown lemurs (Fichtel & Kappeler, 2001; $N = 8$). In the context of predation two call types were produced. According to their acoustic structure they were tentatively classified as "growl" and "roars." Growls were emitted in response to aerial (*Polyboroides radiatus*) and terrestrial predators (*Cryptoprocta ferox, Canis familaris*), but also in non-predator social situations such as group encounters, whereas roars were given exclusively toward the aerial predator. An acoustic analysis of alarm calls given to the different predator species is lacking. Playback experiments were performed with growls as the terrestrial alarm call and roars as the aerial alarm call. When growls were played back, none of the subjects responded vocally, whereas five of eight individuals produced roars in response to roars. Like red-fronted brown lemurs, sifakas looked down more often and climbed up in response to terrestrial predators and terrestrial alarm calls, whereas they looked up and climbed down in response to aerial alarm calls and aerial predator calls.

Verreaux's sifakas showed also evidence for heterospecific alarm call recognition. Thus, they responded to the aerial alarm call of red-fronted brown lemurs by aerial alarm calls (roar) at the Kirindy field site. They looked up more often in response to the aerial than to the terrestrial alarm call, whereas they looked down more often in response to terrestrial than to aerial alarm calls of red-fronted lemurs (Fichtel, 2004; $N = 8$). At the Berenty field site, where Verreaux's sifakas live sympatrically with ring-tailed lemurs, they showed a predator-specific escape response to aerial and terrestrial alarm calls of ring-tailed lemurs. More individuals looked up in response to ring-tailed aerial alarm calls and more individuals moved up in response to the terrestrial alarm call (Oda, 1998; $N = 11$).

In summary, Fichtel & Kappeler (2001) emphasized that Verreaux's sifakas use aerial predator calls referentially and terrestrial alarm calls according to the urgency of predator threat. Furthermore, Verreaux's sifakas demonstrate heterospecific alarm call recognition.

Black and White Ruffed Lemurs (*Varecia variegata variegata*)

One group of ruffed lemurs was studied under semi-free conditions in a large outdoor enclosure at the Duke Primate Center, USA (Macedonia, 1990, 1993). Predator–lemur interaction was observed toward naturally occurring predators as well as in experimental confrontations of lemurs with two aerial and a terrestrial predator (aerial predators: stuffed museum specimen of a perched red-shouldered hawk (*Buteo lineatus*) or a great-horned owl (*Bubo viginianus*); terrestrial predator: living dog; Macedonia, 1993). Four different call types were evoked by these predators, tentatively classified as "abrupt roars," "growls," "growl snorts," and "pulsed squawk" according to their acoustic structure. Abrupt roars were used as mobbing calls to aerial predators. They were continued after the potential threat visually disappeared. With increasing arousal abrupt roars were combined with wails. Ruffed lemurs on the ground, detecting an aerial predator, assumed a threatening posture and emitted roars that induced "scan and roar" behavior in nearby group members. When they were in the tree, they climbed toward the treetop and emitted roars in the direction of the predator. Growls, growl snorts, and pulsed squawks were produced in response to the dog as a potential terrestrial predator while lemurs showed mobbing behavior. Calls may integrate into each other with increasing arousal. In a playback study, abrupt roars were diffused as the aerial alarm call and pulsed squawks as the terrestrial alarm call. Ruffed lemurs responded to playbacks of the aerial alarm call by producing roars and to playbacks of terrestrial alarm calls by producing pulsed squawks. Growls and growl snorts were produced in response to both call types, but significantly more often to terrestrial than aerial alarm calls. Lemurs showed more sky scanning and roaring behavior in response to the aerial alarm call than to the terrestrial alarm call. In response to terrestrial alarm calls, adults on the ground ran up into trees more often. Adults in the tree did not show a specific escape response, whereas immatures in the trees climbed higher more often in response to terrestrial than to aerial alarm calls. In summary, ruffed lemurs did mob both aerial and terrestrial predators, and consequently did not show a strong predator-specific response. Macedonia (1993) argued furthermore that the pulse squawk (terrestrial alarm call) lies at one end of a structurally graded acoustic continuum and the wail of the abrupt roars (aerial alarm call) at its other end. This acoustic continuum coincides with predictions for an urgency-based alarm call system.

Ring-Tailed Lemurs (*Lemur catta*)

Ring-tailed lemurs were studied under the same conditions as ruffed lemurs at the Duke Primate Center (Macedonia, 1990, 1993) under similar semi-free conditions in the Izu Cactus Park in Shizuoka, Japan, and additionally in the field in the Berenty Reserve in southern Madagascar (Oda & Masataka, 1996). Direct interactions with naturally occurring predators as well as experimental confrontations of lemurs with two aerial predators and a terrestrial predator (aerial predators: stuffed museum specimen of a perched red-shouldered hawk (*Buteo lineatus*), or a great-horned owl (*Bubo viginianus*); terrestrial predators: living dog; Macedonia, 1993) were observed. Ring-tailed lemurs produced seven acoustically different call types, which were tentatively classified, based on their acoustic structure, as "gulps," "rasps," "shrieks," "chirps," "clicks," "closed mouth click," "open mouth click," and "yaps." In response to aerial and terrestrial predators subjects first emitted gulps. When a large moving bird approached a lemur group subjects emitted rasps grading into a shriek chorus when all group members detected it. During aerial predator detection subjects looked skyward and tracked the flight of the predator or ran into cover. Group relocation was accompanied by chirp calls. In response to terrestrial predators ring-tailed lemurs produced clicks, closed mouth clicks, open mouth clicks, and yaps and usually jumped immediately into the trees. Rasps, shrieks, open mouth clicks, and yaps were only observed in alarm contexts in contrast to gulps, chirps, clicks, and closed mouth clicks. In a playback experiment rasps and shrieks were used as aerial alarm calls, whereas yaps were used as the terrestrial alarm call. In response to aerial alarm calls subjects emitted significantly more chirp calls than in response to the terrestrial alarm calls, whereas subjects emitted significantly more clicks in response to terrestrial than to aerial alarm calls. Furthermore ring-tailed lemurs showed different escape responses to aerial and terrestrial alarm calls. Ring-tailed lemurs on the ground looked up and stood up bipedally in response to aerial alarm calls, subjects in trees climbed lower to dense vegetation. In response to terrestrial alarm calls subjects on the ground ran up into the trees or climbed higher into the dense canopy.

Oda & Masataka (1996) provided evidence for heterospecific alarm call recognition in ring-tailed lemurs at Berenty, where they live sympatrically with Verreaux's sifakas. Ring-tailed lemurs on the ground ($N = 26$) looked skyward in response to playbacks of sifakas' aerial alarm calls and ran up into the trees in response to playbacks of sifakas' terrestrial alarm calls. Ring-tailed lemurs in the trees ($N = 26$) responded also by looking skyward in response to the aerial alarm call and climbed down in response to the terrestrial alarm call. These results coincide well with their reactions to the respective conspecific alarm calls. Experience seems to influence the recognition of heterospecific alarm calls since subjects in captivity did not show such a predator-specific escape response.

An experiment illustrating the effect of learning on acoustic recognition of raptor calls was performed by Macedonia & Yount (1991). They conducted a playback experiment with two ring-tailed lemur groups at the Duke Primate Center. Calls of three avian species–red-tailed hawk (*Buteo jamaicensis*), Madagascar harrier-hawk (*Polyboroides radiatus*), wood thrush (*Hylocicla mustelina*)–and one mammal species–the eastern gray squirrel (*Sciurus carolinensis*)–were used as acoustic stimuli. All subjects were familiar with all acoustic stimuli besides the Madagascar harrier-hawk. Results of this experiment were mixed. No significant difference in bipedal scanning was found after the presentation of the familiar red-tailed hawk and the unfamiliar Madagascar harrier-hawk. However, one of the two tested groups jumped significantly more often onto tree trunks in response to playbacks of the familiar red-tailed hawk than the unfamiliar Madagascar harrier-hawk. Furthermore, more individuals of this group leapt more often onto trees in response to playbacks of red-tailed hawks and of gray squirrels than to playbacks of wood thrushes.

In summary, ringtail lemurs show evidence for predator-specific alarm calls and semantic conspecific alarm call recognition. Furthermore, experience appears to shape heterospecific alarm call recognition as well as acoustic predator recognition.

Discussion

In this paper we have estimated predation risk of lemurs based on a review of current data on the number of predator species. We have related this information to the variation of particular life history traits and ecology to explore whether these traits are shaped by predation. According to our study foraging group size and female maximum reproductive output per year were not related to predation risk for the whole lemur sample, whereas body mass was significantly correlated and could be best illustrated by a logarithmic model. This contradicts findings in anthropoid primates where predation is an important selective force shaping group size and female fecundity (e.g., Anderson, 1986; Hill & Lee, 1998; Hill & Dunbar, 1998; Janson, 2003). Predation risk in our study was estimated solely by the number of predatory species per genus based on direct predator–lemur interactions and indirect cues such as owl pellets, feces, and dietary analyses, and should therefore be treated cautiously. Accordingly, the intensity with which predation and antipredator strategies were studied differs among the different genera. However, the described anecdotal observations and experimental data on lemurs indicate that not only hunting styles of predators (as suggested by current theory, see Seyfarth & Cheney, 2003), but also predation risk linked to body mass seem to have an impact on the evolution of their anti-predator strategies.

Reviewing the lemur literature we found that lemurs, including the nocturnal ones, do not only rely on crypsis to avoid predators as often is suggested by the anthropoid literature (Stanford, 2002; Janson, 1998), but they do show as highly sophisticated anti-predator strategies as anthropoid primates. We found information about anti-predator behaviors for 11 out of 13 genera (no information is available yet for *Daubentonia* and *Indri*) based on direct predator–prey interactions as well as visual and acoustical confrontation experiments. Snake anti-predation behaviors during direct snake–lemur interactions were observed in six lemur genera of the medium to high predation risk class (Table 5.3). In one genus (*Cheirogaleus*) it was observed that a female attacked a snake defending her offspring. Five out of six lemur genera showed mobbing responses toward the snake, supporting the prediction of snake anti-predator strategy outlined at the beginning of this chapter. Mobbing seems to be a universal snake anti-predator strategy, similar to the strategy shown by anthropoid primates. Terrestrial carnivore anti-predator behaviors were observed in five lemur genera belonging exclusively to the medium and low predation risk class. All of them showed a universal escape response by climbing up into the canopy. Furthermore, three of them also displayed mobbing behavior. In these cases the animals were directly confronted with the predator. These findings support the prediction of the terrestrial carnivore anti-predator strategy concerning medium and large bodied lemurs, whereas data for small-bodied lemurs are lacking so far and require further studies. Furthermore, playback experiments with predator calls were conducted for four lemur genera. They indicated that three out of four lemur genera perceived predator calls semantically, as is known for anthropoid primates (e.g., Seyfarth & Cheney, 2003; Zuberbühler, 2003). Aerial anti-predation behaviors were noted in seven lemur genera belonging to the medium to low predation risk class, confirming the predictions of aerial anti-predator strategy. Observations of direct raptor–lemur interactions occurred in two lemur genera. In both cases the animals searched for cover by climbing into the dense canopy. Anti-predator behaviors were experimentally induced in seven lemur genera by playback studies using predator calls and in two by confrontation experiments with potential predators. One genus (*Varecia*) responded with mobbing behavior ("scan and roar behavior"; Macedonia, 1990, 1993). Six out of seven lemur genera showed a universal anti-predator response of scanning the sky and five of them also responded by searching for cover. Results of playback experiments imply that calls of raptors are perceived semantically, as is shown for anthropoid primates. A panic cry anti-predator strategy was observed in three lemur species belonging to the high and medium predation risk class. In all cases the lemurs were already gripped by the predator and emitted panic cries. Panic crying seemed also to be universal across lemur genera. But to date there is no evidence that these calls may startle the predator, lead to mobbing, or attract a larger predator to compete for the prey. Evidence for the presence of the semantic predator recognition strategy as outlined in this paper currently seems to exist for three lemur genera. Two of them responded specifically to the alarm calls of one sympatric lemur genera, whereas one genus, *Propithecus*, showed heterospecific alarm call recognition of two sympatric lemur genera. It is not clear, however, to what extent these heterospecific alarm calls differ in acoustic structure. *Microcebus murinus* also responded to alarm calls of sympatric *M. ravelobensis*.

TABLE 5.3. Overview of anti-predation behavior in response to direct confrontations with predators (marked bold) and predator calls.

Genus	Terrestrial	Snake	Aerial	References
Microcebus		**mobbing**		Schmelting, 2000
Cheirogaleus		**attacking**		Fietz & Dausmann, 2003
Mirza		**mobbing**		Schuelke, 2001
Phaner		**mobbing**		Schuelke, 2001
Lepilemur	**mobbing from conspecifics**[1] **victim fleeing in canopy**[2]		scanning sky[3]	[1]Zimmermann, unpubl. data; [2]Rabesandratana et al., 2005; [3]Karpanty & Grella, 2001
Avahi			looked to sound source	Karpanty & Grella, 2001
Hapalemur			scanning sky climbed into cover	Karpanty & Grella, 2001
Eulemur	**mobbing**[4,5] looked down[7] climbed up[7]	**mobbing**[4]	**scanning sky**[6,7] **climbed into cover**[4,6] climbed down[7]	[4]Colquhoun, 1993; [5]Sussman, 1975; [6]Karpanty & Grella, 2001; [7]Fichtel & Kappeler, 2001
Lemur	climbed up		scanning sky climbed into cover	Macedonia 1990, 1993; Oda & Masataka 1994
Varecia	**mobbing**[9] climbed up[9]		scanning sky[8] climbed into cover[8,9] **mobbing**[9] scan and roar[9]	[8]Karanty & Grella, 2001; [9]Macedonia, 1993
Propithecus	looked down[13] climbed up[13]	**mobbing**[10]	**climbing up into dense canopy**[11] scanning sky[12,13] climbed into cover[12] climbed down[13]	[10]Burney, 2002; [11]Brockmann, 2003; [12]Karpanty & Grella, 2001; [13]Fichtel & Kappeler, 2001

However, in this case calls did not differ statistically in their acoustic structure (Zietemann, 2000).

So far, for most genera, sophisticated quantitative and experimental approaches aimed toward gaining insight into sensory and cognitive abilities of lemurs are lacking. Our overview of our current knowledge nevertheless provides some evidence for the expected anti-predator strategies outlined at the beginning of this paper. Thus, it seems as if all lemur species studied to date, irrespective of their activities and social patterns, have evolved particular anti-predator strategies that minimize the risk of being eaten. Behavior strategies shown under these circumstances appear to be adapted to the perception abilities and hunting styles of three different predator categories as well as to body mass and location of the respective lemur itself. Further direct observations on natural predator–lemur interactions, comparative studies on induced predator–lemur confrontations and playback experiments with lemur alarm and predator calls using the same experimental paradigm and exploring all sensory domains for predator detection are necessary to assess the extent to which signaling and recognition mechanisms in lemurs correspond to those of anthropoid primates or show lineage-specific constraints. They may also shed light on potential universal principles governing communication and cognition.

First research on anti-predator behaviors of predator-naïve lemurs in comparison to experienced ones (see also Oda, 1996; Bunte, 1998; Bunkus et al., 2005; Sündermann et al., 2005) provided some evidence for the influence of experience on predator perception and recognition, a rather neglected area of research in the strepsirrhines. The question on why and how lemurs learn about predators is, however, highly important from an applied perspective. Almost all extant lemurs bear a high risk of extinction (e.g., Mittermeier et al., 2003). Conservation and reintroduction programs are therefore urgently needed, and some are partly established, for the most threatened species. From a variety of bird and mammalian species it is known that reintroduced and translocated individuals are highly vulnerable to predation after release, unfortunately reducing the success of the respective conservation programs (e.g., MacMillon, 1990; Beck, 1994; Wolf et al., 1996). To improve their anti-predator skills and to enhance the efficiency of these programs, pre-release anti-predator training was used in which individuals learned to associate particular predator categories with an unpleasant experience (e.g., Ellis et al., 1977; Miller et al., 1990; Richards, 1998; Griffin et al., 2000; McLean et al., 2000; Griffin et al., 2001). Research on predator learning in lemurs is therefore required not only to get a better understanding on the origin and evolution of primate communication and cognition, but also to deliver appropriate tools for effective management and conservation.

Acknowledgments. We thank Anna Nekaris and Sharon Gursky for the invitation to contribute to this volume. Our own research on lemurs in Madagascar was made possible by logistic support of CAFF/CORE, the Association pour la Gestion des Aires Protégées (ANGAP), and the Faculty of Sciences at the University Antananarivo, Madagascar. We are indebted to the late Prof. Berthe

Rakotosamimanana (GERP), Prof. Olga Ramilijaona, Prof. Noro Raminosoa, and Dr. Daniel Rakotondravony (Départment de Biologie Animale, University of Antananarivo) for all institutional help. This paper has also profited from discussions within the DFG FOR 499 as well as with Kalle Esser, Sabine Schmidt, and Ute Radespiel. We also thank three anonymous referees for valuable comments on an earlier version of this manuscript. Our research was financially supported by the German Research Council (FOR 499 and GK 289), Volkswagenstiftung (Zusammenarbeit mit Entwicklungsländern), and the German Academic Exchange Council (Sandwich programme fellowship to AR).

References

Alcock, J. (1997). *Animal Behavior.* Sunderland: Sinauer Assoc., Inc.

Anderson, C.M. (1986). Primates and primate evolution. *Primates*, 27, 15–39.

Bearder, S.K., Nekaris, K.A.I., and Buzzell, C.A. (2002). Danger in the night: Are some nocturnal primates afraid of the dark? In L.E. Miller (Ed.), *Eat or be eaten: Predator sensitive foraging among primates* (pp. 21–40). Cambridge: Cambridge Univ. Press.

Beck, B.B. (1994). Reintroduction of captive-born animals. In P.J.S. Olney, G.M. Mace, and A.T.C. Feistner (Eds.), *Creative conservation: Interactive management of wild and captive animals* (pp. 265–286). London: Chapman and Hall.

Blumstein, D.T. (1999). Alarm calling in three species of marmots. *Behav.*, 136: 731–757.

Braune, P., Schmidt, S., and Zimmermann, E. (2005). Spacing and group coordination in a nocturnal primate, the golden brown mouse lemur (*Microcebus ravelobensis*). *Behav. Ecol. Sociobiol.*, 58: 587–596.

Brockman, D.K. (2003). *Polyboroides radiatus* predation attempts on *Propithecus verreauxi*. *Folia Primatol.*, 74: 71–74.

Bunkus, E., Scheumann, M., and Zimmermann, E. (2005). Do captive-born grey mouse lemurs (*Microcebus murinus*) recognize their natural predators by acoustic cues? *Primate Report*, 72: 22.

Bunte, S. (1998). Verhaltensbiologische und bioakustische Untersuchungen zum Wachsamkeitsverhalten und zur Räubervermeidung bei Mausmakis (*Microcebus spp.*) in Gefangenschaft. Unpublished thesis.

Burney, D.A. (2002). Sifaka predation by a large boa. *Folia Primatol.*, 73: 144–145.

Charles-Dominique, P., and Petter, J.J. (1980). Ecology and social life of *Phaner furcifer*. In P. Charles-Dominique, H.M. Cooper, A. Hladik, C.M. Hladik, E. Pages, G.F. Pariente., A. Petter-Rousseaux, A. Schilling, and J.J. Petter (Eds.), *Nocturnal Malagasy primates: Ecology, physiology, and behavior* (pp. 75–95). New York: Academic Press, New York.

Clutton-Brock, T.H., and Harvey, P.H. (1980). Primates, brains and ecology. *J. Zool. Lond.*, 190: 309–323.

Colquhoun, I.C. (1993). The socioecology of *Eulemur macaco*: A preliminary report. In P.M. Kappeler and J.U. Ganzhorn (Eds.), *Lemur social systems and their ecological basis* (pp. 11–23). New York: Plenum Press.

Curio, E. (1993). Proximate and developmental aspects of antipredator behavior. *Adv. Study Behav.*, 22: 135–238.

Driver, P.M., and Humphries, D.A. (1969). The significance of the high-intensive alarm call in captured passerines. *Ibis*, 111: 243–244.

Ellis, D.H., Dobrott, S.J., and Goodwin, J.G. (1977). Reintroduction techniques for masked bobwhites. In S.A. Temple (Ed.), *Endangered birds: Management techniques for preserving threatened species* (pp. 345–354). Madison: Univ. of Wisconsin Press.

Fay, R.R. (1988). *Hearing in vertebrates: A psychophysics databook*. Winnetka: Hill-Fay Associates.

Fichtel, C. (2004). Reciprocal recognition of sifaka (*Propithecus verreauxi verreauxi*) and red-fronted lemur (*Eulemur fulvus rufus*) alarm calls. *Anim. Cogn.*, 7: 45–52.

Fichtel, C., and Kappeler, P.M. (2002). Anti-predator behavior of group-living Malagasy primates: Mixed evidence of a referential alarm call system. *Behav. Ecol. Sociobiol.*, 51: 262–275.

Fietz, J., and Dausmann, K.H. (2003). Costs and potential benefits of parental care in the nocturnal fat-tailed dwarf lemur (*Cheirogaleus medius*). *Folia Primatol.*, 74: 246–258.

Gaffney, M.F., and Hodos, W. (2003). The visual acuity and refractive state of American kestrel (*Falco sparverius*). *Vision Res.*, 43: 2053–2059.

Gilbert, M., and Tingay R.E. (2001). Predation of a fat tailed dwarf lemur *Cheirogaleus medius* by a Madagascar harrier-hawk *Polyboroides radiatus*: An incidental observation. *Lemur News*, 6: 6.

Glaw, F., and Vences, M. (1996). *A field guide to the amphibians and reptiles of Madagascar*, (2nd ed.). Köln: Verlags GbR.

Griffin, A.S., Blumstein, D.T., and Evans, C.S. (2000). Training captive bred and translocated animals to avoid predators. *Conserv. Biol.*, 14: 1317–1326.

Griffin, A.S., Evans, C.S., and Blumstein, D.T. (2001). Learning specificity in acquired predator recognition. *Anim. Behav.*, 62: 577–589.

Goodman, S.M. (2003). *Predation on lemurs*. In S.M. Goodman, and J.P. Benstead (Eds.), *The natural history of Madagascar* (pp. 1221–1228). Cambridge: Cambridge Univ. Press.

Goodman, S.M., O'Connor, S., and Langrand, O. (1993). A review of predation on lemurs: Implications for the evolution of social behaviour in small, nocturnal primates. In P.M. Kappeler and J.U. Ganzhorn (Eds.), *Lemur social systems and their ecological basis* (pp. 51–66). New York: Plenum Press.

Goodman, S.M., Ganzhorn, J.U., and Rakotondravony, D. (2003). Introduction to mammals. In S.M. Goodman, and J.P. Benstead (Eds.), *The natural history of Madagascar* (pp. 1159–1186). Cambridge: Cambridge Univ. Press.

Hill, R.A., and Dunbar, R.I.M. (1998). An evaluation of the roles of predation rate and predation risk as selective pressure on primate grouping behaviour. *Behaviour*, 135(4): 411–430.

Hill, R.A., and Lee, P. (1998). Predation risk as an influence on group size in cercopithecoid primates: Implications for social structure. *J. Zool.*, 245: 447–456.

Hogstedt, G. (1983). Adaptation unto death: Function of fear screams. *Am. Nat.*, 121: 562–570.

Isbell, L.A. (1994). Predation on primates: Ecological patterns and evolutionary consequences. *Evol. Anthropol.*, 3(2): 6171.

Janson, C.H. (2003). Puzzles, predation and primates: Using life history to understand selection pressures. In P.M. Kappeler and M.E. Pereira (Eds.), *Primate life histories and socioecology* (pp. 103–131). University of Chicago Press, Chicago.

Janson, C.H. (1998). Testing the predation hypothesis for vertebrate sociality: Prospects and pitfalls. *Behaviour*, 135: 389–410.

Kappeler, P.M., and Pereira, M.E. (2003). *Primate life histories and socioecology*. Chicago: Univ. of Chicago Press.

Karpanty, S.M., and Grella, R. (2001). Lemur response to diurnal raptor calls in Ranomafana National Park, Madagascar. *Folia Primatol.*, 72: 100–103.

Konishi, M. (1973). Locatable and nonlocatable acoustic signals for barn owls. *Am. Nat.*, 107: 775–785.

Langrand, O. (1990). *Guide to the birds of Madagascar*. New Haven: Yale Univ. Press.

Macedonia, J.M. (1990). What is communicated in the antipredator calls of lemurs: Evidence from playback experiments with ring-tailed and ruffed lemurs. *Ethology*, 86: 177–190.

Macedonia, J.M. (1993). Adaption and phylogenetic in the antipredator behaviour of ring-tails and ruffed lemurs. In P.M. Kappeler and J.U. Ganzhorn (Eds.), *Lemur social systems and their ecological basis* (pp. 67–84). New York: Plenum Press.

Macedonia, J.M., and Evans, C.S. (1993). Variation among mammalian alarm call systems and the problem of meaning in animal signals. *Ethology*, 93: 177–197.

Macedonia, J.M., and Polak, J.F. (1989) Visual assessment of avian threat in semi-captive ringtailed lemurs (*Lemur catta*). *Behaviour*, 111: 291–304.

Macedonia, J.M., and Yount, P.L. (1991) Auditory assessment of avian predator threat in semi-captive ring-tailed lemurs (*Lemur catta*). *Primates*, 32(2): 169–182.

MacMillon, B.W.H. (1990). Attempts to re-establish wekas, brown kiwis and red-crowned parakeets in the Waitakere Ranges. *Notornis*, 37: 45–51.

Manser, M.B., Seyfarth, R.M., and Cheney, D.L. (2002). Suricate alarm calls signal predator class and urgency. *Trends Cogn. Sci.*, 6(2): 55–57.

McLean, I.G., Schmitt, N.T., Jarman, P.J., Duncan, C., and Wynne, C.D.L. (2000). Learning for life: Training marsupials to recognise introduced predators. *Behaviour*, 137: 1361–1376.

Miller, D.B., Hicinbothom, G., and Blaich, C.F. (1990). Alarm call responsitivity of mallard ducklings: Multiple pathways in behavioural development. *Anim. Behav.*, 39(6): 1207–1212.

Mittermeier, R.A., Konstant, W.R., and Rylands, A.B. (2003). Lemur conservation. In S.M. Goodman, and J.P. Benstead (Eds.), *The natural history of Madagascar* (pp. 1538–1543). Cambridge: Cambridge Univ. Press.

Müller, A.E., Thalmann, U. (2000). Origin and evolution of primate social organisation: A reconstruction *Biol. Rev.*, 75: 405–435.

Neuweiler, G. (2003). *Neuro- und Sinnesphysiologie*. Berlin: Springer.

Oda, R. (1998). The response of Verreaux's sifakas to anti-predator alarm calls given by sympatric ring-tailed lemurs. *Folia Primatol.*, 69: 357–360.

Oda, R., and Masataka, N. (1996). Interspecific responses of ring-tailed lemurs to playback of antipredator alarm calls given by Verreaux's sifakas. *Ethology*, 102: 441–453.

Owings, D.H., and Hennessy, D.F. (1984). The importance of variation in sciurid visual and vocal communication. In J.O. Murie and G.R. Michener (Eds.), *The biology of ground-dwelling squirrels: Annual cycles, behavioural ecology, and sociality* (pp. 169–200). Lincoln: Univ. of Nebraska Press.

Owings, D.H., and Morton, E.S. (1998). *Animal vocal communication: A new approach*. Cambridge: Cambridge Univ. Press.

Pages, E. (1980). Ethoecology of *Microcebus coquereli* during the dry season. In P. Charles-Dominique, H.M. Cooper, A. Hladik, C.M. Hladik, E. Pages, G.F. Pariente., A. Petter-Rousseaux, A. Schilling, and J.J. Petter (Eds.), *Nocturnal Malagasy primates: Ecology, physiology, and behavior* (pp. 97–116). New York: Academic Press, New York.

Paillette, M., and Petter, J.J. (1978). Vocal repertoire of *Lemur fulvus albifrons*. In D.J. Chivers and J. Herbert (Eds.), *Recent advances in primatology*, Vol. 1. (pp. 831–834). London: Academic Press.

Pereira, M.E., and Macedonia, J.M. (1991). Ring-tailed lemur anti-predator calls denote predator class, not response urgency. *Anim. Behav.*, 41: 543–544.

Pereira, M.E., Seeligson, M.L., and Macedonia, J.M. (1988). The behavioral repertoire of the black-and-white ruffed lemur, *Varecia variegata variegata* (Primates: Lemuridae). *Folia Primatol.*, 51: 1–32.

Petter, J.J., and Charles-Dominique, P.C. (1979). Vocal communication in prosimians. In G.A. Doyle and R.D. Martin (Eds.), *The study of prosimian behavior* (pp. 247–305). London: Academic Press, Inc.

Pollock, J.I. (1975). Field observations on *Indri indri*: A preliminary report. In I. Tattersall and R.W. Sussman (Eds.), *Lemur biology* (pp. 287–311). New York: Plenum Press.

Rabesandratana, Z.A., Raminosoa, N., and Zimmermann, E. (2005). How the Milne Edwards' sportive lemur (*Lepilemur edwardsi*) reacts to a fossa encounter. *Primate Report*, 72: 79.

Radespiel, U., Cepok, S., Zietemann, V., and Zimmermann, E. (1998). Sex-specific usage patterns of sleeping sites in grey mouse lemurs (*Microcebus murinus*) in northwestern Madagascar. *Am. J. Primatol.*, 46: 77–84.

Radespiel, U., Ehresmann, P., and Zimmermann, E. (2003). Species-specific usage of sleeping sites in two sympatric mouse lemur species (*Microcebus murinus* and *M. ravelobensis*) in northwestern Madagascar. *Am. J. Primatol.*, 59: 139–151.

Rainey, H.J., Zuberbühler, K., and Slater, P.J.B. (2004). Hornbills can distinguish between primate alarm calls. *Proc. R. Soc. Lond. B*, 271: 755–759.

Rakotoarison, N., Zimmermann, H., and Zimmermann, E. (1996). Hairy-eared dwarf lemur (*Allocebus trichotis*) discovered in a highland rain forest of eastern Madagascar. In W.R. Lourenço (Ed.), *Biogéographie de Madagascar* (pp. 275–282). OSTROM, Paris.

Rakotondravony, D., Goodman, S. M., and Soarimalala, V. (1998). Predation on *Hapalemur griseus griseus* by *Boa manditra* (Boidae) in the Littoral Forest of Eastern Madagascar. *Folia Primatol.*, 69: 405–408.

Rasoloharijaona, S. (2001). Contribution à l'étude comparative de la communication vocale et de la vie sociale de deux formes de *Lepilemur* (Geoffroy, 1858) (Lepilemuridae, Petter et al. 1977) provenant de la forêt sèche de l'ouest et de la forêt humide de l'est de Madagascar. Doctorat de troisième cycle, Université d'Antananarivo, Antananarivo, Madagascar.

Richard, A.F. (1978). *Behavioral variation: Case study of a Malagasy lemur*. Lewisburg, PA: Bucknell Univ. Press.

Richards, J. (1998). Return of the natives. *Aust. Geographic*, special edition: 91–105.

Safer, A.B., and Grace, M.S. (2004). Infrared imaging in vipers: Differential response of crotaline and viperine snakes to paired thermal targets. *Behav. Brain Res.*, 154: 55–61.

Sauther M.L. (1989). Antipredator behavior in troops of free-ranging Lemur catta at Beza Mahafaly Special Reserve, Madagascar. *Int. J. Primatol.*, 10: 595–606.

Schmelting, B. (2000). Saisonale Aktivität und Reproduktionsbiologie von Grauen Mausmaki-Männchen (*Microcebus murinus*, J.F. Miller 1777) in Nordwest-Madagaskar. Dissertation, Tierärztliche Hochschule Hannover, Hannover.

Schülke, O. (2001). Social anti-predator behaviour in a nocturnal lemur. *Folia Primatol.*, 72: 332–334.

Schülke, O., and Ostner J. (2001). Predation on *Lepilemur* by a harrier hawk and implications for sleeping site quality. *Lemur News*, 6:5.

Seyfarth, R. M., and Cheney, D. L. (2003). Signalers and receivers in animal communication. *Annu. Rev. Psychol.*, 54: 145–173.

Seyfarth, R.M., Cheney, D.L., and Marler, P. (1980). Monkey responses to three different alarm calls: Evidence of predator classification and semantic communication. *Science*, 210: 801–803.

Stanford, C.B (2002). Predation in primates. *Inter. Jour. of Primatol.*, 23: 741–757.

Stanger, K.F. (1995). Vocalizations of some cheirogaleid prosimians evaluated in a phylogenetic context. In L. Alterman, G.A. Doyle, and M.K. Izard (Eds.), *Creatures of the dark: The nocturnal prosimians* (pp. 353–376.). New York: Plenum Press.

Stanger, K.F., and Macedonia, J.M. (1994). Vocalizations of aye-ayes (*Daubentonia madagascariensis*) in captivity. *Folia Primatol.*, 62: 160–169.

Sündermann, D.; Scheumann, M. and Zimmermann, E. (2005). Olfactory predator recognition in predator-naive grey mouse lemurs (*Microcebus murinus*). *Primate Report*, 72: 94.

Sussman, R.W. (1975) A preliminary study of the behavior and ecology of *Lemur fulvus rufus* Audebert 1800. In I. Tattersall and R.W. Sussman (Eds.), *Lemur biology* (pp. 237–258). New York: Plenum Press.

Terborgh, J., and Janson, C.H. (1986). The socioecology of primate groups. *Annu. Rev. Ecol. Syst.*, 17: 111–136.

van Schaik, C.P., and van Hooff, J.A.R.A.M. (1983). On ultimate causes of primate social systems. *Behaviour*, 85: 91–117.

Wolf, C.M., Griffith, B., Reed, C., and Temple, S.A. (1996). Avian and mammalian translocations: An update and reanalyses of 1987 survey data. *Conserv. Biol.*, 10: 1142–1154.

Yoder, A. D. (2003). Phylogeny of the lemurs. In S.M. Goodman, and J.P. Benstead (Eds.), *The natural history of Madagascar* (pp. 1242–1247). Cambridge: Cambridge Univ. Press.

Zietemann, V. (2000). Artendiversität bei Mausmakis: die Bedeutung der akustischen Kommunikation. Dissertation, Universität Hannover, Hannover.

Zimmermann, E. (1995). Acoustic communication in nocturnal primates. In L. Alterman, G.A. Doyle, and M.K. Izard (Eds.), *Creatures of the dark: The nocturnal prosimians* (pp. 311–330). New York: Plenum Press.

Zimmermann, E., and Radespiel, U. (in press). Primate life histories. In W. Henke, H. Rothe, and I. Tattersall (Eds.), *Handbook of palaeoanthropology*, Vol. 2. Heidelberg: Springer-Verlag.

Zimmermann, E., Newman, J.D., and Jürgens, U. (Eds.). (1995). *Current topics in primate vocal communication*. New York: Plenum Press.

Zimmermann, E., Vorobieva, E., Wrogemann, D., and Hafen,T. (2000). Use of vocal fingerprinting for specific discrimination of gray (*Microcebus murinus*) and rufous mouse lemurs (*Microcebus rufus*). *Int. J. Primatol.*, 21(5): 837–852.

Zuberbühler, K. (2000). Referential labelling in Diana monkeys. *Anim. Behav.*, 59: 917–927.

Zuberbühler, K. (2003). Referential signalling in non-human primates: Cognitive precursors and limitations for the evolution of language. *Adv. Study Behav.*, 33: 265–307.

Zuberbühler, K., and Jenny, D. (2002). Leopard predation and primate evolution. *J. Hum. Evol.*, 43: 873–886.

6
A Consideration of Leaping Locomotion as a Means of Predator Avoidance in Prosimian Primates

Robin Huw Crompton and William Irvin Sellers

Introduction

Predator pressure is normally very difficult to assess, and most reports tend to be anecdotal. However, it has been estimated that an annual predation rate of 25% may apply to *Microcebus* populations (Goodman et al., 1993). Such a rate, albeit for a particularly small prosimian, implies strong selective pressure in favor of adaptations that reduce predation, and it seems reasonable to assess adaptations with predation in mind. Predator avoidance by vigilance is usually seen as an attribute of social foragers (see, e.g., Terborgh & Janson, 1986), to which category many of the Lemuridae, and arguably some Indriidae and Lepilemuridae, belong. However, the small body size and nocturnality of those prosimians described as "solitary foragers" are often regarded as facilitating alternative predator avoidance strategy, crypsis (e.g., Clutton-Brock & Harvey, 1977; Stanford, 2002).

A rather more obvious and striking specialization of prosimians, however, is their proclivity for leaping. In this paper we suggest that rather than crypsis, leaping is actually the primary predator-avoidance device in prosimian primates classed as solitary foragers, and indeed may play as important a role as vigilance in many more gregarious taxa. Equally, while no single selective pressure is likely to be uniquely responsible for the widespread adoption of leaping locomotion by prosimian primates, the balance of evidence suggests that as in many non-primate leapers, leaping has indeed been adopted *primarily and originally* as a predator-avoidance device.

Leaping in Prosimians

Among vertebrates, it is the prosimian primates that display the most outstanding saltatory performances. *Galago moholi*, for example, leap distances and heights which are the greatest multiple of body length found in any vertebrate: horizontal leaps of 4 m and height gains of over 2 m may be achieved. Leaping is not only well-developed in prosimians, but it is nearly ubiquitous. In 22 genera of living prosimians, only four (*Nycticebus*, *Loris*, *Perodicticus*, and *Arctocebus*) do not leap at all. All those that do leap use leaping as a substantial element of their locomotor repertoire (reviewed in Walker, 1979; Oxnard et al., 1989). The four exceptions are all tailless, with sub-equal limb lengths, rather than the hindlimb-dominated intermembral indices which Napier and Walker (1967) famously identified as a marker of the locomotor category, vertical clinging and leaping. The four are supposedly all relatively slow moving and have adaptations such as a *rete mirabile*, an enhanced vascular network in the muscles, which permits muscles to remain in contracted state for extended periods. Their predatory behavior has been described as "stealthy" (Walker, 1969): slow movement, it was claimed, that is used to approach prey without disturbing the surrounding vegetation. Stealth may, of course, serve the needs of predator avoidance as well as it may those of predation, and indeed Charles-Dominique (1971) has argued that the slow locomotion of *Loris*, *Arctocebus*, and *Perodicticus* is actually an adaptation for predator-avoidance by crypsis. Walker (1969) contrasted the stealthy strategy of lorises with that of their relatives, the galagos, where speed of predatory movement is served by leaping. Although an apposite characterization of the behavior of *G. alleni* (Charles-Dominique, 1971) and *G. moholi* (Crompton, 1984), this adaptation is even more characteristic of tarsiers, which have recruited the leap as the basis of the predatory pounce from perches on vertical sapling-trunks near ground level (Fogden, 1974; Niemitz, 1979, 1984a; Crompton, 1989; Crompton & Andau, 1986; Oxnard et al., 1989; Jablonski & Crompton, 1994).

Thus, for some species at least (as in the case of the tarsier's predatory pounce) there might be an argument for linking prosimian leaping to hunting (i.e., engaging in, rather than avoiding, predation); but of course many prosimian leapers such as indriids, gentle and sportive lemurs, and ringtail lemurs are not primarily, or even substantially animalivorous (reviewed in, e.g., Hladik, 1979).

(NB: To link "stealth" necessarily to slow speed seems increasingly inappropriate. Anna Nekaris (pers. comm.) has since discovered that in the wild, the 130-g red loris can reach 1.29 m/s, an absolute speed well within the range of human walking speeds. Nekaris (2005) thus describes the Mysore loris's (*Loris lydekkerianus*) locomotion as "stealthy but swift." At least the pygmy slow loris may also be capable of quite high speeds, and the applicability of both Walker's (1969) and Charles-Dominique's (1971) descriptions may thus be quite limited.)

Kinetics and Kinematics of Leaping

Leaping style and leaping mechanism

Three categories of leaping "style" (see e.g., Oxnard et al., 1989) have been defined: static leaping, in which the animal pauses before making a leap; running leaping, in which the animal makes a transition from a run to a leap; and ricochetal leaping, in which the animal links together a succession of individual leaps with no pause or strides between each individual leap. In addition to these *outcome* groupings there are categorizations depending on the *mechanism* (see, e.g., Alexander, 2003) used to generate the power required for leaping: squat leaping, where muscle contraction alone is the motive force; countermovement leaping, in which a previous movement is used to store elastic energy that is released during take-off; catapult leaping, in which a locking mechanism is used to allow muscles slowly to bring about maximum tension, which can then be quickly released during take-off; and vaulting leaping, where a rigid strut is used to alter the direction of movement of the center of mass. All these mechanisms are potentially applicable to all the leap styles (with the probable exception of vaulting combined with ricochetal leaping), but it is most likely that the squat, catapult, and countermovement mechanisms are all used to perform static leaps. Countermovements are also involved in ricochetal leaping. It is striking in the context of a possible predator-avoidance role for leaping that the commonest outcome category in most primates (let alone prosimians), running leaping, is almost certainly brought about by vaulting, where an intrinsic element of change in the direction of movement exists; and this change is of course *sudden*.

While large animals benefit from the absolutely greater length of their limbs, which allows them to apply smaller forces over a longer take-off period, scaling effects also suggest that muscle physiological cross-sectional area will be larger in them compared to body mass in small mammals (see e.g., Demes & Günther 1989), so that even though the reduced take-off distance available to small animals necessitates higher power outputs, relatively more power is indeed available to them. However, Hall-Craggs (1962) noted that the calculated required power output for an observed maximum vertical leap of *Galago senegalensis*, gaining 2.25 m in height, is well in excess of the maximum capacity of vertebrate muscle (Bennet-Clark, 1977), which implies the existence of some means of power amplification. Aerts (1998) made a dynamic analysis of leaping in the lesser galago which lead him to suggest that the required power amplification could be obtained by a sequential recruitment of countermovement, catapult, and squat "with compliant tendons" (Alexander, 1995) mechanisms.

Leaping as a specialization

While leaping always tends to require a higher degree of musculoskeletal specialization than cyclic locomotor modes, not all prosimian leapers are equally specialized. Indeed, they may usefully be divided into specialist and non-specialist leapers. This is not just a matter of the percentage of movements that are leaps or the contribution to each kilometer of travel that is made up by leaping. Although arm swinging is used to extend a series of leaps by sifakas (author Crompton, pers. obs.), it is almost certain that all prosimian leapers power the leap primarily with the hindlimb. Specialist prosimian leapers, indeed, tend both to take off from, and to land on, their hindlimbs. This both maximizes the distance over which the body center of mass can be accelerated before losing contact with the ground and the distance over which it can be decelerated on landing. This in turn implies that specialist leapers require some mechanism for changing body posture in mid-flight. This is accomplished by a tail-flick in *Galago* and *Tarsius*. Given these animals' small body size, the tail-flick presumably must act by changing the rotational inertia of the body, not by means of air resistance (Peters & Preuschoft, 1984). Mid-flight rotation is, however, accomplished by countermovements of the fore-limbs in the large-bodied *Indri* and *Propithecus* (Preuschoft et al., 1998). Whether the use of the arms in these large species is a consequence of the greater air resistance they encounter (Bennet-Clark, 1977) is unclear, but air resistance may be exploited by indriids to increase maximum leap length, since loose skin under the abducted arms might provide a "gliding" effect, albeit at the expense of reducing speed. *Indri*, of course, lacks a tail; but *Propithecus'* tail appears simply to trail the body during leaps. Tail-flicks, and forelimb countermovements can alter orientation during flight. However, only the use of air resistance permits change in direction and/or leap length in mid-flight.

Generalists tend to land forelimb first, which at least in larger species may limit the force they can afford to experience on landing, and may thus also limit leap speed or distance (Oxnard et al., 1989). Choice of a compliant substrate as a landing target will, however, negate this problem, albeit at the cost of increased disturbance to the surrounding vegetation. Thus, for example, while Demes et al. (2005) found that the *Lemur catta* they studied tended to land hindlimb first, their *Eulemur* subjects landed forelimb first. In addition, *Eulemur* forelimb forces on landing were greater than hindlimb forces, although hindlimb forces on take-off were larger still.

Leaping and efficiency of transport

From basic physical principles it has been established that leaping locomotion is not in itself a very efficient way of moving around (Walton & Anderson, 1988). Except for ricochetal leaping, where leaps follow immediately upon each other at a stable resonant frequency, there is little or no possibility of the primate conserving energy between one leap and the next. Energy savings in ricochetal leaping may be served by elastic recoil of tendon and ligament and elastic units in muscles, stretched during landing, to help power the next leap. (There will of course be ecological situations where leaping remains the most efficient locomotor option: e.g., when crossing between trees, where the alternative to leaping from one canopy to the next may be to climb down one trunk and up the next.)

Leaping and musculoskeletal load

Leaping is also associated with high ground reaction forces compared to quadrupedalism (Günther et al., 1991; Demes et al., 1999) and behaviors that result in large forces are likely to influence musculoskeletal morphology (Alexander, 1981). As we have seen, the scale of forces required to be exerted during take-off varies with body size, so that Demes and colleagues (1999) give values of hindlimb take-off force of thirteen times body mass in *G. moholi*, but nine times body mass in *P. verreauxi*. The more striking contrast was, however, with quadrupeds of equivalent size, where forces are only just over twice body mass. Thus, leaping is thus not only energetically expensive as a means of transport compared to quadrupedalism, but also incurs higher musculoskeletal loads, and thus requires a greater degree of musculoskeletal specialization.

Leaping and transport speed

In contrast to popular expectation, leaping is not a particularly fast method of travel. Günther et al. (1991) recorded maximum velocities at take-off of 5.1 m/s for G. moholi leaping from a forceplate, and noted that this compares unfavorably to velocities of 15 m/s or more, which may be attained over short distances by a galloping, cursorial quadruped. The velocity Günther recorded is slightly greater than the 4.4 m/s required by leapers to attain a height of 1 m, irrespective of size (according to Bennet-Clark, 1977). As G. moholi (atleast according to Hall-Craggs, 1964, 1965) can gain 2.25 m in a leap, 5.1 m/s must be an underestimate of actual velocity maxima (although doubling vertical take-off velocity would quadruple height gain (Bennet-Clark, 1977). However, under natural conditions, Crompton recorded only a single record of a 2 m estimated height gain and only 39 of an estimated height gain of over 1 m in 2786 leaps by G. moholi. For Tarsius bancanus he recorded a maximum estimated height gain of 1.5 m, and only eight records of leaps over a 1 m height gain (of a total 1425 observed leaps). These field data tend to suggest that a take-off velocity of 5.1 m/s (Günther et al., 1991) is not substantially less than actual maximum velocities. Moreover, even anatomically specialized leapers do not often attain a velocity of 4.4 m/s (see Bennet-Clarke, 1977) in nature. Thus, most leaping occurs at ground speeds well under a third of the maximum speeds attained by cursorial quadrupeds, and actually rather closer to the speeds reached in arboreal quadrupedalism by Loris.

Since leaping is a ballistic action, we can readily derive predicted performances under different conditions. The ratio of distance travelled to force exerted at take-off varies with take-off angle and in-flight trajectory. Flight time is also dependent on trajectory, and the relative heights of the initial and terminal supports also need to be taken into consideration. Figures 6.1–6.3 show these relationships, and the equations used to derive these curves are given in Appendix 1, so that they may be used to analyze field data.

Figure 6.1 shows the mechanical energy cost of a leap for a set of take-off angles given the relative heights of the initial and terminal support (labeled "slope") for a 1 kg animal leaping 1 m. Figure 6.2 shows the flight time for a range of take-off angles and differing relative heights of initial and terminal supports ("slope"), again for a 1 m/s take-off velocity (the range for any combination of these values are given for equivalent values in Figure 6.3). Flight times for different take-off velocities are simple multiples, so the flight time for a speed of 2 ms would be twice the value given, for 4 ms it is 4 times the value, etc.). Figure 6.3 shows the range of a leap for a set of take-off angles and "slopes" for a 1 m/s take-off velocity. The range distance shown is the length of a line drawn from start to endpoint; the horizontal distance can be obtained by multiplying by the cosine of the "slope"

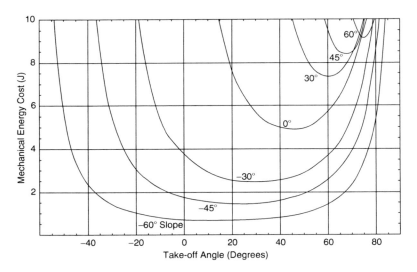

FIGURE 6.1. Mechanical energy cost of a leap for a set of take-off angles and relative heights of initial and terminal supports (slopes) for a 1 kg animal leaping a distance of 1 m

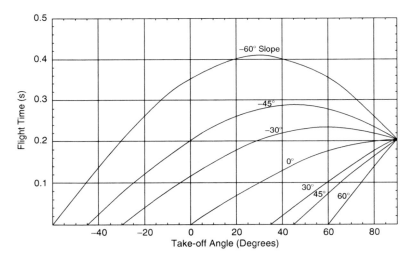

FIGURE 6.2. Flight time for a set of take-off angles and relative heights of initial and terminal supports (slopes) for a 1 ms take-off velocity

angle. Range depends on the square of the velocity, so range quadruples for twice the speed, is sixteen times greater for four times the speed, etc.

For any given combination of support heights, there is thus a take-off angle that will maximize travel distance (or equally minimize the energetic cost of travel). For level leaps at a take-off angle of 45°, distance covered for a given take-off

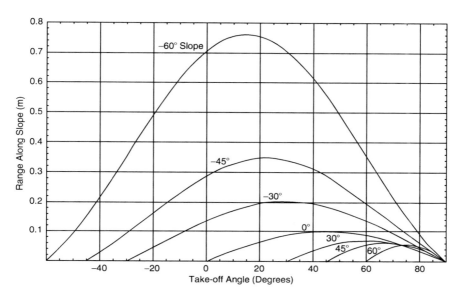

FIGURE 6.3. Range of a leap for a set of take-off angles and relative heights of initial and terminal supports (slopes) for a 1 ms take-off velocity. The range is measured along the slope

force is maximized, but such a leap is relatively slow. By contrast, a low, 20° take-off angle gives lowest costs for a 60° descent, while a take-off angle around 75° is required for maximum efficiency in a 60° ascent. Flatter trajectories cover less distance for the same take-off force, but less time is spent in the air. Very low take-off angles, while minimizing flight time, are always energetically expensive. While, in general, short flight times require low take-off angles, for downward leaps the longest flight times occur with moderate take-off angles. Leaping upward, however, is clearly much more expensive than leaping downward.

Perhaps surprisingly, Crompton et al. (1993) found that of five prosimian leapers studied in the laboratory, only the most anatomically specialized, G. moholi, habitually used the ballistically optimum take-off angle' 45°' at all leap lengths, in level leaps, although the other species tended to use this angle more often as leap distances approached the maximum they performed. This would seem to suggest that most prosimians opt for speed rather than distance in their leaping, or cannot readily adopt an appropriate body posture for a high-angled take-off, as discussed below. Demes et al. (1999) showed that "specialist" leapers, such as the indriids, exert relatively lower take-off and landing forces than less specialized leapers such as G. garnetti. Take-off force did not increase with distance (within the limited range of leap distances the authors could examine). In a study of leaping forces in Indri, P. verreauxi, and P. diadema, Demes and colleagues (1995) found that both take-offs and landings nearly always resulted in tree sway, and that for take-offs, the indriids lost contact with the initial

TABLE 6.1. Support diameters and effective jump distance in *Galago moholi*, *Tarsius bancanus* and *Galago crassicaudatus* (data from Crompton et al., 1993).

Effective Jump Distance	*Galago moholi*[a] Initial Support Diameter	*Tarsius bancanus*[b] Terminal Support Diameter	*Otolemur crassicaudatus*[c] Terminal Support Diameter
0–0.200 m	4.0 cm	1.8 cm	4.2 cm
0.201t–0.400 m	4.1 cm	1.8 cm	4.1 cm
0.401t–0.800 m	4.2 cm	2.4 cm	4.1 cm
0.801t–1.600 m	4.4 cm	2.8 cm	4.1 cm
1.601t–3.200 m	5.3 cm	3.0 cm	4.3 cm
3.200 m +	6.2 cm	3.7 cm	3.1 cm

[a]Diameters for all leap categories above 0.800 m were significantly different ($P < 0.05$) from each of those below (Duncan's multiple range test); [b]diameters for all leap categories above 0.400 m were significantly different ($P < 0.05$) from each of those below (Duncan's multiple range test); [c]diameters for leap category 3.200 m + significantly different ($P < 0.05$) from each of those below (Duncan's multiple range test).

support before rebound occurred, so that energy was lost to the branch at take-off as well as landing. Crompton et al. (1993, and see Table 6.1) however found that in *G. moholi*, leaps over 0.8 m began on larger diameter supports than did shorter ones, suggesting that the risk of energy loss to the substrate might have an effect on the choice of take-off supports. This was not the case in *T. bancanus*, which, on the other hand, tended to land on substantially larger supports for leap lengths over 0.4 m than for leaps up to 0.4 m; *Otolemur*, however, favored smaller supports in leaps over 3.2 m than in all leap lengths below this distance. There is thus no conclusive evidence for a consistent pattern of avoidance of loss of energy to the substrate either on take-off or on landing. But substrate orientation also needs to be taken into consideration, as does trajectory, since it might also be the case that leaps with flatter trajectories, when taken from a horizontal or low-angled support, may exert the greater proportion of take-off force in the strongest direction of the support. Conversely, leaps with higher trajectories might be expected to exert a greater proportion of force in the strongest direction of the support when taken from high-angled supports.

Discussion

Ability to use a high take-off angle requires that the body center of gravity be positioned along, or close as possible to, a line extended at that angle to the take-off support from the propelling limb(s). Assuming, as is appropriate for prosimian primates, it will be the hindlimbs which are primarily responsible for propelling a leap, the implication is that an orthograde trunk posture needs to be adopted. High take-off angles are thus more readily attained from supports at a relatively high angle to the horizontal, although they can be performed even from horizontal supports, as a consideration of the locomotor ecology of ground-foraging genera such as *Tarsius* (and to a lesser extent *Galago moholi*, for example) makes immediately obvious.

If the finding of Crompton et al. (1993)—that the specialist leaper among their five experimental subjects used steeper trajectory leaps in leaps of all lengths, whereas the generalists used steeper trajectories in only their longest leaps—could be generalized, it would then be expected that more specialized leapers, which would be more likely to avoid flat trajectories, would also be more likely to use near-vertical supports. From a pronograde body posture, low take-off angles, and thus low trajectories, with short flight times, can more readily be adopted. Leaping from low-angled supports is more feasible, and more of the thrust may be directed along the strongest axis of the takeoff support, reducing energy loss to the branch and hence branch displacement. Both a short flight time and lack of disturbance of vegetation might be seen as advantageous in predator-avoidance. But mechanical energy costs are inevitably high, and ranges short.

Hence, use of high angled supports for take-off would be expected to be more characteristic of specialist leapers, low-angled supports characteristic of more generalized leapers. This appears generally to be the case, both in comparisons of closely related pairs such as *G. moholi* and *O. crassicaudatus* (Crompton, 1984) and in broader comparisons of the prosimians as a whole (Oxnard et al., 1989). However, comparison of the behavior of *G. moholi* between different seasons shows a greater affinity for vertical supports in a cold, dry season, but lower affinity in a warm, wet season (Crompton, 1984). This would not be expected simply from an association, in nature, of steep trajectories with near-vertical take-off supports, but flat trajectories with low-angled supports. Consideration of height of observation and support availability in the open *Acacia* woodland, which is the natural habitat of *G. moholi*, shows that as *G. moholi* are found much more often low down in the cold, dry season, they will encounter fewer low-angled supports and more high-angled supports. Leap distances are longer; this would be expected *both* from the greater separation of supports nearer ground level and from an hypothesized association of steep trajectory leaps with high-angled supports.

Field data also show that mean leap length in specialist leapers is far below the attainable maximum. In *G. moholi* and *T. bancanus*, while the longest leaps observed in the field were often in excess of 4 m (Crompton, 1980, 1983, 1984; Crompton & Andau, 1986), and while Niemitz (1979) suggests over 6 m may be attained by *T. bancanus* when pursued, Crompton (1980, 1983, 1984) found that the mean leap length was only 0.69 m for *G. moholi*, and Crompton & Andau (1986) obtained a mean of 1.12 m for *T. bancanus*.

This might suggest that under field conditions, these specialist leapers do not use the ballistically optimum take-off angle as regularly as they do in the laboratory, preferring the shorter flight duration and greater unpredictability of a relatively "flat" jump; *or* that they are using asymmetric leaps, again for unpredictability because of the potential for change in direction we have mentioned above; *or* that they are interrupting their leaps by use of air resistance (perhaps less likely in small species) *or* that they often use the ballistic trajectory to gain height by landing early in the trajectory, rather than using climbing for height gain (since it may be even more expensive than height change by leaping). It may be relevant that in field data for *T. bancanus*, *G. moholi*, and even *O. crassicaudatus*, the longest leaps tended to be associated with height gain rather than height loss (Crompton et al., 1993), suggesting that these might be such interrupted ballistic leaps.

Indeed, re-analysis of Crompton's field data shows that even unspecialized arboreal quadrupeds as *O. crassicaudatus* (about 1300 g) regularly attain distances like as *G. moholi* (185 g). Some care must be taken in discussing raw leap lengths, as calculations of the mechanical cost of a leap must take into consideration the height of initial and terminal supports (see, e.g., Crompton et al., 1993; Warren & Crompton, 1998). Further, maximum ranges recorded in the field are difficult to compare, both because *Otolemur* moves much higher (see Crompton, 1984), and can thus lose much more height, and because in unusual circumstances (such as when it is being chased) it can alter leap kinematics. For example, when being chased *Otolemur* can (no doubt at some energetic cost) take off and land hindlimb first, and will then often use vertical or near-vertical supports (Crompton, 1980): This presumably gives high trajectories and therefore increases range. Re-analysis of Crompton's field data, however, shows that means for *level* leaps are very similar (0.63 m and 0.64 m, respectively, not significantly different).

While the frequency of leaping in the folivorous specialist leapers *Avahi occidentalis* and *Lepilemur edwardsi* is similar to that in *Tarsius bancanus*, their mean leap length at the study site of Ampijoroa is greater than in the latter species: 1.5 m and 1.23 m, respectively. (For the lemurs, a *t*-test on a 50% random sample for mechanically effective ranges gave a two-tailed, equal-variance probability of < 0.001 for overall means of 1.38 m (*Lepilemur*) and 1.56 m (*Avahi*), (Warren & Crompton, 1998)). However, this is still considerably less than the mean intertrunk distance (2.55 m, $N = 613$, SE 0.09) at this site, and much less than the maximum leap distance that both species were observed to attain (7 m). The ability of each of these species to cross the wide gap between tree trunks is not often used. Thus, the importance of the ability to leap long distances may rather be that an ability to perform occasional very long leaps is an effective means of avoiding predation in open cover. We must however ask *why* this ability is not often used. The contribution of the mechanical costs of locomotion to the total energy budget was estimated by Charles-Dominique and Hladik (1971) and Hladik and Charles-Dominique (1974) for *Lepilemur mustelinus leucopus* in Didiereaceae bush. Their estimates suggest that the caloric value of dietary intake was insufficient to sustain total energetic costs, and they proposed caecotrophy as a possible means whereby the deficit might be made up. The predicted deficit existed, they argued, notwithstanding the fact that locomotor costs contributed only 10% to the total energy expenditure. Their conclusions have, however, subsequently been challenged by Russell (1977).

It is difficult to reliably predict the metabolic costs of locomotion, unless a forward-dynamic musculoskeletal model is used to estimate the metabolic cost of muscle contraction, as we did recently for walking in *Australopithecus afarensis* (Sellers et al., 2005). Nevertheless a case can be made that the costs of leaping locomotion in species with unusual dietary habits may be such as to bring the total budget close to tolerable limits. This particularly applies to small-bodied species where thermoregulation is highly expensive (Karasov, 1981; Schmidt-Neilsen, 1990). The most rigorous estimate of the contribution of locomotion to total metabolic costs of wild animals (the field metabolic rate, FMR) is that of Kenagy and Hoyt (1990) for golden-mantled ground squirrels. Their estimate of 15% contrasts with the figure of 2% calculated by Nagy and Milton (1979) for mantled howler monkeys. Warren & Crompton (1998) used their field data to estimate the contribution of locomotion to total energy costs for five nocturnal prosimians: four specialist leapers (*L. edwardsi, A. occidentalis, T. bancanus, G. moholi*) and one generalist (*O. crassicaudatus*) and found that *Avahi* had the highest contribution at 3%. But they noted that the contribution of locomotion to FMR is very sensitive to daily movement distances (DMD) (Goszczynski, 1986; cf.) daily path length), which are (notoriously) underestimated in observational studies of ranging behavior. Elastic energy savings through ricochetal leaping are one (untested) means whereby *Avahi* may be able to tolerate its rather high locomotor costs. Warren & Crompton (1998) suggested that *T. bancanus* might also be close to its energetic limits, on the basis of Niemitz's (1985a) and Jablonski and Crompton's (1994) data on dietary intake, and Crompton's (1989) data on DMD in *T. bancanus*. Thus, for a leaping specialist with a long DMD (such as *T. bancanus*); or a particularly low metabolic rate (such as *L. ruficaudatus* (Schmid & Ganzhorn, 1996), the energetic costs of leaping may indeed be critical, and particularly expensive leaps may need to be avoided except in life-threatening situations (of which predation must surely be the most common).

Thus, rather than concluding—as one might from the marked difference between mean and maximum leaps of *Galago, Otolemur*, and *Tarsius* (see above)—that specialist leaping species are "over-specified" in terms of their morphological adaptation to leaping, consideration from a predator-avoidance perspective suggests that the ability to perform long leaps may be selected for primarily by the risk of predation attempts: such attempts are likely to be far less rare than successful predation.

Clearly, if a threatened bushbaby or tarsier performs a leap some four times longer than its mean leap length, this capability would likely confuse a predator familiar with their quotidian performance. However rare, such a capability would be strongly selected for wherever predation pressure was substantial, as the effects of a successful predation on reproductive fitness are uniquely drastic (Lima & Dill, 1990).

Günther et al. (1991) suggest that specialist leapers such as *G. moholi* also tend to use asymmetrical leaping quite often, where one hindlimb applies more force than does the other, so to effect changes in direction; whereas, these authors argue, less specialized leapers do not. Asymmetrical leaping is commonly seen in other vertebrates such as frogs, where leaping is regarded as primarily a predator avoidance strategy since it reduces the predictability of leaping direction (Gans & Parsons, 1966). In invertebrates such as locusts and grasshoppers, escape leaps seem to have a completely random direction. Thus, leaping specialization in prosimians may not be so much an adaptation to maximize the leap length that can be obtained, as an adaptation to maximize the leap length that can be obtained using the force from one hindlimb. In other words, it may be that specialist leapers are adapted to perform well in *asymmetric* leaping rather than *symmetric* leaping. This argument would, however, also be consistent with their specialization serving the ends of unpredictability (and so predator avoidance, and where relevant, predation) rather than locomotor efficiency. It could also be the case in species where energetic budgets are so finely balanced (perhaps including *T. bancanus*, *A. occidentalis*, and *L. edwardsi*) that a high degree of locomotor efficiency is also selected for.

Finally, extra leaping performance may allow leaps to be performed at energetically suboptimal trajectories. This increases the energetic cost of the leap but can reduce the flight time and increase the horizontal speed, or allow reduction in the predictability of the trajectory—all potentially valuable methods of avoiding predators. However, since use of flat trajectories is actually rather commoner in unspecialized rather than specialized leapers, this factor is not likely to be important.

Goodman et al. (1993) provide an excellent review of the anecdotal data we have on predation on lemurs. While snakes appear to be less frequent predators, the fossa (*Cryptoprocta ferox*), and to a lesser extent other viverrids, such as *Galidia elegans*, frequently prey on diurnal and nocturnal lemurs, large and small alike. Owls, such as the barn owl (*Tyto alba*) and the Madagascar long-eared owl (*Asio madagascariensis*) are primarily predators of small-bodied, nocturnal genera such as *Microcebus,* while large raptors, such as the Madagascar harrier hawk (*Polyboroides radiatus*) and the Madagascar buzzard (*Buteo brachypterus*) prey on large-bodied, diurnal genera, including *Propithecus* and *Indri*. Defensive movements made by adult *Indri* at Mantadia when *Polyboroides* is in sight suggest the latter is a predator on young *Indri*. Both *Polyboroides* and *Buteo* elicit alarm calls from *Hapalemur griseus* at Mantadia, are often heard circling *Hapalemur* home ranges, and are likely major predators (Mary Blanchard pers. comm. and authors' pers. obs.). These data imply, and the "short-winged" nomen of the Madagascar buzzard reminds us, that we need to consider the locomotor capabilities of predator species as well as those of their prey. Short-winged birds, such as most owls, are generally more capable of rapid changes of direction (see, e.g., Norberg, 1985), whereas long-winged species may only be capable of taking lemurs from the very top of the canopy.

Cryptoprocta, the fossa, is a large-bodied but short-legged carnivore, den-living but competent arboreally and capable of leaping (see, e.g., Wright et al., 1997; Hawkins, 1998; Dollar, 1999; Dollar et al., this volume; Patel, 2005). The fossa's powerful forelimbs, clawed digits, and short, flexed limbs permit pursuit by climbing on large to medium size tree trunks and branches. Because it is a large predator, we would expect and indeed find that *Cryptoprocta*, in preying on small lemurs, will concentrate on nocturnal species that use nests or tree-hollows for sleeping (and may sleep in groups), rather than risk failure in an active chase. A rare film sequence (an edited version can be seen in BBC Wildlife, *Life of Mammals*) of

Cryptoprocta in pursuit of *Propithecus* show that while the fossa is quite capable of leaps of one to two meters (level) from, and to, vertical supports, it is less agile than a sifaka on smaller, low-angled branches, where body weight deforms the support, but where the fossa's lack of grasping appendages renders it unstable. In the case of the BBC film, however, it appears to have been primarily the sifakas' ability to make repeated leaps with frequent and marked changes of direction that leads to their escape.

Predation by raptors on large lemurs almost inevitably occurs most often at canopy level or in open ground, as long wings and a soaring habit do not permit ready flight in woodland, where frequent changes of direction are required. This may suggest one reason why indris tend to travel from tree to tree just below canopy level, despite the long leaps that are required. *Indri* appear to avoid having to come to the ground (Mary Blanchard, pers. comm.), where they are at a disadvantage with respect to *Cryptoprocta*. Bipedal hopping by *Propithecus* may, however, permit this genus more extensive use of the ground and lower forest levels by permitting confusingly sudden changes in direction when pursued by these predominantly quadrupedal predators.

Predation on galagos was discussed briefly by Bearder (1987) who estimated that 15% of *G. moholi* populations are harvested annually by predators, primarily owls but also, during the day, hawks. During the day, *G. moholi* and its sympatric relative *O. crassicaudatus* are relatively protected by the long thorns of the *Acacia* trees (the gums of which contribute substantially to their diet). At night, *Otolemur* exhibit alarm in the presence of genets (*Genetta tigrina*). The genet is an agile, arboreal species like itself. On the other hand, even young *Otolemur* will approach and touch monitor lizards of considerable (about one meter) size if they are found on a branch (author Crompton pers. obs.). Rapidity in movement seems to be a prerequisite for nocturnal predators on galagos. However, there is no doubt that owls are agile enough to take *Galago* in mid-leap: it happens commonly enough to have been captured on film (BBC Wildlife, *Mara Nights*). In contrast, instances of predation on tarsiers are relatively rare in the literature. MacKinnon and MacKinnon (1980) remark on a lack of any alarm response by tarsiers to the presence of potential predators. On the other hand, Gursky (2001) reported a successful predation on *Tarsius spectrum* by a python and (2005) noted frequent alarm calling and mobbing in response to potential predators, and Susmann (1999) reported that Shekelle has observed a predation event on *T. syrichta* by a monitor lizard. But the Sulawesi and Phillippine forest habitats are relatively open compared to lowland evergreen rainforest, the habitat of the largest species, *T. bancanus*.

Niemitz (1979) working on *T. bancanus* in a forest enclosure at Semongok, Sarawak, observed that this species lacks any obvious alarm response to potential predators introduced into the enclosure. Similarly, Crompton, working at Sepilok in Sabah, did not observe predation or any suggestion of an alarm response during the active period in many hours of close-contact following of free-ranging *T. bancanus*. This species usually forages within the first two meters above the ground. In the normal primary rainforest habitat of this species, little moon- or starlight (and relatively little sunlight) reaches this level. Thus, at night *T. bancanus* must be very difficult for any predator to locate, since its background will always be relatively dark, and lacking a tapetum, light that does reach it will not be reflected back from its eyes. (It does not seem likely that absence of a tapetum is related to a cryptic "strategy." The tarsier's lack of a tapetum is of course amply compensated by eye size and a likely consequence of secondary adoption of nocturnality by its branch of the common haplorhine lineage (Crompton, 1989). Nocturnality serves niche differentiation more directly than it does crypsis). Crompton found *T. bancanus*' habitual response to (human) pursuit to be immediate flight by a rapid series of upward leaps to a height of up to 12 m. Similarly, vine and thorn tangles at 3–4 above ground (well above the normal height of activity) in dense tree fall zones were identified as the commonest diurnal sleeping site. This suggests that diurnal terrestrial predators may be more of a problem for this species.

Conclusions

As we have seen, it has often been proposed that the single greatest advantage conferred by leaping locomotion is the ability to make sudden and unpredictable changes in direction: in anurans (Gans & Parsons, 1966), fleas (Bennet-Clark & Lucey, 1967), and locusts (Bennet-Clark, 1975, 1977). Amongst mammals, a very clear case for this argument is that made for the hopping of pocket mice by Bartholomew and Cary (1954), which rarely use their hopping as a means of travel, preferring to use quadrupedalism unless threatened. It is therefore most economical to conclude that while no single selective pressure is likely to be responsible for the widespread adoption of leaping locomotion by prosimian primates, the balance of the weight of evidence suggests that as in many non-primate leapers, prosimian leaping has been adopted primarily as a predator-avoidance device. As one of the most striking characteristics of prosimians, this in turn suggests that— outside of infancy, dormancy, or the inactive part of the diel cycle, and with the possible exception of the lorises—crypsis, as a predator avoidance strategy, is no more typical of what Bearder (1987) aptly terms "solitary foragers," the small-bodied, nocturnal forms, than it is of the large-bodied, diurnal, social foragers.

Future Directions for Research

In his recent but already classic text, Alexander (2003) observes that a major need in locomotor biology is for studies of the mechanics of arboreal locomotion that take account of the flexibility and uneven spacing of branches. We need more locomotor studies designed to collect biomechanically relevant data, rather than just raw locomotor counts, and to allow integrated analysis of leap length (raw and effective), and initial and terminal support characteristics. This study suggests that we need to understand the decisions made by animals crossing gaps between such supports in terms of the costs and risks (both biomechanical and ecological) that each choice incurs. Among these risks predation must surely be the most adaptively challenging.

Acknowledgments. We thank Anna Nekaris for the invitation to contribute to this volume and Sharon Gursky and four anonymous reviewers for their helpful comments and suggestions. We thank BBC Wildlife for giving us access to unedited footage of a chase by *Cryptoprocta*. Our field and laboratory studies of prosimian locomotion have been funded by the generosity of: The L.S.B. Leakey Foundation; World Wildlife Fund Hong Kong; The Royal Society; the Erna and Victor Hasselblad Foundation, the Natural Environment Research Council, and the Biotechnology and Biological Sciences Research Council.

References

Aerts, P. (1998). Vertical jumping in *Galago senegalensis*; the quest for an obligate mechanical power amplifier. *Phil. Trans. Roy. Soc. Lond.* B, 533: 1299–1308.

Alexander, R. Mc N. (1981). Factors of safety in the structure of animals. *Sci. Prog.*, 67: 109–130.

Alexander, R. Mc N. (1995). Leg design and jumping technique for humans, other vertebrates and insects. *Phil. Trans. Roy. Soc. Lond.* B, 347: 235–248.

Alexander, R. Mc N. (2003). *Principles of animal locomotion.* Princeton, NJ: Princeton Univ. Press.

Bartholomew, G.A., and Cary, G.R. (1954). Locomotion in pocket mice. *J. Mammalogy*, 35: 386–392.

Bearder, S.K. (1987). Lorises, bushbabies and tarsiers: Diverse societies in solitary foragers. In B.B. Smuts, D.L. Cheney, R.M. Seyfarth, R.W. Wrangham, and T.T. Struhsaker (Eds.), *Primate societies* (pp. 11–24). Chicago: Univ. of Chicago Press.

Bennet-Clark, H.C. (1975). The energetics of the jump of the locust *Schistocerca gregaria. J. Exp. Biol.*, 82: 105–121.

Bennet-Clark, H.C. (1977). Scale effects in jumping animals. In T.J. Pedley (Ed.), *Scale effects in animal locomotion* (pp. 185–201), London: Academic Press.

Bennet-Clark, H.C., and Lucey, E.C.A. (1967). The jump of the flea: A study of the energetics and a model of the mechanism. *J. Exp. Biol.*, 47: 59–76.

Clutton-Brock, T.H., and Harvey, P.H. (1977). Primate ecology and social organization. *J. Zool.*, 183: 1–39.

Charles-Dominique, P. (1971). Eco-ethologie des prosimiens du Gabon. *Biologica Gabonica*, 7: 21–228.

Charles-Dominique, P., and Hladik, C.M. (1971). Le *Lepilemur* du sud du Madagascar: Ecologie, alimentation et vie sociale. *Terre et Vie*, 25: 3–66.

Crompton, R.H. (1980). *Galago* locomotion. Unpublished Doctoral thesis. Harvard University.

Crompton, R.H. (1983). Age differences in locomotion of two subtropical Galaginae. *Primates*, 24: 241–259.

Crompton, R.H. (1984). Foraging, habitat structure and locomotion in two species of *Galago.* In P. Rodman and J. Cant, J. (Eds.), *Adaptations for foraging in non-human primates* (pp. 73–111). New York: Columbia Univ. Press.

Crompton, R.H. (1989). Mechanisms for speciation in *Galago* and *Tarsius. Hum Evol.*, 4: 105–116.

Crompton, R.H., and Andau, P.M. (1986). Locomotion and habitat utilization in free ranging *Tarsius bancanus*: A preliminary report. *Primates*, 27: 337–355.

Crompton, R.H., Sellers, W.I., and Günther, M.M. (1993). Energetic efficiency and ecology as selective factors in the saltatory adaptation of prosimian primates. *Proc. Roy. Soc. Lond.*, 254: 41–45.

Demes, B., and Günther, M.M. (1989). Biomechanics and allometric scaling in primate locomotion and morphology. *Folia primatol.*, 52: 58–59.

Demes, B., Jungers, W.L., Gross, T.S., and Fleagle, J.G. (1995). Kinetics of leaping primates: influence of substrate orientation and compliance. *Am. J. Phys. Anthrop.*, 96: 419–429.

Demes, B., Fleagle, J.G., and Jungers, W.L. (1999). Takeoff and landing forces of leaping strepsirhine primates. *J. Hum. Evol.*, 37: 279–292.

Demes, B., Franz, T.M., and Carlson, K.J. (2005). External forces on the limbs of jumping lemurs at takeoff and landing. *Am. J. Phys. Anthrop.*, 128: 348–358.

Dollar, L. (1999). Preliminary report on the status, activity cycle, and ranging of *Cryptoprocta ferox* in the Malagasy rainforest, with implications for conservation. *Small Carnivore Conservation*, 20: 7–10.

Fogden, M. (1974). A preliminary field-study of the western tarsier, *Tarsius bancanus* HORSFIELD. In R.D. Martin, G.A. Doyle, and A.C. Walker. (Eds.), *Prosimian biology* (pp. 151–165). Pittsburgh: Pittsburgh Univ. Press.

Gans, C., and Parsons, T.S. (1966). On the origin of the jumping mechanism in frogs. *Evolution*, 20: 92–99.

Goodman, S.M., O'Connor, S., and Langrand, O. (1993). A review of predation on lemurs: Implications for the evolution of social behaviour in small, nocturnal primates. In P.M. Kappeler and J.U. Ganzhorn (Eds.), *Lemur social systems and their ecological* basis (pp. 51–67). New York: Plenum.

Goszczynski, J. (1986). Locomotor activities of terrestrial predators and their consequences. *Acta Theriol.*, 31: 79–95.

Günther, M.M., Ishida, H., Kumakura, H., and Nakano, Y. (1991). The jump as a fast mode of locomotion in arboreal and terrestrial biotopes. *Z. Morph. Anthrop.*, 78(3): 341–372.

Gursky, S. (2005). Predator mobbing in *Tarsius spectrum*. *Int. J. Primatol.*, 26(1): 207–221.

Gursky, S. (2002). Predation on a wild spectral tarsier (*Tarsius spectrum*) by a snake. *Folia primatol.*, 73(1): 60–62.

Hall-Craggs, E.C.B. (1962). The jump of *Galago senegalensis*. Unpublished doctoral thesis. Univ. of London.

Hall-Craggs, E.C.B. (1964). The jump of the bushbaby—A photographic analysis. *Med. Biol. Ill.*, 14:170–174.

Hall-Craggs, E.C.B. (1965). An analysis of the jump of the lesser galago (*Galago senegalensis*). *J. Zool.*, 147:20–29.

Hawkins, C.E. (1998). The behaviour and ecology of the fossa, *Cryptoprocta ferox* (Carnivora: Viverridae) in a dry deciduous forest in western Madagascar. Unpublished doctoral thesis. Univ. of Aberdeen.

Hladik, C.M., and Charles-Dominique, P. (1974). The behaviour and ecology of the sportive lemur (*Lepilemur mustelinus*) in relation to its dietary peculiarities. In R.D. Martin, G.A. Doyle, and A.C. Walker (Eds.), *Prosimian biology* (pp. 23–37). Pittsburgh: Pittsburgh Univ. Press.

Hladik, C.M. (1979). Diet and ecology of prosimians. In G.A. Doyle, R.D. Martin (Eds.), The study of prosimian behaviour (pp. 307–359). London: Academic Press.

Jablonski, N.G., and Crompton, R.H. (1994). Feeding behavior, mastication and toothwear in the Western Tarsier, *Tarsius bancanus*. *Int. J. Primatol.*, 15(1): 29–59.

Karasov, W.H. (1981). Daily energy expenditure and the cost of activity in a free-living mammal. *Oecologia*, 51: 253–259.

Kenagy, G.J., and Hoyt, D.F. (1990). Speed and time energy budget for locomotion in golden-mantled ground squirrels. *Ecology*, 70:1834–1839.

Lima, S.L., and Dill, L.M. (1990). Behavioral decisions made under the risk of predation: A review and prospectus. *Can. J. Zool.*, 68: 619–640.

MacKinnon, J., and MacKinnon K (1980). The behavior of wild spectral tarsiers. *Int. J. Primatol.*, 1: 361–379.

Nagy, K.A., and Milton, K. (1979). Energy metabolism and food consumption by wild howler monkeys. *Ecology*, 60: 475–480.

Napier, J.R., and Walker, A.C. (1967). Vertical clinging and leaping—A newly recognized category of locomotor behaviour of primates. *Folia Primatol.*, 6: 204–219.

Nekaris, K.A.I. (2005). Foraging behaviour of the slender loris (*Loris lydekkerianus lydekkerianus*): Implications for theories of primate origins. *J. Hum. Evol.*, 49: 289–300.

Niemitz, C. (1979). Results of a study on the western tarsier (*Tarsius bancanus borneanus* HORSFIELD 1821) in Sarawak. *Sarawak Mus. J.*, 48: 171–228.

Niemitz, C. (1979) Outline of the behavior of *Tarsius bancanus*. In G.A. Doyle, R.D. Martin (Eds.), *The study of prosimian behavior*. New York: Academic Press. (pp. 631–660)

Niemitz, C. (1984a). Synecological relationships and feeding behaviour of the genus *Tarsius*. In C. Niemitz (Ed.), *Biology of tarsiers* (pp. 59–75). Stuttgart: Gustav Fischer.

Niemitz, C. (1984b). Activity rhythms and use of space in semi-wild Bornean tarsiers, with remarks on wild spectral tarsiers. In C. Niemitz (Ed.) *Biology of tarsiers* (pp. 85–115). Stuttgart: Gustav Fischer.

Norberg, U.M. (1985). Flying, gliding and soaring. In M. Hildebrand, D.M. Bramble, K. Liem, and D.B. Wake (Eds.) *Functional vertebrate morphology* (pp. 129–158). Cambridge, MA: Belknap.

Norton, F.G.J. (1987). *Advanced mathematics*. London: Pan Books.

Oxnard, C.E., Crompton, R.H., Liebermann, S.S. (1989). *Animal lifestyles and anatomies: The case of the prosimian primates*. Seattle: Washington. Univ. Press. Patel, E.R. (2005). Silky sifaka predation (*Propithecus candidus*) by a fossa (*Cryptoprocta ferox*). *Lemur News*, 10: 25–27.

Peters, A., and Preuschoft, H. (1984). External biomechanics of leaping in *Tarsius* and its morphological and kinematic consequences. In C. Niemitz C. (Ed.), *Biology of tarsiers* (pp. 227–255). Stuttgart: Gustav Fischer.

Preuschoft, H., Günther M.M., and Christian, A. (1998). Size dependence in prosimian locomotion and its implications for the distribution of body mass. *Folia primatol.*, 69: 60–81.

Russell, R. (1977). The behaviour, ecology and environmental physiology of a nocturnal primate, *Lepilemur mustelinus*. Unpublished doctoral thesis. Duke Univ.

Schmid, J., and Ganzhorn, J. (1996). Resting metabolic rates of *Lepilemur mustelinus ruficaudatus*. *Am. J. Primatol.*, 38: 169–174.

Schmidt-Neilsen, K. (1990). *Animal physiology: Adaptation and environment* (ed. 4). Cambridge: Cambridge Univ. Press.

Sellers, W.I., Cain, G.M., Wang, W.J., and Crompton, R.H. (2005). Stride lengths, speed and energy costs in walking of *Australopithecus afarensis*: Using evolutionary robotics to predict locomotion of early human ancestors. *J. Roy. Soc. Interface*, 2: 431–442.

Stanford, C.B. (2002). Avoiding predators: Expectations and evidence in primate antipredator behavior. *Int. J. Primatol.*, 23: 741–757.

Sussman R. (1999). *Primate ecology and social structure, vol 1: Lemurs, lorises and tarsiers.* New York: Pearson Custom Publishing.

Terborgh, J., and Janson, C.H. (1986). The socioecology of primate groups. *Annu. Rev. Ecol. Syst.,* 17: 111–135.

Walker, A. (1969). The locomotion of the lorises with special reference to the potto. *E. Afr. Wildl. J.,* 7: 1–5.

Walker, A. (1979). Prosimian locomotor behaviour. In G.A. Doyle, R.D. Martin (Eds.), *The study of prosimian behaviour* (pp. 543–566). London: Academic Press.

Walton, M., and Anderson, B.D. (1988). The aerobic cost of saltatory locomotion in the Fowler's toad (*Bufo woodhousei fowleri*). *J. Exp. Biol.,* 136: 273–288.

Warren, R.D., and Crompton, R.H. (1997). Locomotor ecology of *Lepilemur edwardsi* and *Avahi occidentalis. Am. J. Phys. Anthrop.,* 104: 471–486.

Warren, R.D., and Crompton, R.H. (1998). Diet, body size and the energy costs of locomotion in saltatory primates. *Folia primatol.,* 69: 86–100.

Wright, P.C., Heckscher, S.K., and Dunham, A.E. (1997). Predation on Milne-Edwards' sifaka (*Propithecus diadema edwardsi*) by the fossa (*Cryptoprocta ferox*) in the rain forest of southeastern Madagascar. *Folia Primatol.,* 68: 34–43.

Appendix 1. Leaping Mechanics

A leaping prosimian can be considered a projectile and the mechanics of projectiles are well understood. The basic equations can be found in most mathematics textbooks and a worked derivation can be found in, for example, Norton (1987).

In the general case (as illustrated in Figure 4) an animal leaps a distance R (measured in meters) at an angle α to the horizontal (α is positive for an upward leap and negative for a downward leap). This angle will be referred to as the *slope* of the leap. The actual horizontal distance is $R \cos \alpha$ and the vertical height change is $R \sin \alpha$. The animal achieves this leap by taking off at a velocity of U m/s at an angle φ to the horizontal. g is acceleration due to gravity: 9.81 m/s. The flight time for a given leap can be calculated using equation (6.1) and examples are shown in Figure 6.2

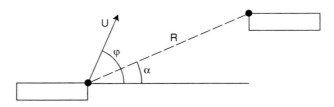

FIGURE 4. Diagram illustrating an animal leaping between supports at different heights from the ground

$$t = -\frac{2U\,\mathrm{Sec}[\alpha]\mathrm{Sin}[\alpha - \phi]}{g} \tag{6.1}$$

Similarly the range for a given leap can be calculated using equation (6.2) and examples are shown in Figure 6.3. For any given combination of support heights (slope) there is a take-off angle that will maximize travel distance (and hence minimize energetic cost of travel). This can be calculated directly using equation (6.3). The range for this maximally efficient leap can be calculated from equation (6.4).

$$R = -\frac{2U^2\mathrm{Cos}[\phi]\mathrm{Sec}[\phi]^2\mathrm{Sin}[\alpha - \phi]}{g} \tag{6.2}$$

$$\phi_{\mathrm{eff}} = \frac{90 + \alpha}{2} \tag{6.3}$$

$$R_{\mathrm{max}} = -\frac{U^2\mathrm{Sec}[\alpha]^2(-1 + \mathrm{Sin}[\alpha])}{g} \tag{6.4}$$

The mechanical energy cost of a general leap can be calculated using equation (6.5) and examples are shown in Figure 6.1. Variable m is mass of the animal in kg.

$$KE = -\frac{1}{4}gm\,R\,\mathrm{Cos}[\alpha]^2\mathrm{Csc}[\alpha - \phi]\mathrm{Sec}[\phi] \tag{6.5}$$

7
Anti-Predator Strategies of Cathemeral Primates: Dealing with Predators of the Day and the Night

Ian C. Colquhoun

Introduction

The entire evolutionary history of the order Primates has occurred in ecological contexts where all primates, like all other animals (Vermeij, 1987; Endler, 1991, p. 176), are, at least, at risk of predation at some point in their lives (Hart & Sussman, 2005). These predator-prey ecological relationships can be conceived of as interspecific, asymmetric "attack-defense" arms races that give rise to diffuse coevolutionary effects (Dawkins & Krebs, 1979; Janzen, 1980). Predators and their prey exhibit asymmetric interactions because the selective pressure of predators on prey species is stronger than the selective pressure of prey species on their predators. The asymmetric nature of these relationships has been termed the "life-dinner principle" (Dawkins & Krebs, 1979): Failure on a predator's part means it has lost a meal, but failure on the prey's part dramatically increases its likelihood of being the meal (e.g., Terborgh, 1983; Vermeij, 1987; Lima & Dill, 1990; Endler, 1991, p. 176; Stanford, 2002).

Despite terminology in the literature such as "act of predation," or "predation event," predation is more accurately regarded as a sequence, or process, involving several stages, or phases, on the part of the predator: search/encounter/detection; identification/approach/pursuit; and, subjugation/consumption (Curio, 1976, p. 98; Taylor, 1984; Vermeij, 1987; Endler, 1991, p. 176). Because predation risk for any prey increases as the predation sequence proceeds from one stage to the next, and because many prey species are subject to predation by more than one predatory species, selection should be greatest for prey defenses that result in early detection of predators. While most predators hunt more than one prey type, predators will also often prey preferentially on the most common prey type(s), a form of frequency-dependent selection known as *apostatic selection* (Clarke, 1962; Maynard Smith, 1970; Curio, 1976, p. 98; Endler, 1991, p. 176). Thus, predators exert stronger selection pressure on individual prey species than any individual prey species can exert on its predators. The "life-dinner principle" and the asymmetric nature of predator-prey arms races mean that, in general, the prey tend to have the advantage in those particular arms races (Dawkins & Krebs, 1979).

Regardless of the advantage prey species might generally have in their arms races with predators, it is not immediately clear how those primate species that have adopted cathemerality deal with arms races with both diurnal or nocturnal predators (Kappeler & Erkert, 2003; Hill, 2006). Previously, I reviewed the effects of predators on the activity patterns of cathemeral lemurids and ceboids (Colquhoun, 2006). The present paper is the obverse of that earlier paper; here, I review the anti-predator behaviors of cathemeral lemurids and ceboids. Cathemeral primates hold some anti-predator strategies in common with diurnal primates (e.g., alarm calling and mobbing). But, in addition to discussing how cathemeral primates use their cathemerality as an anti-predator *strategy of temporal crypticity* (e.g., Wright, 1989, 1994, 1995, 1998, 1999; Wright et al., 1997; Donati et al., 1999; Curtis & Rasmussen, 2002; Colquhoun, 2006; Hill, 2006), I will show how cathemeral primates also make differential use of other behavioral and morphological anti-predator strategies (i.e., behavioral crypticity, social groups, "escape in size", and polymorphism) (Clarke, 1962; Terborgh, 1983; Endler, 1991, p. 176). While cathemeral primates are in arms races with the same types of major predators in both Madagascar and the Neotropics (i.e., carnivores, birds of prey, and constricting snakes), there is a non-convergence of the faunal communities in these two biogeographic regions (Terborgh & van Schaik, 1987; Kappeler & Ganzhorn, 1993; van Schaik & Kappeler, 1993, 1996; Kappeler & Heymann, 1996; Peres & Janson, 1999; Ganzhorn et al., 1999; Kappeler, 1999a, b; Hart, 2000; Colquhoun, 2006). Consequently, the anti-predator strategies of cathemeral lemurids exhibit different emphases from the anti-predator strategies emphasized by cathemeral ceboids. In particular, I propose that the sexual dichromatism that characterizes all species and subspecies of the lemurid genus *Eulemur* may represent a polymorphic anti-predator adaptation to apostatic selection. Thus, this paper will consider both proximate and ultimate anti-predator strategies of cathemeral primates.

Primate Cathemerality and Predation

The Taxonomic Distribution of Primate Cathemerality

Tattersall (1987, 2006) introduced the term "cathemeral" to describe the activity patterns of organisms in which equal or significant amounts of feeding and/or traveling occur "through the day"— that is, through the 24-hour cycle. Over the last 30 years, cathemeral activity has been reported in all species of the lemurid genus *Eulemur* (Colquhoun, 1993, 1997, 1998a; Overdorff & Rasmussen, 1995; Curtis, 1997; Donati et al., 1999; Wright, 1999; Curtis & Rasmussen, 2002; Kappeler & Erkert, 2003; Overdorff & Johnson, 2003; see, also, Table 7.1).

Considerable inter-species variability has also been noted in the cathemerality observed across the genus *Eulemur* (see Table 7.2). Cathemerality has also been reported or suggested for at least some populations of the other lemurid

TABLE 7.1. Taxa of the lemurid genus *Eulemur* for which cathemerality has been reported and the sites where that cathemeral activity was observed.

Lemur Taxon	Common Name	Site(s)	Reference(s)
Eulemur coronatus	crowned lemur	Ankarana; Mt. d'Ambre NP	Wilson et al., 1989; Freed,1996a, 1999
E. fulvus fulvus	brown lemur	Ampijoroa	Rasmussen, 1998a
E. (f.) albifrons	white-fronted brown lemur	Andranobe watershed	Vasey, 1997, 2000
E. f. mayottensis	Mayotte brown lemur	Mavingoni; Mayotte	Tattersall, 1977, 1979; Tarnaud, 2004
E. f. rufus	red-fronted, or rufous, brown lemur	Ranomafana NP; Kirindy Forest; Antserananomby, Tongobato;	Overdorff, 1996; Donati et al., 1999; Kappeler & Erkert, 2003 Sussman, 1972
E. f. sanfordi	Sanford's brown lemur	Ankarana; Mt. d'Ambre NP	Wilson et al., 1989; Freed, 1996a, 1999
E. (f.) albocollaris	white-collared brown lemur	Andringitra NP	Johnson, 2002
E. macaco macaco	black lemur	Ambato Massif; Lokobe	Colquhoun, 1993, 1997, 1998a; Andrews & Birkinshaw, 1998
E. mongoz	mongoose lemur	Ampijoroa; Anjouan; Moheli; Anjamena	Tattersall & Sussman, 1975; Sussman & Tattersall, 1976; Rasmussen, 1998b; Tattersall, 1976; Curtis & Zaramody, 1999; Curtis et al., 1999
E. rubriventer	red-bellied lemur	Ranomafana NP	Overdorff, 1988, 1996

genera besides *Eulemur* (see Table 7.3). Comparative assessment of lemur activity cycle data led Rasmussen (1999) and Curtis & Rasmussen (2002), to recognize three modes of cathemerality. Their "Pattern A" refers to the seasonal shifting from diurnal to nocturnal activity that has only been described in *Eulemur mongoz*. "Pattern B" involves a seasonal shift from diurnal activity to cathemerality, a pattern that has only been described in *E. fulvus fulvus*. "Pattern C" is the year-round cathemerality that has been described in most of the *Eulemur* taxa, as well as the Lac Alaotra gentle lemur (*Hapalemur griseus alaotrensis*) (Mutschler et al., 1998; Mutschler, 2002), and the greater bamboo lemur (*H. simus*) (Tan, 2000; Grassi, 2001).

Comparative data raise questions concerning the activity patterns of the lesser bamboo lemur (*H. griseus*). In eastern Madagascar populations of the gray bamboo lemur (*H. griseus*) exhibiting either diurnality (Overdorff et al., 1997; Tan, 2000; Grassi, 2001) or largely nocturnal activity (Vasey, 1997, 2000) have been reported. At Ambato Massif, the western bamboo lemur (*H. g. occidentalis*) was observed to be diurnal (Colquhoun, 1993, 1998b). Such intraspecific variability in activity cycle is reminiscent of differing activity patterns reported for different populations of *Eulemur mongoz* (e.g., Tattersall, 1976) and the owl monkey

TABLE 7.2. Inter-species variability in cathemeral activity across the lemurid genus *Eulemur*.

Lemur Taxon	Cathemerality (Year-Round or Seasonal)	Nocturnal Activity Independent of Lunar Cycle?	References
Eulemur coronatus	Year-round	Yes	Wilson et al., 1989; Freed, 1996a, 1999
E. fulvus fulvus	Seasonal	No	Harrington, 1975; Rasmussen, 1998a
E. (f.) albifrons	Year-round;	No data	Vasey, 1997, 2000
E. f. mayottensis	Year-round	Yes	Tattersall, 1977, 1979; Tarnaud, 2004
E. f. rufus	Year-round;	No	Overdorff, 1996; Donati et al., 1999; Kappeler & Erkert, 2003
E. f. sanfordi	Year-round	No data	Freed, 1996a, 1999
E. (f). albocollaris	Year-round	No data	Johnson, 2002
E. macaco macaco	Year-round	No	Colquhoun, 1993, 1997, 1998a; Andrews & Birkinshaw, 1998
E. mongoz	Year-round;	Yes	Tattersall, 1976; Curtis, 1997; Curtis et al., 1999;
	seasonal		Curtis et al., 1999; Rasmussen, 1998a,b
E. rubriventer	Year-round	Yes	Overdorff, 1988, 1996

(*Aotus*, see below). Further comparative data are needed to clarify our understanding of *H. griseus* activity patterns (Tan, 2000; Mutschler & Tan, 2003). The remaining genera in the family Lemuridae, the ruffed lemurs (*Varecia*) and the ring-tailed lemur (*Lemur catta*), are usually considered to be strictly diurnal (Tattersall, 1982; Vasey, 2000, 2003; Jolly, 2003). But, there is a single report from Ranomafana National Park that the black and white ruffed lemur (*Varecia variegata*) exhibits cathemerality (Wright, 1999). Similarly, there is also a recent lone report of *L. catta* in Berenty Reserve being cathemeral (Traina, 2001). The extent to which these latter two taxa are indeed cathemeral deserves further attention in the field. In light of Wright's (1999) report, I include *V. variegata* in this review so as to provide as complete a reflection of the literature on primate cathemerality as possible; *L. catta* is the focus of another chapter (Gould & Sauther, this volume), and will not be considered further here.

    In addition to most taxa in the family Lemuridae being cathemeral, cathemerality has also been reported in at least two genera of platyrrhine monkeys (see Table 7.3). The genus *Aotus* (variously known as the owl monkey, night monkey, or douroucouli) is usually noted for being the only nocturnal platyrrhine and the only nocturnal anthropoid (e.g., Moynihan, 1964, 1976; Thorington et al., 1976; Wright, 1978; Garcia & Braza, 1987; Kinzey, 1997). Yet, *Aotus* is also a cathemeral taxon because some populations exhibit diurnal activity (Rathbun & Gache, 1980). *Aotus azarai* in the Paraguayan Chaco has been reported as cathemeral (Wright, 1985, 1989, 1994). *Aotus azarai* in the eastern Argentinean Chaco has also been observed to be cathemeral (Fernandez-Duque & Bravo, 1997; Fernandez-Duque et al., 2001, 2002; Fernandez-Duque, 2003). Finally, the howler monkeys (*Alouatta*) are usually considered strictly diurnal (e.g., Kinzey, 1997). But, there are some data hinting that cathemerality may occasionally occur in the mantled howler monkey (*Alouatta palliata*) in Costa Rica (Glander, 1975) and the

TABLE 7.3. Other primate taxa besides *Eulemur* for which cathemerality has been reported or suggested.

Taxon	Common Name	Site(s)	Nature of Report and References
Hapalemur simus	greater bamboo lemur	Ranomafana NP	"...*H. simus* is cathemeral" "(Tan, 2000 p. iv)"
Hapalemur griseus alaotrensis	Lac Alaotra gentle lemur	Lac Alaotra	"...flexible 24-hour activity cycle... *H. g. alaoternsis* is cathemeral." (Mutschler et al., 1998, p. 329) "Night activity is substantial." (Mutschler, 2002, p.102)
H. griseus	lesser bamboo lemur	Andranobe	"*Hapalemur griseus* is largely nocturnal at Andranobe." (Vasey, 2000, p. 426)
		Ranomafana NP	"*H. griseus* are diurnal." (Overdorff et al., 1997, p. 217) "*H. griseus* and *H. aureus* are diurnal." (Tan, 2000) "*H. griseus* is strictly diurnal." (Grassi, 2001, p. 189)
Varecia variegata	black and white ruffed lemur	Ranomafana NP	"...*Varecia v. variegata* and *Hapalemur griseus alaotrensis* exhibit cathemeral behavior." (Wright, 1999, p. 45)
Lemur catta	ring-tailed lemur	Berenty Reserve	"...link between key sites in the home range of ringtailed lemurs and their day and night activity will be examined." (Traina, 2001, p. 188)
Aotus azarai	red-necked owl monkey	Paraguayan and Argentinean Chaco regions	See text
Alouatta palliata	mantled howler monkey	Hacienda La Pacifica, Guanacaste Prov., Costa Rica	"...traveling through the trees on several nights around midnight, and often began feeding well before dawn." (Glander, 1975, p. 41)
A. pigra	black howler monkey	Cayo District, Belize	"The group was found to become active and feed between three and five hours before sunrise" (Dahl & Hemingway 1988, p. 201)

black howler monkey (*A. pigra*) in Belize (Dahl & Hemingway, 1988). Because these reports are consistent with Tattersall's (1987) definition of cathemerality, and because other researchers (e.g., Kinzey, 1997) have cited these reports as suggesting howler monkey cathemerality, *A. palliata* and *A. pigra* are also included in this review. As with *Varecia* and *Lemur*, however, the aim of future fieldwork on *A. palliata* and *A. pigra* should be to seek to clarify the degree of cathemeral activity in these taxa.

Explaining Primate Cathemerality: Predation and Other Factors

In recent years, several authors have drawn attention to the fact that primate cathemerality is not a unitary phenomenon (Rasmussen, 1999; Curtis & Rasmussen, 2002; Mutschler, 2002; Overdorff & Johnson, 2003; Colquhoun, 2006; Tattersall, 2006). While considerations of primate cathemerality have often stressed single-cause explanations, it is apparent that a full understanding of primate cathemerality will show that both proximate and ultimate factors (Tinbergen, 1963) are involved in the variable and flexible activity patterns of cathemeral primates. Although proximate selective factors may be identified, it complicates matters that these factors do not necessarily provide an explanation for, nor give a reflection of, the ultimate factors that gave rise to cathemerality (Endler, 1986).

The range of possible factors that may contribute to a thorough explanatory model of primate cathemerality includes it being: an ancestral condition; a response to seasonality or availability of food resources; a mechanism for (or result of) reduction of interspecific competition; a response to precipitation, the lunar cycle, and/or ambient temperature, and an anti-predator strategy. Tattersall (1982) proposed that cathemerality represents the ancestral lemurid activity cycle. This is the view that has been taken in several reports of lemurid cathemerality (Overdorff & Rasmussen, 1995; Colquhoun, 1998a; Rasmussen, 1998a; Curtis & Rasmussen, 2002). A similar view was taken by Dahl & Hemingway (1988) in their preliminary report of cathemerality in *A. pigra*; they interpreted it as an activity pattern that provided a good degree of adaptability and considered it a characteristic that traced back to the earliest anthropoids. Engqvist & Richard (1991) suggested that cathemerality was a seasonal response to changes in food availability and/or quality. Some field data provide support for this interpretation (e.g., Overdorff, 1996). However, other field studies have produced results that are inconsistent with this ecologically-based explanation (e.g., Overdorff & Rasmussen, 1995; Colquhoun, 1998a; Kappeler & Erkert, 2003). But, as noted above, an absence of proximate evidence does not rule out seasonality of food resources as a possible ultimate cause of cathemerality. The recent extinctions of the diurnal "giant" lemurs were implicated by van Schaik & Kappeler (1993, 1996) as events that precipitated an "evolutionary disequilibrium." By this model the cathemeral lemur species we observe today are in a transitional stage between nocturnality and diurnality, as they occupy econiches that opened up with the extinctions of the "giant" lemurs. This explanation is also problematic, however, as the retinal and optic foramina morphologies of lemurid species are not consistent with having been nocturnal until only 1,500–500 years ago (e.g., Martin, 1990; Colquhoun, 1998a; Kay & Kirk, 2000; Mutschler, 2002).

At present, the functional explanations most often invoked for the evolution of primate cathemerality focus on it being either a thermoregulatory strategy or a predator avoidance strategy. Thermoregulatory stress has been cited as a likely cause for cathemerality in several species: *Hapalemur griseus alaotrensis*

(Mutschler et al., 1997; Mutschler, 2002); *Eulemur mongoz* (Curtis et al., 1999); and *Aotus azarai* (Fernandez-Duque et al., 2002; Fernandez-Duque, 2003). Data correlating ambient temperature and activity patterns provide strong support for cathemerality in these species having a thermoregulatory basis. The predator avoidance function of cathemerality—a kind of concealment in time, or temporal crypticity—has been cited in regards to the activity cycles of many lemurids (Wright, 1995, 1998, 1999; Wright et al., 1997; Curtis & Rasmussen, 2002; Colquhoun, 2006), including: *Eulemur rubriventer* (Overdorff, 1988); *E. mongoz* (Curtis, 1997; Rasmussen, 1998b; Curtis et al., 1999), and *E. fulvus rufus* (Donati et al., 1999). Colquhoun (2006) proposed that *Eulemur* cathemerality might be an evolutionary stable strategy (ESS) to predation pressure from the fossa (*Cryptoprocta ferox*), a viverrid carnivoran that is also cathemeral and appears to be a lemur-hunting specialist (see also, Hart, 2000; Hart & Sussman, 2005; Hill, 2006). A release from the threat of diurnal raptor predation, together with the nocturnal threat posed by the great horned owl (*Bubo virginianus*), has been suggested as the reason for diurnal activity by *Aotus azarai* in the Paraguayan Chaco (Wright, 1985, 1989, 1994). While the literature on primate cathemerality has seen this consistent implication of predation as a causal factor, it seems an odd incongruity that over the last 20 years several major reviews of predation on primate species made no mention of primate cathemerality being a possible adaptive response to predation (Anderson, 1986; Cheney & Wrangham, 1987; Goodman et al., 1993; Isbell, 1994; Stanford, 2002; Goodman, 2003; Hart & Sussman, 2005).

Other Anti-Predator Strategies of Cathemeral Primates

Behavioral crypticity

Cathemeral primates tend to be relatively small-bodied (under about 2.5 kg), placing them at some risk of predation (Overdorff & Johnson, 2003). The largest primate species for which there are indications of cathemerality (but see above concerning uncertainties about that cathemerality) are the black and white ruffed lemur (*Varecia variegata*, at about 3.0 kg (Tattersall, 1982), and the howler monkeys, *Alouatta palliata* and *A. pigra*, averaging 6.4 kg (Kinzey, 1997). In some cathemeral primate species, the increased risk of predation that comes with small body size translates into cryptic behavior. Several cathemeral primate species also occur in small groups or family groups (Freed, 1999): *H. g. alaotrensis* (Mutschler et al., 1997; Mutschler, 2002); *E. rubriventer* (Overdorff, 1988, 1996); *E. mongoz* (Curtis, 1997; Curtis et al., 1999); and *Aotus azarai* (Rathbun & Gache, 1980; Wright, 1985, 1989, 1994; Stallings et al., 1989). By virtue of their size the small social groups of these species are less likely to be detected by predators. These species are also relatively cryptic when active. For example, at Ranomafana, Overdorff (1996) found that *E. rubriventer* (mean group size = 3 individuals), rested more, traveled less, and were less active at night than the sympatric rufous lemur (*E. f. rufus*), with their larger social groups (mean group size = 8 individuals). Grassi (2001) regards *H. griseus* at Ranomafana as an understory specialist whose ecology has been shaped by predator avoidance strategies. The resting sites used by *H. griseus* tend to be in dense vegetation tangles at a height of about 7 m (which Grassi interprets as an anti-raptor strategy), and sleeping sites are found above a height 15 m in large trees (which Grassi interprets as an anti-carnivore strategy). Grassi also reports that *H. griseus* has distinct alarm calls for snakes (Rakotodravony et al., 1998) and birds of prey; the response shown by group members to the latter call is to become quiet (an example of what Curio (1976, p. 98) terms "adaptive silence"), drop in height in the understory, and stay still (Grassi, 2001). Grassi concludes that *H. griseus* can distinguish between different types of predators and exhibit appropriate predator-specific behaviors.

Similarly, although *Aotus* is a powerful leaper and can move rapidly through the trees (Moynihan, 1964; see also Wright's (1984) observation of an *Aotus* male with an infant on its back narrowly escaping a pursuing, and possibly predatory, male *Cebus* monkey), *Aotus* lives in small family groups and utilizes cryptic sleeping sites in Peru, where they are nocturnal. In Paraguay, diurnal owl monkeys often sleep at night on open branches (Wright, 1985). Wright (1985, 1989, 1994) has interpreted the diurnal activity of *Aotus azarai* in the Paraguayan Chaco as the result of predation release from diurnal raptors, combined with the presence of the great horned owl (*B. virginianus*) as a nocturnal predation threat. However, Wright (1985) notes that while *B. virginianus* is large enough to prey on *Aotus*, it does not specialize on feeding on arboreal prey, unlike the harpy eagle (*Harpia harpyja*) and the Guiana crested eagle (*Morphnus guianensis*); rather, the great horned owl often catches its prey on the ground. Rathbun & Gache (1980, p. 213) describe what they term the "*Aotus* distress vocalization"—a "whoop, whoop, whoop"—but they provide no further information about this call, so the contexts of its use are unknown.

Other cathemeral primate species are not particularly cryptic, despite their relatively small body sizes. For example, when the black lemur (*E. macaco macaco*) is active at night, group progressions are quite noisy and groups often engage in nocturnal loud calling typical of inter-group encounters (Colquhoun, 1998a); the same is true of many nocturnal lemur species, which exhibit noisy behaviors and are highly vocal (e.g., Schulke, 2001). Black lemur resting sites, however, can be cryptic (e.g., dense liana tangles), especially in the dry season when activity levels are dramatically lower than in the wet season (Colquhoun, 1993, 1998a).

Social Groups—Safety in Numbers

Larger primate social groups are known to be better able to detect potential predators (e.g., Terborgh, 1983; van Schaik et al. 1983; Landeau & Terborgh, 1986; Hauser & Wrangham, 1990; Peres, 1993; Sauther, 2002; Hart & Sussman, 2005). Freed (1999) has presented comparative data on average group sizes in lemurid species which show that those species in the genus *Eulemur* that do not form family groups (i.e., *E. coronatus*, *E. fulvus*, and *E. macaco*) all have multi-male, multi-female social groups with group sizes that tend to range between 6–11 individuals. In two of these taxa, *E. macaco* (Colquhoun, 1998a) and *E. f. rufus* (Kappeler & Erkert, 2003), year-round cathemeral activity has been linked to the lunar cycle, with more nocturnal activity occurring on nights with bright moonlight.

Predator alarm calls are well developed in the *Eulemur* species that exhibit multi-male, multi-female social groups (see Table 7.4). In *E. coronatus*, Wilson et al., (1989) reported that detection of the fossa (*Cryptoprocta ferox*) was met with lemurs staring in the direction of the fossa and giving "grunt-shriek" alarm vocalizations; response to the fossa and the "grunt-shriek" vocalizations was to flee upwards. While Freed (1996a) reported often seeing fossas at night, he did not observe what response(s) fossas elicited in either *E. coronatus* or *E. f. sanfordi*. Freed does note, however, that while both lemurs would grunt occasionally on sighting the smaller Malagasy ring-tailed mongoose (*Galidia elegans*), neither gave distinct vocalizations or directed particular behaviors towards *Galidia*. Wilson et al. (1989) report that *E. coronatus* typically responded with evasive behavior on sighting large raptors, fleeing downwards rather than giving alarm vocalizations. Freed (1996a), however, describes both *E. coronatus* and *E. f. sanfordi* as giving loud and distinct vocalizations when either the Madagascar harrier hawk (*Polyboroides radiatus*), or the Madagascar buzzard (*Buteo brachypterus*) were sighted. But, Freed does not give further description of these distinctive vocalizations.

TABLE 7.4. Predator alarm responses of "safety in numbers" primate taxa that are, or may be, cathemeral.

Taxon	Terrestrial Predator alarm	Aerial Predator Alarm	Mobbing	References
Eulemur coronatus	grunt-shriek vocalization	no; loud and distinctive vocalizations	?	Wilson et al., 1989; Freed, 1996a
E. fulvus rufus	generalized alarm call	yes, directed at Madagascar harrier hawk	yes	Fichtel & Kappeler 2002; Fichtel & Hammerschmidt, 2002; Sussman, 1975, 1977; Karpanty & Grella, 2001
E. macaco macaco	generalized huff or hack alarm call	scream-whistle vocalization in response to Madagascar harrier hawk	yes, in response to large boa constrictors and harrier hawks	Colquhoun, 1993, 2001
Varecia variegata	anti-carnivore call	no; yes, but not to all large raptors	?	Macedonia, 1990; Karpanty & Grella, 2001
Alouatta palliata	generalized roars, woofs and barks	generalized roars, woofs and barks	yes	Baldwin & Baldwin, 1976

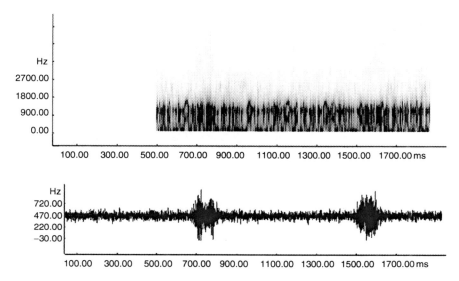

FIGURE 7.1. Spectrogram and waveform of black lemur "huff/hack" alarm vocalizations. The spectrogram of two closely-spaced calls is of poor quality due to load insect buzzing on the recording; regardless, two atonal pulses can discerned at around 700 ms and 1600 ms

Sussman (1975, 1977) reported that *Eulemur fulvus rufus*, *Lemur catta*, and *Propithecus verreauxi* would all move into very dense foliage and each give particular loud calls if a Madagascar harrier hawk was sighted overhead. All three species performed their calls in unison, and Sussman neither recorded the calls being given towards any other species of bird, nor towards the Madagascar fruit bat (*Pteropus rufus*—cf. *E. macaco*, below). Fichtel & Kappeler (2002) and Fichtel & Hammerschmidt (2002) report that at Kirindy Forest (western Madagascar), *E. f. rufus* exhibits a mixed alarm call system. Terrestrial predators (such as the fossa) are met with a generalized alarm call, while aerial predators (specifically, the Madagascar harrier hawk) elicit a specific alarm call. Fichtel & Kappeler (2002) noted that the ultimate cause for this predator alarm call variability is unclear, but they suggest that it might be explained by the so-called evolutionary disequilibrium hypothesis (van Schaik & Kappeler, 1993, 1996). But, in playback experiments at Ranomafana (southeastern Madagascar), Karpanty & Grella (2001) found that calls of the Madagascar serpent eagle (*Eutriorchis astur*), Henst's goshawk (*Accipiter henstii*), and the Madagascar harrier hawk, all elicited general predator alarm calls, dropping in the canopy, and fleeing from the source of the sound. However, none of the raptor calls was responded to by *E. f. rufus* with specific aerial predator alarm calls. Karpanty & Grella (2001) also report that the responses of *E. rubriventer* to the raptor call playbacks were similar to those of *E. f. rufus*.

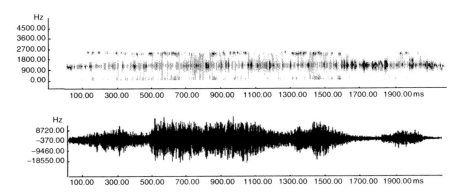

FIGURE 7.2. Spectrogram and waveform of black lemur "bark" vocalization

At Ambato Massif, Colquhoun (1993, 2001) documented a distinctive set of loud calls used by black lemurs (*E. macaco macaco*). Three distinct loud calls were noted. A short, sharp "huff" or "hack" vocalization (Figure 7.1) was heard fairly often in various situations where animals had been surprised or startled (e.g., by a falling tree branch, non-predatory birds suddenly taking flight, or the sudden appearance of humans and/or dogs). While local informants told of the fossa occurring at Ambato Massif, I never sighted a fossa. However, when a black lemur group sighted a domestic dog, they would give one or two "hack" vocalizations, and then move quickly and silently away from the dog. Most commonly heard was a generalized "loud call," or "bark" (Figure 7.2), that was given in many different situations, such as inter-group encounters, the sighting of small to midsize raptors, and the "mobbing" of large boa constrictors (*Acrantophis madagascariensis*) (Colquhoun, 1993). The most distinctive loud call, the "scream-whistle" (Figure 7.3), was noted in only two particular types of situations. During the day it was invariably given upon sighting the Madagascar harrier hawk, the largest raptor species commonly seen at Ambato. Harrier hawks often circled overhead, not far above treetop level, and this would set off scream-whistle vocalizations from the lemurs, followed by urgent evasive behaviors (e.g., diving several meters down into the crowns of trees). On one occasion, as I observed a group of juvenile black lemurs playing on the ground in an open patch of disturbed, low-stature forest, a harrier hawk soared overhead. The juveniles scattered on the scream-whistle vocalization that ensued, leaping up into nearby saplings; one juvenile female, however, found herself closer to a dried palm frond on the ground and dove underneath it, lying flat against the forest floor. On another occasion, I observed a black lemur group mobbing, from below the forest canopy, a pair of Madagascar harrier hawks that were copulating on an exposed dead branch in the canopy. At the same time, the copulating harrier hawks were also mobbed and dive-bombed by a pair of crested drongos (*Dicrurus forficatus*); indeed, it appeared that the mobbing vocalizations of the drongos attracted the attention of the black lemurs and set them mobbing the harrier hawks as well (albeit from a safe distance and

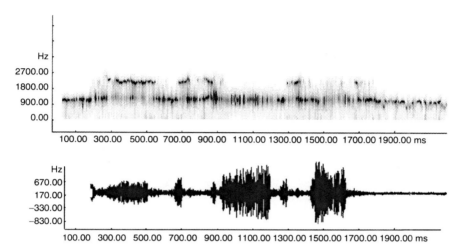

FIGURE 7.3. Spectrogram and waveform of black lemur "scream-whistle" vocalization

an unexposed position). But, I also recorded one occasion—on a brightly moonlit night—when a black lemur group was giving "scream-whistles" in response to swooping fruit bats (*Pteropus rufus*—a species with a similar wingspan to that of the Madagascar harrier hawk). "Scream-whistles" directed at fruit bats could be explained as due to young animals that had not yet learned to reliably identify harrier hawks. This explanation is problematic; however, as no generalized use of "scream-whistles" toward all large raptor species was ever heard. Rather than being a vocal signal solely symbolizing the harrier hawk, a more parsimonious explanation of the "scream-whistle" is that it carries ordinal information, signaling not just that something large has been sighted overhead but that something extremely large (and potentially dangerous) is overhead.

Black-and-white ruffed lemurs (*Varecia variegata*) have a variable, fission-fusion social community structure and organization (e.g., Freed, 1999; Vasey, 2005), but still give vocalizations in response to both raptors and carnivores (Table 7.4). Macedonia (1990) describes the vocal signal given by *Varecia* in response to sighting raptors as also occurring in other generalized, high-arousal contexts that don't involve predators. Thus, he suggests that this call does not represent either predator class or signal a situation requiring urgent response, but rather indicates an aggressive/defensive demeanor. By contrast, the anti-carnivore call of *Varecia* is interpreted by Macedonia as a high-urgency signal for the group to reaggregate. But, in the playback experiments of Karpanty & Grella (2001), only *Varecia*, along with the diurnal *Propithecus* (*diadema*) *edwardsi*, gave aerial alarm predator calls in response to the calls of the Madagascar serpent eagle and Henst's goshawk; interestingly, *Varecia* did not give aerial predator alarm calls in response to the Madagascar harrier hawk.

Both mantled howler monkeys (*Alouatta palliata*) and black howler monkeys (*A. pigra*) live in groups that exhibit variable social organization. Mantled howler monkeys are the smaller of the two species (females: 3.1–7.6 kg.; males: 4.5–9.8 kg; Ford & Davis, 1992); the multi-male, multi-female social groups vary widely in size, from 4–23 individuals (Kinzey, 1997). Black howler monkeys are larger (females: 6.4 kg; males: 11.35 kg; Ford & Davis, 1992), and tend to be found in smaller groups (4–10 individuals; Horwich & Lyon, 1990; Kinzey, 1997; Treves & Brandon, 2005); social organization may be either one male and multiple females, or one to four adult males and one to four adult females (Horwich & Lyon, 1990; Kinzey, 1997; Treves & Brandon, 2005).

An assortment of ecological data are consistent with the interpretation that *Alouatta palliata* seems to, at least in some contexts, use protection in numbers as an anti-predator strategy (see Table 7.4). Baldwin & Baldwin (1976) found that mantled howler monkeys gave "roar" vocalizations in response to "danger stimuli", such as large birds, dogs, a fallen infant, proximity to humans, and gunfire. In some areas, howler monkeys "roar" at humans, in others they don't, a function of whether or not humans are associated with danger. Where the stimuli are less intense or less dangerous, these same situations can also evoke "woofs" or "barks" from the monkeys. "Roars" do not necessarily function as an anti-predator strategy. While "roars" may interrupt the activities of some potential predators, or cause them to move off, "roars" can also have the reverse effect and attract potential predators (e.g., dogs and humans). Howler monkeys will "woof" when moving towards targets of group mobbing, but that they will "roar" when not moving towards the target; on this evidence, Baldwin & Baldwin (1976) suggest that "roars" may serve to signal avoidance or withdrawal responses, rather than an approach response. Terborgh (1983) noted that howler monkey groups spend much of their time resting on exposed perches in the forest canopy, making no efforts to be inconspicuous; he suggested that such behavior would seem to make it possible to detect predators at a distance. More recently, Gil-da-Costa et al. (2003) found that a safety in numbers response to "predator assessment" vocalizations by harpy eagles was critical in determining whether the harpy would attempt a predatory attack or not. When a mantled howler monkey group responded to the predator assessment calls of a harpy in a coordinated manner (i.e., vigilance by all group members, females collecting their infants and moving into dense foliage, males moving distally on canopy branches, often with alarm calls being given), the harpy either delayed its attack or moved off to seek other prey. If, however, a mantled howler monkey group was inattentive to the harpy's "predator assessment" calls, or reacted in an unorganized and chaotic manner, the harpy would either attack or move closer (see, also, Gil-da-Costa, this volume).

"Escape in Size"

As several researchers have pointed out, in the life history course predation risk is greater for young individuals since they are potential prey for a larger range of predators; as individuals grow and mature, they often become "protected" from predation by certain predators simply because they are too big to be captured (e.g., Sauther, 1989; Endler, 1991, p. 176; Csermely, 1996). Because of the relatively small body size of most cathemeral lemurids and *Aotus*, adults are still potential prey for multiple predators. With an adult body weight of about 3 kg, *Varecia* is the largest of the lemurs for which there are indications of cathemerality. Even so, it is not so large that it cannot be preyed upon by the fossa (e.g., Britt et al., 2001, 2003). The playback results reported by Karpanty & Grella (2001) indicate that *Varecia* also regard large raptors as potential threats. Certainly, given that the Madagascar harrier hawk can capture sifakas that are heavier than adult *Varecia* (Karpanty & Goodman, 1999; Brockman, 2003), adult *Varecia* may fall prey to harrier hawks on occasion. However, it may be that adult *Varecia* have "escaped in size" from the majority of Madagascar's extant raptors.

Asensio and Gomez-Marin (2002) observed a group of four adult tayras (*Eira barbara*, Mustelidae) display aggressive behavior towards a group of mantled howler monkeys; two adult female howler monkeys approached the tayras, causing the tayras to retreat. Asensio and Gomez-Marin (2002) also note that a successful predation of a primate by a tayra has never been observed; they, thus, conclude that unlike the jaguar (*Panthera onca*) and harpy eagle (Eason, 1989; Peres, 1990; Sherman, 1991; Peetz et al., 1992), the tayra is not a serious threat to the howler monkey (see also, Terborgh, 1983; cf. reports of observed unsuccessful attacks by *E. barbara* against, or anti-predator responses to *E. barbara* by, various callitrichid species: black-mantled tamarins (*Saguinus nigricollis*) (Izawa, 1978); buffy-headed marmosets (*Callithrix flaviceps*) (Ferrari & Lopes Ferrari, 1990); saddle-back and moustached tamarins (*S. fuscicollis* and *S. mystax*) (Peres, 1993); golden lion tamarins (*Leontopithecus rosalia*) (Stafford & Ferreira, 1995). In other words, it would seem that though the tayra appears to be a potential predator of small Neotropical primates (Colquhoun, 2006), *Alouatta* avoids threat from this particular predator on the basis of size.

Predator Confusion by Polymorphism

It is a notable fact that, although sexual dichromatism is rare across Order Primates, all taxa of the cathemeral genus *Eulemur* are, to one degree or another, sexually dichromatic (Tattersall, 1982; Mittermeier et al., 1994; Overdorff & Johnson, 2003). Clarke (1962) introduced the concept of apostatic polymorphisms, pointing out that such polymorphisms can be features that favor the prey in the arms races with their predators, especially if those predators employ a "search image" manner of hunting. If a particular predator species hunts a prey species in an apostatic manner (i.e., preying on the most commonly encountered form), a polymorphism in the preferred prey (i.e., dichromatism in the case of genus *Eulemur*), produces a selective advantage for those phenotypes that do not match the search image of their potential predators. Endler (1991, p. 176) subsequently expanded on the concept of apostatic selection (see also, e.g., Maynard Smith, 1970; Curio, 1976, p. 98), noting that polymorphisms in prey species produce what he termed "apparent rarity", reducing the predation risk per individual because the apostatic predator encounters two (or more) rarer prey forms as opposed to one common prey form. Further, the apparent rarity of the polymorphic prey may lead the apostatic predator to switch predation effort to an apparently more common monomorphic prey species. Finally, Endler (1991, p. 176) notes that a predator need not prey apostatically for polymorphism to be advantageous to a prey species. A predator may find itself confused by having to select among apparently different prey, particularly if the different morphs are seen at the same time by the predator (as would likely be the case when a *Eulemur* social group was encountered during the day by a raptor or a fossa). Functionally, this is the same effect achieved by mixed-species groups (see Landeau & Terborgh, 1986). Confusion or hesitation on the part of the predator could be the chance that the potential prey targets would need to elude the predator. Given that the relatively small-bodied *Eulemur* species are at risk of predation from both large raptors (Karpanty & Goodman, 1999; Karpanty & Grella, 2001; Karpanty, 2003; Colquhoun, 2006), and the cathemeral fossa (Hart, 2000; Goodman, 2003; Hart & Sussman, 2005; Colquhoun, 2006), if dichromatism lowers the probability of predation on individuals and/or increases the chances of eluding predatory attacks, it would certainly confer a selective advantage.

Discussion

In very broad terms, primate social organization has been characterized as enabling two general anti-predator strategies: (i) social groups that are relatively large and conspicuous, but that can detect, and even deter, potential predators; or (ii) social groups that are relatively small in size and primarily employ cryptic behaviors to avoid many potential predators (e.g., Cheney & Wrangham, 1987; Janson, 1998; Gautier et al., 1999). Among cathemeral primates, predator avoidance can be thought of as consisting of at least a two-track strategy that operates in parallel. One track centers on being cathemeral and exercising temporal crypticity (Donati et al., 1999; Colquhoun, 2006; Hill, 2006); the second track centers on the strong association between group size and whether a species practices behavioral crypticity or safety in numbers. Overall, family groups, or small groups, of cathemeral primates are more likely to be behaviorally cryptic. The association of relatively small body size (i.e., adult weight < 2.5 kg), small group size and crypticity (temporal, and in some cases behavioral) in some cathemeral lemurids accords well with the general pattern reported among small-bodied diurnal primates. For example, with average adult body weights ranging between 120–600 g (Ford & Davis, 1992; Kinzey, 1997), the New World callitrichids are the smallest anthropoids, and the cryptic nature of much of their behavior is well-documented (e.g., Izawa, 1978; Dawson, 1979; Sussman & Kinzey, 1984; Kinzey, 1997); their typical response to the threat of raptor predation is to rapidly seek the protective cover of thick vegetation, often dropping several meters in the forest canopy to do so, and then remain motionless as long as the raptor continues to be a threat (e.g., Ferrari & Lopes Ferrari, 1990; Heymann, 1990; Peres, 1993; Searcy & Caine, 2002).

This reaction to aerial predators is very similar to that described for *Hapalemur griseus* (Grassi, 2001).

But, while there may be a general association between small body size and cryptic behavior, particularly in response to raptors (e.g., Terborgh, 1983), other possibilities exist. That is, a range of anti-predator strategies across primate species in response to various predators is to be expected. For example, Terborgh (1983) notes that the squirrel monkey (*Saimiri sciureus*) employs a safety in numbers strategy, despite its relatively small body size. Similarly, Peres (1993) found that among Amazonian tamarins, the formation of large, stable mixed-species groups between *Saguinus fuscicollis* and one of its larger congeners (*S. mystax*, *S. imperator*, or *S. labiatus*) provided enhanced predator defense through safety in numbers for both species in the mixed-species groups, particularly for the smaller-bodied *S. fuscicollis* (see also Landeau & Terborgh, 1986). While at least some cathemeral lemurids also form mixed-species groups (i.e., crowned lemurs (*Eulemur coronatus*) and Sanford's brown lemurs (*E. fulvus sanfordi*) – Freed, 1996a, b), the relative predator defense benefits to each species are not clear at present. In some situations, even small-bodied primates will mob potential predators (e.g., marmosets mobbing scansorial carnivores: *Callithrix flaviceps* mobbing a tayra (Ferrari & Lopes Ferrari, 1990); *C. jacchus* mobbing a margay (*Felis wiedii*) (Passimani, 1995); or Coquerel's dwarf lemurs (*Mirza coquereli*) mobbing a boa constrictor (*Acrantophis madagascariensis*) (pers. obs.). Likewise, at Ambato Massif, family groups of *Hapalemur griseus occidentalis* traveling low in the forest occasionally encountered me as I observed one of my black lemur study groups; these chance encounters often resulted in the *Hapalemur* mobbing me (a potential terrestrial predator) at close range with staccato "ah-ah-ah-ah-ahhhhhh" vocalizations. Conversely, when faced with a formidable predator, even primates in relatively large social groups may opt for cryptic behavior (e.g., see the report by van Schaik & van Noordwijk (1989) of the responses of wild capuchin monkeys, *Cebus albifrons* and *C. apella*, to presentation of a harpy eagle (*Harpia harpyja*) model and harpy eagle vocal playbacks).

Cathemeral lemurid species that live in larger social groups are not necessarily behaviorally cryptic and seem to rely more on safety in numbers. Along with the strategy of safety in numbers in some cathemeral lemurids, well-developed predator alarm call systems also occur (Zuberbühler et al., 1999). These alarm call systems appear to be particularly fine-tuned to aerial predators; specific aerial predator calls in *Eulemur fulvus rufus* and *E. macaco* are associated with immediate evasive behaviors and the seeking of cover. By itself, however, group living may not be sufficient protection against attack from nocturnal (or cathemeral) predators (Peetz et al., 1992; Wright et al., 1997; Wright, 1998). Risk from nocturnal attack is especially high on moonless nights or nights with gusting winds, the noise of which provides an acoustic screen for a stalking predator (Terborgh, 1983; Bearder et al., 2002, 2006). Thus, living in multi-male, multi-female social groups combined with cathemerality may produce a heightened anti-predator strategy— that is, a two-track anti-predator strategy of safety in numbers coupled with predator avoidance through time (Donati et al., 1999; Colquhoun, 2006; Hill, 2006).

Apostatic selection is regarded as a commonly occurring predator-prey phenomenon (e.g., Hubbard et al., 1982; Endler, 1986; Gendron, 1987; Allen, 1988; but see, e.g., Sherratt & MacDougall, 1995, for conditions where anti-apostatic selection may occur). The selective advantage of polymorphism as an adaptation to (and even for) apostatic predation has been documented in studies of a wide range of non-primate taxa; e.g., "predation" of dichromatic bait by passerine birds (Allen & Clarke, 1968; Allen, 1972); predation of dimorphic bugs (*Sigara distincta*) by rudd (a fish, *Scardinius eryophthalmus*) (Elton & Greenwood, 1970); "predation" of dichromatic bait by Japanese quail (*Coturnix coturnix japonica*) (Cook & Miller, 1977); predation risk in normal and melanistic morphs of the adder (*Vipera berus*) (Andrén & Nilson, 1981); differential predation of the polymorphic aquatic isopod *Idotea baltica* by perch (*Perca fluviatilis*) (Jormalainen et al., 1995); "predation" of computer-generated polymorphic moth images by blue jays (*Cyanocitta cristata*) (Bond & Kamil, 2002); predation of mammal species by raptor species exhibiting plumage polymorphism (Roulin & Wink, 2004). The suggestion made here that the sexual dichromatism found across the cathemeral genus *Eulemur* is a polymorphic anti-predator strategy against apostatic predation is new, but it represents an extension of a well-established concept. While the concept of polymorphism as an adaptation to apostatic predation pressure has not heretofore been applied to nonhuman primates, there is no a priori reason why it could not be. Interestingly, in reviews of color polymorphisms in birds, Galeotti et al., (2002) and Galeotti & Rubolini (2004) found that the greatest expression of color polymorphism occurred in avian species that were active during both day and night (i.e., were cathemeral). These authors noted the selective importance of varying light levels affecting the detectability of the organisms and that this could be a key mechanism in maintaining color polymorphism.

Among primates, Bicca-Marques & Calegaro-Marques (1998) previously considered the evolution of sexual dichromatism in the black and gold howler monkey (*Alouatta caraya*). This species exhibits striking sexual dichromatism similar to that of *Eulemur macaco*: adult males are entirely black, while adult females are golden brown (see Rowe, 1996). Bicca-Marques & Calegaro-Marques found that despite the strong sexual dichromatism, as well as pronounced sexual dimorphism (adult males weigh 4.0–9.6 kg, adult females 3.8–5.4 kg) (Ford & Davis, 1992), there were no male-female differences in the thermoregulatory behavior of *A. caraya*. They thus concluded that sexual dichromatism in the black and gold howler monkey might better be explained as a result of sexual selection (their analysis did not include apostatic selection). Similarly, a thermoregulatory function would not seem to sufficiently explain the sexual dichromatism in *Eulemur* species. For example, the highly dichromatic black lemur (*E. macaco macaco*), showed no apparent sex differences in microhabitat preferences (Colquhoun, 1997).

Sexual selection was also the paradigm Cooper and Hosey (2003) employed to analyze sexual dichromatism in *Eulemur fulvus* subspecies, as well as *E. (f.) collaris* and *E. (f.) albocollaris*. Although their experimental results were consistent with the interpretation that sexual dichromatism in these *Eulemur* taxa was the evolutionary outcome of females exhibiting mating preference for brightly colored males, like Bicca-Marques & Calegaro-Marques (1998), Cooper & Hosey (2003) did not consider apostatic selection. Just as the evolution of cathemerality was likely influenced by multiple factors, (Tattersall, 2006) the same may be said of the evolution of sexual dichromatism. While it is worth noting that all *Eulemur* species exhibit strict seasonal breeding (e.g., Tattersall, 1982; Wright, 1999), both sexual selection and apostatic selection could drive the evolution of sexual dichromatism. Apostatic selection would favor sexual dichromatism outside the breeding season, and sexual selection could provide a breeding season advantage, thus enhancing the adaptive significance of this characteristic.

There are several aspects of the anti-predator behavior of cathemeral primates that deserve further research. Specifically, to test the possible impact of apostatic selection on the genus *Eulemur*, future field research on *Eulemur* species should pay attention to how the sexual dichromatism and the social organizations of these taxa may affect, or be affected by, ecological relationships with potential aerial and terrestrial predators (e.g., In any given lemur ecological community, does *Eulemur* represent the most numerous potential prey for these predators?). For *Eulemur* taxa to have possibly evolved sexual dichromatism as an adaptive anti-predator response to apostatic selection, one would have to predict that, in fact, *Eulemur* do represent a rather abundant potential prey pool for large raptors and the fossa. Detailed research on the predators of *Eulemur* taxa is also needed to augment the few available data and to try to establish how heavily different predatory species rely on these cathemeral lemurids (e.g., Do predators of *Eulemur* taxa prey on them in an apostatic fashion and, if so, which predators are responsible for any apostatic selective pressure?). On a more general level, additional field data on the activity patterns of *Varecia*, *Lemur catta*, *Alouatta palliata*, and *A. pigra* are needed to clarify the extent to which these taxa might be, or are, cathemeral. As is the case in general with studies of predation and predators, further data on the ecological interactions between the cathemeral primate species and their predators will allow us to better conceptualize and understand these predator-prey arms races and the processes involved therein.

Acknowledgments. My thanks to Sharon Gursky and Anna Nekaris for inviting me to contribute this paper to the volume. The careful reading of, and detailed comments on, an earlier version of this paper by two anonymous reviewers, and one not-so-anonymous reviewer, is greatly appreciated. Inclusion of the black lemur vocalization spectrographic figures would not have been possible without the assistance of my wife, Sonia Wolf, who made the multiple between-program graphic translations that were necessary. Thanks must also go to Katie Whitaker, my research assistant, for the eagerness with which she helped me track down reference sources.

References

Allen, J.A. (1972). Evidence for stabilizing and apostatic selection by wild blackbirds. *Nature*, 237: 348–349.

Allen, J.A. (1988). Frequency-dependent selection by predators. *Philosophical Transactions of the Royal Society of London, Series B, Biological Sciences*, 319: 485–503.

Allen, J.A., and Clarke, B. (1968). Evidence for apostatic selection by wild passerines. *Nature*, 220: 501–502.

Andrén, C., and Nilson, G. (1981). Reproductive success and risk of predation in normal and melanistic colour morphs of the adder, *Vipera berus. Biological Journal of the Linnean Society*, 15: 235–246.

Anderson, C.M. (1986). Predation and primate evolution. *Primates*, 27(1): 15–39.

Andrews, J.R., and Birkinshaw, C.R. (1998). A comparison between the daytime and night-time diet, activity and feeding height of the black lemur, *Eulemur macaco* (Primates: Lemuridae), in Lokobe forest, Madagascar. *Folia Primatologica*, 69(Suppl. 1): 175–182.

Asensio, N., and Gomez-Marin, F. (2002). Interspecific interaction and predator avoidance behavior in response to tayra (*Eira barbara*) by mantled howler monkeys (*Alouatta palliata*). *Primates*, 43(4): 339–341.

Baldwin, J.D., and Baldwin, J.L. (1976). Vocalizations of howler monkeys (*Alouatta palliata*) in southwestern Panama. *Folia Primatologica*, 26(2): 81–108.

Bearder, S.K., Nekaris, K.A.I., and Buzzell, C.A. (2002). Dangers in the night: Are some nocturnal primates afraid of the dark? In L.E. Miller (Ed.), *Eat or be eaten: Predator sensitive foraging among primates* (pp. 21–43). Cambridge: Cambridge Univ. Press.

Bearder, S.K., Nekaris, K.A.I., and Curtis, D.J. (2006). A re-evaluation of the role of vision in the activity and communication of nocturnal primates. *Folia Primatologica*, 77(1–2): 50–71.

Bicca-Marques, J.C., and Calegaro-Marques, C. (1998). Behavioral thermoregulation in a sexually and developmentally dichromatic Neotropical primate, the black-and-gold howling monkey (*Alouatta caraya*). *American Journal of Physical Anthropology*, 106(4): 533–546.

Bond, A.B., and Kamil, A.C. (2002). Visual predators select for crypticity and polymorphism in virtual prey. *Nature*, 415: 609–612.

Britt, A., Welch, C., and Katz, A. (2001). The impact of *Cryptoprocta ferox* on the *Varecia v. variegata* reinforcement project at Betampona. *Lemur News*, 6: 35–37.

Britt, A., Welch, C., and Katz, A. (2003). Can small, isolated primate populations be effectively reinforced through the release of individuals from a captive population? *Biological Conservation*, 115: 319–327.

Brockman, D.K. (2003). *Polyboroides radiatus* predation attempts on *Propithecus verreauxi. Folia Primatologica*, 74(2): 71–74

Cheney, D.L., and Wrangham, R.W. (1987). Predation. In B.B. Smuts, D.L. Cheney, R.M. Seyfarth, R.W. Wrangham, and T.T. Struhsaker (Eds.), *Primate societies* (pp. 227–239). Chicago: Univ. of Chicago Press.

Clarke, B. (1962). Balanced polymorphism and the diversity of sympatric species. In D. Nichols (Ed.), *Taxonomy and geography* (pp. 47–70). London: The Systematics Association.

Colquhoun, I.C. (1993). The socioecology of *Eulemur macaco*: A preliminary report. In P.M. Kappeler and J.U. Ganzhorn (Eds.), *Lemur social systems and their ecological basis* (pp. 11–23). New York: Plenum Press.

Colquhoun, I.C. (1997). A predictive socioecological study of the black lemur (Eulemur macaco macaco) in northwestern Madagascar. Doctoral dissertation. Washington University, St. Louis, MO.

Colquhoun, I.C. (1998a). Cathemeral behavior of *Eulemur macaco macaco* at Ambato Massif, Madagascar. *Folia Primatologica*, 69(Suppl. 1): 22–34.

Colquhoun, I.C. (1998b). The lemur community of Ambato Massif: An example of the species richness of Madagascar's classified forests. *Lemur News*, 3: 11–14.

Colquhoun, I.C. (2001). Spectroscopic analysis of black lemur "loud calls": Evidence for ordinal-level communication in a prosimian primate. *Canadian Association for Physical Anthropology Newsletter*, 2001(1): 20–21.

Colquhoun, I.C. (2006). Predation and cathemerality: comparing the impact of predators on the activity patterns of lemurids and ceboids. *Folia Primatologica*, 77(1–2): 143–165.

Cook, L.M., and Miller, P. (1977). Density-dependent selection on polymorphic prey—some data. *American Naturalist*, 111: 594–598.

Cooper, V.J., and Hosey, G.R. (2003). Sexual dichromatism and female preference in *Eulemur fulvus* subspecies. *International Journal of Primatology*, 24(6): 1177–1188.

Curio, E. (1976). *The ethology of predation*. Berlin: Springer-Verlag.

Curtis, D.J. (1997). *The mongoose lemur (Eulemur mongoz): A study in behavior and ecology*. Doctoral dissertation. Universitat Zurich, Zurich.

Curtis, D.J., and Zaramody, A. (1999). Social structure and seasonal variation in the behaviour of *Eulemur mongoz*. *Folia Primatologica*, 70: 79–96.

Curtis, D.J., and Rasmussen, M.A. (2002). Cathemerality in lemurs. *Evolutionary Anthropology*, (Suppl. 1): 83–86.

Curtis, D.J., Zaramody, A., and Martin, R.D. (1999). Cathemerality in the mongoose lemur, *Eulemur mongoz*. *American Journal of Primatology*, 47: 279–298.

Csermely, D. (1996). Antipredator behavior in lemurs: Evidence of an extinct eagle on Madagascar or something else? *International Journal of Primatology*, 17(3): 349–354.

Dahl, J.F., and Hemingway, C.A. (1988). An unusual activity pattern for the mantled howler monkey of Belize. *American Journal of Physical Anthropology*, 75(2): 201.

Dawkins, R., and Krebs, J.R. (1979). Arms races between and within species. *Proceedings of the Royal Society of London, Series B, Biological Sciences*, 205: 489–511.

Dawson, G.A. (1979). The use of time and space by the Panamanian tamarin, *Saguinus oedipus*. *Folia Primatologica*, 31: 253–284.

Donati, G., Lunardini, A., and Kappeler, P.M. (1999). Cathemeral activity of red-fronted brown lemurs (*Eulemur fulvus rufus*) in the Kirindy forest/CFPF. In B. Rakotosamimanana, H. Rasamimanana, J.U. Ganzhorn, and S.M. Goodman (Eds.), *New directions in lemur studies* (pp. 119–137). New York: Kluwer Acadamic/Plenum Publishers.

Eason, P. (1989). Harpy eagle attempts predation on adult howler monkey. *The Condor*, 91(2): 469–470.

Elton, R.A, and Greenwood, J.J.D. (1970). Exploring apostatic selection. *Heredity*, 25(4): 629–633.

Endler, J.A. (1986). *Natural selection in the wild*. Monographs in Population Biology 21. Princeton, NJ: Princeton Univ. Press.

Endler, J.A. (1991). Interactions between predators and prey. In J.R. Krebs and N.B. Davies (Eds.), *Behavioral ecology: An evolutionary approach* (pp. 169–196). Oxford: Blackwell Scientific Publications.

Engqvist, A., and Richard, A. (1991). Diet as a possible determinant of cathemeral activity patterns in primates. *Folia Primatologica*, 57: 169–172.

Fernandez-Duque, E. (2003). Influences of moonlight, ambient temperature, and food availability on the diurnal and nocturnal activity of owl monkeys (*Aotus azarai*). *Behavioral Ecology and Sociobiology*, 54: 431–440.

Fernandez-Duque, E., and Bravo, S.P. (1997). Population genetics and conservation of owl monkeys (*Aotus azarai*) in Argentina: A promising field site. *Neotropical Primates*, 5(2): 48–50.

Fernandez-Duque, E., Rotundo, M., and Sloan, C. (2001). Density and population structure of owl monkeys (*Aotus azarai*) in the Argentinean Chaco. *American Journal of Primatology*, 53(3): 99–108.

Fernandez-Duque, E., Rotundo, M., and Ramirez-Llorens, P. (2002). Environmental determinants of birth seasonality in night monkeys (*Aotus azarai*) of the Argentinean Chaco. *International Journal of Primatology*, 23(3): 639–656.

Ferrari, S.F., and Lopes Ferrari, M.A. (1990). Predator avoidance behaviour in the buffy-headed marmoset, *Callithrix flaviceps*. *Primates*, 31(3): 323–338.

Fichtel, C., and Kappeler, P.M. (2002). Anti-predator behavior of group-living Malagasy primates: Mixed evidence for a referential alarm call system. *Behavioral Ecology and Sociobiology*, 51: 262–275.

Fichtel, C., and Hammerschmidt, K. (2002). Responses of red fronted lemurs to experimentally modified alarm calls: Evidence for urgency-based changes in call structure. *Ethology*, 108: 763–777.

Ford, S.M., and Davis, L.C. (1992). Systematics and body size: Implications for feeding adaptations in New World monkeys. *American Journal of Physical Anthropology*, 88(4): 415–468.

Freed, B.Z. (1996a). Co-occurrence among crowned lemurs (Lemur coronatus) and Sanford's lemur (Lemur fulvus sanfordi) of Madagascar. Doctoral dissertation. Washington University, St. Louis, MO.

Freed, B.Z. (1996b). Habitat use and mixed-species associations of crowned lemurs and Sanford's lemurs. *American Journal of Physical Anthropology*, (Suppl. 22): 106.

Freed, B.Z. (1999). An introduction to the ecology of daylight-active lemurs. In P. Dolhinow and A. Fuentes (Eds.), *The nonhuman primates* (pp. 123–132). Mountain View, CA: Mayfield Publishing Co.

Galeotti, P., and Rubolini, D. (2004). The niche variation hypothesis and the evolution of colour polymorphism in birds: A comparative study of owls, nightjars and raptors. *Biological Journal of the Linnean Society*, 82: 237–248.

Galeotti, P., Rubolini, D., Dunn, P.O., and Fasola, M. (2003). Color polymorphism in birds: Causes and functions. *Journal of Evolutionary Biology*, 16(4): 635–646.

Ganzhorn, J.U., Wright, P.C., and Ratsimbazafy, J. (1999). Primate communities: Madagascar. In J.G. Fleagle, C.H. Janson, and K.E. Reed (Eds.), *Primate communities* (pp. 75–89). Cambridge: Cambridge Univ. Press.

Garcia, J.E., and Braza, F. (1987). Activity rhythms and use of space of a group of *Aotus azarae* in Bolivia during the rainy season. *Primates*, 28(3): 337–342.

Gautier, J.P., Vercauteren Drubbel, R., Fleury, M.C., and Martinez, B. (1999). Difficulties of observing terrestrial forest dwelling primates as a consequence of their anti-predator strategies. *Folia Primatologica*, 70(4): 211–212.

Gendron, R.F. (1987). Models and mechanisms of frequency-dependent predation. *American Naturalist*, 130: 603–623.

Gil-da-Costa, R. (this volume). Howler monkeys and harpy eagles: A communication arms race. In S. Gursky, and K.A.I. Nekaris (Eds.), *Primate anti-predator strategies* (pp. 287–305). New York: Springer.

Gil-da-Costa, R., Palleroni, A., Hauser, M.D., Touchton, J., and Kelley, J.P. (2003). Rapid acquisition of an alarm response by a Neotropical primate to a newly introduced avian predator. *Proceedings of the Royal Society of London Series B, Biological Sciences*, 270: 605–610.

Glander, K.E. (1975). Habitat description and resource utilization: A preliminary report on mantled howling monkey ecology. In R.H. Tuttle (Ed.), *Socioecology and psychology of primates* (pp. 37–57). The Hague: Mouton Publishers.

Goodman, S.M. (2003). Predation on lemurs. In S.M. Goodman and J.P. Benstead (Eds.), *The natural history of Madagascar* (pp. 1221–1228). Chicago: Univ. of Chicago Press.

Goodman, S.M., O'Connor, S., and Langrand, O. (1993). A review of predation on lemurs: Implications for the evolution of social behavior in small nocturnal primates. In P.M. Kappeler and J.U. Ganzhorn (Eds.), *Lemur social systems and their ecological basis* (pp. 51–66). New York: Plenum Press.

Gould, L., and Sauther, M.L. (this volume). Anti-predator strategies of a diurnal prosimian. In S. Gursky and K.A.I. Nekaris (Eds.), *Primates anti-predators strategies* (pp. 273–286). New York: Springer.

Grassi, C. (2001). The behavioral ecology of *Hapalemur griseus griseus*: The influence of microhabitat and population density on this small-bodied prosimian folivore. Doctoral dissertation. The University of Texas at Austin.

Harrington, J.E. (1975). Field observations of social behavior of *Lemur fulvus fulvus* (E. Geoffroy 1812). In I. Tattersall and R.W. Sussman (Eds.), *Lemur biology* (pp. 259–279). New York: Plenum Press.

Hart, D.L. (2000). Primates as prey: Ecological, morphological and behavioral relationships between primate species and their predators. Doctoral dissertation. Washington University, St. Louis, MO.

Hart, D.L., and Sussman, R.W. (2005). *Man the hunted: Primates, predators and human evolution*. New York: Westview Press.

Hauser, M.D., and Wrangham, R.W. (1990). Recognition of predator and competitor calls in nonhuman primates and birds: A preliminary report. *Ethology*, 86: 116–130.

Heymann, E.W. (1990). Reactions of wild tamarins, *Saguinus mystax* and *Saguinus fuscicollis* to avian predators. *International Journal of Primatology*, 11(4): 327–337.

Hill, R.A. (2006). Why be diurnal? Or, why not be cathemeral? *Folia Primatologica*, 77(1–2): 72–86.

Horwich, R.H., and Lyon, J. (1990). *A Belizean rain forest: The community baboon sanctuary*. Gays Mills, WI: Orangutan Press.

Hubbard, S.F., Cook, R.M., Glover, J.G., and Greenwood, J.J.D. (1982). Apostatic selection as an optimal foraging strategy. *Journal of Animal Ecology*, 51(2): 625–633.

Isbell, L.A. (1994). Predation on primates: Ecological patterns and evolutionary consequences. *Evolutionary Anthropology*, 3: 61–71.

Izawa, K. (1978). A field study of the ecology and behavior of the black-mantle tamarin (*Saguinus nigricollis*). *Primates*, 19(2): 241–274.

Janzen, D.H. (1980). When is it coevolution? *Evolution*, 34(3): 611–612.

Janson, C.H (1998). Testing the predation hypothesies for vertebrate sociality: Prospects and pitfalls. *Behaviour*, 135: 389–410.

Johnson, S.E. (2002). Ecology and speciation in brown lemurs: White-collared lemurs (*Eulemur albocollaris*) and hybrids (*Eulemur albocollaris X Eulemur fulvus rufus) in southeastern Madagascar*. Doctoral dissertation. The University of Texas at Austin.

Jolly, A. (2003). *Lemur catta*, ring-tailed lemur, *Maky*. In S.M. Goodman and J.P. Benstead (Eds.), *The natural history of Madagascar* (pp. 1320–1331). Chicago: Univ. of Chicago Press.

Jormalainen, V., Merilaita, S., and Tuomi, J. (1995). Differential predation on sexes affects colour polymorphism of the isopod *Idotea baltica* (Pallas). *Biological Journal of the Linnean Society*, 55: 45–68.

Kappeler, P.M. (1999a). Lemur social structure and convergence in primate socioecology. In P.C. Lee (Ed.), *Comparative primate socioecology—Cambridge studies in biological and evolutionary anthropology*, vol. 22. (pp. 273–299). Cambridge: Cambridge Univ. Press.

Kappeler, P.M. (1999b). Convergence and nonconvergence in primate social systems. In J. Fleagle, C. Janson, and K. Reed (Eds.), *Primate communities* (pp. 158–170). Cambridge: Cambridge Univ. Press.

Kappeler, P.M., and Ganzhorn, J.U. (1993). The evolution of primate communities and societies in Madagascar. *Evolutionary Anthropology*, 2(5): 159–171.

Kappeler, P.M., and Heymann, E.W. (1996). Nonconvergence in the evolution of primate life history and socio-ecology. *Biological Journal of the Linnean Society*, 59: 297–326.

Kappeler, P.M., and Erkert, H. (2003). On the move around the clock: Correlates and determinants of cathemeral activity in wild redfronted lemurs (*Eulemur fulvus rufus*). *Behavioral Ecology and Sociobiology*, 54: 359–369.

Karpanty, S.M. (2003). Behavioral and ecological interactions of raptors and lemurs in southeastern Madagascar: A multiple predator approach. Doctoral dissertation. The State University of New York at Stony Brook.

Karpanty, S.M., and Goodman, S.M. (1999). Diet of the Madagascar harrier-hawk, *Polyboroides radiatus*, in southeastern Madagascar. *Journal of Raptor Research*, 33(4): 313–316.

Karpanty, S.M., and Grella, R. (2001). Lemur responses to diurnal raptor calls in Ranomafana National Park, Madagascar. *Folia Primatologica*, 72: 100–103.

Kay, R.F., and Kirk, E.C. (2000). Osteological evidence for the evolution of activity pattern and visual acuity in primates. *American Journal of Physical Anthropology*, 115(2): 235–262.

Kinzey, W.G. (Ed.). (1997). *New World primates: Ecology, evolution, and behavior*. New York: Aldine de Gruyter, Hawthorne.

Landeau, L., and Terborgh, J. (1986). Oddity and the 'confusion effect' in predation. *Animal Behaviour*, 34: 1372–1380.

Lima, S.L., and Dill, L.M. (1990). Behavioral decisions made under risk of predation: A review and prospectus. *Canadian Journal of Zoology*, 68(4): 619–640.

Macedonia, J.M. (1990). What is communicated in the antipredator calls of lemurs: Evidence from playback experiments with ringtailed and ruffed lemurs. *Ethology*, 86: 177–190.

Martin, R.D. (1990). *Primate origins and evolution*. London: Chapman and Hall.

Maynard Smith, J. (1970). The causes of polymorphism. In R.J. Berry and H.N. Southern (Eds.), *Variation in mammalian populations. Symposia of the Zoological Society of London*, 26: 371–383.

Mittermeier, R.A., Tattersall, I., Konstant, W.R., Meyers, D.M., and Mast, R.M. (1994). *Lemurs of Madagascar*. Washington, DC: Conservation International.

Moynihan, M. (1964). Some behavior patterns of platyrrhine monkeys. I. The night monkey (*Aotus trivirgatus*). *Smithsonian Miscellaneous Collections*, 146(5): 1–84.

Moynihan, M. (1976). *The New World primates: Adaptive radiation and the evolution of social behavior, languages, and intelligence.* Princeton, NJ: Princeton Univ. Press.

Mutschler, T. (2002). Alaotran gentle lemur: Some aspects of its behavioral ecology. *Evolutionary Anthropology*, (Suppl. 1): 101–104.

Mutschler, T., Feistner, A.T.C., and Nievergelt, C.M. (1998). Preliminary field data on group size, diet and activity in the Alaotran gentle lemur *Hapalemur griseus alaotrensis. Folia Primatologica*, 69: 325–330.

Mutschler, T., and Tan, C.L. (2003). *Hapalemur*, bamboo or gentle lemurs? In S.M. Goodman and J.P. Benstead (Eds.), *The natural history of Madagascar* (pp. 1324–1329). Cambridge: Cambridge Univ. Press.

Overdorff, D.J. (1988). Preliminary report on the activity cycle and diet of the red-bellied lemur (*Lemur rubriventer*) in Madagascar. *American Journal of Primatology*, 16(2): 143–153.

Overdorff, D.J. (1996). Ecological correlates to activity and habitat use of two prosimian primates: *Eulemur rubriventer* and *Eulemur fulvus rufus* in Madagascar. *American Journal of Primatology*, 40(4): 327–342.

Overdorff, D.J., and Johnson, S. (2003). *Eulemur*, true lemurs. In S.M. Goodman and J.P. Benstead (Eds.), *The natural history of Madagascar* (pp. 1320–1324). Cambridge: Cambridge Univ. Press.

Overdorff, D.J., and Rasmussen, M.A. (1995). Determinants of nighttime activity in "diurnal" lemurid primates. In L. Alterman, G.A. Doyle, and M.K. Izard (Eds.), *Creatures of the dark: The nocturnal prosimians* (pp. 61–74). New York: Plenum Press.

Overdorff, D.J., Strait, S.G., and Telo, A. (1997). Seasonal variation in activity and diet in a small-bodied folivorous primate, *Hapalemur griseus*, in southeastern Madagascar. *American Journal of Primatology*, 43(3): 211–223.

Passamani, M. (1995). Field observations of a group of Geoffroy's marmosets mobbing a margay cat. *Folia Primatologica*, 64(3): 163–166.

Peetz, A., Norconk, M.A., and Kinzey, W.G. (1992). Predation by jaguar on howler monkeys (*Alouatta seniculus*) in Venezuela. *American Journal of Primatology*, 28(3): 223–228.

Peres, C.A. (1990). A harpy eagle successfully captures an adult red howler monkey. *Wilson Bulletin*, 102(3): 560–561.

Peres, C.A. (1993). Anti-predation benefits in a mixed-species group of Amazonian tamarins. *Folia Primatologica*, 61: 61–76.

Peres, C.A., and Janson, C.H. (1999). Species coexistence, distribution, and environmental determinants of Neotropical primate richness: A community-level zoogeographic analysis. In J.G. Fleagle, C.H. Janson, and K.E. Reed (Eds.), *Primate communities* (pp. 55–74). Cambridge: Cambridge Univ. Press.

Rakotondravony, D., Goodman, S.M., and Soarimalala, V. (1998). Predation on *Hapalemur griseus* by *Boa manditra* (Boidae) in the littoral forest of eastern Madagascar. *Folia Primatologica*, 69: 405–408.

Rasmussen, M.A. (1998a). Variability in the cathemeral activity cycle of two lemurid primates at Ampijoroa, northwest Madagascar. *American Journal of Physical Anthropology*, (Suppl. 26): 183.

Rasmussen, M.A. (1998b). Ecological influences on cathemeral activity in the mongoose lemur (*Eulemur mongoz*) at Ampijoroa, northwest Madagascar. *American Journal of Primatology*, 45(2): 202.

Rasmussen, M.A. (1999). Ecological influences on activity cycle in two cathemeral primates, *Eulemur mongoz* (mongoose lemur) and *Eulemur fulvus fulvus* (common brown lemur). Doctoral dissertation. Duke University, Raleigh, NC.

Rathbun, G.B., and Gache, M. (1980). Ecological survey of the night monkey, *Aotus trivirgatus*, in Formosa Province, Argentina. *Primates*, 21(2): 211–219.

Roulin, A., and Wink, M. (2004). Predator-prey relationships and the evolution of colour polymorphism: A comparative analysis in diurnal raptors. *Biological Journal of the Linnean Society*, 81: 565–578.

Rowe, N. (1996). *The pictorial guide to the living primates*. East Hampton, NY: Pogonias Press.

Sauther, M.L. (1989). Antipredator behavior in groups of free-ranging *Lemur catta* at Beza Mahafaly Special Reserve, Madagascar. *International Journal of Primatology*, 10: 595–606.

Sauther, M.L. (2002). Group size effects on predation sensitive foraging in wild ring-tailed lemurs (*Lemur catta*). In L.E. Miller (Ed.), *Eat or be eaten: Predator sensitive foraging among primates* (pp. 107–125). Cambridge: Cambridge Univ. Press.

Schulke, O. (2001). Social anti-predator behaviour in a nocturnal lemur. *Folia Primatologica*, 72(6): 332–334.

Searcy, Y.M., and Caine, N.G. (2003). Hawk calls elicit alarm and defensive reactions in captive Geoffroy's marmosets (*Callithrix geoffroyi*). *Folia Primatologica*, 74: 115–125.

Sherman, P.T. (1991). Harpy eagle predation on a red howler monkey. *Folia Primatologica*, 56(1): 53–56.

Sherratt, T.N., and MacDougall, A.D. (1995). Some population consequences of variation in preference among individual predators. *Biological Journal of the Linnean Society*, 55: 93–107.

Stafford, B.J., and Ferreira, F.M. (1995). Predation attempts on callitrichids in the Atlantic coastal rain forest of Brazil. *Folia Primatologica*, 65: 229–233.

Stallings, J.R., West, L., Hahn, W., and Gamarra, I. (1989). Primates and their relation to habitat in the Paraguayan Chaco. In K.H. Redford and J.F. Eisenberg (Eds.), *Advances in Neotropical mammalogy* (pp. 425–442). Gainesville, FL: The Sandhill Crane Press, Inc.

Stanford, C.B. (2002). Avoiding predators: Expectations and evidence in primate antipredator behavior. *International Journal of Primatology*, 23(4): 741–757.

Sussman, R.W. (1972). An ecological study of two Madagascan primates: *Lemur fulvus rufus* (Audebert) and *Lemur catta* (Linnaeus). Doctoral dissertation. Duke University, Durham, NC.

Sussman, R.W. (1975). A preliminary study of the behavior and ecology of *Lemur fulvus rufus* (Audebert 1800). In I. Tattersall and R.W. Sussman (Eds.), *Lemur biology* (pp. 237–258). New York: Plenum Press.

Sussman, R.W. (1977). Feeding behavior of *Lemur catta* and *Lemur fulvus*. In T.H. Clutton-Brock (Ed.), *Primate ecology: Studies of feeding and ranging behaviour in lemurs, monkeys and apes* (pp. 1–36). London: Academic Press, Inc.

Sussman, R.W., and Kinzey, W.G. (1984). The ecological role of the Callitrichidae: A review. *American Journal of Physical Anthropology*, 64: 419–449.

Sussman, R.W., and Tattersall, I. (1976). Cycles of activity, group composition, and diet of *Lemur mongoz mongoz* Linnaeus 1766 in Madagascar. *Folia Primatologica*, 26: 270–283.

Tan, C.L. (2000). Behavior and ecology of three sympatric bamboo lemur species (Genus: *Hapalemur*) in Ranomafana National Park, Madagascar. Doctoral dissertation. The State University of New York at Stony Brook.

Tarnaud, L.D. (2004). Cathemerality in the Mayotte brown lemur: Seasonality and food quality. *Folia Primatologica*, 75(Suppl. 1): 420.

Tattersall, I. (1976). Group structure and activity rhythm in *Lemur mongoz* (Primates, Lemuriformes) on Anjouan and Moheli islands, Comoro Archipelago. *Anthropological Papers of the American Museum of Natural History*, 53(4): 367–380.

Tattersall, I. (1977). Ecology and behavior of *Lemur fulvus mayottensis* (Primates, Lemuriformes). *Anthropological Papers of the American Museum of Natural History*, 54(4): 421–482.

Tattersall, I. (1979). Patterns of activity in the Mayotte lemur, *Lemulr fulvus mayottensis*. *Journal of Mammalogy*, 60(2): 314–323.

Tattersall, I. (1982). *The primates of Madagascar*. New York: Columbia Univ. Press.

Tattersall, I. (1987). Cathemeral activity in primates: A definition. *Folia Primatologica*, 49: 200–202.

Tattersall, I. (2006). The concept of cathemerality: History and definition. *Folia Primatologica*, 77(1–2): 7–14.

Tattersall, I., and Sussman, R.W. (1975). Observations of the ecology and behavior of the mongoose lemur (*Lemur mongoz mongoz*) Linnaeus (Primates, Lemuriformes), at Ampijoroa, Madagascar. *Anthropological Papers of the American Museum of Natural History*, 52(4): 193–216.

Taylor, R.J. (1984). Predation. In M.B. Usher and M.L. Rosenzweig (Eds.), *Population and community biology series*. New York: Chapman and Hall.

Terborgh, J. (1983). *Five New World primates: A study in comparative ecology*. Princeton, NJ: Princeton Univ. Press.

Terborgh. J., and van Schaik, C.P. (1987). Convergence vs. nonconvergence in primate communities. In J.H.R. Gee and P.S. Giller (Eds.), *Organization of communities past and present: The 27th Symposium of the British Ecological Society* (pp. 205–226). Oxford: Blackwell Scientific Publications.

Thorington, R.W., Muckenhirn, N.A., and Montgomery, G.G. (1976). Movements of a wild night monkey (*Aotus trivirgatus*). In R.W. Thorington, and P.G. Heltne (Eds.), *Neotropical primates: Field studies and conservation. Proceedings of a symposium on the distribution and abundance of Neotropical primates* (pp. 32–34). Washington, DC: National Academy of Sciences.

Tinbergen, N. (1963). On the aims and methods of ethology. *Zeitschrift fur Tierpsychologie*, 20: 410–433.

Traina, A. (2001). Activity pattern and feeding behaviour of ringtailed lemurs (*Lemur catta*) at Berenty Reserve in Madagascar during the day and night. *Folia Primatologica*, 72(3): 188.

Treves, A., and Brandon, K. (2005). Tourist impacts on the behavior of black howling monkeys (*Alouatta pigra*) at Lamanai, Belize. In J.D. Paterson and J. Wallis (Eds.), *Commensalism and conflict: The human-primate interface (Special topics in primatology,* vol 4, J. Wallis, series ed.) (pp. 146–167). Norman, OK: The American Society of Primatologists.

van Schaik, C.P., and Kappeler, P.M. (1993). Life history, activity period and lemur social systems. In P.M. Kappeler and J.U. Ganzhorn (Eds.), *Lemur social systems and their ecological basis* (pp. 241–260). New York: Plenum Press.

van Schaik, C.P., and Kappeler, P.M. (1996). The social systems of gregarious lemurs: Lack of convergence with anthropoids due to evolutionary disequilibrium? *Ethology*, 102: 915–941.

van Schaik, C.P., and van Noordwijk, M.A. (1989). The special role of male *Cebus* monkeys in predation avoidance and its effect on group composition. *Behavioral Ecology and Sociobiology*, 24: 265–276.

van Schaik, C.P., van Noordwijk, M.A., Warsono, B., and Sutriono, E. (1983). Party size and early detection of predators in Sumatran forest primates. *Primates*, 24(2): 211–221.

Vasey, N. (1997). Community ecology and behavior of *Varecia variegata rubra* and *Eulemur fulvus albifrons* on the Masoala Peninsula, Madagascar. Doctoral dissertation. Washington University, St. Louis, MO.

Vasey, N. (2000). Niche separation in *Varecia variegata rubra* and *Eulemur fulvus albifrons*: Interspecific patterns. *American Journal of Physical Anthropology*, 112(3): 411–431.

Vasey, N. (2003). *Varecia*, ruffed lemurs. In S.M. Goodman, and J.P. Benstead (Eds.), *The natural history of Madagascar* (pp. 1332–1336). Cambridge: Cambridge Univ. Press.

Vasey, N. (2005). New developments in the behavioral ecology and conservation of ruffed lemurs (*Varecia*). *American Journal of Primatology*, 66(1): 1–6.

Vermeij, G.J. (1987). *Evolution and escalation: An ecological history of life*. Princeton, NJ: Princeton Univ. Press.

Wilson, J.M., Stewart, P.D., Ramangason, G-S., Denning, A.M., and Hutchings, M.S. (1989). Ecology and conservation of the crowned lemur, *Lemur coronatus*, at Ankarana, N. Madagascar, with notes on Sanford's lemur, other sympatrics and subfossil lemurs. *Folia Primatologica*, 52(1): 1–26.

Wright, P.C. (1978). Home range, activity pattern, and agonistic encounters of a group of night monkeys (*Aotus trivirgatus*) in Peru. *Folia Primatologica*, 29: 43–55.

Wright, P.C. (1984). Biparental care in *Aotus trivirgatus* and *Callicebus moloch*. In M.F. Small (Ed.), *Female primates: Studies by women primatologists* (pp. 59–75). New York: Alan R. Liss, Inc.

Wright, P.C. (1985). The costs and benefits of nocturnality for *Aotus trivirgatus* (the night monkey). Doctoral dissertation. The City University of New York.

Wright, P.C. (1989). The nocturnal primate niche in the New World. *Journal of Human Evolution*, 18: 635–658.

Wright, P.C. (1994). The behavior and ecology of the owl monkey. In J.F. Baer, R.E. Weller, and I. Kakoma (Eds.), *Aotus: The owl monkey* (pp. 97–112). San Diego: Academic Press.

Wright, P.C. (1995). Demography and life history of free-ranging *Propithecus diadema edwardsi* in Ranomafana National Park, Madagascar. *International Journal of Primatology*, 16(5): 835–854.

Wright, P.C. (1998). Impact of predation risk on the behaviour of *Propithecus diadema edwardsi* in the rain forest of Madagascar. *Behaviour*, 135: 483–512.

Wright, P.C. (1999). Lemur traits and Madagascar ecology: Coping with an island environment. *Yearbook of physical anthropology*, 42: 31–72.

Wright, P.C., Heckscher, S.K., and Dunham, A.E. (1997). Predation on Milne-Edwards' sifaka (*Propithecus diadema edwardsi*) by the fossa (*Cryptoprocta ferox*) in the rain forest of southeastern Madagascar. *Folia Primatologica*, 68: 34–43.

Zuberbühler, K., Jenny, D., and Bshary, R. (1999). The predator deterrence function of primate alarm calls. *Ethology*, 105(6): 477–490.

8
Moonlight and Behavior in Nocturnal and Cathemeral Primates, Especially *Lepilemur leucopus*: Illuminating Possible Anti-Predator Efforts

Leanne T. Nash

Introduction

What factors affect the behavioral decisions individuals make to forage, travel, rest, or engage in sociality, including reproductive behaviors, at a particular time or place? Decisions to engage in any of these activities should depend on trade-offs between gains in nutrients and, ultimately, in reproductive success, and predation risk (Lima & Dill, 1990). Lima & Dill argue that since most of the components of predation risk are potentially assessable by prey, prey may make behavioral decisions that could reduce the risks. In the case of foraging, a decision concerning foraging when a predation risk exists may differ from a decision based only upon energy considerations. However, Lima & Dill also review the variability of predator behavior, which may limit the prey's ability to assess risk and the consequent cues prey could use to assess risk. For the observer, the difficulty is one of assessing the animal's perceived risk of predation. Perceived risk is likely to be more important in understanding prey behavior than predation rate, since rate of predation is what we see after the evolution of anti-predator adaptations in morphology or behavior (Hill, 1998; Janson, 1998; Stanford, 2002).

Much of the discussion about the effects of predation in the primatological literature has focused on the role (or lack thereof) of predation in the evolution of sociality (e.g., van Schaik, 1983; van Schaik & van Hooff, 1983; Cheney & Wrangham, 1987; Isbell, 1994; Boinski & Chapman, 1995; Hill, 1998; Janson, 1998; Treves, 1999; Stanford, 2002). Among nocturnal primates, anti-predator strategies available to diurnal primates such as shared vigilance, early detection with warning calls, and group living *per se* may be less available. It is argued that nocturnal primates use crypsis, solitary foraging, and selective use of range habitat or microhabitat to minimize risk (Stanford, 2002). There is relatively little information available on anti-predator activities in "nongregarious" nocturnal species, though both mobbing and warning calls do occur in some. *Galagoides zanzibaricus* gives warning calls and will mob genets and snakes, i.e., puff adders and an arboreal snake, probably a mamba (Nash, pers. obs.). *Galago moholi* (Bearder, 1987; Bearder et al., 2002) behaves similarly. Schülke (2001) reports two examples involving a sub-adult and an adult fork-marked lemur (*Phaner furcifer*) accompanied by a Coquerel's dwarf lemur (*Mirza coquerli*) engaged in simultaneous mobbing of a boa with alarm calls. Gursky (2005) used natural observations and experiments to show that male *Tarsius spectrum* from multiple social groups would join to mob a snake.

This paper will examine in some detail the response of a nocturnal Malagasy lemur, *Lepilemur leucopus* from southwestern Madagascar, to differing moonlight conditions and relate its response to potential predation risk. All primates, even nocturnal ones, are defined as a taxon in terms of their "visual orientation" (Cartmill, 1974, 1992; Fleagle, 1999). Thus, at night, the level of ambient light is expected to influence their ability to detect their food, conspecifics, and predators. Moonlight can be 100 or even 1000 times brighter than starlight alone, and the brightness of the full moon is 10 times that of the first quarter moon (Pariente, 1979, 1980). Also, there is marked light quality variation between quarter and full moon, with a relative decrease in spectral energies beyond 750 nm (i.e., longer wavelengths) at quarter moon. Given the variability in strepsirrhine spectral sensitivities, perception of color in the environment may vary across species under differing moonlight conditions (Dominy et al., 2001).

Responses to Moonlight—Nocturnal Primates and Other Animals: Lunar Philia, Lunar Phobia, Lunar Neutrality

Responses of nocturnal animals to moonlight vary among birds and mammals depending on species and behaviors assessed. Table 8.1 presents a non-exhaustive sampling for birds and non-primate mammals and as many sources as I could find for nocturnal or cathemeral primates. Most of these animals are prey and several are both predators and prey of others. In general, most studies on birds and non-primate mammals indicate lunar phobic behavior, i.e., activity is reduced under brighter moonlight conditions. This can also include altering microhabitat use so as to increase the use of cover. As indicated in Table 8.1, this is usually attributed to assumed or (occasionally) known changes in predation risk. In contrast, where an effect of moonlight has been found in primates it is almost always one of lunar philia, most commonly expressed as increased movement or vocalizations in the presence of more light, but in some cases an increase seeking of cover also has been noted. In general, authors cited in Table 8.1 have suggested that this is because primates, as an order, are particularly visually oriented mammals (Allman, 1977; Bearder et al., 2006) and may be able to detect predators better in bright light. An alternative hypothesis, one relating to predators, including primate insectivores, is that the prey may become more active in bright moon conditions requiring more activity by predators to catch them. Since these ideas are not mutually exclusive notions, both merit testing (Gursky, 2003a). Though less surprising, it can be noted that activity of diurnal primates may also be increased in bright

TABLE 8.1. Summary of moonlight effects on behavior in nocturnal and cathemeral primates and selected birds and other mammals[a].

Species	With Increased Moonlight[b]	Method of Comparing Moonlight Levels[c]	Hypothesized Causes	Sources[d]
Galagos, pottos, lorises, and tarsiers				
Galago moholi, South Africa	↑ speed, night range length, time on ground in males; ↔ females' night range	Focal, H (2), N (3)	Avoid predation	1
Galagoides zanzibaricus[e], *Otolemur garnettii*, Kenya	↑ travel (both species); calling only in *G. zanzibaricus*	Focal, H (2): all occurrences, N (2)	(1) Improve vision during movement; (2) avoid predators (detect, alarm calling, mob)	2
G. demidovii, Euoticus elegantulus, G. alleni, Perodicticus potto, Arctocebus calabarensis, Gabon	↑ activity for galago species; no information for pottos	Qualitative	Avoid predation	3
Nycticebus coucang, captive indoors	↑ "low activity" (e.g., resting); ↓ "high activity" behaviors (e.g., locomotion); ↔ medium activity (eating, grooming, and social grooming)	Group scans, N (4)	Avoid predation	4
Loris tardigradus, Dindigul, India	↑ parked infants seek cover, adults in time spent foraging, travel between trees; ↓ calling; ↔ time on ground, travel speed, time resting, rate of prey capture, or feeding location	Focal, N (2), H (2)	(1) Locate prey; (2) avoid predation ("parked" infants) but low risk at this site	5
L. tardigradus, Dindigul, India	↑ time inactive; ↓ time in exploration; ↔ locomotion, feeding; some differences in effects by age-sex classes	Focal, N (2)	Avoid predation	6
Tarsius spectrum, Sulawesi, Indonesia	↑ foraging, travel time, path length, range size, calling, time animals travel together, territorial disputes; ↓ resting, scent marking	Focal, N (5)	(1) Locate prey; (2) avoid predation, (3) enhance gregariousness	7
Malagasy lemurs				
Lepilemur leucopus, Berenty	↑ calling	Qualitative	—	8
L. leucopus, Microcebus murinus, Beza Mahafaly	↔ encounter rates (both species) and *Lepilemur* calling incidence	Census, N (2)	—	9
Eulemur macaco macaco at Ambato Massif	Cathemeral; ↑ activity	Focal, H (2)	Improved vision in general (since lack of tapetum)	10

(Continued)

175

TABLE 8.1. (Continued).

Species	With Increased Moonlight[b]	Method of Comparing Moonlight Levels[c]	Hypothesized Causes	Sources[d]
E. rubriventer and E. fulvus rufus at Ranomafana	Cathemeral; ↔ general behavior	Qualitative	—	11
E. fulvus fulvus at Ampijoroa, E. mongoz at Anjamena and Ampijoroa	Cathemeral; for E. mongoz ↑ activity with light level (only when correct for cloud cover); E. f. fulvus, ↔ activity	Focal, N (correlation with nightly index of illumination)	—	12
E. fulvus rufus Kirindy	Cathemeral; ↑ movement	Qualitative (eclipse); Group scan, N (2); H (correlation)	(1) Improved vision in general; (2) avoid predation	13
E. fulvus rufus Kirindy	Cathemeral; ↑ movement	Focal (activity logging collar), H (4)	(1) Improved vision in general; (2) avoid predation	14
Propithecus v. verreauxi, Kirindy	Diurnal; ↑ nocturnal activity (slightly) during the first 3 hrs of the night (only during long-day warm, wet season)	Focal (activity logging collar), N (4)	Carryover of daytime arousal	15
New World monkeys				
Aotus trivirgatus, captive indoors	↑ movement at about 0.1 lux vs. both lower and higher light intensities	Experimental varying of night light levels; focal (automated recording of activity), H (visual comparison of plots)	—	16
A. trivirgatus in Peru and A. azarai in Paraguay	↑ nightly path length, time active (travel, feed, socialize) at both sites, diurnal activity in site without diurnal predators (Paraguay)	Focal, N (2), H (qualitative)	(1) Improved vision in general; (2) avoid predation	17
A. azarai in Argentina	Cathemeral; ↑ activity at night; ↓ diurnal activity in day following a full-moon night	Group scans, N (4)	(1) Improved vision in general; (2) avoid predation	18
Non-primate mammals and birds				
Onychomys leucogaster breviauritus, grasshopper mouse, captive outdoors	↓ activity, time out of nest	Automated activity recorder of individuals, N (4), H (visual comparison of plots)	(1) Locate prey; (2) avoid predation	19
Dipodomys spectabilis, bannertail kangaroo rats, wild and captive outdoors	Wild: ↓ activity at least at some seasons of year; captive: ↓ activity in pens but not in running wheels	Automated activity recorders, capture rates of wild, N(2), H (2, 4)	(1) Locate prey; (2) avoid predation (trade-off increased risk and higher food payoff)	20

D. nitratoides, Fresno kangaroo rats, wild and captive outdoors	↔ activity	See above	n.a.; need wild data to confirm	21
Desert rodents, wild	↑ use of cover; ↓ use of open spaces	Live capture rates, N (3)	(1) changes in resource availability; (2) avoid predation	22
Desert rodent communities, wild	↓ proportion of animals trapped in open areas, especially for species more vulnerable to predation	Live capture rates, N (2, with lanterns manipulating light)	Avoid predation	23
Desert rodents (aggregate of various); Dipodomys merriami, Merriam's kangaroo rat, wild	↓ seed removal from feeders in open microhabitats	Activity measured as food seed harvested in 3 microhabitats, N (2)	Avoid predation	24
Peromiscus polionotus, old-field mouse, wild	↓ time in open areas	Activity measure as tracks, H (8)	Avoid predation	25
Gerbillus allenby, G. pyramidum, gerbils, captive outdoors	↑ capture rate by owls: ↓ foraging activity, especially with less cover, but less strongly in G. pyramidum	Aviary experiment with gerbils and owls; activity measured as food seed harvested, N (2, with lanterns manipulate light)	Avoid predators; G. pyramidum may be less vulnerable to owls (larger, better hearing ability)	26
Dipodomys merriami, Merriam's kangaroo rat, wild	↑ crepuscular activity, predation by diurnal predators during crepuscular activity; ↓ distance from burrow, time out of burrow, predation by nocturnal predators	Focal follows of radio tracked animals, N (2 or 3)	(1) Avoid predation; (2) trade-off: compensate for decreased foraging in full moon by increasing crepuscular activity	27
Peromyscus leucopus, white-footed mice, wild	↑ perceptual ability, though less than in twilight; ability to view a goal in higher light but move toward it later in the dark	Tracks left by animals; N (2)	Avoid predation	28
Lepus mericanus, snowshoe hare, wild	↓ movement; ↔ use of cover	Tracks left by animals to feeders in different cover; N (2)	(1) Avoid predation; (2) other unspecified seasonal effects	29
Meles meles, badgers, wild	↑ time to emerge from daytime resting place; ↓ travel speed, distance, night range size	Focal, H (4)	—	30
Setonix brachyurus, quokka (nocturnal macropod marsupial), wild	↓ activity	Automatic activity recorder, H (2)	Avoid predation	31
African insectivorous bats, wild	↑ forage in more closed areas; ↓ activity	Focal follow light tagged animals, N (2, visual comparison of plots), H (2, visual comparison of plots)	Avoid predation	32

(Continued)

TABLE 8.1. (Continued).

Species	With Increased Moonlight[b]	Method of Comparing Moonlight Levels[c]	Hypothesized Causes	Sources[d]
Artibeus jamaicensis, Neotropical fruit bat, wild	↓ foraging, locating of new foraging trees	Focal follow with radio tracking, H in week (2. visual comparison of plots)	Avoid predation; (by sit-and-wait predators by minimizing cast shadows)	33
Lophostoma silvicolum, Neotropical perch-hunting insectivorous bat, wild	↓ nocturnal flying, spent more time in roosts	Focal follow with radio tracking, N (3)	Increase foraging efficiency on prey which also ↓ activity with moonlight	34
Oceanodroma leucorhoa, Leach's storm-petrels, wild	↓ nocturnal flying	All occurrences counts of birds at a point, H (correlational)	Avoid predation	35
Ptychoramphus aleuticus, Cassin's auklet, wild	↑ number of dead auklets found in nesting colony; ↓ time spent on land	Counts of dead, observations of activity, N (3)	(1) Locate prey; (2) avoid predation	36
Phalaenoptilus nuttallii, common poorwill, wild	↑ foraging (when moon is higher but not "more full"); ↔ hatching rate	Focal follow with radio tracking, N (5 % of moon face illuminated at midnight), N (4, height of moon in sky); for nesting cycle, days since full moon	Increased foraging efficiency	37

[a] Localities given for primates studied to facilitate intrageneric comparisons (Madagascar is country for all lemurs). When different studies are in same country, site is given. For non-primates, locale is wild or captive (indoors, light levels manipulated, or outdoors exposed to moon, with or without light levels manipulated with lamps).

[b] ↑: measure increases with increasing light; ↓: measure decreases with increasing light; ↔: no change on measure with increasing light.

[c] Sampling method, time scale of samples, i.e., N: nightly, H: time blocks within nights taking moon rise/set times into account. Number in parentheses indicates where ANOVA-like test was used. The number 2 generally represents new vs. full moon, but it might also indicate that light levels were aggregated in other ways; alternately, correlation tests used; see cited references for details.

[d] Sources: 1 (Bearder et al., 2002; Bearder et al. (2006), 2 (Nash, 1986), 3 (Charles-Dominique, 1977), 4 (Trent et al., 1977), 5 (Bearder et al., 2002; Bearder et al., (2006), 6 (Radhakrishna & Singh, 2002), 7 (Gursky, 2003a), 8 (Charles-Dominique & Hladik, 1971), 9 (Nash, 2000), 10 (Colquhoun, 1988), 11 (Overdorff, 1988; Overdorff & Rasmussen, 1995), 12 (Curtis et al., 1999; Curtis & Rasmussen, 2002), 13 (Donati, 1999; Donati et al., 2001), 14 (Kappeler & Erkert, 2003), 15 (Erkert & Kappeler, 2004), 16 (Erkert, 1974, 1976), 17 (Wright, 1978; Wright, 1989; Wright, 1996), 18 (Fernandez-Duque & Rotundo, 2003), 19 (Jahoda, 1973), 20 (Lockard & Owings 1974b, a), 21 (Lockard & Owings, 1974a), 22 (Price et al., 1984), 23 (Kotler, 1984), 24 (Bowers, 1988,1990), 25 (Wolfe & Tan Summerlin, 1989), 26 (Kotler et al., 1991), 27 (Daly et al. 1992), 28 (Zollner & Lima, 1999), 29 (Gilbert & Boutin, 1991), 30 (Cresswell & Harris, 1988), 31 (Packer, 1965), 32 (Fenton et al., 1977), 33 (Morrison, 1978), 34 (Lang et al., 2006), 35 (Watanuki, 1986), 36 (Nelson, 1989), 37 (Brigham & Barclay, 1992)

[e] now called *Galagoides cocos* (Grubb et al., 2003)

moonlight. Anderson (1984, 2000) reviews moonlight associated with cases of increased sensitivity to stimuli, group movements and foraging, increased vocalizations, delayed retiring, and earlier departure from sleeping sites; predation risk may increase at such times (Peetz et al., 1992). However, the differences in lunar philic and lunar phobic responses do not exactly match the dichotomy of primates versus non-primates. Reasons for this will be discussed below, after I show that *Lepilemur leucopus* does not show the increase in activity with bright moon that is found in most other primates examined.

At Berenty, in southeastern Madagascar, *L. leucopus* is reported to vocalize more during times of the night when the moon is up compared to the same times of the night when the moon is down (Charles-Dominique & Hladik, 1971). However, only three nights of data are presented in support of this conclusion. In contrast, observations during census walks at Beza did not find a moonlight effect on vocalization rate (Nash, 2000). During those censuses there was also no moonlight effect on encounter rate in *Lepilemur* or *Microcebus* (though there was a trend to more encounters for *Lepilemur* in full moon), but sample sizes for census walks were small so statistical power was low. Consequently, I will investigate in more detail the effect of moonlight on both the "when" and the "where" of behaviors, i.e., on vocalizations, activity budget, and canopy heights used by *L. leucopus* at Beza. I will use a variety of approaches to test for such effects to see if they give congruent results. Finally, the response of *Lepilemur* to moonlight will be placed in the broader context of responses by other nocturnal and cathemeral primates and nonprimates. These comparisons will lead to a number of further questions about the causes of species variation in response to moonlight, questions that researchers would find fruitful to explore.

Predators of Nocturnal Primates and Their Anti-Predator Behavior

As is the case for most primates, much of the evidence for predation of nocturnal primates is indirect or anecdotal, but all nocturnal genera have some predators (Treves, 1999; Stanford, 2002; Goodman, 2003). In many cases the actual predators are not known to the researcher and the risk of predation must be assumed from examination of (1) the potential predators in a given geographic area and (2) apparent anti-predator behavior. The mobbing responses described above are examples of such indirect evidence. Galagos in Gabon also respond to possible predators (i.e., children who hunted them, the observer) with warning calls and mobbing, though no information was available on other predators (Charles-Dominique, 1977). Charles-Dominique pointed out the very different response to threats by the galagos, who could leap away quickly, compared to the sympatric pottos (*Perodicticus, Arctocebus*). The lorids slow, quiet locomotion seemed designed to help the animals avoid detection by predators, and, like galagos, they seemed to use vision to spot a predator (e.g., the observer, if he was not very still). *Perodicticus* apparently had few predators. Their remains were not found in owl pellets but they were attacked by viverrids, the palm civet (*Nandinia binotata*), and a mongoose (*Bdeogale nigripes*). If it can, *Perodicticus* will "flee" slowly and quietly. If attacked, it curls up and bites under its arm and butts its "scapular shield" at the threat. As a last resort, it will drop from the trees. *Arctocebus calabarensis* also uses slow, noiseless locomotion and concealment; it will also drop to the ground when threatened at close range. *Galago moholi*, from southern Africa, is known to be preyed on by genets and possibly by owls, domestic cats, jackals, and snakes. In Asia, *Loris tardigradus* is sympatric with owls, wild and domestic cats, and snakes. As with the galagos from Gabon, *G. moholi* gives warning calls but *Loris* does not, though it will flee swiftly (Bearder et al., 2002). In Kenya I observed *G. senegalensis* foraging on the ground "startle" and jump into a bush when it noticed an owl perched nearby (Nash, pers. obs.).

Known predators on *Lepilemur* include nocturnal raptors i.e., barn owl (*Tyto alba*), Madagascar long-eared owl (*Asio madagascariensis*), diurnal raptors, i.e., Madagascar buzzard, (*Buteo brachypterus*), Madagascar harrier hawk (*Polybroides radiatus*), and Henst's goshawks (*Accipiter henstii*), and mammalian carnivores, i.e., the fossa (*Cryptoprocta fossa*), and possibly a snake, the Madagascar tree boa (*Sanzinia madagascariensis*) (Charles-Dominique & Hladik, 1971; Goodman et al., 1993c; Rasoloarison et al., 1995; Goodman et al., 1997; Goodman et al., 1998; Ratsirarson & Emady, 1999; Goodman, 2003). In addition, other nocturnal lemurs of similar or smaller size have been preyed on by other nocturnal and diurnal predatory birds, mongooses, domestic dogs, and possibly domestic cats (Fietz & Dausmann, 2003; Goodman, 2003). Finally, even though many predators may now be extinct, those extant in the recent past may have influenced current behavioral responses of prey (Goodman et al., 1993c).

Predation by diurnal hawks indicates the importance of appropriate sleeping sites but does not indicate what behavioral modifications during the night may be related to avoidance of such predation (Gilbert & Tingay, 2001; Schülke & Ostner, 2001). Observations on predation of *Lepilemur*, as well as other nocturnal primates, are usually made during the day (e.g., Schülke & Ostner, 2001; Fietz & Dausmann, 2003), or from indirect means such as examination of scats, regurgitated owl pellets, or discarded remains found around raptor's nests (Goodman et al., 1993a; Goodman et al., 1993c; Goodman et al., 1997; Goodman, 2003). Consequently, the predator's activity at the time of a predation attempt or success is rarely known. Thus, hypotheses on how moonlight might influence anti-predator activities are based mainly on suppositions about what will affect the predator's hunting success. These include the movements of the prey, light levels and hence visibility and ease of capture (cover versus lack thereof). Genet predation on *G. moholi* always occurred on dark nights and some owls may have more success at moonlit times (Clark, 1983; Bearder et al., 2002) as discussed further below. In the Neotropics *Aotus* face predation mainly from diurnal raptors, though owls, snakes, and felids eat them occasionally (Wright, 1994).

Methods

As reviewed in Table 8.1, most of the behavioral variables assessed have to do with activity budgets, and more rarely, vocalizations, as will be the case discussed here. However, animals may also shift some behaviors to greater "cover"; in the case of this study, this approach will be examined vis-à-vis the height in the trees that prey used for various activities. As is also apparent in Table 8.1, there are a variety of methods used to assess moonlight effects on prey animals' behavior. For example, phases of the lunar month may be grouped into as few as two or as many as eight levels (Nash, 1986; Gursky, 2003a). Also, on any night other than the night of the full moon, times when the moon is above the horizon can be contrasted with times it is not visible. Light levels are rarely directly measured, but moonrise and -set times and moon phase can be found in astronomical tables for any date. Light level can sometimes be varied experimentally, even in the wild (Kotler, 1984). For cathemeral animals, e.g., *Aotus*, it is even important to test for moonlight effects during diurnal activity on a day that follows nocturnal activity (Fernandez-Duque, 2003).

Site and Subjects

The study was done in Parcel 1 of the Beza Mahafaly Special Reserve in south-western Madagascar. The site is a xerophytic strip of riverine deciduous and semi deciduous vegetation dominated by *Tamarindus indica* (Sauther, 1989; Richard et al., 1991; Sauther, 1991; Sussman, 1991; Sussman & Rakotozafy, 1994). This is a highly seasonal environment in the driest region of Madagascar. Annual rainfall is highly variable and most rain falls between October and March. The area around the reserve averages 720 mm per year. During the night, the austral summer minimum temperatures are as low as 21°C, and temperatures sink to as low as 3°C in the austral winter. The total rain from October through May during the study was 906 mm, considerably more than in other years. The bird predators present at Beza Mahafaly that might have been a threat to *Lepilemur* include five nocturnal raptors and nine hawks (one a boreal winter migrant) (Langrand, 1990; Goodman et al., 1993b). Four are known to prey on *Lepilemur* at Beza or elsewhere (see above). Other potential predators include the fossa, domestic cats and dogs, and snakes.

  Lepilemur leucopus is a small, folivorous, nocturnal lemur. At night it is a relatively inactive species overall (Nash, 1998) and during the day sleeps in nest holes or thick tangles of branches at this site. When it uses nest holes in the daytime *L. leucopus* will often sit at the opening of the hole in a rather exposed position. However, at the slightest disturbance, it will retreat into the hole. The seven subjects included six males and one adult female. Weights of three captured females were 585 g, 475 g, and 740 g. The first listed was the female radio-tracked subject in the study. The low weight female was likely an immature and the high weight was a near term pregnant female. Weights of the six males ranged from 520 g to 675 g, with a median of 635 g. The lightest male may have been a juvenile, an observation based on the date of capture (March 1, after a November–December birth season) and his low weight (the next smallest male ever trapped weighed 600 g). However, his testes size was in the range of the heavier males. See Nash (1998) for further details.

All subjects' ranges were in the part of Parcel 1 nearest the river, where trees were the tallest. Neither identification of an animal's sex nor individual identification would have been possible without trapping and artificial marking. Subjects were trapped by blow-darting or removing them by hand from a sleeping place during the morning daylight hours. While sedated (Telozol or Ketamine were used), each was fitted with a collar carrying a Telonics radio-tracking transmitter. At this time, body weight and morphometrics measurements were made and the animal was examined for indices of age, reproductive maturity (i.e., testes size in males, nipple length in females), and, for females, reproductive condition was noted (patent or sealed vagina, palpable fetus). Animals were returned to their trapping site within four to six hours of being caught. The entire collar-transmitter package weighed 15–20 g, less than 4% of the animal's body weight.

Observation Procedures and Behavior Variables

Observations reported here were conducted between December 3, 1992, and June 27, 1993. Radio-tracking permitted the author and an assistant to locate the subject's sleeping site during the day and follow it from there during the night using headlamps. The animals very quickly habituated to being followed and observed. By the second night of being followed they did not seem overly disturbed. Focal animal follows were carried out from the time the animal left its sleeping place at dusk until midnight. Observations consisted of eight to nine follows for one female and four males, about 50 hrs per subject, and three follows for two other males, about 12 hrs per subject. Observations were frequently disrupted by the higher than usual rainfall, which commonly fell at night.

During focal individual follows, the animal's activity was recorded with instantaneous point samples at 5-min intervals (Martin & Bateson, 1993). The activities were categorized in this analysis as REST (animal alone, not moving, eyes open or closed), LOCOMOTE (animal moving within, or, more usually, between crowns), FEED (picking, handling, ingesting, or clearly chewing food), SELF-GROOM (scratching self or use of toothcomb on own body), or OTHER (all else, including all social behavior, vocalizing, and otherwise, mainly excreting). Also, at each interval the animal's substrate was recorded for height (nearest m), diameter (nearest cm), and angle (four categories: vertical, horizontal, angled head up, or angled head down). In this analysis, scoring for activity and substrate started with the last interval of rest before an animal left the sleeping site.

Because of the difficulty in seeing the items being fed upon at each 5-min interval, feeding was also scored in a one-zero (1/0) fashion within each interval (Martin & Bateson, 1993). In addition, as an alternative assessment of locomotor activity, moving to a new tree was scored 1/0 within each 5-min interval. Characteristic loud vocalizations, which may have a territorial function (Charles-Dominique & Hladik, 1971) and often occur in bouts exchanged between neighboring animals, were difficult to count as individual calls. Scoring of vocalizations was done in two ways. First, the focal subject's vocalizations were scored 1/0 within each 5-min interval. Second, an index of "total calls in vicinity" was created summing a 1/0 score for each interval for calls heard from the area around the focal animal (and not by it) in five "vicinities," each direction (N, E, S, W), plus undetermined (for calls heard but not easily localized), and at each of three estimated distances (from loudness: near, medium, far). Thus, this score could, in principal, range from zero to a maximum of 15. For added details of site, subjects, and procedures see Nash (1998).

Though light levels were not measured during observations, all animals were followed during periods when the moon was in different phases and during times when the moon was set and when it was not. The choice of nights on which an individual was followed was influenced by a variety of logistical factors. Consequently, differing amounts of observation intervals in different moonlight conditions were available for each subject. The specific amount of observation time for each subject varied with the differing analyses of the data, which are described below.

Data Analysis

Analysis of data was done in four ways to deal with two issues and to ask slightly different questions. The two issues were these: first, there are various ways to separate observations into moonlight levels; second, since data were taken as sequential 5-min observation intervals or as sequential 5-min 1/0 scores, the problem of temporal autocorrelation of data was present within each follow. Analyses proceeded to establish "moon phase" for each interval by collapsing intervals into times defined as "dark" and times defined as "light". First, each 5-min interval was assigned as "dark" if the moon had set or was not yet risen, or if the moon was up and new, waxing but less than one quarter; or waning and less than one quarter (i.e., crescent, new, and not up). "Light" moon phase was assigned to an interval if the moon was up and full or if it was waxing and greater than one quarter or if it was waning and greater than one quarter (i.e., full, gibbous, and quarter, respectively). During waning moon phase, the moon rises later after sunset on succeeding nights (approximately one hour later for each night following full moon), so most of a follow at this time of the month was in a dark moon condition. During waxing moon, the moon rises prior to sunset, so it was up during these follows.

In the first analysis, called "interval analysis," total data for each subject were cross-tabulated for the behavioral variable at an interval and moon phase for that interval. This analysis examined the possibility of moment to moment changes in behavior with light level as it varied due to moonrise and -set times and due to moon phase. The proportion of time a subject performed a behavior within a moon phase was the subject's "score."

The second analysis was similar to the first, but all intervals on a night were assigned the moon's phase for the night, regardless of whether the moon was up or not. Thus, the number of light moon intervals was overestimated in this approach, called "night's interval data." This analysis addressed the possibility that it is the night within the lunar cycle that is important, not whether or not the moon is risen at a given moment (Morrison, 1978; Donati et al., 2001). This type of analysis was not applied to the 1/0 scored behaviors. In both of these types of analyses, a matched-pairs analysis was performed on seven subject's scores from dark and light intervals using a Wilcoxson's signed rank test (Sokal & Rohlf, 1995). Descriptive statistics presented will be the median and range across animals. The total observation time in the dark conditions was 175.1 hrs (median across subjects, 27.8 hrs, range 5.3–40.2 hrs) and the total observation time in light conditions was 90.6 hrs (median, 15.5 hrs, range 4.1–19.9 hrs).

The third analysis took selected nights of extremes in moon phase (light levels) as the unit of analysis. These "selected extreme nights" were chosen from all nights of observations to be within a seven-day period centered on the night of the full moon (light) or the night of the new moon (dark). In addition, since just after a night of full moon the moon does not rise until after dusk, intervals prior to moonrise were omitted from analysis for the light moon condition. For most tests there were 10 dark nights and 8 light nights, except for the test on "total calls in vicinity," where there were only 8 dark and 7 light nights available with data. Subjects contributed varying numbers of nights to the sample of observations (two subjects for one night each, one subject for two nights, two subjects each for three and four nights). However, observations on a single subject were separated by between 11 and 102 days (median = 30 days), so in this study are considered independent. Dark and light night's data were compared with a Mann-Whitney U test (Sokal & Rohlf, 1995). The total observation time in the dark conditions was 53.3 hrs (median across 10 nights, 6.0 hrs, range 2.9–6.5 hrs) and the total observation time in light conditions was 36.4 hrs (median across 8 nights, 4.4 hrs, range 2.7–5.8 hrs). Descriptive statistics presented will be the median and range across nights.

Because of the possibility that for the selected nights there was a confound of moon phase with seasonal effects (Nash, 1998), a test of association of the selected nights for dark versus light moon with warm season (12/20/1992–4/15/1993) or cool season (4/16/1993–6/22/1993) was made. There was no association of season with moonlight level for the selected nights ($N = 18$, Yates corrected chi-square = 1.25, df = 1, NS).

Finally, in the fourth type of analysis "pooled data" were compared using a chi-square test of association between behavior and moon phase at an interval (Sokal & Rohlf, 1995) after intervals were selected that were separated at a length of time which produced no temporal autocorrelation of successive data points. In this situation statistical sample size is this number of data points. The method of Cant et al. (2001) was followed to establish independent data points for analysis of pooled data, though it did not solve the problem of repeated measures on the same individual with the pooled data. Depending on the behavioral measure, varying time periods were required to produce independent data (described below). This analysis was not applicable to the behavior "total calls in vicinity." For more on issues of data pooling see Leger & Didrichsons (1994) and Jenkins (2002).

All results will be presented as the proportion of good observation data for each variable, i.e., intervals where observations were not missed. In no case was there a significant difference by moonlight level in the proportion of missed observations for a particular variable for an interval. Depending on the particular measure and method of contrasting dark and light moon nights, the range of missed values was 4–16% for dark moon observations and 3–13% for light moon observations. However, as a comment on the effect of moonlight on the observer, the proportion of missed data on dark moon nights did exceed the proportion on light moon nights in every type of comparison. Since there was some difference by variable in the proportion of missing data, the actual number of intervals that were available for analysis and that occurred under either moonlight condition varied (Table 8.2). However, in general, there were more observations made during dark moon conditions than during light moon conditions since there were fewer opportunities for the latter.

Results

Activity Budget

None of the various measures of activity level showed a significant association with moonlight level. For activity budgets (Table 8.2), depending on the measure used, the median percentage of time feeding was 29–36% (dark: 31–36%, light: 29–33%), of locomoting 12–13% (dark: 12–13%, light: 11–13%), of resting 37–45% (dark: 39–41%, light 37–45%), of self-grooming 5–10% (dark: 7–10%, light: 5–10%) and for other 3–6% (dark: 4–6%, light: 3–4%). The general picture of activity budgets did not vary greatly by the type of measure used. Using the "interval data," the p values associated with the Wilcoxson matched pairs signed rank test, $N = 7$, ranged from 0.05–0.81 for the five behaviors. The lowest value was associated with "self-groom." However, if the proportions were rounded to two decimals, there was one tie and the difference was no longer significant. Using the "night's interval data" with the same statistic, the *p* values ranged from 0.32–1.00 across the five behaviors. For the "selected extreme nights," using the selected dark ($N = 10$) and light ($N = 8$) nights, the Mann-Whitney U test values ranged from 41–49 with associated *p* values of 0.965–0.460, respectively, across the five behaviors. Note that since these are medians, and not scores from an individual, across these three measures, the time in all activities for any one moon phase within a measure will not necessarily add to 1.00 (the same will apply to time in all height classes, below). Finally, in the pooled analysis, activity budget was limited to the use of intervals spaced at 15 min to produce temporally independent samples, and there was no association between moonlight and activity for those intervals ($N = 979$, df $= 4$, chi-square $= 5.69$, NS).

For the variables assessing activity level with 1/0 scores for each interval, there were also no associations with moonlight level for movement to a new tree or for feeding (Table 8.2). Depending on the measure, the percentage of intervals in which animals moved to a new tree ranged from 29–38% (dark: 29–32%, light: 31–38%). Using the "interval data" and the Wilcoxson matched pair signed rank test, there was no difference in movement with moon phase ($W+ = 10$, $W- = 18$, $N = 7$, $p = 0.5781$). There was also no difference

TABLE 8.2. Proportions of observations made at each moonlight level, <u>median</u> (and range) of proportion of time in dark or light in each measure of behavior, for each test type.

Measures		Interval Data Moon phase		Night's Interval Data Moon phase		Selected Extreme Nights Moon phase		Pooled Data[1] Moon phase	
		dark	light	dark	light	dark	light	dark	light
Total intervals§ — activities		2944		2944		986		979	
Proportion time observed for activities		0.588	0.413	0.492	0.508	0.578	0.422	0.654	0.346
Activity budget	feed	<u>0.320</u> (0.163–0.411)	<u>0.330</u> (0.103–0.478)	<u>0.328</u> (0.186–0.408)	<u>0.298</u> (0.098–0.473)	<u>0.363</u> (0.091–0.489)	<u>0.318</u> (0.146–0.478)	0.311	0.289
	locomote	<u>0.130</u> (0.091–0.252)	<u>0.130</u> (0.057–0.199)	<u>0.122</u> (0.089–0.265)	<u>0.117</u> (0.065–0.172)	<u>0.119</u> (0.054–0.244)	<u>0.124</u> (0.068–0.369)	0.122	0.130
	rest	<u>0.390</u> (0.159–0.609)	<u>0.372</u> (0.283–0.770)	<u>0.386</u> (0.153–0.599)	<u>0.367</u> (0.273–0.772)	<u>0.410</u> (0.133–0.672)	<u>0.403</u> (0.215–0.750)	0.414	0.454
	self-groom	<u>0.092</u> (0.065–0.142)	<u>0.091</u> (0.034–0.128)	<u>0.096</u> (0.051–0.157)	<u>0.087</u> (0.033–0.128)	<u>0.073</u> (0.022–0.153)	<u>0.053</u> (0.023–0.154)	0.093	0.097
	other	<u>0.045</u> (0.026–0.112)	<u>0.044</u> (0.022–0.050)	<u>0.046</u> (0.026–0.122)	<u>0.044</u> (0.018–0.058)	<u>0.038</u> (0.000–0.121)	<u>0.036</u> (0.000–0.095)	0.059	0.029
Total intervals§ — 1/0³ new tree		3124		NA[2]		1045		1035	
Proportion time observed for 1/0 new tree		0.655	0.345			0.584	0.416	0.654	0.346
1/0 activity—new tree		<u>0.319</u> (0.266–0.707)	<u>0.333</u> (0.180–395)			<u>0.285</u> (0.132–0.694)	<u>0.376</u> (0.091–0.522)	0.301	0.310
Total intervals§ — 1/0³ feed		2841		NA		948		703	
Proportion time observed for 1/0 feed		0.651	0.349			0.578	0.422	0.653	0.347
1/0 activity — feed		<u>0.505</u> (0.298–0.570)	<u>0.546</u> (0.238–0.644)			<u>0.507</u> (0.233–0.698)	<u>0.496</u> (0.205–0.644)	0.486	0.496
Total intervals§ — heights		2836		2836		957		359	
Proportion time observed for heights		0.654	0.346	0.494	0.506	0.588	0.412	0.641	0.359

Heights (m)							
0–5	0.114 (0.046–0.280)	0.122 (0.015–0.055)	0.104 (0.055–0.353)	0.133 (0.000–0.585)	0.158 (0.000–0.438)	0.109	0.171
6–9	0.210 (0.176–0.470)	0.233 (0.179–0.444)	0.330 (0.134–0.582)	0.317 (0.138–0.529)	0.234 (0.107–0.932)	0.261	0.287
10–14	0.436 (0.318–0.623)	0.423 (0.280–0.680)	0.442 (0.267–0.679)	0.325 (0.154–0.828)	0.457 (0.068–0.821)	0.474	0.488
15+	0.127* (0.000–0.308)	0.104 (0.000–0.325)	0.069 (0.000–0.127)	0.125* (0.000–0.328)	0.008* (0.000–0.047)	0.157	0.054
Total intervals§ – 1/0 vocalize	3188		NA	1076		797	
Proportion time observed for 1/0 vocalize	0.659	0.341		0.595	0.405	0.659	0.341
1/0 activity — vocalize	0.107 (0.061–0.221)	0.088 (0.043–0.145)		0.069 (0.013–0.204)	0.096 (0.000–0.246)	0.097	0.085
Total intervals§ – Total calls in vicinity	2856		NA	978		NA	
Proportion time observed for total calls in vicinity	0.658	0.342		0.596	0.404		
Total calls in vicinity[4]	1.409* (1.128–2.008)	2.134* (1.596–2.907)		1.755* (1.232–2.527)	2.246* (1.938–3.781)		

* Significant difference of light vs. dark for an analysis of a variable at p < 0.05 (see text)

§ "Total intervals" refers to the number of 5-min intervals available across all subjects, within each moon phase, available for analysis. The sample sizes for statistical tests were: (1) for the "interval data" and "night's interval data," 7 animals at each moon phase (matched pairs) and the median and range is across these 7 animals; (2) for "selected extreme nights" there were 10 dark nights and 8 light nights (except for "total calls in vicinity," where it was 8 and 7 nights, respectively) and the median and range is across these nights; and (3) for the pooled data N does represent the number of intervals analyzed by chi-square (after deletion of data points to correct for temporal autocorrelation). See text for more details.

[1] Only overall proportion provided

[2] "Night moon" not done for 1/0 scores or total vocalizations in vicinity (see text)

[3] 1/0 data are presented as proportion of intervals with behavior present any time during the interval

[4] This variable is an index not a proportion; see text.

contrasting "selected extreme nights" for movement between trees (Mann-Whitney $U = 41.5$, $N = 10, 8$, $p = 0.897$). The pooled data analysis for move to a new tree was limited to intervals spaced at 15 min and there was no association between moonlight and move or not for those intervals ($N = 1035$, df $= 1$, Yates corrected chi-square $= 0.08$, NS). The percentage of intervals in which animals fed ranged from 49–55% (dark: 49–51%, light: 50–55%). Using the "interval data" and the Wilcoxson matched pair signed rank test, there was no difference in feed with moon phase ($W+ = 16$, $W- = 12$, $N = 7$, $p = 0.8125$). There was also no difference contrasting "selected extreme nights" for feed (Mann-Whitney $U = 43$, $N = 10, 8$, $p = 0.829$). The pooled data analysis for feed was limited to intervals spaced at 20 min and there was no association between moonlight and feed or not for those intervals ($N = 703$, df $= 1$, Yates corrected chi-square $= 0.06$, NS).

Heights Used

Height-use classes were created by examination of a histogram of use at 1-m intervals and breaking classes around peaks in the histogram. Higher moonlight levels were associated with a decrease in time spent at the highest levels in the trees during light moon phase but there was not a clear pattern at other canopy levels (Table 8.2). Depending on the measure used, the percentage of time at 15 m or more was 2–16% (dark: 10–16%, light: 2–5%), at 0–5 m it was 10–17% (dark: 11–13%, light: 10–17%), at 6–9 m it was 21–33% (dark: 21–32%, light: 23–33%), and at 10–14 m it was 32–49 % (dark: 32–47%, light: 44–49%). In all cases time spent at the lowest levels and highest levels represented the smallest time, and time spent at 10–14 m was the largest. Using the "interval data," the p values associated with the Wilcoxson matched pairs signed rank test, $N = 7$, was significant for the 15-m or more level ($W+ = 26$, $W- = 2$, $p < 0.05$). For the other three levels, p values ranged from 0.38–0.94. Using the "night's interval data" with the same statistic, however, there were no significant differences; the p values ranged from 0.15–0.93 across the four levels (with the smallest p for the 15-m or more level). For the "selected extreme nights," using the selected dark ($N = 10$) and light ($N = 8$) nights, the Mann-Whitney test produced a significant difference for the 15-m or more level ($U = 70$, $p = 0.006$). For the other three levels there were no differences; U values ranged from 41–43 with associated p values of 0.965–0.829, respectively. Finally, in the pooled analysis, height was limited to using intervals spaced at 45 min to produce temporally independent samples and thus severely reduced the sample size to about 13% of the available intervals. That test still found that there was an association between moonlight and height level for those intervals ($N = 359$, df $= 3$, chi-square $= 9.87$, $p < 0.05$). The greatest deviations from expected values in the test were in the cells for the 15-m or more level.

Trade-offs between height in a tree and activity budget were examined qualitatively, using all available data, because when temporal autocorrelation in both variables was avoided sample sizes were too severely restricted for statistical testing. When height and activity budget were examined collapsed over all moonlight

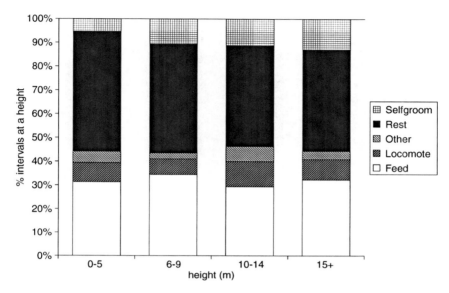

FIGURE 8.1. Percent of good observation time at each height spent in each activity

conditions there was little difference in activity by height (Fig. 8.1). In particular, if time spent in the more active behaviors vs. less active behaviors was lumped, feed plus locomote was 39–41% of the time at all heights, and rest plus self-groom was always 54–56% of the time.

When the effect of moonlight was incorporated (Fig. 8.2) there were shifts in activity with moonlight level at each height. This was especially apparent at the 15+ m and the 6–9 m levels. Although less time was spent at 15+ m at moonlit times, when animals were at this level in such times they spent more time feeding, locomoting, and self-grooming than when they were at this level in dark times. There was little difference in activity budget at the 10–14 m height in the two moonlight conditions. At the lower two height levels, even though the total time spent at those levels did not differ significantly by moon phase, there does appear to be some shift in activities with moon phase. The shift is in the opposite direction to that found at the highest canopy level—that is, in moonlit conditions there was less feeding and more resting.

Vocalizations

Relative frequency of vocalizations in relation to moonlight levels, as assessed by the 1/0 calling scores of the focal individuals, did not differ between moonlight levels (Table 8.2). Depending on the measure used, the percentage of 5-min intervals during which an animal called at least once ranged from 7–11% (dark: 7–11%, light: 9–10%). Using the "interval data" and the Wilcoxson matched pair signed rank test, there was no difference in vocalizing with moon phase ($W+ = 7$, $W- = 21$, $N = 7$, $p = 0.2969$). There was also no difference

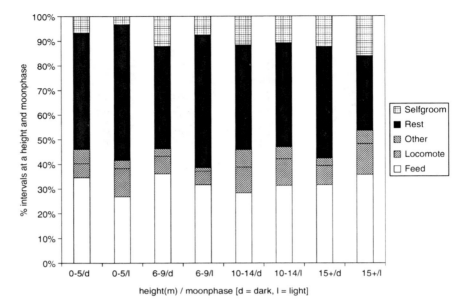

FIGURE 8.2. Percent of good observation time at each combination of height and moon phase spent in each activity

contrasting "selected extreme nights" for calling (Mann-Whitney $U = 50$, $N = 10, 8$, $p = 0.408$). The pooled data analysis for vocalizing was limited to intervals spaced at 20 min and there was no association between moonlight and calling for those intervals ($N = 797$, df $= 1$, Yates corrected chi-square $= 0.35$, NS).

In contrast, when the score assessing "total calling in vicinity" was tested, it did show higher median calling indeces during light moon conditions (Table 8.2) when using the "interval" data ($W+ = 0$, $W- = 21$, $N = 6$, $p = 0.0312$) even though there was one less subject available for this test than in other "interval" analyses. Similarly, there was also a significant difference contrasting "selected extreme nights" for "total calling in vicinity," despite there being a smaller sample of nights available (Mann-Whitney $U = 47$, $N = 8, 7$, $p = 0.029$).

Given that there was only one female subject, it was not possible to test for any sex differences. However, in all the tests done, the female's value was at or near the median score in almost all of the tests done. She was the lowest score for time at a height of 15 m or more in some cases, but in all those cases she was tied with a male's score. Thus, she was not a remarkable outlier in her scores.

Discussion

The results of analyses on *L. leucopus* at Beza can be summarized as follows. Moonlight had little effect on activity budgets, but at moonlit times there was a reduction of time spent in the highest parts of the canopy. When activity budget and height were examined together qualitatively, when at that highest level during bright moon times, animals increased feeding, locomoting, and self-grooming compared to dark moon times. When examined as all vocalizations in the vicinity of focal animals, there were more vocalizations during light moon times. Although none of the measures of an individual focal's vocal behavior showed a significant increase with moonlight, the measure showing the largest difference ("selected extreme nights") was in that direction.

I suggest that the minimal shift in activity budget may be due to the constraints, i.e., features not under immediate control of the individual (Lima & Dill, 1990), of digestive processes for a very small-bodied folivore (Nash, 1998). Although activity budget does vary seasonally, there may be limits on the "elasticity" of activity budgets to vary with moon phase. Animals might have made the shift away from the highest levels of the trees during light moon to improve cover to avoid predation by owls. The alteration of activity budget at the highest level might seem counterintuitive in that animals were more active in moonlight. However, since overall they spent less time at that height, they might have been going there only when they had to feed or move. That activity increase, nevertheless, fits with the more common pattern in other primates of increasing activity with more moonlight. It is possible they were more nervous if self-grooming, as scratching, is a sign of anxiety (Troisi et al., 1991). However, much self-grooming involves putting the head down, which would reduce visual vigilance, and at lower levels of the forest, self-grooming was reduced during light moon periods. Possibly at lower levels other predators, such as domestic dogs or cats, are a greater danger, so animals stay "heads up" more in the dimmer light near the forest floor.

Previously, I reported that there was little association of height and activity (Nash, 1998); this observation is replicated here as seen in the overall data. The 1998 study found that (1) there was an association of more feeding on the smallest diameter supports (up to 5 cm) with less resting and locomoting on such supports, and that (2) feeding occurred more on horizontal than on other support orientations while locomoting and resting occurred slightly less on horizontal supports. I do not examine moonlight effects on support angle and diameter here in detail. However, a very preliminary analysis suggested little change in support angle with moon phase, but an increase in the proportion of time on the smallest diameter supports in light moon phase. Given that feeding was proportionately more frequent at high levels and on small diameter supports, and that feeding at highest levels increased in light moon phase, the association with feeding on small diameters is the likely reason for a moonlight effect on substrate diameter use. In retrospect, it would have been useful to record the degree of "cover" the animal was under at each height, but this was not done.

The calls recorded here seem to have a social, possibly territorial, function and do not appear to be alarm calls (Charles-Dominique & Hladik, 1971; Warren & Crompton, 1997). This analysis partly confirms previous observations of *Lepilemur* calling more in more moonlit conditions (Charles-Dominique & Hladik, 1971). Perhaps a measure that is more sensitive than a 1/0 index would have been better (Altmann, 1974). If calling is a form of territorial display, it is likely initiated by visual contact with conspecifics (as well as hearing them call), which would be easier under light moon conditions. In future, it would be interesting to see if locations from which calling is done shift with moonlight conditions to afford greater protection from owl's attacks, e.g., if locations change more under cover or internal to a crown.

Different measures of moonlight show effects in a given circumstance, species, and seasons of the year (Lockard & Owings, 1974a; Gilbert & Boutin, 1991; Brigham & Barclay, 1992). The analyses presented here, as well as the variety of behaviors measured and methods for contrasting moonlight levels reviewed in Table 8.1, show the importance of looking in various ways for moonlight effects. For example, previously, within the same field project, I found no effect of moonlight on calling during censuses on *Lepilemur* (Nash, 2000), but a different measure (this study) did find an effect. Within this study different analytic approaches tended to produce the same result for each of the three sets of behavioral variables examined. However, the value of the four different tests was that it allowed me to either look at results from different time scales or to consider the different units of analysis (individual animals, nights, parts of night) to cross-check the robustness of results and to see which measure might be the most sensitive to moonlight effects. The pooled data were relatively insensitive due to problems of temporal autocorrelation in a species that rests a great deal. Packer (1965) points out issues complicating analysis: (1) strong circadian behavioral rhythms may override moonlight effects unless analysis looks at comparable time periods through the night; and (2) climatic conditions (cloud cover) may sometimes mask moonlight effects (see also Lang et al., 2006).

For a good source for practical information on measurement of light levels and the issues that go with it, see Erkert (2003). For more on the complexities of design of studies of moonlight effects and the analyses of data, see Daly et al. (1992) and Zollner & Lima (1999).

Interspecific Comparisons

Moonlight effects are part of the complex relationships between predators and their prey and understanding them requires knowledge of the visual systems of both the predators and their primate prey. In addition, if the primates or other species are also themselves predators, we need to know how moonlight affects the behavior of their prey in order to distinguish causal hypotheses about predator avoidance versus feeding behavior (Lang et al., 2006). For primates and other species with primarily plant-based diets, we also need to understand how diet, e.g., fruit vs. foliage, as well as other factors, might constrain their activity and affect their responses to moonlight. As indicated above, the importance of vision to primates (Bearder et al., 2006) may be a reason that they tend to be more lunar philic than other species (Table 8.1). However, given the variability among primates and non-primates, factors other than phylogeny may be important in influencing a species' response to moonlight. These include the type of predators a species faces, its methods of avoiding predators (including locomotor agility), its diet (including what it preys upon and how it detects food), and possible energy constraints.

Moon Effects on Predators

A number of birds that are predators may increase activity with moonlight. Barn owls can hunt in complete darkness, but other owls may be more visually oriented (Goodman et al., 1993a). Owls hunting rodents do have better success on moonlit nights in some cases (Clark, 1983; Kotler et al., 1991). Amongst birds, gulls are more successful preying on petrels and auklets during moonlit times (Watanuki, 1986; Nelson, 1989). However, amongst nocturnal and crepuscular predatory birds, lunar philia was not related to foraging strategy in a simple way (Brigham & Barclay, 1992). Whippoorwills (*Caprimulgus vociferous*), a sit-and-wait predator, showed increased locomotion, vocalization, and nest visits with moonlight and the nestling period occurring during the two-week period of most moonlight. In contrast, common nighthawks (*Chordeiles minor*) forage by flying continuously and only crepuscularly, do not alter foraging with moon phase, and do not synchronize hatching dates with lunar cycles. In a third pattern, the common poorwill (*Phalaenoptilus nuttallii*), also a sit-and-wait predator, increased foraging with increasing moon height, but not with increasing proportion of the moon phase illuminated, and it did not synchronize breeding with moon phase.

Although behavior of some insects may be altered by moonlight (e.g., fewer come to light traps on moonlit nights), some insectivorous bats do not seem to experience an effect on their foraging, as the types of prey they consumed does not differ (Fenton et al., 1977). However, these bats are at risk from the visually oriented bat hawk, whose predation improves with bright moonlight, and the bats seek more cover in foraging locations on moonlit nights (Fenton et al., 1977). In contrast, for an under story-foraging species of bat, activity of both the bat and its insect prey decreased on moonlit nights (Lang et al., 2006). In this case, effects of the moon on foraging efficiency were interpreted to be more important than effects on predation risk.

These examples indicate how complex the behavior of predators may be in response to moonlight. Also, animals that are both predators and prey must trade off benefits and risks of altering behavior with moonlight.

Moon Effects on Prey

Given that we have relatively little idea about how the actual risk of predation varies with moonlight for most nocturnal and cathemeral primates, we have to assume that most of their predators gain an advantage in locating their prey on the nights when light is brighter. Most of the non-primates reviewed in Table 8.1— those that decrease activity with moonlight (mostly rodents, bats, and smaller mammals)—probably detect predators more through hearing than vision. It is hypothesized here that primates may be able to increase activity in moonlight, even if it makes them more visible to predators, because primates can and do visually detect the predator. Birds that are prey may have poorer vision than their avian predators.

However, exceptions among primates to this pattern indicate that visual ability alone does not produce the species differences between primates and nonprimates, nor should we expect such a simple relationship between responses to moonlight and phylogeny. Agility to escape the attack may also be important. Most lorisines and *Lepilemur* are the exceptions. Lorisines do not have the agile leaping ability found in galagos, tarsiers, and possibly the other primates that are lunar philic. The need for high visual ability coupled with agility is suggested by the lunar phobic response of bats. While they are certainly agile, they may have poorer visual ability than their most likely predators (owls and falcons) and it has been suggested that their silhouettes while flying in moonlight may expose them to sit-and-wait visual predators (Fenton et al., 1977; Morrison, 1978; Heffner et al., 2001; Ortega & Castro-Arellano, 2001).

While *Lepilemur* can and does move quickly at times, it may be energetically constrained from adjusting its movements with moon phase. In the most food-rich season (or perhaps at other locations), *Lepilemur* might be predicted to respond more like a tarsier or galago and thus show the least moonlight effect in the most food-constrained season. Unfortunately, my study did not encompass the end of the cold dry season or the beginning of the rainy season, which might be periods when extremes of feeding conditions occur. However, this prediction would only be so if there were some advantage to moving more, and the folivorous diet of *Lepilemur* might not provide an advantage to increasing movement when more light is available. Improved understanding of how *Lepilemur* detects its food patches would be helpful. It would also be helpful to know if *Avahi*, which is similar to *Lepilemur* in locomotion and in being a small-bodied folivore, responds to moonlight like most other primates or more like *Lepilemur*. In order to test these ideas an examination of moonlight effects in existing data on other lorisines— e.g., on *Nycticebus*, on *Lepilemur* at other sites, and on *Avahi*—would be useful (Warren & Crompton, 1997; Wiens & Zitzmann, 1999; Thalmann, 2002; Rasoloharijaona et al., 2003; Wiens & Zitzmann, 2003a, b; Zinner et al., 2003; Ganzhorn et al., 2004).

Among the nocturnal primates studied to date for moonlight effects there are no particularly clear associations of the presence or absence of lunar philia with diet. Galagos, tarsiers, and *Aotus*, which show the strongest lunar philic responses for activity, are mainly leapers and insectivores, though *Otolemur garnetti* and *Aotus* also incorporate considerable fruit in their diet (Harcourt & Nash, 1986; Wright, 1994; Gursky, 2000; Bearder et al., 2002). However, slow movers like the slender loris show less effect and are also insectivores (Nekaris & Rasmussen, 2003). *Lepilemur*, which also shows less moonlight effect on activity, is a folivore (Nash, 1998). In addition, species do not always show congruent effects on measures of both activity and vocal behavior. A similar difficulty exists for interpreting differences among bats in reaction to moonlight (Erkert, 2000). It is clear that "lunar philia" and "lunar phobia" are really not unitary concepts and individual behavior variables must be examined.

Further complexity is found in the responses of some prey to predation and moonlight. Although most researchers theorize that small rodents are more vulnerable in moonlight, some results are opposite to what is expected, i.e., some kangaroo rats are less vulnerable to owls during moonlit periods (Price et al., 1984). For Merriam's kangaroo rats (Daly et al., 1992) there is a complex pattern of predation risk from diurnal vs. nocturnal predators and a pattern of rats compensating for decreased foraging activity in full moon with increased crepuscular foraging on full moon days. Consequently, predation by nocturnal predators was most frequent when rats were most active, in the new moon period. Thus, effects of moonlight can extend to other times of the animal's active period.

In rodents, which may detect nocturnal predators by hearing rather than sight, vulnerability may be related to body size and hearing ability (Kotler et al., 1991), and such species differences may shape communities through different microniche usage. However, vision may also be more important to some rodents, and visual ability, moonlight, predation risk, and activity cycle interact in complex ways. Experimental data show that nocturnal white-footed mice (*Peromyscus leucopus*) have limited ability to perceive a distant forest on dark nights and full moonlight extends their perceptual range, but travel on full moon nights increases their risk of predation. However, they are able to gather information about more distant goals under twilight conditions and act on that information later when it is darker, doing so in a "look now and move later" strategy (Zollner & Lima, 1999). To my knowledge, there are no studies of how differences in vulnerability to predators might interact with responsiveness to moonlight levels in shaping nocturnal primate communities or whether primates might also use a "look now and move later" strategy.

The increases in calling with moonlight seen in tarsiers, *G. zanzibaricus*, and *Lepilemur* might reveal them to predators (Table 8.1). Again, agility might be the key, as the only other species for which calling was investigated—the slender loris—is the only lorisine that is particularly "noisy," and it decreases calling with more moonlight. Most of the nonprimates in Table 8.1 are not very vocal and none had measures of vocalization relative to moonlight reported. Another factor that might interact with agility and vocal crypticity in altering primate reactions to moonlight could be the degree of gregariousness combined with the usefulness of mobbing in different species. Slender lorises and spectral tarsiers may be more gregarious than *G. moholi*, but tarsiers, *G. moholi*, and *G. zanzibaricus* may mob a predator (Bearder et al., 2002; Gursky, 2002, 2005). Gregariousness was low in *Lepilemur* and mobbing was not noted (Nash, pers. obs.). We need better quantitative comparative data on both mobbing patterns and gregariousness of more species of nocturnal primates as well as their reactions to moonlight.

An examination of perceptual system differences among nocturnal primates might help to sort out some of the differences in responsiveness to moonlight (Allman, 1977; Pariente, 1980). Differences in color vision may be one of the factors, since the apparently secondarily nocturnal primates *Tarsius* and *Aotus* differ from each other in genes for photopigment opsins, and there is also recently discovered variability among prosimians in these genes (Jacobs, 1995; Jacobs, 2002). Clearly more information about the qualities of light under different forest conditions would be helpful. For insectivores, as mentioned, light level might increase insect activity and increase foraging ability. For frugivores or folivores, might moonlight levels influence perception and thus location of food? Or are other senses more important (Dominy et al., 2001; Dominy, 2004; Dominy et al., 2004)?

Nocturnal Primates, Moonlight, and Modeling of Predation's Role in Primate Sociality

Many models of how predation rate and risk influence sociality in primates have usually "lumped" all but a few nocturnal animals as "solitary" and assigned a group size of 1 in analyses (Hill & Dunbar, 1998). The problem is that this treats all nocturnal primates as the same and ignores social cohesion differences, e.g., *Tarsius* (Gursky, 2005) and *Phaner* (Schülke & Kappeler, 2003), as well as social responses of nocturnal primates to predators (see above). Authors often assume all nocturnal primates emphasize crypsis to the exclusion of other responses. In fact, neither predation (Treves, 1999) nor crypsis (Bearder et al., 2002) are "unitary phenomena." When the effects of moonlight are incorporated, results of analyses of crypsis become complex, as discussed above.

Janson (1998) developed a model contrasting species that deal with predators by crypsis and so should live in small groups versus those that use warning calls and so should live in larger groups. He assumed, but could not prove with a model, that these are mutually exclusive strategies. Among his predictions were (1) insectivores would be more likely to use early warning systems than species with other diets but with similar body sizes; (2) that nocturnal primates should favor cryptic strategies; and (3) that within any species, individuals when solitary should become more cryptic upon detection of a predator. His discussion did not address effects of moonlight. It is not clear that the array of nocturnal primates reviewed here fits nicely into his model. As Janson did, I leave it to future researchers to elaborate his model and incorporate the influence of moonlight levels.

Future Research

The major impediment to understanding species variation in the responses of nocturnal primates to moonlight levels is the limited number of species for which we have data. As is the usual plaint, we need more information on the predators of primates (Isbell, 1994). However, Boinski (1995) recommends that experimental approaches are likely to be as important as, or more important than, studies of predators. Gursky (2002, 2003b, 2003a, 2005) has shown the value of such experimentation in her work on tarsiers for understanding predation's effects and has also looked at moonlight. An approach combining both moonlight information and predator model experiments can be imagined for a variety of nocturnal primates, possibly incorporating some of the approaches reviewed here for rodents.

For medium and larger size nocturnal primates the greatest predation risk might be in species where an infant is "parked" while the mother forages (e.g. tarsiers, most galagos, lorises) (Charles-Dominique, 1977; Bearder, 1987; Gursky, 1994; Kappeler, 1998; Bearder et al., 2002; Nekaris, 2003). It would be interesting though difficult to discover if females alter when or where they park infants with respect to moonlight levels. Infant slender lorises seek more cover on moonlit nights (Bearder et al., 2002).

Other possible data sets where researchers might look for moon phase effects on primate behavior are available for a number of Malagasy lemurs, in addition to those mentioned above. Since many species with different body sizes (and thus potential predation risk), diets, and degrees of locomotor agility are often sympatric, an examination of their responses to moonlight might allow testing of various hypotheses about moonlight effects while controlling for differences in habitat features. These species include *Cheirogaleus* (Fietz, 1999a,b; Fietz & Ganzhorn, 1999; Müller, 1999b,a,c), *Phaner* (Schülke, 2003b,a; Schülke & Kappeler, 2003), *Daubentonia* (Sterling, 1993; Sterling & Richard, 1995), *Mirza* (Kappeler, 1997), and *Microcebus*, (Schmid & Kappeler, 1998; Atsalis, 1999; Schmid, 1999; Atsalis, 2000; Radespiel et al., 2001; Eberle & Kappeler, 2002; Radespiel et al., 2003; Rendigs et al., 2003; Eberle & Kappeler, 2004b; Eberle & Kappeler, 2004a; Schwab & Ganzhorn, 2004; Weidt et al., 2004). Most of these data are from radio-tracking studies of individuals using focal follows. However, even trap-retrap studies, e.g., of *Microcebus* (Randrianambinina et al., 2003), might be examined to discover if "trappability" varied with moon phase (Plesner-Jensen & Honess, 1995). In addition, new information indicating that some populations of *Hapalemur* are cathemeral means that they might also be studied as a folivore that might, like *Lepilemur*, be constrained for moon phase effects on activity budgets (Mutschler, 1999; Mutschler, 2002; Mutschler & Tan, 2003). Further analysis for more nocturnal and cathemeral primates will be illuminating with regard to why many nocturnal primates are lunar philic in many behaviors, some lunar phobic in the same or other behaviors, and, for some behaviors, some species don't seem to care.

Acknowledgments. The fieldwork on *Lepilemur* was supported by the Wenner Gren Foundation, the National Geographic Society, and donations of supplies from Nails Pro and 3M Company. It was carried out under an approved animal research protocol reviewed by Arizona State University's Institutional Animal Care and Use Committee. I thank the Government of Madagascar, M. Berthe Rakotsamimana, and P. Rakotomanga for permission to work at Beza Mahafaly Special Reserve. R. Sussman, A. Richard, P. Wright, B. Andriamihaja, M. Nash, and S. O'Connor provided logistic help. For help and companionship in field: R. Randriambololona, T. Bertrand, L. Gould, J. Ratsimbaza, and the entire staff at Beza, especially Enafa. M. Nash and L. Bidner made helpful comments on earlier drafts. Thanks to Sharon Gursky and Anna Nekaris for the invitation to participate in this volume, their patience, and to Gursky's paper on lunar philia in tarsiers, which inspired much of this paper. During the final revision stage, Simon Bearder provided a key copy of the Bearder, Nekaris & Curtis unpublished manuscript, which converged on several of the ideas in this paper. Comments of reviewers significantly improved the paper.

References

Allman, J. (1977). Evolution of the visual system in the early primates. *Prog. Psychob. Physiol.*, 7: 1–53.

Altmann, J. (1974). Observational study of behaviour: Sampling methods. *Behaviour*, 49: 227–267.

Anderson, J.R. (1984). Ethology and ecology of sleep in monkeys and apes. *Advances in the Study of Behavior*, 14: 165–229.

Anderson, J.R. (2000). Sleep-related behavioural adaptations in free-ranging anthropoid primates. *Sleep Medicine Reviews*, 4: 355–373.

Atsalis, S. (1999). Seasonal fluctuations in body fat and activity levels in a rain-forest species of mouse lemur, *Microcebus rufus*. *Int. J. Primatol.*, 20: 883–910.

Atsalis, S. (2000). Spatial distribution and population composition of the brown mouse lemur *Microcebus rufus* in Ranomafana National Park, Madagascar, and its implications for social organization. *Am. J. Primatol.*, 51: 61–78.

Bearder, S.K. (1987). Lorises, bushbabies, and tarsiers: Diverse societies in solitary foragers. In B. Smuts, D. Cheney, R. Seyfarth, R. Wrangham, and T. Struhsaker (Eds.), *Primate societies* (pp. 11–24). Chicago: Univ. of Chicago Press.

Bearder, S.K., Nekaris, K.A.I., and Buzzell, C.A. (2002). Dangers in the dark: Are some nocturnal primates afraid of the dark? In L.E. Miller (Ed.) *Eat or be eaten: Predator sensitive foraging among primates* (pp. 21–40). Cambridge: Cambridge Univ. Press.

Bearder, S.K., Nekaris, K.A.I., Curtis, D.J. (2006). A re-evaluation of the role of vision in the activity and communication of nocturnal primates. *Folia Primatol.*, 77: 50–71.

Boinski, S., and Chapman, C.A. (1995). Predation on primates: Where are we and what's next? *Evol. Anthropol.*, 4: 1–3.

Bowers, M.A. (1988). Seed removal experiments on desert rodents: The microhabitat by moonlight effect. *J. Mammal.*, 69: 201–204.

Bowers, M.A. (1990). Exploitation of seed aggregates by Merriam's kangaroo rats: Harvesting rates and predatory risk. *Ecology*, 71: 2334–2344.

Brigham, M., Barclay, R. (1992). Lunar influence on foraging and nesting activity of common poorwills. *The Auk*, 109: 315–320.

Cant, J.G.H., Youlatos, D., Rose, M.D. (2001). Locomotor behavior of *Lagothrix lagothricha* and *Ateles belzebuth* in Yasuní National Park, Ecuador: General patterns and nonsuspensory modes. *J. Hum. Evol.*, 41: 141–166.

Cartmill, M. (1974). Rethinking primate origins. *Science*, 184: 436–443.

Cartmill, M. (1992). New views on primate origins. *Evol. Anthropol.*, 1: 105–111.

Charles-Dominique, P. (1977). *Ecology and behaviour of nocturnal prosimians*. London: Duckworth.

Charles-Dominique, P., Hladik, C.M. (1971). Le *Lepilemur* du sud de Madagascar: Ecologie, alimentation et vie sociale. *Terre et Vie*, 25: 3–66.

Cheney, D., Wrangham, R. (1987). Predation. In B. Smuts, D. Cheney, R. Seyfarth, R. Wrangham, and T. Struhsaker (Eds.), *Primate societies* (pp. 227–239). Chicago: Univ. of Chicago Press.

Clark, J.A. (1983). Moonlight's influence on predatory prey interactions between short-eared owls (*Asio flammeus*) and deermice (*Peromyscus manicularus*). *Behav. Ecol. Sociobiol.*, 13: 205–209.

Colquhoun, I. (1988). Cathemeral behavior of *Eulemur macaco macaco* at Ambato Massif, Madagascar. *Folia Primatol.*, 69(Suppl.1): 22–34.

Cresswell, W., and Harris, S. (1988). The effects of weather conditions on the movement and activity of badgers (*Meles meles*) in a suburban environment. *J. Zool.*, 215: 187–194.

Curtis, D.J., and Rasmussen, M.A. (2002). Cathemerality in lemurs. *Evol. Anthropol.*, 11: 83–86.

Curtis, D.J., Zaramody, A., and Martin, R.F. (1999). Cathemerality in the mongoose lemur. *Eulemur mongoz. Am. J. Primatol.*, 47: 279–298.

Daly, M., Behrends, P.R., Wilson, M.I., and Jacobs, L.F. (1992). Behavioural modulation of predation risk: Moonlight avoidance and crepuscular compensation in a nocturnal desert rodent, *Dipodomys merriami. Anim. Behav.*, 44: 1–10.

Dominy, N.J. (2004). Fruits, fingers, and fermentation: The sensory cues available to foraging primates. *Integr. Comp. Biol.*, 44: 295–303.

Dominy, N.J., Lucas, P.W., Osorio, D., and Yamashita, N. (2001). The sensory ecology of primate food perception. *Evol. Anthropol.*, 10: 171–186.

Dominy, N.J., Ross, C.F., and Smith, T.D. (2004). Evolution of the special senses in primates: Past, present, and future. *Anat. Rec.*, 281: 1078–1082.

Donati, G. (1999). Cathemeral activity of red-fronted brown lemurs (*Eulemur fulvus rufus*) in the Kirindy forest/CFPF. In B. Rakotosamimanana, H. Rasamimanana, J.U. Ganzhorn, and S.M. Goodman (Eds.), *New directions in lemur studies* (pp. 119–138). New York: Plenum.

Donati, G., Lunardini, A., Kappeler, P.M., Tarli, B. (2001). Nocturnal activity in the cathemeral red-fronted lemur (*Eulemur fulvus rufus*), with observations during a lunar eclipse. *Am. J. Primatol.*, 53: 69–78.

Eberle, M., Kappeler, P.M. (2002). Mouse lemurs in space and time: A test of the socioecological model. *Behav. Ecol. Sociobiol.*, 51: 131–139.

Eberle, M., Kappeler, P.M. (2004a). Selected polyandry: Female choice and inter-sexual conflict in a small nocturnal solitary primate (*Microcebus murinus*). *Behav. Ecol. Sociobiol.*, 57: 91–100.

Eberle, M., Kappeler, P.M. (2004b). Sex in the dark: Determinants and consequences of mixed male mating tactics in *Microcebus murinus*, a small solitary nocturnal primate. *Behav. Ecol. Sociobiol.*, 57: 77–90.

Erkert, H.G. (1974). Der Einfluss des Mondlichtes auf die Aktiviatsperiodik machtaktiver Saugetuere. *Oecologia*, 14: 269–287.

Erkert, H.G. (1976). Beleuchtungsabhängiges aktivitätsoptimum bie Nachtaffen (*Aotus trivirgatus*). *Folia Primatol.*, 25: 186–192.

Erkert, H.G. (2000). Bats: Flying nocturnal mammals. In S. Halle and N.C. Stenseth (Eds.), *Activity patterns in small mammals: An ecological approach* (pp. 253–272). Berlin: Springer.

Erkert, H.G. (2003). Chronobiological aspects of primate research. In J.M. Setchell and D.J. Curtis (Eds.), *Field and laboratory methods in primatology: A practical guide* (pp. 252–270). Cambridge: Cambridge Univ. Press.

Erkert, H.G., Kappeler, P.M. (2004). Arrived in the light: Diel and seasonal activity patterns in wild Verreaux's sifakas (*Propithecus v. verreauxi*; Primates: Indriidae). *Behav. Ecol. Sociobiol.*, 57: 174–186.

Fenton, M.B., Boyle, N.G.H., Harrison, T.M., and Oxley, D.J. (1977). Activity patterns, habitat use, and prey selection by some African insectivorous bats. *Biotropica*, 9: 73–85.

Fernandez-Duque, E. (2003). Influences of moonlight, ambient temperature, and food availability on the diurnal and nocturnal activity of owl monkeys (*Aotus azarai*). *Behav. Ecol. Sociobiol.*, 54: 431–440.

Fernandez-Duque, E., Rotundo, M. (2003). Field methods for capturing and marking Azarai night monkeys. *Int. J. Primatol.*, 24: 1113–1120.

Fietz, J. (1999a). Demography and floating males in a population of *Cheirogaleus medius*. In B. Rakotosamimanana, H. Rasamimanana, J.U. Ganzhorn, and S.M. Goodman (Eds.), *New directions in lemur studies* (pp. 159–172). New York: Plenum.

Fietz, J. (1999b). Monogamy as a rule rather than exception in nocturnal lemurs: The case of the fat-tailed dwarf lemur, *Cheirogaleus medius*. *Ethology*, 105: 255–272.

Fietz, J., and Dausmann, K.H. (2003). Costs and potential benefits of parental care in the nocturnal fat-tailed dwarf lemur (*Cheirogaleus medius*). *Folia Primatol.*, 74: 246–258.

Fietz, J., and Ganzhorn, J.U. (1999). Feeding ecology of the hibernating primate *Cheirogaleus medius*: How does it get so fat? *Oecologia*, 121: 157–164.

Fleagle, J.G. (1999). *Primate adaptation and evolution* (2nd ed.). New York: Academic Press, Inc.

Ganzhorn, J.U., Pietsch, T., Fietz, J., Gross, S., Schmid, J., and Steiner, N. (2004). Selection of food and ranging behaviour in a sexually monomorphic folivorous lemur: *Lepilemur ruficaudatus*. *J. Zool.*, 263: 393–399.

Gilbert, B.S., and Boutin, S. (1991). Effect of moonlight on winter activity of snowshoe hares. *Artic Alpine Res.*, 23: 61–65.

Gilbert, M., and Tingay, R.E. (2001). Predation of a fat-tailed dwarf lemur *Cheirogaleus medius* by a Madagascar harrier-hawk *Polyboroides radiatus*: An incidental observation. *Lemur News*, 6: 6.

Goodman, S.M. (2003). Predators on lemurs. In S.M. Goodman and J.P. Benstead (Eds.), *The natural history of Madagascar* (pp. 1221–1228). Chicago: Univ. of Chicago Press.

Goodman, S.M., Langrand, O., and Rasolonandrasana, B.P.N. (1997). The food habits of *Cryptoprocta ferox* in the high mountain zone of the Andringitra Massif, Madagascar (Carnivora: Viverridae). *Mammalia*, 61: 185–192.

Goodman, S.M., Langrand, O., and Raxworthy, C.J. (1993a). The food-habits of the barn owl *Tyto alba* at 3 sites on Madagascar. *Ostrich*, 64: 160–171.

Goodman, S.M., Langrand, O., and Raxworthy, C.J. (1993b). Food-habits of the Madagascar long-eared owl *Asio madagascariensis* in 2 habitats in southern Madagascar. *Ostrich*, 64: 79–85.

Goodman, S.M., O'Connor, S., and Langrand, O. (1993c). A review of predation on lemurs: Implications for the evolution of social behavior in small, nocturnal primates. In P.M. Kappeler and J. Ganzhorn (Eds.), *Lemur social systems and their ecological basis* (pp. 51–66). New York: Plenum Press.

Goodman, S.M., Rene de Roland, L.A., and Thorstrom, R. (1998). Predation on the eastern wooly lemur (*Avahi laniger*) and other vertebrates by Henst's goshawk (*Accipiter henstii*). *Lemur News*, 3: 14–15.

Grubb, P., Butynski, T.M., Oates, J.F., Bearder, S.K., Disotell, T.R., Groves, C.P., and Struhsaker, T.T. (2003). An assessment of the diversity of African primates. *Int. J. Primatol.*, 24: 1301–1357.

Gursky, S. (1994). Infant care in the spectral tarsier (*Tarsius spectrum*) Sulawesi, Indonesia. *Int. J. Primatol.*, 15: 843–853.

Gursky, S. (2000). Effect of seasonality on the behavior of an insectivorous primate, *Tarsius spectrum*. *Int. J. Primatol.*, 21: 477–496.

Gursky, S. (2002). Determinants of gregariousness in the spectral tarsier (Prosimian: *Tarsius spectrum*). *J. Zool.*, 256: 401–410.

Gursky, S. (2003a). Lunar philia in a nocturnal primate. *Int. J. Primatol.*, 24: 351–367.

Gursky, S. (2003b). Predation experiments on infant spectral tarsiers (*Tarsius spectrum*). *Folia Primatol.*, 74: 272–284.

Gursky, S. (2005). Predator mobbing in *Tarsius spectrum*. *Int. J. Primatol.*, 26: 207–221.

Harcourt, C.S., and Nash, L.T. (1986). Species differences in substrate use and diet between sympatric galagos in two Kenyan coastal forests. *Primates*, 27: 41–52.

Heffner, R.S., Koay, G., and Heffner, H.E. (2001). Sound localization in a new-world frugivorous bat, *Artibeus jamaicensis*: Acuity, use of binaural cues, and relationship to vision. *Journal of the Acoustical Society of America*, 109: 412–421.

Hill, R.A. (1998). Predation risk as an influence on group size in cercopithecoid primates: Implications for social structure. *J. Zool.*, 245: 447–456.

Hill, R.A., and Dunbar, R.I.M. (1998). An evaluation of the roles of predation rate and predation risk as selective pressures on primate grouping behaviour. *Behaviour*, 135: 411–430.

Isbell, L.A. (1994). Predation on primates: Ecological patterns and evolutionary consequences. *Evol. Anthropol.*, 3: 61–71.

Jacobs, G.H. (1995). Variations in primate color vision: Mechanisms and utility. *Evol. Anthropol.*, 3: 196–205.

Jacobs, G.H. (2002). Progress toward understanding the evolution of primate color vision. *Evol. Anthropol.*, 11: 132–135.

Jahoda, J. (1973). The effect of the lunar cycle on the activity pattern of *Onychomys leucogaster breviauritius*. *J. Mammal.*, 54: 544–549.

Janson, C.H. (1998). Testing the predation hypothesis for vertebrate sociality: Prospects and pitfalls. *Behaviour*, 135: 389–410.

Jenkins, S.H. (2002). Data pooling and Type I errors: A comment on Leger & Didrichsons. *Anim. Behav.*, 63: F9–F11.

Kappeler, P.M. (1997). Intrasexual selection in *Mirza coquereli*: Evidence for scramble competition polygyny in a solitary primate. *Behav. Ecol. Sociobiol.*, 41: 115–127.

Kappeler, P.M. (1998). Nests, tree holes, and the evolution of primate life histories. *Am. J. Primatol.*, 46: 7–33.

Kappeler, P.M., and Erkert, H.G. (2003). On the move around the clock: Correlates and determinants of cathemeral activity in wild redfronted lemurs (*Eulemur fulvus rufus*). *Behav. Ecol. Sociobiol.*, 54: 359–369.

Kotler, B.P. (1984). Risk of predation and the structure of desert rodent communities. *Ecology*, 65: 689–701.

Kotler, B.P., Brown, J.S., and Hasson, O. (1991). Factors affecting gerbil foraging, behavior and rates of owl predation. *Ecology*, 72: 2249–2260.

Lang, A., Kalko, E., Römer, H., Bockholdt, C., and Dechmann, D. (2006). Activity levels of bats and katydids in relation to the lunar cycle. *Oecologia*, 146: 659–666.

Langrand, O. (1990). *Guide to the birds of Madagascar*. New Haven: Yale Univ. Press.

Leger, D.W., and Didrichsons, I.A. (1994). An assessment of data pooling and some alternatives. *Anim. Behav.*, 48: 823–832.

Lima, S.L., and Dill, L.M. (1990). Behavioral decisions made under the risk of predation A review and prospectus. *Can. J. Zool.*, 68: 619–640.

Lockard, R., Owings, D. (1974a). Moon-related surface activity of bannertail (*Dipodomys spectabilis*) and Fresno (*D. nitratoides*) kangaroo rats. *Anim. Behav.*, 22: 262–273.

Lockard, R., and Owings, D. (1974b). Seasonal variation in moonlight avoidance by bannertail kangaroo rats. *J. Mammal.*, 55: 189–193.

Martin, P., and Bateson, P. (1993). *Measuring behavior.* Cambridge: Cambridge Univ. Press.

Morrison, D.W. (1978). Lunar phobia in a neotropical fruit bat, *Artibeus jamaicensis* (Chiroptera: Phyllostomidae). *Anim. Behav.*, 26: 852–855.

Müller, A.E. (1999a). Aspects of social life in the fat-tailed dwarf lemur (*Cheirogaleus medius*): Inferences from body weights and trapping data. *Am. J. Primatol.*, 49: 265–280.

Müller, A.E. (1999b). The social organisation of the fat-tailed dwarf lemur, *Cheirogaleus medius* (Lemuriformes, Primates). Doctoral dissertation. University of Zurich, Zurich, Switzerland.

Müller, A.E. (1999c). Social organization of the fat-tailed dwarf lemur (*Cheirogaleus medius*) in northwestern Madagascar. In B. Rakotosamimanana, H. Rasamimanana, J.U. Ganzhorn, and S.M. Goodman (Eds.), *New directions in lemur studies* (pp. 139–156). New York: Plenum.

Mutschler, T. (1999). Folivory in a small-bodied lemur: The nutrition of the Alaotran gentle lemur (*Hapalemur griseus alaotrensis*). In B. Rakotosamimanana, H. Rasamimanana, J.U. Ganzhorn, and S.M. Goodman (Eds.), *New directions in lemur studies* (pp. 221–238). New York: Plenum.

Mutschler, T. (2002). Alaotran gentle lemur: Some aspects of its behavioral ecology. *Evol. Anthropol.*, 11: 101–104.

Mutschler, T., and Tan, C.L. (2003). *Hapalemur*, bamboo or gentle lemurs. In S.M. Goodman and J.P. Benstead (Eds.), *The natural history of Madagascar* (pp. 1324–1329). Chicago: Univ. of Chicago Press.

Nash, L.T. (1986). Influence of moonlight level on traveling and calling patterns in two sympatric species of *Galago* in Kenya. In D. Taub and F. King (Eds.) *Current perspectives in primate social dynamics* (pp. 357–367). New York: Van Nostrand Reinhold Co.

Nash, L.T. (1998). Vertical clingers and sleepers: Seasonal influences on the activities and substrate use of *Lepilemur leucopus* at Beza Mahafaly Special Reserve, Madagascar. *Folia Primatol.*, 69(Suppl.1): 204–217.

Nash, L.T. (2000). Encounter rate estimates on *Lepilemur leucopus* and *Microcebus murinus* at Beza Mahafaly Special Reserve, southwestern Madagascar. *Lemur News*, 5: 38–40.

Nekaris, K.A.I. (2003). Observations of mating, birthing and parental behaviour in three subspecies of slender loris (*Loris tardigradus* and *Loris lydekkerianus*) in India and SriLanka. *Folia Primatol.*, 74: 312–336.

Nekaris, K.A.I., and Rasmussen, D.T. (2003). Diet and feeding behavior of Mysore slender lorises. *Int. J. Primatol.*, 24: 33–46.

Nelson, D. (1989). Gull predation on Cassin's auklet varies with the lunar cycle. *The Auk*, 106: 495–497.

Ortega, J., and Castro-Arellano, I. (2001). *Artibeus jamaicensis. Mammalian Species*, 662: 1–9.

Overdorff, D.J. (1988). Preliminary report on the activity cycle and diet of the red-bellied lemur (*Lemur rubriventer*) in Madagascar. *Am. J. Primatol.*, 16: 143–153.

Overdorff, D.J., and Rasmussen, M.A. (1995). Determinants of nighttime activity in "diurnal" lemurid primates. In L. Alterman, G.A. Doyle and M.K. Izard (Eds.), *Creatures of the dark: The nocturnal prosimians* (pp. 61–74). New York: Plenum Press.

Packer, W.C. (1965). Environmental influences on daily and seasonal activity in *Setonix brachyurus* (Quoy and Gaimard) (Marsupialia). *Anim. Behav.*, 13: 270–283.

Pariente, G.F. (1979). The role of vision in prosimian behavior. In G.A. Doyle and R. Marin (Eds.), *The study of prosimian behavior* (pp. 411–459). New York: Academic Press, Inc.

Pariente, G.F. (1980). Quantitative and qualitative study of the light available in the natural biotope of Malagasy prosimians. In P. Charles-Dominique, H.M. Cooper, A. Hladik, C.M. Hladik, E. Pagès, G.F. Pariente, A. Petter-Rousseaux, A. Schilling, and J.J. Petter (Eds.), *Nocturnal Malagasy primates: Ecology, physiology, and behavior* (pp. 117–134). New York: Academic Press, Inc.

Peetz, A., Norconk, M.A., and Kinzey, W.G. (1992). Predation by jaguar on howler monkeys (*Alouatta seniculus*) in Venezuela. *Am. J. Primatol.*, 28: 223–228.

Plesner-Jensen, S., and Honess, P. (1995). The influence of moonlight on vegetation height preference and trappability of small mammals. *Mammalia*, 59: 35–42.

Price, M., Waser, N., and Bass, T. (1984). Effects of moonlight on microhabitat use by desert rodents. *J. Mammal.*, 65: 211–219.

Radespiel, U., Lutermann, H., Schmelting, B., Bruford, M.W., and Zimmermann, E. (2003). Patterns and dynamics of sex-biased dispersal in a nocturnal primate, the grey mouse lemur, *Microcebus murinus. Anim. Behav.*, 65: 709–719.

Radespiel, U., Sarikaya, Z., Zimmermann, E., and Bruford, M.W. (2001). Sociogenetic structure in a free-living nocturnal primate population: Sex-specific differences in the grey mouse lemur (*Microcebus murinus*). *Behav. Ecol. Sociobiol.*, 50: 493–502.

Radhakrishna, S., and Singh, M. (2002). Social behaviour of the slender loris (*Loris tardigradus lydekkerianus*). *Folia Primatol.*, 73: 181–196.

Randrianambinina, B., Rakotondravony, D., Radespiel, U., and Zimmermann, E. (2003). Seasonal changes in general activity, body mass and reproduction of two small nocturnal primates: A comparison of the golden brown mouse lemur (*Microcebus ravelobensis*) in northwestern Madagascar and the brown mouse lemur (*Microcebus rufus*) in eastern Madagascar. *Primates*, 44: 321–331.

Rasoloarison, R.M., Rasolonandrasana, B.P.N., Ganzhorn, J.U., and Goodman, S.M. (1995). Predation on vertebrates in the Kirindy forest, western Madagascar. *Ecotropica*, 1: 59–65.

Rasoloharijaona, S., Rakotosamimanana, B., Randrianambinina, B., and Zimmermann, E. (2003). Pair-specific usage of sleeping sites and their implications for social organization in a nocturnal Malagasy primate, the Milne Edwards' sportive lemur (*Lepilemur edwardsi*). *Am. J. Phys. Anthropol.*, 122: 251–258.

Ratsirarson, J., and Emady, R.J. (1999). Prédation de *Lepilemur leucopus* par *Buteo brachpterus* dans la forêt galarie de Beza Mahafaly. *Working Group on Birds in the Madagascar Region Newsletter*, 9: 14–15.

Rendigs, A., Radespiel, U., Wrogemann, D., and Zimmermann, E. (2003). Relationship between microhabitat structure and distribution of mouse lemurs (*Microcebus spp.*) in northwestern Madagascar. *Int. J. Primatol.*, 24: 47–64.

Richard, A.F., Rakotomanga, P., and Schwartz, M. (1991). Demography of *Propithecus verreauxi* at Beza Mahafaly, Madagascar: Sex ratio, survival, and fertility, 1984–1988. *Am. J. Phys. Anthropol.*, 84: 307–322.

Sauther, M.L. (1989). Antipredator behavior in troops of free-ranging *Lemur catta* at Beza Mahafaly Special Reserve, Madagascar. *Int. J. Primatol.*, 10: 595–606.

Sauther, M.L. (1991). Reproductive behavior of free-ranging *Lemur catta* at Beza Mahafaly Special Reserve, Madagascar. *Am. J. Phys. Anthropol.*, 84: 463–477.

Schmid, J. (1999). Sex-specific differences in activity patterns and fattening in the gray mouse lemur (*Microcebus murinus*) in Madagascar. *J. Mammal.*, 80: 749–757.

Schmid, J., and Kappeler, P.M. (1998). Fluctuating sexual dimorphism and differential hibernation by sex in a primate, the gray mouse lemur (*Microcebus murinus*). *Behav. Ecol. Sociobiol.*, 43: 125–132.

Schülke, O. (2001). Social anti-predator behaviour in a nocturnal lemur. *Folia Primatol.*, 72: 332–334.

Schülke, O. (2003a). Living apart together: Patterns, ecological basis, and reproductive consequences of life in dispersed pairs of fork-marked lemurs (*Phaner furcifer*, Primates). Doctoral dissertation. Universität Würzburg, Würzburg, Germany.

Schülke, O. (2003b). To breed or not to breed—Food competition and other factors involved in female breeding decisions in the pair-living nocturnal fork-marked lemur (*Phaner furcifer*). *Behav. Ecol. Sociobiol.*, 55: 11–21.

Schülke, O., and Kappeler, P.M. (2003). So near and yet so far: Territorial pairs but low cohesion between pair partners in a nocturnal lemur, *Phaner furcifer*. *Anim. Behav.*, 65: 331–343.

Schülke, O., and Ostner, J. (2001). Predation on *Lepilemur* by a harrier hawk and implications for sleeping site quality. *Lemur News*, 6: 5.

Schwab, D., and Ganzhorn, J.U. (2004). Distribution, population structure and habitat use of *Microcebus berthae* compared to those of other sympatric cheirogalids. *Int. J. Primatol.*, 25: 307–330.

Sokal, R.R., and Rohlf, F.J. (1995). *Biometry*. New York: W.H. Freeman and Company.

Stanford, C.B. (2002). Avoiding predators: Expectations and evidence in primate antipredator behavior. *Int. J. Primatol.*, 23: 741–758.

Sterling, E.J. (1993). Patterns of range use and social organization in aye-ayes (*Daubentonia madagascariensis*) on Nosy Mangabe. In P.M. Kappeler and J. Ganzhorn (Eds.), *Lemur social systems and their ecological basis* (pp. 1–10). New York: Plenum Press.

Sterling, E.J., and Richard, A.F. (1995). Social organization in the aye-aye (*Daubentonia madagascariensis*) and the perceived distinctiveness of nocturnal primates. In L. Alterman, G.A. Doyle, and M.K. Izard (Eds.), *Creatures of the dark: The nocturnal prosimians* (pp. 439–451). New York: Plenum Press.

Sussman, R.W. (1991). Demography and social organization of free-ranging *Lemur catta* in the Beza Mahafaly Reserve, Madagascar. *Am. J. Phys. Anthropol.*, 84: 43–58.

Sussman, R.W., and Rakotozafy, A. (1994). Plant diversity and structural analysis of a tropical dry forest in southwestern Madagascar. *Biotropica*, 26: 241–254.

Thalmann, U. (2002). Contrasts between two nocturnal leaf-eating lemurs. *Evol. Anthropol.*, 11: 105–107.

Trent, B.K., Tucker, M.E., and Lockard, J.S. (1977). Activity changes with illumination in slow loris *Nycticebus coucang*. *Appl. Anim. Ethol.*, 3: 281–286.

Treves, A. (1999). Has predation shaped the social systems of arboreal primates? *Int. J. Primatol.*, 20: 35–68.

Troisi, A., Schino, G., D'Antoni, M., Pandolfi, N., Aureli, F., and D'Amato, F.R. (1991). Scratching as a behavioral index of anxiety in macaque mothers. *Behav. Neural Biol.*, 56: 307–313.

van Schaik, C.P. (1983). Why are diurnal primates living in groups? *Behaviour*, 87: 120–143.

van Schaik, C.P., and van Hooff, J.A.R.A.M. (1983). On the ultimate causes of primate social systems. *Behaviour*, 85: 91–117.

Warren, R.D., and Crompton, R.H. (1997). A comparative study of the ranging behaviour, activity rhythms and sociality of *Lepilemur edwardsi* (Primates, Lepilemuridae) and *Avahi occidentalis* (Primates, Indriidae) at Ampijoroa, Madagascar. *J. Zool.*, 243: 397–415.

Watanuki, Y. (1986). Moonlight avoidance behavior in Leach's storm-petrels as a defense against slaty-backed gulls. *The Auk*, 103: 14–22.

Weidt, A., Hagenah, N., Randrianambinina, B., and Radespiel, U. (2004). Social organization of the golden brown mouse lemur (*Microcebus ravelobensis*). *Am. J. Phys. Anthropol.*, 123: 40–51.

Wiens, F., and Zitzmann, A. (1999). Predation on a wild slow loris (*Nycticebus coucang*) by a reticulated python (*Python reticulatus*). *Folia Primatol.*, 70: 362–364.

Wiens, F., and Zitzmann, A. (2003a). Social dependence of infant slow lorises to learn diet. *Int. J. Primatol.*, 24: 1007–1021.

Wiens, F., and Zitzmann, A. (2003b). Social structure of the solitary slow loris *Nycticebus coucang* (Lorisidae). *J. Zool.*, 261: 35–46.

Wolfe, J.L., and Tan Summerlin, C. (1989). The influence of lunar light on nocturnal activity of the old-field mouse. *Anim. Behav.*, 37: 410–414.

Wright, P.C. (1978). Home range, activity pattern, and agonistic encounters in a group of night monkeys (*Aotus trivirgatus*) in Peru. *Folia Primatol.*, 29: 43–55.

Wright, P.C. (1989). The nocturnal primate niche in the New World. *J. Hum. Evol.*, 18: 635–658.

Wright, P.C. (1994). The behavior and ecology of the owl monkey. In J.F. Baser, R.E. Weller, and I. Kakoma (Eds.), *Aotus: The owl monkey* (pp. 97–112). San Diego: Academic Press.

Wright, P.C. (1996). The neotropical primate adaptation to nocturnality: Feeding in the night (*Aotus nigriceps* and *A. azarae*). In M.A. Norconk, A.A. Rosenberger, and P.A. Garber (Eds.), *Adaptive radiations of Neotropical primates* (pp. 369–382, 546–547). New York: Plenum Press.

Zinner, D., Hilgartner, R.D., Kappeler, P.M., Pietsch, T., and Ganzhorn, J.U. (2003). Social organization of *Lepilemur ruficaudatus*. *Int. J. Primatol.*, 24: 869–888.

Zollner, P.A., and Lima, S.L. (1999). Illumination and the perception of remote habitat patches by white-footed mice. *Anim. Behav.*, 58: 489–500.

9

A Comparison of Calling Patterns in Two Nocturnal Primates, *Otolemur crassicaudatus* and *Galago moholi* as a Guide to Predation Risk

Simon K. Bearder

Introduction

Predation on nocturnal primates is rarely witnessed, but a strong indication of predation risk can be inferred from the reactions of potential prey species to various sources of danger. In these circumstances, some nocturnal primates exhibit a rich array of calls whereas others are relatively silent. But predators are not the only cause of anxiety, and identical calls may sometimes be given when an animal is threatened by a conspecific. Individuals also call when in vulnerable situations, for example, if they are out of touch with companions, on crossing an open space, or when unable to leap between branches. For these reasons, calling patterns provide not only a means to assess the influence of potential predators, but also the extent of other dangers faced by individuals and the mechanisms they use in avoiding them.

Here I make a detailed functional comparison of the calls made in the context of exposure to all kinds of danger for two highly vocal galago species (bushbabies) that were each studied for over 12 months in both the field and in captivity. Captivity provides ideal conditions for hearing faint or subtle calls, whereas a more natural frequency and pattern of use is more likely in the wild. Careful monitoring of the full context in which calls occur and the responses they evoke from others of the same or different species help to determine the functional significance of each call (Charles-Dominique & Petter, 1979; Zimmermann, 1985; Zimmermann, 1995a,b). This monitoring also indicates whether or not calls carry specific information about the evoking stimulus (referential signaling) or whether they reflect the state of arousal of the caller and the fact that it is aware of the danger (perception advertisement or pursuit deterrent signals) (Seyfarth et al., 1980; Macedonia & Evans, 1993; Hasson, 1991). Building on case studies of these two species, I also discuss what is known about the calling patterns of other galagos when exposed to danger.

A useful definition of "communication" is this: It is the process whereby actors use specifically designed signals or displays to modify the behavior of reactors (Krebs & Davies, 1993). Animals are expected to call if it brings advantages to either themselves or their kin (Harvey & Greenwood, 1978). Zimmermann et al. (1988) divide the calls of *Galago senegalensis* and *G. moholi* into three functional categories: (1) social cohesion and spacing; (2) agonistic encounters; and (3) attention and alarm. Considerable attention has been paid to the first category, which usually represents up to 75–95% of all calling bouts heard in the wild (Honess, 1996; Ambrose, 1999) and includes some of the loudest calls in the repertoire, which are used to advertise the presence of the caller to companions and rivals (Zimmermann, 1995a,b; Bearder et al., 1995). This paper, however, is restricted to the last two categories and includes all calls that are given when an individual perceives a threat. Relatively few studies of these calls have been published in the case of galagos, although they are often included in MPhil and PhD theses (Andersson, 1969; Bearder, 1974; Honess, 1996; Ambrose, 1999; Wallace, 2005). Notable exceptions are Charles-Dominique (1977), Petter & Charles-Dominuque (1979), Zimmermann (1985) and Becker et al. (2003).

Study Animals and Methods

Continuous follows of focal individuals of large-eared greater bushbabies (*Otolemur crassicaudatus*) and southern lesser galagos (*Galago moholi*) (Grubb et al., 2003) were conducted over a period of 663 and 1016 hrs respectively in the Northern Province and KwaZulu, South Africa, and Zimbabwe (Bearder, 1969; Bearder, 1974; Bearder & Martin, 1980). Following a cautious approach, the study animals became habituated to the observer within one hour, and the use of binoculars and red light from a headlamp facilitated observations from a distance of 2–15 m (Charles-Dominique & Bearder, 1979). In the case of *G. moholi*, some individuals were also radio tagged and calls noted when they were trapped and handled (Bearder & Martin, 1980). Observations were made during every month of the year. Additional records (130 hrs) were made of the same species in captivity at the University of the Witwatersrand in Johannesburg (Bearder & Doyle, 1974). Calls were given descriptive labels and noted using ad libitum sampling, together with relevant information about the context of the call, the state of the caller and any apparent responses (Altmann, 1974). Discrete calls are relatively distinct in structure and easily separated, while the parts of a graded (or continuous) series merge into one another via a series of intermediates, which make the naming of divisions somewhat arbitrary (Hockett, 1960; Marler, 1961). However, in addition to structural variation (syntactic), continuous patterns also have functional variation (pragmatic), which can be judged from differences in the context in which the calls occur and the responses of other animals (Altmann, 1967). Mixed calls are combinations of different calls that are given in a single bout of calling.

Zimmermann (1995) has pointed to the difficulty of comparing the size and shape of vocal repertoires between species, especially when different investigators study these repertoires with different methods. These include: (1) differences in the quality of recording and analyzing techniques; (2) differences in sampling accuracy in captivity compared to the wild, or in calls that are relatively quiet; and (3) different ways of categorizing each call, which result in different numbers of calls recognized for the same species. Bias is minimized here by restricting comparisons to functional aspects of calls distinguished by a single observer at close range and irrespective of how many times they were heard.

The species populations differ in body size (maximum *O. crassicaudatus* 2,000 g versus *G. moholi* 330 g) and habitat preferences (riparian forest vs. acacia savannah), and they are sympatric in some areas. Both species sleep either alone or in small groups, but *O. crassicaudatus* is unusual amongst galagos in forming cohesive groups at night (a mother and up to three offspring), particularly when fruit is abundant (Bearder, 1987). In southern Africa there is evidence of predation on these two species at night by genets (*Genetta tigrina*), jackals (*Canis mesomelas*), domestic dogs and cats, eagle owls (*Bubo africanus*), and snakes (*Naja naja, Naja nigricillis, Dendroapsis polylepis, Python sebae*). During the day galagos are at risk from birds of prey (*Aquila verreauxi, Polemaetus bellicosus*), monitor lizards (*Varanus spp.*) and snakes (Pullen, 2000; Bearder et al., 2002). Humans are probably the major cause of danger through their disturbing of habitat and hunting for bushmeat or for the pet trade (Charles-Dominique & Bearder, 1979).

Results

The repertoires of calls used by each species in situations of potential danger are examined in detail below. Three aspects are noted for each call: (1) the situations evoking the call, (2) observed responses, and (3) associated calls. This illustrates the complex interplay between calls with changes in context and levels of arousal.

Case Study 1: Otolemur crassicaudatus

In threatening circumstances adults of this species give calls that have been divided into one continuous series consisting of seven parts (calls 1–7, Table 9.1). Continuous calls are generally given in bouts of varying duration up to 1 hr 30 min. They may be mixed with six discrete calls (calls 8–13) of shorter duration. Additional calls are given by infants and juveniles in distress (calls 14–16). *O. crassicaudatus* gives a further two calls, which are associated with social cohesion: "cries" and "clucks." They are not listed here (Bearder, 1974).

1. *Sniff*: When faced with a situation that evokes curiosity (e.g., the observer) adults sometimes react by approaching hesitantly, staring and pointing their noses toward the object of interest while giving a "sniff." Other members of the foraging group may then stare briefly. Juveniles reacted in this way to resting wood owl (*Ciccaba woodfordii*). This very low intensity call is sometimes accompanied by drawn out "moans" (a discrete call) and can grade to or from "knocks."

TABLE 9.1. A comparison of related calls in the continuous "mobbing" sequence of *O. crassicaudatus*, showing increasing intensity from left to right.

Feature	Type of Vocalization						
	Sniff	Knock	Creak	Squawk	Whistled-yap	Whistle	Chirp
Loudness	faint	faint to intermediate	intermediate	intermediate to loud	loud	loud	loud
Pitch	low	low	high	intermediate	high	high	very high
Quality	nasal	breathy to guttural	squeaky	harsh	shrill	shrill to queaky	rasping to squeaky
Units/min.	<60	42–108	78–108	24–54	54–114	78–114	>114
Maximum duration	2 min.	40 min	5 min	20 min	1 hr 30 min	25 min	1 hr
Frequency of occurrence	rare	frequent	occasional	frequent	frequent	occasional	rare
Associated discrete calls	moans	moans, spits, screams	moans	chatter, screams	chatter	chatter	moans

2. *Knock*: A low intensity alarm call. "Knocks" are made following any disturbance when an animal is in a vulnerable position or when a bushbaby first sees
a potential danger, such as the observer or a minor potential predator (e.g.,
cat or genet). Knocks are also common around the time of mating and when
females are accompanied by infants. Such females are abnormally agitated by
the presence of an observer and they continue to call long after the youngsters
have lost interest. The caller may stare intently, move hesitantly, urinate and
approach, or move away from the evoking stimulus. Up to three animals may
call together. Knocks become louder and faster when the danger is close by,
but trail off or stop as the distance increases, and can be interspersed by occasional "moans." They sometimes become faster and higher pitched, grading
into "creaks," or they become louder and harsher, passing imperceptibly into
"squawks."
3. *Creak*: "Creaks" precede or follow high intensity alarm induced by potential
predators and occur during mild agonistic social interactions or following a
disturbance such as that caused by an observer; however, this call usually occurs
when danger is not imminent or when the stimulus has ceased to attract the full
attention of the caller; its calling subsiding as it moves away. Creaks can grade
to or from "knocks" or "whistles" and are often accompanied by "moans." The
response of others is unknown.
4. *Squawk*: "Squawks" are given when a bushbaby is startled or threatened by the
movements of a large potential predator (observer, bushpigs, domestic dogs, or
jackals) or by any major change in the environment such as tree felling. They
were once given toward sleeping monkeys (*Cercopithecus aethiops*) and during
agonistic encounters. The caller often moves to a vantage point and remains
still, staring at the disturbance. Alternately, it moves away from the stimulus,
particularly when this is a rival bushbaby. Squawks are loud calls that may
start abruptly, or build up from and subside to knocks. At higher levels of
arousal they grade into "whistled-yaps" or "whistles." When highly agitated,
squawks are occasionally interspersed by "chatters" and may be accompanied
by "screams" when interacting with a conspecific. Other bushbabies in the
vicinity may remain still and stare, join in with a 'squawk or whistled-yap,
or continue their activity after briefly looking toward the call from a distance.
5. and 6. *Whistled-yap and whistle*: These two calls occur in similar situations to
the squawk but, by contrast, they may persist long after the evoking stimulus
has gone (up to 1 hr 30 min). They are given toward potential predators (e.g.,
cat, genet, eagle owl) and unfamiliar objects following major disturbance by the
observer and during and after agonistic encounters. The caller gives the impression of considerable agitation, moving back and forth, staring fixedly or looking
in all directions while the call increases in intensity, subsides or stops intermittently. The whistled-yap often develops from a squawk, varying considerably in
tone, pitch, and speed of repetition, becoming a whistle or subsiding to a creak.
Whistled-yaps and whistles may be interspersed with short bouts of a discrete
call, a rapid chatter. Other bushbabies may look toward this call from a distance

or join in chorus with two or three animals beginning and ending the call almost simultaneously.

7. *Chirp*: This call was made during an interaction between an adult male bushbaby and a large snake (species unidentified). It was associated with knocks and moans as the bushbaby kept the snake in view, approaching it to within one meter and following it when it moved. The calling increased in intensity as the bushbaby advanced. The interaction lasted for over an hour at dawn, causing the bushbaby to return to its sleeping place 25 minutes later than usual.

8. *Moan*: "Moans" are evoked in many potentially dangerous situations, e.g., by strange animals (snake or genet) or objects (photographic equipment), during social encounters and when an animal is in a vulnerable position low down in a tree or moving between trees. The caller shows signs of wariness, approaching hesitantly, staring or moving away from the stimulus. During social encounters moans are usually made by the individual that is being approached, irrespective of its size, age, or sex relative to the approaching animal. The exception is adult males, which sometimes give moans when approaching a female. When it is used in a social context the call accompanies chasing and grappling and is usually followed by allogrooming, particularly when a male joins a female with offspring after a long period of separation. The moan appears to represent a mood of mild anxiety, which may act as a qualifier to other anxiety/alarm calls (sniff, knock) or those involved with agonistic contact (spit, scream). The call induces others to stare at or move away from a danger or to approach the caller.

9. *Chatter*: The "chatter" is intimately linked to the high intensity calls of the alarm series without grading into them. It interrupts bouts of whistles and whistled-yaps or, occasionally, squawks. The behavior of the caller indicates that it is intensely excited, but the exact context is not clear.

10. *Hack*: "Hacks" are made spontaneously by juveniles or adults when they are startled, for example, by the noise of a camera when they are being photographed or by the sudden approach of another animal such as a swooping owl. The caller retreats rapidly and then remains tense and alert. The same call is given during contact between bushbabies while grooming or playing.

11. *Spit*: Spitting occurs in situations of close contact between bushbabies (during play, grooming, fighting, competing for food, and prior to sleeping together) or with another species (a genet), and when handled by the observer after trapping. It is accompanied by lunging or cuffing at the protagonist or by retreat of the caller, and it usually serves to avert or at least subdue the approach of another animal. Spits are given by all age classes in association with knocks, moans, and, occasionally, screams.

12. *Scream*: "Screams" occur during agonistic contact between bushbabies involved in a skirmish or fight. Screams are particularly common during the mating season when they are often preceded or followed by an advertising call ("cry"). In one instance an adult female screamed when she was merely held by a male, after which the two sat quietly together. Screams are associated with spits, knocks, moans, and squawks.

13. *Yell*: This blood-curdling series of loud cries was heard only once in the field when a guard dog caught a bushbaby in a garden. The same call was made when the observer handled the bushbaby. On one occasion when an adult gave this call on being handled in captivity, his "yell" instantly provoked squawks from his cage mates, who pounced aggressively onto the cage mesh next to the handler.

14. and 15. *Squeak and click*: Young bushbabies make a number of clicking, crackling, and squeaking notes. "Squeaks" are given by struggling infants of a few days old while being carried by the mother. "Clicks" can be heard in infants up to the age of seven months. They occur in situations of distress, particularly after the offspring loses physical contact with the mother. As soon as infants are capable of following the mother there is a sharp increase in the amount of clicking. Older infants and juveniles move around actively while calling, showing signs of agitation. These calls evidently act as contact signals when the infant counter-calls the mother, but they also occur in situations that evoke alarm calls in older animals (e.g., after falling to the ground or before making an awkward jump). Clicks are no longer heard by the time the young are late juvenile or sub-adult, at which time they are replaced by "buzzes." They evoke a retrieval response in the mother.

16. *Buzz*: Juveniles give bouts of "buzzes" that are intermittently loud and soft. Adults rarely give buzzes, but do so when members of a foraging group become separated. Buzzes may be directly preceded by loud advertising calls ("cries") or interspersed with clicks. Buzzing is accompanied by searching movements that can be frantic. The caller moves back and forth rapidly, stopping now and again to stare into the distance or to sniff a branch. The caller responds to sounds of moving branches or the calls of its fellows by going immediately toward these sounds. Answering buzzes were rare, but in general individuals that had been left behind soon rejoined their group and their calling ceased.

Case Study 2: Galago moholi

When faced with danger, southern lesser galagos give two continuous series of anxiety and alarm calls, each grading from sneezes (call 1) into three and five parts, respectively (calls 2–4 and 5–9, Table 9.2). There are also seven discrete calls (calls 10–16) and infants make a further two distress calls (17–18). Three other calls, associated with social cohesion, "barks," "hoots," and "coos," are not considered here (Andersson, 1969; Zimmermann et al., 1988).

1. *Sneeze*: Curiosity is signaled by bouts of "sneezes" that an animal can make almost continuously or sporadically when on the move, particularly when an object or animal evokes interest. These calls start after any sudden disturbance (e.g., by the observer), on landing after a jump, or during social interactions. They vary in intensity and speed of repetition, with single sneezes replaced by double and triple sneezes at a faster rate as attention increases. Sneezes can induce sneezes from galagos nearby, or cause them to jump away from a novel

TABLE 9.2A. A comparison of related calls in the first anxiety/alarm sequence of *G. moholi*, showing increasing intensity from left to right.

Feature	Type of Call			
	Sneeze	Explosive cough	Whistle	Sob
Loudness	faint	intermediate	loud	loud
Pitch	low	intermediate	high	high
Quality	nasal	explosive	plaintive	mournful
Units/min	–	12–20	12–20	12–20
Maximum duration	15	3	40	2
Frequency of occurrence	frequent	frequent	occasional	rare
Associated discrete calls		shivering-stutter, spits, spit-grunts, chatter, screams	shivering-stutter, spits, spit-grunts, chatter, screams	shivering-stutter, spits, spit-grunts, chatter, screams

TABLE 9.2B. A comparison of related calls in the second anxiety/alarm sequence of *G. moholi*, showing increasing intensity from left to right.

Feature	Type of Call					
	Sneeze	Gerwhit	Cluck	Yap	Yap-alarm	Wuff/wail
Loudness	faint	intermediate	intermediate	loud	very loud	very loud
Pitch	low	low	intermediate	intermediate	high	high
Quality	nasal	subdued	intermediate	strident	strident	shrill
Units/min	–	60–70	70–90	90–100	110–180	200
Maximum duration	15	3	3	60	5	2
Frequency of occurrence	frequent	occasional	occasional	frequent	occasional	rare
Associated discrete calls		shivering-stutter, spits, spit-grunts, chatter, screams	shivering-stutter, spits, spit-grunts, chatter, screams	shivering-stutter, spits, spit-grunts, chatter, screams	shivering-stutter, caws	shivering-stutter, caws

object. They can develop into either explosive coughs or "gerwhits" that are the starting points of the two-graded call series associated with anxiety and alarm.

2. *Explosive cough*: Any situation that causes anxiety may be associated with "explosive coughs," with the number and speed of repetition of coughs increasing as the danger intensifies. A mother gives this call toward strange objects or potential predators when her infant falls from the tree. Subordinate animals almost invariably give explosive coughs before and during social interactions, even when the caller is 20–30 m from an approaching galago. Coughs grade into "whistles" and accompany "grunts," "spits," and "chatters" during agonistic interactions.

3. *Whistle*: "Whistles" occur in the presence of domestic cats and jackals, when a galago is disturbed by an observer, and during agonistic encounters. They are frequently associated with yaps when the animal is alarmed and may grade into "sobs" during social interactions (chasing and grappling). Other individuals may join in with whistles or yaps or stare towards the disturbance.

4. *Sob*: "Sobs" reflect high levels of anxiety in subordinate animals during intense bouts of chasing and fighting between bushbabies. This call may be mixed with other elements of the anxiety and alarm series. No obvious response is shown by others.

5. and 6. *Gerwhit and cluck*: These calls may develop from sneezes at low levels of arousal from dangers such as the presence of the observer or a potential predator, and they also may accompany mildly agonistic social interactions, which also evoke explosive-coughs, grunts, spits and chatter. "Gerwhits" grade into "clucks" and then to "yaps" if the cause of disturbance increases, and they replace yaps following high intensity alarm situations as the animal calms down.

7. and 8. *Yap and yap-alarm*: As an animal becomes more alarmed it will give "yaps," grading into "yap-alarms" towards cats, genets, eagle owls, jackals, or the observer from a place of safety, sometimes interspersed with whistles, when it is anxious, and caws, when it is highly alarmed. These calls are also made when the animal is being chased by a conspecific, at which time the calls are combined with grunts, spits, and chatter. Others may join in and mob a predator, approaching it while calling. The call attracts the mother when given by youngsters, and females with infants will give yaps towards predators that they would otherwise ignore (e.g., a genet in a tree as opposed to on the ground).

9. *Wuff/wail*: The highest intensity expressions of alarm during bouts of yapping are "wuffs" and "wails" that grade to and from yap-alarm and may be associated with caws and sobs. The caller shows signs of extreme agitation toward a predator or an aggressive conspecific.

10. *Shivering stutter*: This is a discrete call that reflects mild anxiety as a qualifier to all other calls noted above. It is often given on first contact with an observer or a minor predator and causes a startle reaction in conspecifics nearby who may join in with the call. It is given by males around the time of mating when several males surround a single female and during coitus.

11. *Caws*: "Caws" are discrete calls linked to the highest levels of alarm and mixed with other alarm calls in complex sequences. Other individuals freeze when they hear the call.

12. *Grunts*: "Grunts" are made by subordinate animals at the close approach of an assertive conspecific or the observer. The call is a frequent component of agonistic interactions. Both adults and infants give this call if they are hiding in a tree hole during the day, being disturbed by the observer or by a potential predator (monitor lizard). Reverberations from the grunt give the impression of a larger animal inside the hole.

13., 14., and 15. *Spits, chatters, and screams*: If grunts do not deter physical conspecifics or potential predators, then subordinate animals will "spit," "chatter," and "scream." These calls are mixed with calls from the two alarm sequences during all agonistic interactions. The speed of calling and rate at which different calls are mixed together results in a cacophony of sound that represents changes in the level of aggression during the interaction and increase or decrease in proximity between the two individuals. Spits and biting movements are given when an individual is handled and are likely to be associated with the close approach of a predator, although this was not witnessed.

16. *Rasps*: "Rasps" are heard very rarely and are difficult to classify. They are the only calls made by an attacking individual in circumstances that are the reverse of normal dominance/subordination relationships, e.g., when a female with infants chases a male, or when an animal in captivity attacks a caretaker.

17. and 18. *Squeaks and clicks*: These calls are made by infants and juveniles in distress situations or when they are approaching or following the mother.

Discussion

The repertoires of *O. crassicaudatus* and *G. moholi* associated with various sources of disturbance are divided into16 and 18 calls respectively (Table 9.3). Moynihan (1964, 1967) notes that the night monkey (*Aotus trivirgatus*) uses a series of relatively discrete calls. He suggests, in accordance with Marler (1965, 1967) and Altmann (1967), that discrete calls help this animal avoid ambiguity and might therefore be expected in nocturnal and arboreal species. Clearly, this is not the case with *O. crassicaudatus* and *G. moholi* , which have one and two continuous call series, respectively, giving them the capacity to express fine gradations of mood as well as more discrete signals.

The typical response to low levels of danger is to give one or two types of call that grade from one sound to another, becoming faster, louder and more intense as the level of arousal increases and vice verse. Such continuous calls can be mixed with, or replaced by, discrete calls in complex and rapidly changing sequences. For example, when a subordinate *G. moholi* is chased by another galago before a fight it will give yaps interspersed with whistles. It descends to the ground and gives a mixture of spits, grunts, chatter, and explosive coughs when pursued, depending on the proximity of the assailant, passing into intense chatter and screams when attacked. Once it escapes, the calls subside to yaps and then stop when the opponent moves away. It is only the subordinate animal that calls as it retreats, indicating its level of arousal, while the approaching individual always remains silent. Similarly, when faced by a potential predator, galagos will give many of the same calls that are associated with disturbance by a conspecific. The overriding impression gained from observing behavior during bouts of calling that result from any kind of disturbance is that calls are motivated by different types and levels of anxiety and alarm.

Studies of calls associated with situations of danger have been published for 13 other species of galagos (Honess, 1996; Zimmermann et al., 1988; Ambrose, 1999), but the very different ways of dividing the spectrum of calls make meaningful comparisons almost impossible. At the broadest level, a survey of nocturnal primates suggests that differences in both the frequency of calling and the size and composition of the call repertoire between species, and between populations, corresponds to the vulnerability of the caller; calling is frequent when individuals are faced with dangers that can be avoided but stops when potential

TABLE 9.3. A comparison of calls associated with all kinds of danger in *O. crassicaudatus* and *G. moholi* showing functionally analogous calls and those with no functional equivalent in the other species. Other calls are also listed.

Species	
O. crassicaudatus	*G. moholi*
Functionally analogous calls	
knock	shivering stutter
creak	gerwhit/cluck
squawk	whistles and variants
whistled-yap	yap
whistle	yap alarm
chirp	wuff/wail
chatter	caw
moan	explosive cough
spit	spit-grunt
scream	scream/fighting chatter
click	click
squeak	squeak
Calls with no functional equivalent	
sniff	
buzz	
hack	
yell	
	sneeze
	sob/moan
	grunt
	rasp
Other calls	
cry	bark
cluck	coo
	hoot

costs are high. Several observations support this conclusion. First, the most active and agile species tend to be the most vocal (galagos versus lorises and pottos; slender lorises versus slow lorises) (Charles-Dominique, 1977; Pimley, this volume; Nekaris et al., this volume), and among these the smallest species (which presumably attract a greater number of potential predators) give a higher proportion of alarm calls (65–78% of the full repertoire in *Galagoides demidovii*, *G. thomasi* and *G. rondoensis* compared with 10–15% in other species) (Honess, 1996; Ambrose, 1999). Second, within each species, animals rarely give calls when they descend to the ground where they are less able to escape. Third, calls are only heard under the cover of darkness and rarely or never during daylight. Finally, individuals will cease to call if this attracts a potential predator into a position from which it could launch an attack.

Calls are often directed, or addressed, toward another animal by staring at, approaching it, and interacting with a companion or opponent. When calls are directed toward a predator such as a genet, snake, or perching owl (mobbing), the caller is frequently alone (approximately 80% of occasions). This observation helps us to exclude a number of functional explanations for why the calls are given, such as safety in numbers, communal vigilance, and communal deterrence (Rohwer et al., 1976; Harvey & Greenwood, 1978). Calling at these times cannot benefit other individuals and may act to deter the predator, although how this works in the case of snakes that cannot hear is not clear. It has been argued that agile species that are capable of avoiding a predator once it has been spotted will benefit from advertising their awareness of the danger (perception advertisement hypothesis). This might work because it causes the predator to give up its pursuit, and perhaps it even reduces the likelihood that this predator will attempt to hunt it in the future (Zavahi, 1993; Hasson, 1991; Zuberbühler et al., 1997; 1999). Nonetheless, the same calls used by lone animals when pestering a predator do sometimes act as recruitment calls to others, which approach the danger and call toward it, resulting in communal mobbing. This is equivalent to the mobbing of tarsiers, fork-marked lemurs, and many other species (e.g., Gursky, 2005; Schuelke, 2001; Scheumann et al., this volume), and it can be concluded that these calls also alert others to danger if they are nearby.

The fact that the same calls are given toward threats coming from predators and members of the same species, and that the calls vary according to the level of threat, strongly indicates that they convey information about the degree of risk the caller experiences (Robinson 1980; Blumstein, 1995; Weary & Kramer, 1995; Blumstein & Armitage, 1997), rather than the type of predator detected (referential signaling) (Seyfarth et al., 1980; Sherman, 1985; Cheney & Seyfarth, 1990; Pereira & Macedonia, 1991; Macedonia & Evans, 1993; Zuberbühler, 2000). Nevertheless, the calls of galagos are situationally specific (Blumstein, 1995; Blumstein & Armitage, 1997) in three ways. First, they produce acoustically distinctive calls at different levels of arousal. Second, they also vary the speed of calling (units per minute). Third, the intensity or volume of calling reflects the intensity of the situation, building up to a cacophony when the animal is extremely disturbed.

The extent to which calling enables other galagos to judge the nature of the threat depends upon both production specificity and perception specificity (Marler, 1992; Macedonia & Evans, 1993; Blumstein & Arnold, 1995; Blumstein & Armitage, 1997). Production specificity is indicated by the extent to which calls are limited to specific situations. This is only weakly developed among galagos, to the extent that some calls are restricted to low levels of anxiety while others typify alarm. During intraspecific agonistic interactions only the subordinate individual gives these calls and thereby act as signals of relative status. The ability of galagos to mix graded and discrete calls in a rapidly changing sequence with an almost infinite number of possible combinations lends itself to expression of fine gradations in mood and level of arousal. Perception specificity is indicated by the extent to which listeners behave appropriately on hearing the calls alone. This is more difficult to judge without the use of playback experiments, but some degree of perception specificity is shown by specific reactions to certain calls, such as the yell of *O. crassicaudatus* that precipitates attack, or the caw of *G. moholi* that induces others to freeze. These aspects remain to be explored in further detail.

Summary and Conclusion

The comparative study of the calling patterns of galagos in the context of danger demonstrates the following facts: (1) Galagos are capable of making a wide range of discrete and graded calls when facing danger (up to 23 structurally distinct sounds), which may be mixed together in complex and rapidly changing sequences with a high potential to convey a variety of mood messages. (2) While the calls that nocturnal primates use for social cohesion and territorial advertising are well studied there is a paucity of published information about calls associated with dangers. (3) Galago calls that are directed toward potential predators are often given when the caller is alone. Such calling cannot benefit other galagos but may inform the predator that it has been detected and can be outsmarted (perception advertisement and pursuit deterrent signals). (4) Some alarm calls from one individual may induce a mobbing response in others, and calls given when an animal is in the grip of a predator (e.g., the yell of *O. crassicaudatus*) can result in an aggressive reaction from others nearby, pointing to communal defense. Given that close associates will often be genetic relatives, these observations suggest that at least some calling benefits kin. Furthermore, females are more likely to become alarmed when they have infants. (5) The same calls that are given toward a predator can be made when a galago is agitated or alarmed during agonistic interactions with a conspecific. There is no evidence that such calls indicate a specific predator or category of predator. Instead they reflect the level and type of arousal of the caller. (6) Despite the lack of any link between a particular call and a particular threat, the context in which galagos usually give calls and call combinations, and the graded structure of some calls, may be learned by listeners and thereby provide information about the type and imminence of potential dangers. (7) Calls made during agonistic interactions between conspecifics are almost invariably made by the subordinate individual (the one that retreats) and may function to appease the aggressor by signaling the level and type of anxiety or alarm.

References

Altmann, J. (1974). Observational study of behaviour: Sampling methods. *Behav.*, 49: 227–36.

Altmann, S.A. (1967). The structure of primate social communication. In S.A. Altmann (Ed.), *Social communication among primates* (pp. 325–362). Chicago: Univ. of Chicago Press.

Ambrose, L. (1999). Species diversity in West and Central African galagos (Primates, Galagonidae): The use of acoustic analysis. Doctoral thesis. Oxford Brookes University, Oxford.

Andersson, A.B. (1969). Communication in the lesser bushbaby (*Galago senegalensis moholi*). MSc dissertation. University of the Witwatersrand, Johannesburg.

Bearder, S.K. (1969). Territorial and intergroup behaviour of the lesser bushbaby, *Galago senegalensis moholi* (A. Smith), in semi-natural conditions and in the field. M.Sc. dissertation. University of the Witwatersrand, Johannesburg.

Bearder, S.K. (1974). Aspects of the ecology and behavior of the thick-tailed bushbaby *Galago crassicaudatus*. Doctoral thesis. University of the Witwatersrand, Johannesburg.

Bearder, S.K. (1999). Physical and social diversity among nocturnal primates: A new view based on long term research. *Primates*, 40: 267–282.

Bearder, S.K., and Doyle, G.A. (1974). Field and laboratory studies of social organisation in bushbabies (*Galago senegalensis*). *J. Hum. Evol.*, 3: 37–50.

Bearder, S.K., and Martin, R.D. (1980). The social organisation of a nocturnal primate revealed by radio tracking. In C.J. Amlaner and D.W. MacDonald (Eds.), *A handbook on biotelemetry and radio tracking* (pp. 633–648). London: Pergamon Press.

Bearder, S.K., Nekaris, K.A.I., and Buzzell, C.A. (2001). Dangers in the night: Are some nocturnal primates afraid of the dark? In L.E. Miller (Ed.), *Eat or be eaten: Predator sensitive foraging among primates* (pp. 21–43). Cambridge: Cambridge Univ. Press.

Bearder, S.K., Nekaris, K.A.I., and Curtis, D.J. (2005). A re-evaluation of the role of vision in the activity and communication of nocturnal primates. *Folia Primatol.*, 50–71.

Becker, M.L., Buder, E.H., and Ward, J.P. (2003). Spectrographic description of vocalizations in captive *Otolemur garnettii*. *Int. J. Primatol.*, 24: 415–446.

Blumstein, D.T. (1995). Golden-marmot alarm calls: I. The production of situationally specific vocalisations. *Ethology*, 100: 113–125.

Blumstein, D.T., and Armitage, K.B. (1997). Alarm calling in yellow-bellied marmots: I. The meaning of situationally variable alarm calls. *Anim. Behav.*, 53: 143–171.

Blumstein, D.T., and Arnold, W. (1995). Situational-specificity in Alpine-marmot alarm communication. *Ethology*, 100: 1–13.

Charles-Dominique, P. (1977). *Ecology and behaviour of nocturnal primates*. London: Duckworth.

Charles-Dominique, P., and Bearder, S.K. (1979). Field studies of lorisid behavior: Methodological aspects. In G.A. Doyle and R.D. Martin (Eds.), *The study of prosimian behavior* (pp. 567–629). London: Academic Press, Inc.

Cheney, D.L., and Seyfarth, R.M. (1990). *How monkeys see the world*. Chicago: Univ. of Chicago Press.

Curio, E. (1993). Proximate and developmental aspects of antipredator behavior. *Adv. Study Behav.*, 22: 135–238.

Evans, C.S., Evans, L., and Marler, P. (1993). On the meaning of alarm calls: Functional reference in an avian vocal system. *Anim. Behav.*, 39: 785–796.

Frankenberg, E. (1981). The adaptive significance of avian mobbing IV. "Alerting others" and "perception advertisement" in blackbirds facing an owl. *Zeitshrift Tierpsychol.*, 55: 97–118.

Grubb, P., Butynski, T.M., Oates, J.F., Bearder, S.K., Disotell, T.R., Groves, C.P., and Struhsaker, T.T. (2003). Assessment of the diversity of African primates. *Int. J. Primatol.*, 14: 1301–1357.

Gursky, S. (2005). Predator mobbing in *Tarsius spectrum*. *Int. J. Primatol.*, 26: 207–221.

Harvey, P.H., and Greenwood, P.J. (1978). Anti-predator defense strategies: Some evolutionary problems. In J.R. Krebs and N.B. Davies (Eds.), *Behavioural ecology: An evolutionary approach* (pp. 129–151). Sunderland, MA: Sinauer Associates.

Hasson, O. (1991). Pursuit-deterrent signals: Communication between prey and predators. *Trends Ecol. Evol.*, 6: 325–329.

Hockett, C.F. (1960). Logical considerations in the study of animal communication. In W.E. Layon and W.N. Tavolga (Eds.), *Animal sounds and communication* (pp. 392–430). Washington, DC: American Institute of Biological Sciences.

Honess, P.E. (1996). Speciation among galagos (Primates, Galagidae) in Tanzanian forests. Doctoral thesis. Oxford Brookes University, Oxford.

Krebs, J.R., and Davies, N.B. (1993). *An introduction to behavioural ecology*. Oxford: Blackwells.

Macedonia, J.M., and Evans, C.S. (1993). Variation among mammalian alarm call systems and the problem of meaning in animal signals. *Ethology*, 93: 177–197.

Manser, M.B. (2001). The acoustic structure of suricates alarm calls varies with predator type and the level of response urgency. *Proc. Biol. Sci.*, 268: 2315–2324.

Marler, P. (1961). The logical analysis of animal communication. *J. Theoret. Biol.*, 1: 295–317.

Marler, P. (1965). Communication in monkeys and apes. In I. DeVore (Ed.), *Primate behavior: Field studies of monkeys and apes* (pp. 544-584). New York: Holt, Rinehart and Winston.

Marler, P. (1968). Aggregation and dispersal: Two functions in primate communication. In P. Jay (Ed.), *Primates: Studies in adaptation and variability* (pp. 420–438. New York: Holt, Rinehart and Winston.

Marler, P., Evans, C.S. and Hauser, M.D. (1992). Animal signals: Motivational, referential or both? In H. Papousek, U. Jurgens, and M. Papousek (Eds.), *Nonverbal vocal communication: Comparative and developmental approaches* (pp. 66–86). Cambridge: Cambridge Univ. Press.

Moynihan, M. (1964). Some behaviour patterns of platyrrhine monkeys: I. The night monkey (*Aotus trivirgatus*). *Smithson. Misc. Coll.*, 146: 1–146.

Moynihan, M. (1967). Comparative aspects of communication in New World primates. In D. Morris (Ed.), *Primate ethology* (pp. 236–266). London: Weidenfeld and Nicolson.

Pereira, M.E., and Macedonia, J.M. (1991). Ring-tailed lemur anti-predator calls denote predator class, not response urgency. *Anim. Behav.*, 41: 543–544.

Perkin, A.W., Bearder, S.K., Butynski, T, Bytebier, B., and Agwanda, B. (2002). The Taita mountain dwarf galago G*alagoides spp.*: A new primate for Kenya. *J. East African Natural History*, 91: 1–13.

Petter, J.J., and Charles-Dominuque, P. (1979). Vocal communication in prosimians. In G.A. Doyle and R.D. Martin (Eds.), *The study of prosimian behaviour* (pp. 247–306). London: Academic Press, Inc.

Pimley, E.R. (2002). The behavioural ecology and genetics of the potto (*Perodicticus potto edwardsi*) and Allen's bushbaby (*Galago alleni cameronensis*). Doctoral thesis. University of Cambridge, Cambridge.

Pullen, S.L. (2000). Behavioural and genetic studies of the mating system in a nocturnal primate: The lesser galago (*Galago moholi*). Doctoral thesis. University of Cambridge, Cambridge.

Robinson, S.R. (1980). Antipredator behaviour and predator recognition in Belding's ground squirrels. *Anim. Behav.*, 28: 840–852.

Schulke, O. (2001). Social anti-predator behaviour in a nocturnal lemur. *Folia. Primat.*, 72: 332–334.

Sherman, P.W. (1985). Alarm calls of Belding's ground squirrels to aerial predators: Nepotism or self-preservation? *Behav. Ecol. Sociobiol.*, 17: 313–323.

Shields, W.M. (1984). Barn swallow mobbing: Self defence, collateral kin defence or parental care? *Anim. Behav.*, 32: 132–148.

Seyfarth, R.M., Cheney, D.L., and Marler, P. (1980). Monkey responses to three different alarm calls: Evidence of predator classification and semantic communication. *Science*, 210: 802–803.

Wallace, G.E. (2005). Identification, abundance and behaviour of galagos (Galagidae) in the Shire Highlands, Malawi. MSc dissertation. Oxford Brookes University.

Weary, D. M., and Kramer, D.L. (1995). Response of eastern chipmunks to conspecific alarm calls. *Anim. Behav.*, 49: 81–93.

Zahavi, A. (1993). The fallacy of conventional signaling. *Phil. Trans. R. Soc. Lond. B*, 338: 227–230.

Zimmermann, E. (1985). The vocal repertoire of the adult Senegal bushbaby (*Galago senegalensis*). *Behav.*, 94: 211–233.

Zimmermann, E. (1995a). Acoustic communication in nocturnal prosimians. In L. Alterman, M.K. Izard, and G.A. Doyle (Eds.), *Creatures of the dark: The nocturnal prosimians* (pp. 311–330). New York: Plenum Press.

Zimmermann, E. (1995b). Loud calls in nocturnal prosimians: Structure, evolution and ontogeny. In E. Zimmermann, J.D. Newman, and U. Jurgens (Eds.), *Current topics in primate vocal communication* (pp. 47–72). New York: Plenum Press.

Zimmermann, E., Bearder, S.K., Doyle, G.A., and Anderson, A.B. (1988). Variations in vocal patterns of Senegal and South African lesser bushbabies and their implications for taxonomic relationships. *Folia Primatol.*, 51: 87–105.

Zuberbühler, K., Jenny, D., and Bshary, R. (1999). The predator deterrence function of primate alarm calls. *Ethology*, 105: 477–490.

Zuberbühler, K. (2000). Referential labelling in Diana monkeys. *Anim. Behav.*, 59: 917–927.

10
Predator Defense by Slender Lorises and Pottos

K. Anne-Isola Nekaris, Elizabeth R. Pimley, and Kelly M. Ablard

Introduction

Crypsis is argued to be the most widely used anti-predator strategy amongst nocturnal primates, wrought in its extreme form amongst the Asian lorises (Lorisinae: *Loris* and *Nycticebus*) and African pottos (Perodicticinae: *Arctocebus* and *Perodicticus*) (van Schaik & van Hoof, 1983; Terborgh & Janson, 1986; Cheney & Wrangham, 1987; Stanford, 2002; Wiens, 2002). Lorises and pottos are classically characterized by relatively slow, non-saltatory locomotion (Walker, 1969; Sellers, 1996). Silent movement, combined with cryptic coloration, small group size, discrete use of vocalizations, and increased olfactory communication are said to camouflage these primates (Petter & Hladik, 1970; Charles-Dominique, 1977). Much support for these notions has been offered by past studies of lorises and pottos.

The most compelling evidence of cryptic adaptations is provided by the unusual morphological adaptations of lorises and pottos, particularly their locomotor anatomy. Charles-Dominique (1977) contended that silent locomotion without abrupt transition, as seen in both lorises and pottos, is the ultimate behavioral adaptation to evade visually and auditorally directed predators (Petter & Hladik, 1970; Wiens, 2002). Morphological specializations of the post cranial anatomy of pottos and lorises allow them to remain still until a potential threat has passed (Rasmussen & Nekaris, 1998); this is exhibited by pottos and angwantibos in Gabon (Charles-Dominique, 1990). Though some captive settings yield freezing in *Nycticebus* and *Loris* (Fitch-Snyder & Schulze, 2001), the only comparable behavior exhibited by wild animals (*L. l. lydekkerianus*) occurred before they were observed crossing open ground (Bearder et al., 2002).

In addition to specializations that may aid lorises and pottos in escaping localization by predators, these animals also have evolved morphological strategies for coping with any predators they may encounter. Pottos are equipped with a scapular shield covered by thick skin and bristles of sensory hair, which they use to engage in active combat (Fig. 10.1); ultimately, a potto may escape by falling to the ground under conditions of extreme danger (Charles-Dominique, 1977). Although

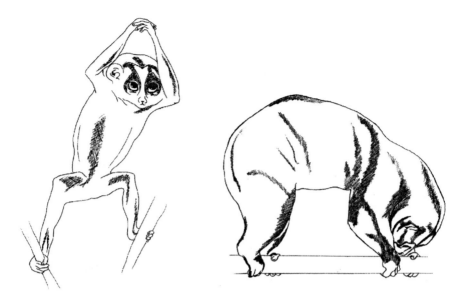

FIGURE 10.1. Two ways in which lorises may defend against predators after being detected: a grey slender loris in a cobra-like pose (drawn from video); a potto assuming a head-butting posture (drawn from a photograph from Charles-Dominique, 1977)

slender lorises have thickened skin in their nuchal area (Schulze & Meier, 1995), they have not been seen to combat potential predators in a way comparable to pottos (Bearder et al., 2002). Instead, slender lorises may ward off or startle predators with a form of mimicry that imitates a cobra (Fig. 10.1). By raising its slender arms near its ears or above its head, swaying its body in a serpentine fashion, and emitting a cobra-like hiss, the slender loris has been seen to ward off a cat (Still, 1905) and to challenge conspecifics (Schulze & Meier, 1995).

Differing strategies of concealment are revealed when observers examine the degree to which pottos and slender lorises use vocal advertisement. Charles-Dominique (1977) found that pottos (*P. potto edwardsi*) and angwantibos (*A. aureus*) not only used vocalizations discretely throughout the night (if at all), but also remained virtually silent in the face of predators, only on occasion omitting a barely audible "wheet." Mysore slender lorises (*L. l. lydekkerianus*), to the contrary, made loud calls throughout the night, with a rate of calling similar to or greater than other nocturnal primates (e.g., lesser bushbabies (*Galago moholi*), fork marked lemurs (*Phaner furcifer*), and spectral tarsiers (*Tarsius spectrum*)) (Schülke, 2001; Bearder et al., 2002; Gursky, 2003). Furthermore, on some occasions, they whistled or screamed when face to face with a potential predator (Bearder et al., 2006). A similar pattern is currently being revealed amongst red slender lorises (*L. t. tardigradus*) (Bernede & Nekaris, unpub. data).

Olfactory communication via scent marking, common amongst both pottos and lorises (Schilling, 1979; Fisher et al., 2003a), has been described as discrete in the extreme (Charles-Dominique, 1974, 1977). It has also been argued that most arboreal predators, including raptors and carnivores, rely on visual and auditory cues for hunting their prey (Charles-Dominique, 1990). However, the olfactory processing capabilities of predators are now known to be profound (Perrot-Sinal, 1999; Shivik, 1999; Gutzke, 2001; Koivula et al., 1997; Koivula & Korpimaki, 2001; Roberts & Gosling, 2001; Wyatt, 2003). Furthermore, strong evidence argues that chemo-communication, rather than being the ultimate form of crypsis, may instead play an important role in anti-predatory strategies (e.g., Jackson et al., 1990; Alterman, 1995; Chivers, 1995; Banks et al., 2000; Rohr & Madison, 2001; Banks et al., 2002; Hagey et al., 2006).

The above précis suggests that rather than being wholly cryptic, lorises and pottos use combined strategies of advertisement and active combat to varying degrees to cope with potential predators. In this contribution we provide information from two new field studies regarding the ways in which slender lorises (*L. l. nordicus* and *L. tardigradus tardigradus*) and pottos (*P. p. edwardsi*) confront actual and potential predators. As is common with studies of predation (Hill & Dunbar, 1998), few observations of actual attack were observed. Thus, we describe the degree to which lorises and pottos use vocal communication when faced with predators, describe how they modify behavior in the face of potential predators, and explore the function of olfactory behavior. We then reassess the anti-predator strategies of these primates.

Methods

Slender Lorises

Nekaris in Sri Lanka carried out fieldwork on slender lorises; details of the study sites and data collection methodology are provided elsewhere (Nekaris, 2001, 2003; Nekaris & Jayewardene 2003, 2004). Terminology for vocalizations follows Schulze & Meier (1995) and for olfactory behavior follows Osman Hill (1938), Ilse (1955), Andrew & Klopman (1974), Manley (1974), Ehrlich & Musicant (1977), and Rasmussen (1986). Data on the northern Ceylon grey slender loris (*L. l. nordicus*), hereafter grey loris, were collected over two field seasons from 2001–2002 at Polonnaruwa and Minneriya-Giritale Sanctuary yielding approximately 190 hrs of direct observation over 446 hrs of nocturnal field effort; additional information regarding predation comes from the sites of Trincomalee, Mihintale, Ritigala, and Anuradhapura. The Southwestern Ceylon red slender loris (*L. tardigradus tardigradus*), hereafter red loris, was studied over six field seasons from 2001–2004 at Masmullah Proposed Forest Reserve; additional information comes from Kanneliya Forest Reserve, Kakanadura Forest Reserve, Bangamukande Estate, and Dandeniya Forest Reserve, yielding 210 hrs of observation over a period covering 519 hrs of nocturnal field effort. All data were entered into SPSS v11.0; the analyses presented here are descriptive statistics.

Pottos

Fieldwork on Milne-Edwards' potto (*P. potto edwardsi*), hereafter potto, was carried out by Pimley from February 1998 to December 2000 at WWF Mount Kupe Forest Reserve in Bakossiland, southwestern Cameroon (Pimley et al., 2005a,b; Pimley & Bearder, in press). During radio tracking of 11 individuals a form of instantaneous point sampling was used (Altman, 1974) whereby the observer recorded whether a behavior occurred or not at the end of each 5-minute sample point. Continuous recording was used for detailed accounts of complex behavioral sequences (Pimley et al., 2005a). Only point samples related to vocalizations and olfaction are presented here (for others, see Pimley, 2002; Pimley et al., 2005a,b). Vocalization terminology follows Bearder et al. (1995) and Ambrose (1999, 2003). Olfactory behavior is divided into smelling and scent marking with genital glands or urine (Evans & Schilling, 1995). "Scent marking" was defined as marking a substrate or conspecifics with scent glands or urine. "Sniffing" involved the nose in contact with or near a substrate or conspecifics or having the head raised while the nose moved. For the purpose of describing olfactory behavior, "substrates" include all non-animal elements of the environment: branches, lianes, climbers, vines, ground and air. "Signaling" involved marking conspecifics or substrates with scent gland or genital secretions, while "receiving" refers to olfactory investigation of a conspecific or substrate.

Data from behavioral observations (consisting of 5-min sample intervals) were not normally distributed, so they were transformed by the arcsine square-root (percentages) and square-root (association indices) to normalize the data distribution (Kolmogorov-Smirnov test, $p > 0.1$), enabling the data to be analyzed with parametric statistics (Motulsky, 1995). T-tests were used to compare transformed data sets for adult males and females. All tests were two-tailed with a significance level of $p \leq 0.05$.

Results

Vocalizations and Displays: Slender Lorises

Both red and grey lorises made calls in proximity to potential predators. A grey loris emitted four sequential single whistles in the direction of a ring-tailed civet (*Viverricula indica majori*) moving on the ground. Another loris emitted a series of single whistles when a fishing cat (*Felis viverrinas*) passed on the ground. On only one occasion did lorises whistle near their sleeping site: A group of red lorises whistled singly six times while a golden palm civet (*Paradoxurus zeylonensis*) was in the vicinity; whistles ceased when the civet moved away.

Reactions to being caught by the researcher may lend a clue as to how lorises would behave when seized by an actual predator (Charles-Dominique, 1977).

Adult and juvenile grey and red lorises growled at observers while twisting their body around the researcher's wrist to bite the opposite side of the hand. One adult and one juvenile loris (both males) engaged in a cobra imitation. The adult engaged in this behavior on a tree branch, while the juvenile stood on the palm of the researcher's hand. Both raised their arms over their heads in the form of a cobra's hood, and made typical growling vocalizations with intermittent spits (Fig. 10.1).

Noisy displays were observed on 38 occasions. An estrous female who was being pursued by males not only whistled loudly, but also loudly thrashed branches at her pursuers. On twelve occasions when males were observed to follow estrous females, they were observed to abandon fluid locomotion for noisy branch scrambling. On 26 different occasions, lorises were observed to "self-play." This behavior involved shaking branches, as well as tumbling and twisting over them while making a play face. On three occasions the researcher detected the loris by hearing the noise it was making.

Vocalizations and Displays: Pottos

In the field no vocalizations were heard between a potto and a predator or between pottos, even when two animals were together. On two occasions pottos temporarily housed together called to one another. Three males housed together whistled to each other when one animal came too close (<1 m) to another. When the researcher entered the enclosure within 50 cm of a potto, three males whistled at a higher frequency than before. One male potto in temporary captivity made a grunting and hissing noise followed by simulated bites when the researcher attempted to pick the animal up. Pottos growled when trapped, suggesting this vocalization may be used towards an enemy.

Evading Predators by Freezing or Fleeing: Slender Lorises

Slender lorises frequently moved near potential predators without showing any sign of fear. The only predator that elicited immobility in the lorises was an Indian krait (*Bangarus caeruleus*). An adult female red loris that was foraging 7 m from her parked juvenile abruptly rejoined it; they entered a dense tangle and both stared down at the ground where the snake was passing, moving into the foliage and were no longer visible to observers. Carnivores and birds were encountered many times and yielded limited behavioral responses from lorises (Table 10.1). Despite the lorises seeming ambivalence to felids, Department of Wildlife officials reported that feral cats and jungle cats catch and kill the animals.

Evading Predators by Freezing or Fleeing: Pottos

Potential carnivore predators on Mt. Kupe included the serviline genet (*Genetta servilina*), African palm civet (*Nandina binotata*), African civet (*Civetticus civetta*), and West African linsang (*Poiana leightoni*). Infrequent sightings of the

TABLE 10.1.

Species	Potential Predator	Predator Behavior	Loris Reaction
grey	ring-tailed civet (*Viverricula indica majori*)	vocalizing in same tree as loris	travel past and ignore
grey	ring-tailed civet (*V. i. majori*)	move on ground within 10 m	ignore
grey	ring-tailed civet (*V. i. majori*)	move within 1.5–15 m, licks lips	forage for 15 minutes in close proximity whilst carrying twins, as low as 1 m
grey	ring-tailed civet (*V. i. majori*)	move within 2 m without stopping or looking	adult female walked past while carrying singleton
grey	ring-tailed civet (*V. i. majori*)	within <1–15 m	forage for 25 minutes in close proximity
red	ring-tailed civet (*V. i. majori*)	civet travels 4 m underneath a loris in tree	ignore and travel
grey	Indian palm cat (*Paradoxurus h. hermaphroditus*)	move 10 m in tree	ignore while grooming, continually wiping hands and arms over face
grey	Indian palm cat (*P. h. hermaphroditus*)	move within 15 m in tree	seemed to detect civet and descended into undergrowth
grey	Indian palm cat (*P. h. hermaphroditus*)	lie draped over branch 20 m away	ignore and forage
red	Indian palm cat (*P. h. hermaphroditus*)	move within 5 m	ignore and travel
red	Indian palm cat (*P. h. hermaphroditus*)	leaping and traveling	stop and stare in the direction of traveling cat
red	golden palm civet (*Paradoxurus zeylonensis*)	move within 20 m	foraging without hesitation
grey	fishing cat (*Felis viverrinas*)	move on ground within 5 m on five separate occasions	forage; ignore cat; meet with social partner for grooming; forage; ignore
grey	jungle cat (*Felis chaus kelaarti*)	move on ground within 5 m and stare	female carrying twins ignores cat
grey	rusty spotted cat (*Pronailurus rubignosa*)	move on ground within 4 m	female and parked juvenile ignore cat
red	serpent eagle (*Spilornis cheela*)	called near loris	moved to dense lianes
red	brown fish owl (*Ketupa zeylonensis*)	called near loris on 18 occasions	continued with normal behavior
grey	barn owl (*Tyto alba*)	landed in same tree	whistled and moved to cover

golden cat (*Profelis aurata*), leopard (*Panthera pardus*), and black-legged mongoose (*Bdeogale nigripes*) have been reported.

Although no active hunting by the 2–3.5-kg *N. binotata* was observed, strong evidence supports that these viverrids are a primary predator of pottos. A sub-adult male potto and his radio collar were eaten by an adult palm civet. Initial evidence for this came from the altered ranging patterns and change in frequency of the collar, indicating that it was inside another animal. Prior to its being eaten by the civet, the potto appeared to be in healthy condition, suggesting it was a predatory rather than a scavenging event. On occasions when palm civets entered an area of forest inhabited by a potto, the potto would either move away from the civet or hide inside dense vegetation until the palm civet had moved on ($n = 5$).

No encounters were observed between pottos and any other potential predators. The arboreal *G. servilina* at 1–2 kg may have some difficulty tackling a similarly sized adult potto, but it would be able to catch a young one. *Bdeogala nigripes*, although predominantly terrestrial, was once seen moving rapidly through the trees, and is a known predator of pottos at another site (Charles-Dominique, 1977). At 2–3.5 kg, this carnivore would be capable of dealing with an adult potto. Terrestrial carnivores such as *Panthera pardus* (50–60 kg), *Profelis aurata* (5.5–18 kg), and domestic dogs would all be capable of surprising a potto on the ground.

Like the slender loris, the potto showed defensive reactions to its human captors. When highly alarmed, a potto would repeatedly bite the substrate in front of it in between hissing and grunts before lunging at the potential attacker.

Chemical Communication: Slender Lorises

Dense rainforest conditions made it impossible to quantify slender loris olfactory communication. Qualitative observations still provide an insight into this behavior. Slender lorises used both direct (urine is directly deposited by the genitalia) and indirect (urine is deposited on a substrate with aid of another body part) modes of urine marking, and they also marked with scent glands. Odors produced were pungent and were often used by researchers to locate animals in the forest. Rhythmic micturition was the most common direct scent marking method for both taxa, occurring during traveling, foraging, and social interactions, and occurring on branches of all sizes and orientations, and during all activities other than resting. Intra-group countermarking of branches at the center of the group range was observed; scents were often inspected by the countermarker with the tongue (with and without Flehmen) before rhythmic micturition took place. Females often used concentrated rhythmic micturition at single prominent points, coating a small surface with a thick layer of urine, keeping one leg raised. Urine marking was observed in proximity to a sleeping site on only one occasion by a grey loris.

Urine washing was common, and seen most often in relation to consumption of noxious food items. Both loris taxa urine-washed before catching and/or after consuming unpalatable prey or when stung by an insect. Urine washing was also conducted before grooming of infants, which were to be parked for the night (grey lorises) or for a period of 2–3 hrs (red lorises).

Olfactory behavior in a social context was pronounced. Passing-over occurred during grooming bouts between red and grey loris males and females. Anogenital sniffing almost always began a grooming bout in both taxa, but usually was directed by males to females. Naso-muzzle sniffing occurred in both taxa when the female accepted the male. It was also directed by the male to infants clinging to the mother. Bouts of grooming sometimes were interrupted by naso-muzzle sniffing.

The specialized brachial gland of *Loris* was used in both taxa. During allogrooming, it was licked and rubbed mainly over the face, but also over the body. Both red and grey males were seen to press their brachial glands against a female, particularly if she rejected a grooming attempt. Mothers were seen to rub this gland over infants during normal grooming sessions. When animals were caught, the brachial gland exuded a pungent sticky substance. Although *Loris* lacks the specialized genital glands of pottos, two grey loris females exhibited a secretion from their vulva. Two male red lorises exhibited a thick pungent secretion on their testes.

Chemical Communication: Pottos

Pottos engaged in either indirect or direct modes of olfactory communication 162 times, or an average of 25.06% (± 0.06) of observations. The relative frequencies of behavior associated with olfaction, namely smelling and/or scent marking conspecifics and/or substrates, are illustrated in Figure 10.2. Pottos spent significantly more time in direct olfactory communication than indirect forms of olfactory behavior (paired t test: $t = 10.43, p = 0.001, \mathrm{df} = 9$). Substrates were both smelled and scent-marked significantly more frequently than conspecifics were either smelled or scent-marked (paired t test: $t = -9.76, p = 0.001$, $\mathrm{df} = 9$. No significant difference was found in the amount of sample points spent by females and males in scent-marking (independent t test: $t = -0.94, p = 0.38, \mathrm{df} = 7$, NS). Olfactory behaviors tended to be more common in males than females, although this difference was not significant (independent t test: $t = 1.04, p = 0.34, \mathrm{df} = 7$, NS). No significant difference was found between the number of sample points the sexes smelled and scent-marked conspecifics (independent t test $= 2.70, p = 0.80, \mathrm{df} = 9$, NS) or substrates (independent t test: $t = 0.37, p = 0.72, \mathrm{df} = 7$, NS).

Pottos marked the substrates with urine by gently lowering the penis or clitoris onto the substrate during locomotion (rhythmic micturition). Genital secretions of pottos were deposited by the female wiping her vulval glands and the male his scrotal glands along the substrate. Scent-marking of conspecifics was another form of indirect communication and was observed between paired male and female pottos, during allogrooming and copulation. A male and female potto would rub their genital glands and then touch the fur of the conspecific, thereby transferring scent, termed genital-scratching marking. One male potto was observed wiping his scrotal gland on his female partner and straddling her and passing-over. Pottos engaged in marking their bodies with their own odors, and were observed

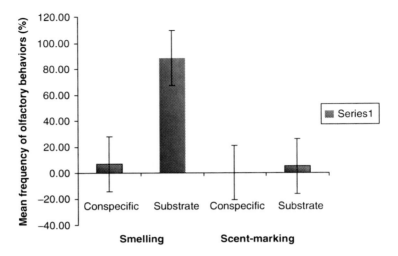

FIGURE 10.2. Mean percentage frequency of olfactory behaviors used by pottos on Mt Kupe, Cameroon. Smelling substrates includes branches, lianes, leaves, ground and air. Scent-marking includes marking with urine, genital glands, passing-over, genital-scratching and rhythmic micturition

marking themselves with genital grand secretions during autogrooming by genital-scratching marking.

Direct modes of olfactory communication were witnessed in pottos when two familiar individuals met. This occurred between paired males and females, and between an adult and a sub-adult male. The two animals faced each other and grasped the shoulders of the conspecific, then sniffed the cheeks and muzzle of each other in turn. Grooming of the head, neck, and shoulder region for several minutes normally followed. When the meeting preceded copulation, the male initiated olfactory investigation of the female's genital region.

Pottos were commonly observed to travel with their noses close to the substrate (nose-down searching). A male potto, after smelling the branch he was on, which was in the home range of another female, engaged in Flehmen, where he opened his mouth, rolled back his lips and raised his nose to sniff the air. Pottos often paused during locomotion to sniff the air or the surrounding substrates.

Discussion

Reactions to potential predators varied between slender lorises and pottos in this study. The most profound difference between the taxa is the common use of vocal communication by lorises, but virtually none pottos. Although all taxa engage in noisy displays, these are more common amongst the lorises. Pottos were seen face to face with predators less often than lorises, which in general ignored them. The greatest similarity amongst the taxa was their profuse use of olfactory communication; whether or not this communication mode is "discrete in the extreme" is discussed further here.

Slender lorises were much more vocal than pottos. In addition to the array of whistles discussed here, lorises frequently utter social "chitters," "kriks," and soft squeaks when engaged in social interactions (Nekaris & Jayewardene, 2003), suggesting that vocal communication is an important aspect of the behavioral repertoire of both taxa. Slender lorises in India were found to be similarly vocal (Bearder et al., 2002). Interestingly, slender lorises are relatively silent with regard to whistles in the captive setting, despite their using milder social calls here (Schulze & Meier, 1995), suggesting that loud whistles provide long distance messages that are less useful in captivity (Zimmerman, 1985, 1995).

Whistles are probably used in many contexts, including contact, spacing, aggression, and appeasement (Schulze & Meier, 1995; Bearder et al., 2002). Our data show that the whistle may also serve a predatory warning function. Both taxa emitted single whistles in the face of potential predators. This short whistle may serve as a quick warning to conspecifics that a predator is in the area (Hersek & Owings, 1993) or may serve as a pursuit deterrence signal, alerting the predator that it has been spotted (Woodland et al., 1980; Hasson, 1991).

A previous study showed that red lorises produced more multi-syllabic calls than grey lorises, with their calls characterized by greater frequency modulation than those of grey lorises (Coultas, 2002). Frequency modulation may function to minimize interference caused by dense vegetation; as vegetation absorbs and scatters sound, multi-syllabic calls may provide an advantage to red slender lorises living in dense rain forest (Wiley & Richards, 1978). Another possibility is that narrow frequency range and long duration of whistles means that no matter how many are emitted, predators would have difficulty localizing them (Daschbach et al. 1981). An ongoing study of red loris vocalizations will, it is hoped, shed additional light on the function of loris calls (Bernede, unpub. data).

Pottos in this study were only heard to vocalize in the presence of potentially threatening humans, accompanied by audible biting of branches. Similar observations were made of the same taxon in Gabon (Charles-Dominique, 1977). When attacked by a palm civet, a potto produced a strident vocalization ("groan" or "heee"), while audibly striking its jaw against a branch. This display had the effect of driving the palm civet away. At the same study site, pottos infrequently made a tsic contact call and a distress wheet call (Charles-Dominique, 1977). Such threat calls were also observed infrequently in the sympatric perodicticine, *Arctocebus aureus*. This relative silence coincides with the classic description of perodicticines as cryptic.

Silence also is a feature meant to characterize the locomotion of lorises and pottos. Charles-Dominique (1977) described discrete locomotion as the primary predator defense of pottos. In general, both lorises and pottos moved silently through their environment. Slender lorises, however, engaged in loud displays with branches. Such displays, lasting as long as 3 hrs, have also been seen in wild *L. l. lydekkerianus* and captive *L. l. nordicus* during self-play, and during social interactions where twig rattling may be part of a dominance display or a sign of stress (Schulze & Meier, 1995; Nekaris, 2001). The benefits of noisy displays are more obvious in a mating context (Lima & Dill, 1990; Andersson, 1994); the function of noisy self-play remains unclear.

Freezing or taking cover in the presence of potential predators was more applicable to pottos than to slender lorises. Lorises rarely modified their behavior in the face of a potential predator, concurring with observations of Mysore slender lorises (Bearder et al., 2002). The most noticeable reaction was freezing in the presence of a snake. A reticulated python consumed a greater slow loris (*N. coucang*) in Malaysia (Wiens & Zitzmann, 1999), and snakes are known to kill other nocturnal primates (Gursky, 2002). Slow lorises also made no reaction when owls or palm civets passed in close proximity, and were most cautious on the ground (Wiens, 2002), again mirroring Mysore slender lorises (Nekaris, 2001). It is possible that lorises are unpalatable to some predators and, thus, are avoided by them (see below).

Viverrids were, however, confirmed predators to pottos; in this study a palm civet consumed a potto, and pottos took cover when civets were present in their range. These viverrids were also a threat to Milne-Edwards' pottos in Gabon. When a palm civet attempted to attack an adult potto, the potto assumed the defense posture, thrusting at the civet with its scapular shield and striking at it by moving its body forward with teeth exposed. The potto remained immobile after this episode, until a second civet approached it, which the potto succeeded in knocking off the branch (Charles-Dominique, 1977). The black-legged mongoose was also identified as a potential predator at Mt. Kupe. In Gabon this mongoose launched an attack on a potto; the potto escaped by moving into impenetrable foliage, a behavior also exhibited by animals in our study. A dog was seen to kill a potto moving on the ground. Both a dead snake and genet experimentally presented to pottos elicited such fear that the animals fell to the ground to evade the danger (Charles-Dominique, 1974). These numerous examples indeed suggest that pottos may be at a higher predation risk than lorises if detected, and thus masterful silence may be of greater benefit to them.

This study further elucidates the vital importance of olfactory communication to slender lorises and pottos. Slender lorises and pottos employ complex social networks that are maintained by olfactory communication to varying degrees (Seitz, 1969; Charles-Dominique, 1977; Schilling, 1979; Fisher et al., 2003a; Wiens & Zitzmann, 2003). Pottos display well-developed anogenital glands, lack brachial glands, and do not urine wash, whereas slender lorises urine wash, have brachial glands, and lack anogenital glands (Schilling, 1979; Rasmussen & Nekaris, 1998). These anatomical differences may confer different functions in terms of both social and anti-predator behavior.

As in the present study, Charles-Dominique (1974, 1978) found that gregariousness amongst pottos occurs primarily via urine marking and/or secretions from their anogenital glands. Females produce a vaginal discharge which is mixed with urine during marking at the time of estrous, inducing a strong attraction for males, informing them of her reproductive status while simultaneously arousing sexual behavior (Epple, 1974).

These chemo-signals can elicit a response in conspecifics and are certainly used in various social contexts (Schilling, 1979; Perret, 1995; Palagi et al., 2002). Odors produced by conspecifics may convey information such as sex, species, social status, or reproductive state (Clark, 1982a, 1982b; Petrulis, et al. 2000). Scent marks from anogenital glands have been found to convey subsets of information, such as intersexual communication, self-advertisement, territorial demarcation, and to incite male-male competition (Heymann, 2000; Heymann, 2001; Smith & Gordon, 2002; Wolff et al., 2002; Lewis, 2004; Braune et al., 2005). Although pottos forage near to one another, they rarely move cohesively (Pimley et al., 2005a); thus, chemo-signals may form a social bridge between individuals who rarely encounter each other (Epple, 1974; Johnston, 1999). The information non-gregarious mammals relay via scent-marking has also been observed in species that live commensally (Kotenkova & Naidenko, 1999; Solomon, 1999; Humphries et al., 2001), reiterating the importance of scent in the social lives of pottos.

The social functions of olfactory behavior cited are also applicable to slender lorises. *Loris*, however, may more actively use chemical communication in anti-predatory behavior by using a pheromone to ward off predators or announce them to conspecifics, as in the case of *Nycticebus* (Nekaris, 2002; Hagey et al., this volume). Both *Loris* and *Nycticebus* emit an oily secretion from their brachial glands when confronted with similar physiological stressors (Alterman, 1995). This exudate, when mixed with saliva, has been determined to be highly volatile and it releases a pungent odor, suggesting that it may function as an alarm pheromone (Schilling, 1979; Alberts, 1992; Hagey et al., this volume). Alarm pheromones are relatively ubiquitous, predominantly not species-specific, and are emitted when a predator is detected, possibly to repel it (Mathis et al., 1995; Wyatt, 2003). Pheromones can elicit a flight or fight response and may contain compounds that can make flesh unpalatable or toxic (Wyatt, 2003). Conspecifics who detect an alarm pheromone may decrease their activity (Chivers, 1995).

This chemical might function in *Loris* in several ways. First, during the cobra imitation lorises were seen to rub their arms on the head, perhaps transferring chemicals from their brachial gland to a vulnerable part of their body. A similar behavior by *Nycticebus* results in the secretion drying and crystallizing on the top of the head (Hagey et al., this volume). In this solid form, the odors are prolonged and can also be detected visually by conspecifics, serving as a potential alarm (Gosling & Roberts, 2001; Roberts & Gosling, 2001). As has been suggested for *Nycticebus* and *Perodicticus* (Alterman, 1995), urine washing in combination with grooming of infants may serve as an additional effective olfactory barrier from predators via a pungent odor (e.g., an alert signal), a perilous chemical signal (e.g., an alarm pheromone), or as a form of predator mimicry (Wyatt, 2003).

Mimicry may also be used by lorises in another manner. During this and previous studies, urine washing was frequent before and after eating or being stung by noxious insects (Nekaris, 2001; Nekaris & Rasmussen, 2003). Such insects form a large part of the loris dietary repertoire and may serve as an exogenous source of a toxic substance that may later be secreted by lorises during urine washing (Darst et al., 2005). Urine washing before capture of toxic insects may be indicative of a self-defense mechanism by the way it mimics the scents of prey items to facilitate their capture without triggering alarm responses (Caldwell, 1996). Urine washing did also occur in other contexts and might serve additional purposes, such as enhancing the grip, as it does in galagos (Welker, 1973; Harcourt, 1981).

Finally, numerous studies point to the ability of potential predators to detect olfactory signals, suggesting that this form of communication may not be cryptic. Scent-marking with glands, urine, or feces containing pheromones such as kairomones, allomones, and allelochemicals can be utilized by predators to either mimic pheromones of prey or to deceive prey (Watson et al., 1999; Heymann, 2000; Wyatt, 2003). Predators can eavesdrop on pheromone trails, which can help them predict the location and movement of prey, thereby enhancing their hunting strategy (Gosling & Roberts, 2001). Predators of lorisines not only employ their ability to analyze chemicals with their vomeronasal organs (e.g., Alterman, 1995), but some avian predators can detect and follow the pheromone trails or marks of prey that are visible as ultraviolet light (Wyatt, 2003). Fisher et al. (2003b) have shown that the urine of *Nycticebus pygmaeus* has UV properties; the pungent odor of slender loris urine indicating that it contains molecules of low volatility suggests it is similar, and again suggests strategies of advertisement that are different from pottos. Further studies should consider patterns of scent-marking amongst the taxa and how these might be used to confound predators (Gosling, 1982).

Conclusions

This study has shown that although in many ways lorises and pottos are adapted to avoid predator detection they differ greatly in other behavioral mechanisms that usually fall under the category of crypsis. Pottos are less vocal, less gregarious, and more cautious in the face of potential predators than slender lorises. Both pottos and slender lorises have evolved independent means to contend with predators upon detection—the potto with its defensive shield and the loris with loud vocalizations, and potentially with the use of alarm pheromones. Olfactory communication is of vital importance to both of these lorisiforms, and its role in predator avoidance and defense should be considered in future studies. This study further elucidates that despite similar anatomical adaptations for slow climbing quadrupedalism, the behavioral repertoire of lorises and pottos is characterized by more variability than previously acknowledged.

Acknowledgments. K.A.I. Nekaris and K.M. Ablard would like to thank the Ball family of Hen Dafarn, Gelliwen, Wales, for making the writing of this paper possible. R. Swaisgood and H. Schulze provided incredibly insightful reviews.

L. Birkett drew the images for this paper. T. Wyatt provided invaluable information and resources regarding olfactory communication.

K.A.I. Nekaris thanks the Department of Wildlife Conservation and Forest Department of Sri Lanka for authorizing this study. Logistical aid and/or field assistance for slender loris studies was provided by M. Abare, D. Coultas, W. Dittus, S. Gamage, A. Gunawardene, K. Hart, J. Jayewardene, E. Leavitt, W. Liyanage, A. Perera, C. Perera, R. Perera, V. Perera, H. Schulze, J. Stokes, and S. Wimalasuriya. Loris studies were funded by grants to Nekaris from Margot Marsh Primate Biodiversity Fund, Columbus Zoological Park and Aquarium, Primate Conservation, Inc., People's Trust for Endangered Species, One with Nature of the Philadelphia Zoo, Petzl Headlamp Company, and the School of Social Sciences and Law of Oxford Brookes University. E.R. Pimley thanks her research assistants: G. Ngole, G. Ekwe, P. Mesumbe, A. Ekinde, O. Mokoko, M. Etube, A. Ekwoge, H. Enongene, O. Mesumbe, and D. Menze for their invaluable support with fieldwork. She also acknowledges the staff of WWF Cameroon and members of the Sub-Department of Animal Behaviour, Departments of Zoology and Biological Anthropology, University of Cambridge, for logistical support and advice. Support for the field study of pottos came from the UK Medical Research Council.

References

Alberts, A.C. (1992). Constraints on the design of chemical communication systems in terrestrial vertebrates. *The American Naturalist*, 139: S62–S69.

Alterman, L. (1995). Toxins and toothcombs: Potential allospecific chemical defenses in *Nycticebus* and *Perodicticus*. In L. Alterman, G.A. Doyle, and M.K. Izard (Eds.), *Creatures of the dark: The nocturnal prosimians* (pp. 413–424). New York: Plenum Press.

Altman, J. (1974). Observational study of behavior: Sampling methods. *Behavior*, 49: 227–265.

Ambrose, L. (1999). Species diversity in West and Central African galagos (Primates, Galagonidae): The use of acoustic analysis. Doctoral thesis. Oxford Brookes University, Oxford.

Ambrose, L. (2003). Three acoustic forms of Allen's galagos (Primates; Galagonidae) in the Central African region. *Primates*, 44: 25–39.

Andersson, M. (1994). *Sexual selection*. Princeton, NJ: Princeton Univ. Press.

Andrew, R.J., and Klopman, R.B. (1974). Urine-washing: Comparative notes. In R.D. Martin, G.A. Doyle, and A.C. Walker (Eds.), *Prosimian biology* (pp. 303–312). London: Duckworth.

Banks, P.B., Norrdahl, K., and Korpimäki, E. (2000). Nonlinearity in the predation risk of prey mobility. *Proceedings of the Royal Society Biological Sciences Series B*, 267: 1621–1625.

Banks, P.B., Norrdahl, K., and Korpimäki, E. (2002). Mobility decisions and the predation risks of reintroduction. *Biological Conservation*, 103: 133–138.

Bearder, S.K., Honess, P.E., and Ambrose, L. (1995). Species diversity among galagos with special reference to mate recognition. In L. Alterman, G. Doyle, and M.K. Izard (Eds.), *Creatures of the dark: The nocturnal prosimians* (pp. 331–352). New York: Plenum Press.

Bearder, S.K., Nekaris, K.A.I., and Buzzell, C.A. (2002). Dangers of the night: Are some primates afraid of the dark? In L.E. Miller (Ed.), *Eat or be eaten: Predator sensitive foraging in primates* (pp. 21–43). Cambridge: Cambridge Univ. Press.

Bearder, S.K., Nekaris, K.A.I., and Curtis, D.J. (2006). A re-evaluation of the role of vision in the activity and communication of nocturnal primates. *Folia Primatologica*, 77 (1–2): 50–71.

Braune, P., Schmidt, S., and Zimmermann, E. (2005). Spacing and group coordination in a nocturnal primate, the golden brown mouse lemur (*Microcebus ravelobensis*). *Behavioural Ecology and Sociobiology*, 58(6): 587–596.

Caldwell, J. (1996). The evolution of myrmecophagy and its correlates in poison frogs (Family: Dendrobatidae). *Journal of Zoology*, 240: 75–100.

Charles-Dominique, P. (1974). Vie sociale de *Perodicticus potto* (Primates: Lorisides). Étude de terrain en forêt equatorial de l'ouest africain au Gabon. *Mammalia*, 38: 355–379.

Charles-Dominique, P. (1977). *Ecology and behaviour of nocturnal primates*. London: Duckworth.

Charles-Dominique, P. (1978). Solitary and gregarious prosimians: Evolution of social structures in primates. In D.J. Chivers and K.A. Joysey (Eds.), *Recent advances in primatology*, Volume 3 (pp. 139–149). New York: Academic Press.

Charles-Dominique, P. (1990). Ecological adaptations related to locomotion in primates: An introduction. In F.K. Jouffroy, M.H. Stack, and C. Niemitz (Eds.), *Gravity, posture and locomotion in primates* (pp. 19–31). Sedicesimo: Editrice II.

Cheney, D., and Wrangham, R.W. (1987). Predation. In B.B. Smuts, D.L. Cheney, R.M. Seyfarth, R.W. Wrangham, and T.T. Struhsaker (Eds.), *Primate societies* (pp. 227–239). Chicago: Univ. of Chicago Press.

Chivers, D.P., Brown, G.E., and Smith, J.F. (1995). Chemical alarm signals: Predator deterrents or predator attractants? *The American Naturalist*, 145: 994–105.

Clark, A.B. (1982a). Scent marks as social signals in *Galago crassicaudatus*. I. Sex and reproductive status as factors in signals and responses. *Journal of Chemical Ecology*, 8(8): 1133–1151.

Clark, A.B. (1982b). Scent marks as social signals in *Galago crassicaudatus*. II. Discrimination between individuals by scent. *Journal of Chemical Ecology*, 8(8): 1153–1165.

Coultas, D.S. (2002). Bioacoustic analysis of the loud call of two species of slender loris (*Loris tardigradus* and *L. lydekkerianus nordicus*) from Sri Lanka. MSc thesis. Oxford Brookes University, Oxford.

Darst, C.R., Menéndez-Guerrero, P.A., Coloma, L.A., and Cannatella, D.C. (2005). Evolution of dietary specialization and chemical defense in poison frogs (Dendrobatidae): A comparative analysis. *The American Naturalist*, 165: 56.

Daschbach, N.J., Schein, M.W., and Haines, D.E. (1981). Vocalizations of the slow loris, *Nycticebus coucang* (Primates, Lorisidae). *Inter. Jour. of Primatol.*, 2, 71–80.

Ehrlich, A., and Musicant, A. (1977). Social and individual behaviors in captive slow lorises (*Nycticebus coucang*). *Behaviour*, 60: 195–220.

Epple, G. (1974). Primate pheromones. In M.C. Birch (Ed.) *Pheromones* (pp. 366–385). New York: Elsevier.

Evans, C., and Schilling, A. (1995). The accessory (vomeronasal) chemoreceptor system in some prosimians. In L. Alterman, G.A. Doyle, and M.K. Izard (Eds.), *Creatures of the dark: The nocturnal prosimians* (pp. 393–411). New York: Plenum Press.

Fisher, H.S., Swaisgood, R.R., and Fitch-Snyder, H. (2003a). Odor familiarity and female preferences for males in a threatened primate, the pygmy loris *Nycticebus pygmaeus*: Applications for genetic management of small populations. *Naturwissenschaften*, 90(11): 509–512.

Fisher, H.S., Swaisgood, R.R., and Fitch-Snyder, H. (2003b). Countermarking by male pygmy lorises (*Nycticebus pygmaeus*): Do females use odor cues to select mates with high competitive ability? *Behav. Ecol. and Sociobiol.*, 53(2): 123–130.

Fitch-Snyder, H., and Schulze, H. (2001). *Management of lorises in captivity: A husbandry manual for Asian lorisines*. San Diego: Zoological Society of San Diego, Center for Reproduction of Endangered Species Press.

Gosling, L.M. (1982). A reassessment of the function of scent marking in territories. *Zeitschrift für Tierpsychologie*, 60: 89–118.

Gosling, L.M., and Roberts, S. (2001). Scent-marking by male mammals: Cheat-proof signals to competitors and mates. *Advances in the Study of Behavior*, 30: 169–217.

Gursky, S. (2002). Predation on a wild spectral tarsier (*Tarsius spectrum*) by a snake. *Folia Primatol.*, 73: 60–62.

Gursky, S. (2003). Predation experiments on infant spectral tarsiers (*Tarsius spectrum*). *Folia Primatol.*, 74(5–6): 272–284.

Gutzke, W.H.N. (2001). Field observations confirm laboratory reports of defense responses by snakes to the odors of predatory snakes. In A. Marchlewska-Koj, J. Lepri, and D. Muller-Schwarze (Eds.), *Chemical signals in vertebrates* (9th ed.). (pp. 285–288). New York: Kluwer Academic/Plenum Publishers.

Hagey, L.R., Fry, B.G., and Snyder, H. (2006). Talking defensively: A dual use for the brachial gland exudate of slow and pygmy lorises. In S. Gursky (Ed.), this volume (pp. xx–yy). New York: Kluwer/Academic Press.

Harcourt, C.S. (1981). An examination of the function of urine washing in *Galago senegalensis. Zeitschrift für Tierpsychologie*, 55: 119–128.

Hasson, O. (1991). Pursuit-deterrent signals: Communication between prey and predator. *Trends in Ecology and Evolution*, 6: 325–329.

Hersek, M.J, and Owings, D.H. (1993). Tail flagging by adult California ground squirrels: A tonic signal that serves different functions for males and females. *Animal Behaviour*, 46: 129–138.

Heymann, E.W. (2000). Spatial patterns of scent marking in wild moustached tamarins, *Saguinus mystax*: No evidence for a territorial function. *Animal Behaviour*, 2000: 723–730.

Heymann, E.W. (2001). Interspecific variation of scent-marking behaviour in wild tamarins, *Saguinus mystax* and *Saguinus fuscicollis. Folia Primatol.*, 72: 253–267.

Hill, R.A., and Dunbar, R.I.M. (1998). An evaluation of the roles of predation rate and predation risk as selective pressures on primate grouping behaviour. *Behaviour*, 135 (4): 411–430.

Humphries, R.E., Robertson, D.H.L., Nevison, C.M., Beynon, R.J., and Hurst, J.L. (2001). The role of urinary proteins and volatiles in competitive scent marking among male house mice. In A. Marchlewskha-Koj, J. Lepri, and D. Muller-Schwarze (Eds.), *Chemical signals in vertebrates* (9th ed.). (pp. 353–360). New York: Kluwer Academic/Plenum Publishers.

Ilse, D.R. (1955). Olfactory marking of territory in two young male lorises kept in captivity in Poona. *British Jour. of Animal Behav.*, 3: 118–120.

Jackson, B.D., Morgan, E.D., and Billen, J.P.J. (1990). A note on pygidial glands of primitive Australian ants: A new source of odorous chemicals. In A.R. McCaffery and I.D. Wilson (Eds.), *Chromatography and isolation of insect hormones and pheromones* New York: Plenum Press. p. 335–341.

Johnston, R.E. (1999). How do hamsters know whose scent is on top and why should it matter? In R. Johnston, D. Muller-Schwartz, and P. Sorenson (Eds.), *Advances in chemical signals in vertebrates* (pp. 227–238). New York: Kluwer Academic/Plenum Publishers.

Koivula, M., Korpimaki, E. and Viitala, J. (1997). Do Tengmalm's owls see vole scent marks visible in ultraviolet light? *Animal Behaviour*, 54: 873–877.

Kotenkova, E.V., and Naidenko, S.V. (1999). Discrimination of con- and heterospecific odors in different taxa of the *Mus musculus* species group. In R. Johnston, D. Muller-Schwartz, and P. Sorenson (Eds.), *Advances in chemical signals in vertebrates* (pp. 299–208). New York: Kluwer Academic/Plenum Publishers.

Lewis, R.J. (2004). Sex differences in scent-marking in Sifaka: Mating conflict or male services? Unpublished doctoral dissertation. University of Texas at Austin, Texas.

Lima, S. L. and Dill, L. M. (1990). Behavioral decisions made under the risk of predation: A review and prospectus. *Canadian Journal of Zoology*, 68: 619–640.

Manley, G. (1974). Functions of the external genital glands of *Perodicticus* and *Arctocebus*. In R.D. Martin, G.A. Doyle, and A.C. Walker (Eds.), *Prosimian biology* (pp. 313–329). London: Duckworth.

Mathis, A., Chivers, D.P., and Smith, J.F. (1995). Chemical alarm signals: Predator deterrents or predator attractants. *The American Naturalist*, 145(6): 994–1005.

Motulsky, H. (1995). *Intuitive biostatistics*. Oxford: Oxford Univ. Press.

Nekaris, K.A.I. (2001). Activity budget and positional behavior of the Mysore slender loris (*Loris tardigradus lydekkarianus*): Implications for "slow climbing" locomotion. *Folia Primatol.*, 72: 228–241.

Nekaris, K.A.I. (2002). Slender in the night. *Natural History*, 2(02): 54–59.

Nekaris, K.A.I. (2003). Observations on mating, birthing and parental care in three taxa of slender loris in India and Sri Lanka (*Loris tardigradus* and *Loris lydekkerianus*). *Folia Primatol.*, 74: 312–336.

Nekaris, K.A.I., and Jayewardene, J. (2003). Pilot study and conservation status of the slender loris (*Loris tardigradus* and *Loris lydekkerianus*) in Sri Lanka. *Primate Conservation*, 19: 83–90.

Nekaris, K.A.I., and Jayewardene, J. (2004). Distribution of slender lorises in four ecological zones in Sri Lanka. *Journal of Zoology*, 262: 1–12.

Nekaris, K.A.I., and Rasmussen, D.T. (2003). Diet of the slender loris. *Inter. Jour. of Primatol.*, 24(1): 33–46.

Osman Hill, W.C. (1938). A curious habit common to lorisoid and platyrrhine monkeys. *Ceylon Journal of Science B*, 21(1): 65.

Palagi, E., Gregorace, A., and Borgognini Tarli, S.M. (2002). Development of olfactory behavior in captive ring-railed lemurs (*Lemur catta*). *Inter. Jour. of Primatol.*, 23(3): 587–599.

Perret, M. (1995). Chemocommunication in the reproduction function of mouse lemurs. In L. Alterman, G.A. Doyle, and M.K. Izard (Eds.), *Creatures of the dark: The nocturnal prosimians* (pp. 372–392). New York: Plenum Press.

Perrot-Sinal, T., Kavaliers, M., and Ossenkopp, P. (1999). Changes in locomotor activity following predator odor exposure are dependent on sex and reproductive status in the meadow vole. In R. Johnston, D. Muller-Schwartz, and P. Sorenson (Eds.), *Advances in chemical signals in vertebrates* (pp. 497–504). New York: Kluwer Academic/Plenum Press.

Petrulis, A., Peng, M., and Johnston, R.E. (2000). The role of the hippocampal system in social odor discrimination and scent-marking in female golden hamsters (*Mesocricetus auratus*). *Behavioural Neuroscience*, 114(1): 184–195.

Petter, J.J., and Hladik C.M. (1970). Observations sur le domaine vital et la densité de population de *Loris tardigradus* dans les forêts de Ceylon. *Mammalia*, 34: 394–409.

Pimley, E.R. (2002). The behavioural ecology and genetics of the potto (*Perodicticus potto edwardsi*) and Allen's bushbaby (*Galago alleni cameronensis*). Doctoral thesis. University of Cambridge, Cambridge.

Pimley, E.R., and Bearder, S.K. (In press). Potto (*Perodicticus*). In J. Kingdon, D. Happold, and T. Butysnki (Eds.), *Mammals of Africa*, Vol. 1. (pp. xx–yy). Cambridge: Cambridge Univ. Press.

Pimley, E.R., Bearder, S.K., and Dixson, A.F. (2005a) Examining the social organization of the Milne-Edwards' potto *Perodicticus potto edwardsi*. *Amer. Jour. of Primatol.*, 66(4): 317–330.

Pimley, E.R., Bearder, S.K., and Dixson, A.F. (2005b). Home range analysis of *Perodicticus potto edwardsi* and *Sciurocheirus cameronensis*. *Inter. Jour. of Primatol.*, 26(1): 191–206.

Rasmussen, D.T. (1986). Life history and behavior of slow lorises and slender lorises. Doctoral thesis. Duke University, Durham, NC.

Rasmussen, D.T., and Nekaris, K.A.I. (1998). Evolutionary history of the lorisiform primates. *Folia Primatol.*, 69: 250–285.

Roberts, S.C., and Gosling, L.M. (2001). The economic consequences of advertising scent mark location on territories. In A. Marchlewsha-Koj, J. Lepri, and D. Schwarze (Eds.), *Chemical signals in vertebrates* (9th ed.). (pp. 11–17). New York: Kluwer Academic/Plenum Press, New York.

Rohr, J.R., and Madison, D.M. (2001). A chemically mediated trade-off between predation risk and mate search in newts. *Animal Behaviour*, 62: 863–869.

Schilling, A. (1979). Olfactory communication in prosimians. In G.A. Doyle and R.D. Martin (Eds.), *The study of prosimian behaviour* (pp. 461–542). London: Academic Press, Inc.

Schülke, O.(2001). Social anti-predator behaviour in a nocturnal lemur. *Folia Primatologica*, 72(6): 332–334.

Schulze, H., and Meier, B. (1995). Behaviour of captive *Loris tardigradus nordicus*: A qualitative description including some information about morphological bases of behavior. In L. Alterman, M. Doyle, and M.K. Izard (Eds.), *Creatures of the dark: The nocturnal prosimians* (pp. 221–250). New York: Kluwer Academic/Plenum Publishers.

Seitz, E. (1969). Die Bedeutung gerüchlicher Orientierung beim Plumplori *Nycticebus coucang* Boddaert 1785 (Prosimii, Lorisidae). *Zeitschrift für Tierpsychologie*, 26: 73–103.

Sellers, W. (1996). A biomechanical investigation into the absence of leaping in the locomotor repertoire of the slender loris (*Loris tardigradus*). *Folia Primatol.*, 67: 1–14.

Shivik, J.A., and Clark, L. (1999). The development of chemosensory attractants for brown tree snakes. In R. Johnston, D. Muller-Schwartz, and P. Sorsenson (Eds.), *Advances in chemical signals in vertebrates* (pp. 649–654). New York: Kluwer Academic/Plenum Publishers.

Smith, E.T., and Gordon, J.S. (2002). Sex differences in olfactory communication in *Saguinus labiatus*. *Inter. Jour. of Primatol.*, 23(2): 429–441.

Solomon, N.G. (1999). The functional significance of olfactory cues in the pine vole (*Microtus pinetorum*). In R. Johnston, D. Muller-Schwartz, and P. Sorsenson (Eds.), *Advances in chemical signals in vertebrates* (pp. 407–419). New York: Kluwer Academic/Plenum Publishers.

Stanford, C. (2002). Avoiding predators: Expectations and evidence in primate antipredator behavior. *Inter. Jour. of Primatol.*, 23(4): 741–757.

Still, J. (1905). On the loris in captivity. *Spolia Zeylanica*, 3: 155–157.

Terborgh, J., and Janson, C. (1986). Socioecology of primate groups. *Annual Review of Ecological Systematics*, 17: 111–135.

van Schaik, C., and van Hoof, J. (1983). On the ultimate causes of primate social systems. *Behaviour*, 5: 91–117.

Walker, A.C. (1969). The locomotion of the lorises, with special reference to the potto. *East African Wildlife Journal*, 7: 1–5.

Watson, S.L., Ward, J.P., David, K.B., and Stavisky, R.C. (1999). Scent-marking and cortisol response in the small-eared bushbaby (*Otolemur garnettii*). *Physiology & Behavior*, 66(4): 695–699.

Welker, C. (1973). Ethological significance of the urine washing by *Galago crassicaudatus* E. Geoffroy, 1812 (Lorisiformes: Galagidae). *Folia Primatol.*, 20: 429–452.

Wiens, F. and Zitzmann, A. (1999). Predation on a wild slow loris (*Nycticebus coucang*) by a reticulated python (*Python reticulatus*). *Folia Primatol.*, 70: 362–364.

Wiens, F. (2002). Behavior and ecology of wild slow lorises (Nycticebus coucang): Social organisation, infant care system and diet. Doctoral thesis. Bayreuth University, Bayreuth (Germany).

Wiens, F., and Zitzmann, A. (2003). Social structure of the solitary slow loris *Nycticebus coucang* (Lorisidae). *Journal of Zoology*, 261(1): 35–46.

Wiley, R.H., and Richards, D.G. (1978). Physical constraints on acoustic communication in the atmosphere: Implications for the evolution of animal vocalizations. *Behav. Ecol. and Sociobiol.*, 3: 69–94.

Wolff, J.O., Mech, S.G., and Thomas, S.A. (2002). Scent marking in female prairie voles: A test of alternative hypotheses. *Ethology*, 108: 483–494.

Woodland, D.J., Jaafar, Z., and Knight, M.-L. (1980). The "pursuit deterrent" function of alarm signals. *American Naturalist*, 115: 748–753.

Wyatt, T.D. (2003). *Pheromones and animal behavior: Communication by smell and taste.* Cambridge: Cambridge Univ. Press.

Zimmerman, E. (1985). Vocalisations and associated behaviours in adult slow loris (*Nycticebus coucang*). *Folia Primatol.*, 44: 52–64.

Zimmermann, E. (1995). Acoustic communication in nocturnal prosimians. In: Alterman, L., Doyle, G. A. and Izard, M. K. (eds.), *Creatures of the Dark: The nocturnal prosimians* (pp. 311–330). New York: Plenum Press.

11

The Response of Spectral Tarsiers Toward Avian and Terrestrial Predators

Sharon L. Gursky

Introduction

Predation pressure has an overwhelming influence on the behavior of nonhuman primates (Cheney & Wrangham, 1986; Isbell, 1994; Hill & Dunbar, 1998; Janson, 1998; Stanford, 1998; Treves, 1998; Wright, 1998; Bearder et al., 2002). In response to this powerful selective pressure, primates have evolved a variety of adaptations to thwart predators including concealment, vigilance, flight, alarm calls, group living, and polyspecific associations (Caine, 1984; Cheney & Wrangham, 1986; Cords, 1990; Ferrari & Ferrari, 1990; Baldellou & Henzi, 1992; Cowlishaw, 1994; Lima, 1995; Gould, 1996; Iwamoto et al., 1996; Bshary & Noë, 1997; Treves, 1998; Bearder et al., 2002). The anti-predator strategy used by a prey often depends on the particular circumstances of a threatening situation (Ydenberg 1986; Welton et al., 2003). In other words, to avoid unnecessary and costly responses, prey animals assess the degree of risk and tailor their responses accordingly (Lima & Dill, 1990; Janson, 1998; Wright, 1999; Henni, 2005). In particular, predators or situations that are unlikely to be successful in their attack can be responded to differently than can predators or situations that are likely to be successful (Welton et al., 2003).

A crucial step in determining the risk of predation is predator identification. Studies of birds and mammals, including primates, indicate that they are often capable of discriminating between different types of predators (Cheney et al., 1980; Hauser, 2000; Gursky, 2003; Ramakrishnan et al., 2005). For example, it is well known that vervet monkeys emit acoustically distinct alarm calls, functionally referential signals, in response to different classes of predator (Cheney et al., 1980). According to Hauser (2000), signals are "referential" if they create in listeners some kind of mental picture of the object or event eliciting the signal. Thus, a prey's alarm call after it sees a leopard evokes a mental image of the leopard in the minds of other individuals. Similarly, it has been observed that wild bonnet macaques (*Macaca radiata*) can distinguish between predatory and non-predatory snakes. They emit alarm calls and stand bipedally in response to a python, but not to other snakes; the python is the only known snake that preys on them (Ramakrishnan et al., 2005).

In addition to producing referential signals, vervet monkeys are also known to exhibit distinct non-vocal behaviors in response to the presence of the different predator types (Cheney et al., 1980). Vervets run under bushes upon seeing an eagle or climb up into the trees upon seeing a leopard. Several other mammals are known to modify behavioral response according to the type of predator encountered (MacWhirter, 1992; Hanson, 1995). For example, during encounters with avian predators, ground squirrels immediately dash into their burrows or other hiding places (Hanson & Coss, 1997). In contrast, upon detecting terrestrial predators, they typically mount promontories where they can continue to monitor the activity of the ground predator. Similarly, while bonnet macaques are known to stand bipedally and emit alarm calls in response to predatory pythons, they respond differently to the presence of venomous snakes. Following detection of a cobra, a well-known venomous snake, bonnet macaques immediately move, running, in the opposite direction from the snake. This response is in keeping with the threat that this predator poses. Predator danger depends on many factors including its identity, which in turn determines its speed and hunting style, as well as the prey's own speed and distance from its refuge.

Referential signaling is also suspected in spectral tarsier infants (Gursky, 2003). Gursky (2003) found that infants consistently emitted a twittering alarm call in response to raptor models and raptor vocalizations, whereas they emitted a harsh, loud call three times in rapid succession in response to model snakes. Thus, based on observations of their response to vocalizations, spectral tarsier infants can distinguish different classes of predator. The next obvious question is whether there are distinct anti-predator behaviors that the spectral tarsiers exhibit in response to different predator classes. Specifically, do spectral tarsiers respond differently to avian than to terrestrial predators, and likewise, do their responses vary in response to mammalian, reptilian, and avian predators?

Despite several studies exploring the non-vocal behavioral response of spectral tarsiers to snakes (Gursky, 2003, 2005), there has not yet been a study comparing the non-vocal behavioral response of these primates to other types of predators. To some extent, this is a reflection of the fact that potential predators are rarely observed when the researchers conduct focal follows. A perusal of the literature (MacKinnon & MacKinnon, 1980; Niemitz, 1984; Gursky, 2003, 2005; Shekelle, 2003) indicates that the primary predators of spectral tarsiers are: monitor lizards (*Varanus indicus*), snakes (e.g., *Python reticulatus*), the Malaysian civet (*Viverra tangalunga*), and various birds of prey such as falcons (*Falco spp.*). Given the rarity with which the spectral tarsiers are observed interacting with potential predators, sample sizes are too small to effectively evaluate differences in the tarsier's response to different types of predators. Thus, it is necessary for studies to experimentally create predator/prey interactions using model predators in order to determine if prey respond differently to distinct classes of predators. In this paper, I explore the influence of predator type on the non-vocal anti-predator behavior of adult spectral tarsiers using wooden model predators.

Methods

Field Site

Sulawesi is located to the east of Borneo and northwest of Australia-New Guinea (125° 14′ East longitude, 1° 34′ North latitude). Sulawesi is the eleventh largest island in the world. It is also the largest and most central island of the biogeographical region of Wallacea, where the Australian and Asian zoogeographical regions meet. Sulawesi exhibits a blend of Asian and Australian elements in its fauna and flora, in addition to very high numbers of endemic taxa (Whitten et al., 1987).

This study was conducted at Tangkoko Nature Reserve on the easternmost tip of the northern arm of the island. The reserve is approximately 3,000 ha and exhibits a full range of forest types, including beach-formation forest, lowland forests, submontane forests, and mossy cloud forests on the summit of Mt. Tangkoko (MacKinnon & MacKinnon, 1980; Whitten et al., 1987; Gursky, 1997). The reserve is far from pristine due to heavy selective logging and encroaching gardens along its borders. The forest canopy is very discontinuous and contains a high proportion of *Ficus* trees (Gursky, 1997, 1998). Rainfall averages approximately 2300 m annually (World Wildlife Fund, 1980; Gursky, 1997). Additional details concerning the habitat type at Tangkoko Nature Reserve can be found in Gursky (1997).

Data Collection

In 2003–2004 nine groups were intensely observed for a 12-month study period. The size and composition of these groups prior to actual data collection (i.e., excludes births, dispersals, and disappearances) are presented in Gursky (2005).

I spent one month prior to data collection locating, trapping, radio-collaring, and habituating groups. Groups were located from the vocalizations each individual tarsier emits upon returning to its sleeping site each morning (MacKinnon & MacKinnon, 1980; Niemitz, 1984; Nietsch & Niemitz, 1992; Gursky, 1997). These vocalizations were given for 3–5 min and were heard from 300–400 m. The age and sex of each group members sharing a sleeping site was recorded. Sex was determined based on the sex-specific vocal calls given by all group members (MacKinnon & MacKinnon, 1980; Niemitz, 1984; Gursky, 1997). Age was estimated by relative body size (Gursky, 1997). Mist nets were then set up at the sleeping site(s) of the study groups approximately one hour before dusk and were continually monitored (Bibby et al., 1992). Upon capture, individuals were placed in a cloth bag and weighed with a portable scale providing an accuracy of ±1 g. An SM1 radio collar (manufactured by AVM Instrument Co., Livermore, CA), weighing 4.0 g was attached to the neck of all group members except infants by covering the folded-back thermoplastic band with heat-shrink tubing (Gursky, 1998b). Previous studies of the effect of radio collars on the activity patterns, mobility, foraging, body weight, and survival of spectral tarsiers suggest that the wearing of low-weight radio collars does not significantly affect this species

(Gursky, 1998b). The groups were each identified according to the location of their respective sleeping tree within the trail system. Upon completion of each research project, the radio collars were removed from all but three study animals. These three individuals proved elusive to recapture by either hand or net.

A radio receiver using 151-MHz frequency and a three-element collapsible Yagi antenna was used to determine the location of each individual radio frequency. An Indonesian assistant and I conducted behavioral focal follows. Initially, we conducted focal follows together on a single individual until approximately 99% of the data recorded by both of us was identical. Subsequently, I collected data on one member of the male-female pair, while the assistant collected data on the other member of the male-female pair. Once each month thereafter, the assistant and I conducted an inter-observer reliability test to determine if we remained consistent in our data recording. I found that our data recording was at least 98% consistent during each inter-observer reliability test.

Over the course of my research, I conducted over 2,000 hours of focal follows. The data set presented here is a subset of the total data collected. I utilized three primary methods of data collection (Altmann, 1974): (1) focal follows, (2) locational positions, and (3) predator experiments. While conducting nightly focal follows, I also recorded all occurrences of mobbing and alarm calling ad libitum. Additional details of the behavioral methods utilized are presented in Gursky (1997, 2002, 2003, and 2005).

During each night I conducted two different predator experiments on the nine groups that were trapped. I conducted 66 experiments using wooden monitor lizard models, 73 experiments using wooden civet models, and 70 experiments using bird of prey models. The model predators were lifelike in both size and color. In addition, there were numerous encounters with natural live predators (n = 25). When conducting these experiments, I remained behind a blind of large *Livistonia* leaves. Two conditions were necessary prior to beginning the experiment. First, the focal animal had to be no greater than 5 m in height. Second, the focal animal could not be near any group member. This was determined through general observation as well as scanning for other group member's radio signals with the radio attenuator on high. The predator experiment was conducted on each adult individual with a radio collar. Thus, focal individuals were restricted to individuals with radio collars. The predator model was attached to a bamboo stick and was exposed to the focal individual from either the top or the side of the blind. Focal animals were exposed to the predator models for 20 min, after which the predator model was removed and replaced behind the blind. While the predator was present, I observed the behavior of the focal animal using both instantaneous sampling as well as ad libitum sampling (Altmann, 1974). Ad libitum data was collected to record all behaviors for the 20 min the predator was present as well as the 20 min directly following the removal of the predator. In addition, while the predator was present, at 1-min intervals I recorded the approximate distance of the adult individuals from the predator. "Mobbing" was defined as the uttering of alarm calls while the animal alternately approached and retreated from the predator. Occasionally, mobbing involved physically biting the predator.

Data were analyzed using Statistica 6.0 (Statsoft Inc.). I used Repeated Measures ANOVA to explore how the spectral tarsiers responded to different types of potential predators (Sokal & Rohlf, 1981; Snedcor & Cochran, 1987).

Results

The results of this study indicate that the spectral tarsiers respond differently to different types of model predators such as snakes, civets, monitor lizards, and birds of prey (ANOVA RM $F = 97.3333$; df $= 4, 284$; $p = 0.0000$). The results for the snakes have been presented elsewhere (Gursky, 2005, 2006) and will not be repeated in the results presented here. However, the tarsier's response to the snakes will be compared to their response to these other predators in the discussion.

In response to the presence of a model falcon (Figure 11.1) the spectral tarsiers frequently (53%) froze their physical movements. Freezing lasted between 8–33 min and averaged 21 min (SD 6.3 min). There were no differences between sexes in the frequency ($X^2 = 7.24$, df $= 2$, $p = 0.9999$) or duration ($t = 1.9677$, df $= 2$, $p = 0.4955$) of freezing by adult males and adult females. The spectral tarsiers were more likely to freeze in response to the falcon model than they were for any of the other predator types (civets, monitor lizards, snakes) (ANOVA RM $F = 34.1535$; df $= 4, 27$; $p = 0.0000$). As a consequence of this anti-predator strategy there was no difference in the distance of the focal

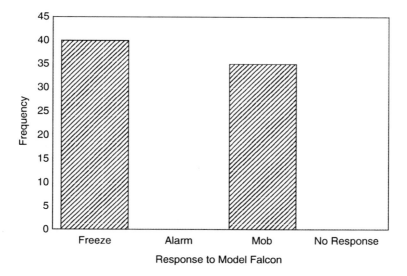

FIGURE 11.1. The variation in the response of spectral tarsiers to a model falcon

individual following exposure to the predator model and its distance 5 min after exposure. The mean height that spectral tarsiers utilized during the scan prior to the falcon presentation was 4.4 m (SD 2.9 m) and ranged from 0–16 m. However, following the presentation of the model falcon, spectral tarsiers decreased the mean height that they used to 1.3 m (SD 0.8 m) and only ranged 0–3 m. After being exposed to the model falcon the spectral tarsiers regularly returned to their sleeping tree, a behavior they rarely exhibited on nights when they were not exposed to a model falcon ($t = 48.4355$, df $= 138$, $p = 0.0000$). Similarly, following the presentation of the falcon models, the mean distance between adult group members significantly increased to 52 m, suggesting that the tarsiers' antipredator strategy, peculiar to this predator, results in decreased sociality.

In addition to freezing, the main anti-predator strategy exhibited by spectral tarsiers in response to the presence of falcons was alarm calling and mobbing behavior. On 47% of the tarsiers' encounters with the falcon model the spectral tarsier was observed alarm calling and mobbing. Mobbing involved lunging and retreating at the model. These two behaviors were always observed together. The tarsiers never alarm called in response to the falcon without mobbing it. The alarm call given in response to the falcon was a twittering alarm call. Mobbing lasted on average 33 min (SD 4.2 min), slightly longer than the duration that the predator model was present. When the model was removed and placed behind the blind, mobbers generally continued to mob the location where the falcon model had been placed. The mean number of mobbers was 5 tarsiers.

In response to the presence of the model civets (Figure 11.2) the tarsiers regularly emitted a harsh alarm call, three times in rapid succession, as well as moved away from the predator model. The mean amount of time from presentation of the model civet to the emission of the alarm call was 3.6 sec. Alarm

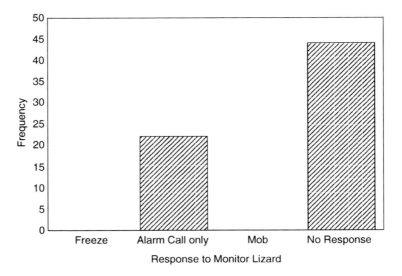

FIGURE 11.2. The variation in the response of spectral tarsiers to monitor lizards

calling lasted between 2 and 26 min and averaged 8 min. There were no statistical differences in the frequency ($X^2 = 5.21$, df $= 144$, $p = 0.8732$) or duration ($t = 1.6221$, df $= 144$, $p = 0.5445$) of alarm calling by adult males and adult females. The mean amount of time from the presentation of the model civet to tarsier traveling was a mere 6 sec. In less than 10 sec after their exposure to the model civet, the tarsiers alarm called and moved away from the model predator. The mean height that spectral tarsiers utilized during the scan prior to the predator presentation was 4.4 m (SD 2.9 m) and ranged 0–16 m. However, following the presentation of the wooden civet, spectral tarsiers increased their mean height to 9.3 m (SD 1.8 m) and only ranged 7–10 m. After exposure to the potential predator, the mean distance between adult group members also significantly decreased to 8 m, providing support for the hypothesis that predation pressure leads to increased sociality. The tarsiers never returned to their sleeping tree following exposure to a wooden civet. In addition to alarm calling upon seeing a civet, the tarsiers occasionally mobbed the wooden model. Unlike their encounters with the falcon model, alarm calling in response to the civet was done independent of mobbing. The mean length of time tarsiers spent mobbing a civet was 28 min (SD 5.7 min). The mean number of individuals at a civet-mobbing event was 4 tarsiers.

The response of the spectral tarsiers to wooden model monitor lizards was the least pronounced (Figure 11.3). They were never observed mobbing the wooden monitor lizard, they never froze their physical movements, and they only occasionally alarm called upon seeing the potential predator; but more often than not they completely ignored its presence. On the occasions in which the tarsiers alarm called, the alarm calls lasted between 1 and 6 min and averaged 2 min. There were no differences in the frequency ($X^2 = 4.27$, df $= 144$, $p = 0.7723$) or

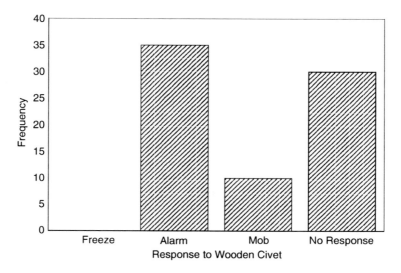

FIGURE 11.3. The variation in the response of spectral tarsiers to civets

duration ($t = 0.6177$, df $= 144$, $p = 0.8766$) of alarm calling by adult males and adult females. The mean amount of time from the presentation of the model monitor lizard to traveling was 38 sec and ranged 18–62 sec. Prior to the presentation of any of the predators, the mean distance between adult group members was 38 m and ranged between 0 and 140 m. Following the presentation of the wooden monitor lizards, the mean distance between adult group members significantly decreased to 24 m.

Discussion

Comparison of tarsier response to the various predators demonstrates a substantial difference between predator types. To begin with, the spectral tarsiers were more likely to freeze in response to the falcon model than with any of the other predator models. Freezing usually lasted the entire time that the falcon was visible. This result suggests that the falcon relies on movement to discern prey, so the tarsier's anti-predator strategy is to minimize movement. Freezing is a typical cryptic strategy exhibited by many prosimian primates when they are exposed to predators. The observation that the spectral tarsiers never ignored the presence of the falcon, but always acknowledged it by either freezing or mobbing, suggests that the falcon may have been the most dangerous predator (of the three predators tested) for this species. Behaviors like freezing that serve to quickly and efficiently avoid predatory threats are often interpreted as urgent responses. Similarly, the observation that spectral tarsiers regularly returned to their sleeping tree as soon as immediate threat of raptor predation was over reinforces the interpretation that falcons are a real threat to spectral tarsiers. The discovery by Gursky of a radio collar in a tree hole regularly used by a falcon also supports this suggestion.

The results of this study also indicate that spectral tarsiers were more likely to mob the falcon than either of the other predator models. One possible explanation for this difference may simply relate to the hunting strategy of the different predator types: Falcons are aerial predators while both the civet and the monitor lizard are more terrestrial predators. More specifically, falcons hunt using a stoop and wait strategy while civets and monitor lizards have a more active pursuit strategy.

Based on tarsier response to predator models in this study, I conclude that while the falcon may be one of the most dangerous predators for spectral tarsiers, the monitor lizard may be the least dangerous. The monitor lizards were regularly ignored, they occasionally caused the emission of alarm calls, but they never caused freezing or mobbing. This result is somewhat surprising as local villagers regularly report seeing monitor lizards consuming tarsiers. I, too, have observed them preying on domestic cats and rats. The fact that I have also observed monitor lizards throughout the tarsier's habitat makes the tarsier's lack of response to this carnivore even more surprising.

The response of the spectral tarsiers to the wooden civet was intermediate to that of the falcon and the monitor lizard. Once again, tarsiers' responses to the model civets were mixed. They did not behave consistently toward the model predator.

Tarsiers never froze their movements upon encountering a civet. The civet models were occasionally ignored, regularly acknowledged with alarm calls, and occasionally mobbed.

The variation observed in the response of the tarsier to the three types of predators probably represents the fact that tarsiers either can identify different predator species or they can recognize the degree of threat posed by different predators or by the same predator under differing circumstances. For example, Vitale (1989) found that free-ranging adult rabbits showed different predator avoidance strategies when presented with aerial or terrestrial predators. More specifically, he found that they mainly froze when faced with an aerial predator (goshawk), but this reaction was negligible when faced with the terrestrial model fox. He interpreted this as indicating that rabbits are capable of identifying different predator species. In contrast, Robinson (1980) found that Belding's ground squirrels are not only capable of discriminating between harmless birds and raptors, they discriminate between more and less dangerous interactions with the same predator species.

The fact that the spectral tarsiers did not behave consistently toward a predator type is suggestive, but not indicative, that tarsiers do not respond to different predator types (Pereira & Macedonia, 1991). Rather, these data suggest that tarsiers are responding to differences in perceived urgency (Macedonia & Evans, 1993; Blumstein, 1995; Blumstein & Armitage, 1997). Although I did not purposefully create differences that might reflect different degrees of urgency, the natural variation in the experimental environment, such as actual variation in the distance of the predator model to the focal animal (both in height and lineal distance), might have created differences in the tarsier's sense of urgency. For example, the tarsiers had to be 5 m below a predator at exposure. However, the tarsier's sense of urgency appeared to have varied depending on whether it was 1 m below or 5 m below the predator upon observing it.

Although this study has demonstrated the different anti-predator strategies of the spectral tarsier toward different types of predators, there are some additional limitations with the research design that should be mentioned. In particular, this study only explores the tarsier's ability to visually perceive predators. It completely ignores olfactory cues as well as auditory cues or any movements. Tarsiers are nocturnal, albeit secondarily nocturnal, and probably rely more on their olfactory and auditory senses to perceive predators than on their visual sense. Thus, future research should utilize olfactory and auditory cues for discerning potential predators. In a similar vein, this study used stationary models. Once the models were removed from the blind they were not moved again. This detail is of particular concern when studying tarsiers, a primate that is known for consuming only live, moving prey (Haring & Wright, 1987). They may be better able to sense miniscule movements by predators. Once again, future research is needed to more fully evaluate this aspect.

Acknowledgments. Funding for this research was provided by Wenner Gren Foundation, Fulbright Foundation, Primate Conservation, Inc., and Texas A&M University. The author thanks the Indonesian Institute of Sciences (LIPI), the

Directorate General for Nature Preservation and Forest Protection (PHPA) in Manado, Bitung, Tangkoko and Jakarta, SOSPOL, POLRI, the University of Indonesia, Jatna Supriatna and Noviar Andayani for their sponsorship while in Indonesia. Special thanks go to my field assistants for their help in collecting the data (Franz, Ben, Lende and Felik). The research protocols for this research were reviewed and approved by the Texas A&M University IACUC committees.

References

Altmann, J. (1974). Observational study of behavior: Sampling methods. *Behavior*, 49: 227–267.

Baldellou, M., and Henzi, S. (1992). Vigilance, predator detection and the presence of supernumary males in vervet monkey troops. *Anim. Behav.*, 43: 451–461.

Barros, M., Alencar, C., and Tomaz, C. (2004). Differences in aerial and terrestrial visual scanning in captive black tufted-ear marmosets (*Callithrix penicillata*) exposed to a novel environment. *Folia Primatol.*, 75(2): 85–92.

Bearder, S., Nekaris, K.A.I., and Buzzell, C. (2002). Dangers in the night: Are some nocturnal primates afraid of the dark? In L.E. Miller (Ed.), *Eat or be eaten: Predator sensitive foraging among primates* (pp. 21–40). Cambridge: Cambridge Univ. Press.

Bibby, R., Southwood, T., and Cairns, P. (1992). *Techniques for estimating population density in birds*. New York: Academic Press.

Blumstein, D. (1995). Golden-marmot alarm calls: I. The production of situationally specific vocalisations. *Ethology*, 100: 113–125.

Blumstein, D., and Armitage, K. (1997). Alarm calling in yellow-bellied marmots: I. The meaning of situationally variable alarm calls. *Anim. Behav.*, 53: 143–171.

Bshary, R., and Noë, R. (1997). Anti predation behavior of red colobus monkeys in the presence of chimpanzees. *Behav. Ecol. Sociobiol.*, 41: 321–333.

Caine, N. (1984). Visual scanning by tamarins. *Folia Primatol.*, 43: 59–67.

Cheney, D., and Wrangham, R. (1986). Predation. In B.B. Smuts, D.L. Cheney, R.M. Seyfarth, R.W. Wrangham, and T.T. Struhsaker (Eds.), *Primate societies* (pp. 227–239). Chicago: Univ. of Chicago Press.

Cheney D., and Seyfarth, R. (1990). *How monkeys see the world*. Chicago: Univ. of Chicago Press.

Cords, M. (1990). Vigilance and mixed-species association of some East African forest monkeys. *Behav. Ecol. Sociobiol.*, 26: 297–300.

Cowlishaw, G. (1994). Trade-offs between foraging and predation risk determine habitat use in a desert baboon population. *Anim. Behav.*, 53: 667–686.

Ferrari, S., and Ferrari, M. (1990). Predator avoidance behavior in the buffy headed marmoset *Callithrix falviceps*. *Primates*, 31: 323–338.

Fishman, M. (1999). Predator inspection: Closer approach as a way to improve assessment of potential threats. *Jour. of Theoretical Biology*, 196: 225–235.

Gould, L. (1996). Vigilance behavior during the birth and lactation season in naturally occurring ring-tailed lemurs (*Lemur catta*) at the Beza-Mahafaly Reserve, Madagascar. *Inter. Jour. of Primatol.*, 17: 331–347.

Gursky, S. (1997). Modeling maternal time budgets: The impact of lactation and gestation on the behavior of the spectral tarsier, *Tarsius spectrum*. Doctoral dissertation. SUNY Stony Brook.

Gursky, S. (1998a). The conservation status of the spectral tarsier, *Tarsius spectrum*, in Sulawesi Indonesia. *Folia Primatol.*, 69: 191–203.

Gursky, S. (1998b). The effect of radio transmitter weight on a small nocturnal primate: Data on activity time budgets, prey capture rates, mobility patterns and weight loss. *Am. Jour. Primatol.*, 46: 145–155.

Gursky, S. (2002). Predation on a wild spectral tarsier (Tarsius spectrum) by a snake. Folia Primatol., 73: 60–62.

Gursky, S. (2003). Predation experiments on infant spectral tarsiers, *Tarsius spectrum. Folia Primatol.*, 74(5–6): 272–284.

Gursky, S. (2005). Predator mobbing in spectral tarsiers. *Int. Jour. Primatol.*, 26: 207–221.

Hanson, M. (1995). The development of the California ground squirrels mammalian and avian anti-predator behavior. *Dissertation Abstracts International*, 56(9): 4696B.

Hanson, M., and Coss, R. (1997). Age differences in the response of California ground squirrels (*Spermophilus beecheyi*) to avian and mammalian predators. *Jour. of Comparative Psych.*, 111(2): 174–184.

Haring, D., Wright, P. (1989). Hand raising a Philippine tarsier, *Tarsius syrichta. Zoo Biology*, 8: 265–274.

Hauser, M. (2000). *The evolution of communication.* Cambridge, MA: MIT Press.

Hill, R., and Dunbar, R.I.M. (1998a). An evaluation of the roles of predation rate and predation risk as selective pressures on primate grouping behavior. *Behaviour*, 135: 411–430.

Isbell, L. (1994). Predation on primates: Ecological patterns and evolutionary consequences. *Evolutionary Anthropology*, 3: 61–71.

Iwamoto, T., Mori, A., Kawai, M., and Bekele, A. (1996). Anti predator behavior of gelada baboons. *Primates*, 37: 389–397.

Janson, C. (1998). Testing the predation hypothesis for vertebrate sociality: Prospects and pitfalls. *Behaviour*, 135: 389–410.

Karpanty S., and Grella, R. (2001). Lemur responses to diurnal raptor calls in Ranomafana National Park, Madagascar. *Folia Primatol.*, 72(2): 100–103.

Lima, S. (1995). Back to the basics of anti-predatory vigilance: The group size effect. *Anim. Behav.*, 49: 11–20.

Lima, S., and Dill, M. (1990). Behavioral decisions made under the risk of predation: A review and prospectus. *Canadian Jour. of Zoology*, 68: 619–640.

Macedonia, J., and Evans, C. (1993). Variation among mammalian alarm call systems and the problem of meaning in animal signals. *Ethology*, 93: 177–197.

MacKinnon, J., and MacKinnon, K. (1980). The behavior of wild spectral tarsiers. *Int. Jour. Primatol.*, 1: 361–379.

MacWhirter, R. (1992). Vocal and escape responses of Columbian ground squirrels to simulated terrestrial and aerial predator attacks. *Ethology*, 91: 311–325.

Mitani, J., Sanders, W., Lwanga, J., and Windfelder, T. (2001). Predatory behavior of crowned hawk-eagles (*Stephanoaetus coronatus*) in Kibale National Park, Uganda. *Behav. Ecol. and Sociobiol.*, 49: 187–195.

Niemitz, C. (1984). *Biology of tarsiers.* Stuttgart: Gustav Fischer.

Nietsch, A., and Niemitz, C. (1992). Indication for facultative polygamy in free ranging *Tarsius spectrum*, supported by morphometric data. *Int. Primatol. Soc. Abst.*, Strasbourg, 1992: 318.

Pereira, M., and Macedonia, J. (1991). Ring-tailed lemur anti-predator calls denote predator class, not response urgency. *Anim. Behav.*, 41: 543–544.

Ramakrishnan, U., Coss, R., Schank, J., Dharawat, A., and Kim, S. (2005). Snake species discrimination by wild bonnet macaques (*Macaca radiata*). *Ethology*, 111: 337–356.

Randall, J., Rogovin, K., and Shier, D. (2000). Anti-predator behavior of a social desert rodent: Foot drumming and alarm calling in the great gerbil, *Rhombomys opiums. Behav. Ecol. and Sociobiol.*, 48: 110–118.

Robinson, S. (1980). Antipredator behaviour and predator recognition in Belding's ground squirrels. *Anim. Behav.*, 28: 840–852.

Shekelle, M. (2003). Taxonomy and biogeography of Eastern tarsiers. Doctoral thesis. Washington University, St. Louis, MO.

Snedcor, G., and Cochran, W. (1989). *Statistical methods*. Iowa: Iowa State Univ. Press.

Sordahl, T. (2004). Field evidence of predator discrimination abilities in American avocets and black necked stilts. *J. Field Ornithol.*, 75(4): 376–385.

Sokal, R., and Rohlf, J. (1981). *Biometry*. New York: Freeman & Co.

Stanford, C. (1998). Predation and male bonds in primate societies. *Behaviour*, 115: 513–533.

Treves, A. (1998). The influence of group size and neighbors on vigilance in two species of arboreal monkeys. *Behaviour*, 115: 453–481.

Vitale, A. (1989). Changes in the antipredator responses of wild rabbits, *Oryctolagus cunicilus* with age and experience. *Behaviour*, 110: 47–60.

Welton, N., McNamara, J., and Houston, A. (2003). Assessing predation risk: Optimal behaviour and rules of thumb. *Theoretical Population Biology*, 64(4): 417–430.

Whitten, T., Mustafa, M., and Henderson, G. (1987). *The ecology of Sulawesi*. Yogyakarta: Gadja Mada Univ. Press.

World Wildlife Fund. (1980). *Cagar Alam Gunung Tangkoko Dua Saudara Nature Reserve Sulawesi Utara management plan 1981–1986*. Bogor: Indonesia.

Wright, P. (1998). Impact of predation risk on the behavior of *Propithecus diadema edwardsi* in the rain forest of Madagascar. *Behaviour*, 115: 483–512.

Wright, P. (1999). Lemur traits and Madagascar ecology: Coping with an island environment. *Yearbook of Physical Anthropology*, 42: 31–72.

Ydenberg, R., and Dill, L. (1986). The economics of escaping from predators. *Advances in the Study of Behavior*, 16: 229–249.

12
Talking Defensively, a Dual Use for the Brachial Gland Exudate of Slow and Pygmy Lorises

Lee R. Hagey, Bryan G. Fry, and Helena Fitch-Snyder

Introduction

On the ventral side of the elbow of the both the slow (*Nycticebus bengalensis*, *N. coucang*) and pygmy (*N. pygmaeus*) lorises, one can perceive a slightly raised, fairly hair-free but barely visible swelling, termed the brachial gland (Figure 12.1). Observers of captive lorises have found that when the animal is disturbed during capture and handling, the gland secretes about 10 microliters (μl) of a clear, strong-smelling liquid in the form of an apocrine sweat. Typically, male and female lorises assumed a defensive position with head bent downward between uplifted forelegs, like a miniature prize fighter in a clinch, while imparting gland exudate to the head and neck (Fitch-Snyder, 1996). The lorises frequently licked their own brachial gland regions, and also wiped these glands against their heads. The gland is active in lorises as young as 6 weeks (Fitch-Snyder, unpubl. data).

Early observers of the loris concluded that the gland contains a form of toxin, basing this conclusion primarily on reports from individuals on the receiving end of painful, slow-healing bites. Lorises have a specialized needle-like, oral tooth comb used for grooming, and the close association with this comb and the licking of the gland made it a natural assumption to visualize the comb as a device for injecting brachial gland derived toxic secretions. Although the design of advanced venom delivery architecture (injecting poison hypodermically through a sharp-pointed tube) has evolved numerous times in vertebrates (i.e., gila monsters, stingrays, stonefish, snakes), offensive and defensive toxins are seldom found in mammals. The platypus and water shrew are the mammals most well documented in the use of this strategy. Thus, the loris would be a singular example of a primate that uses a toxin, has a specialized device for injecting it, and that loads its "sting" in a secondary manner by retrieving the toxin from a part of the body not associated with the toothcomb. Specialized teeth on the lower jaw of the loris have been shown to be effective in conducting liquid upward (Alterman, 1995).

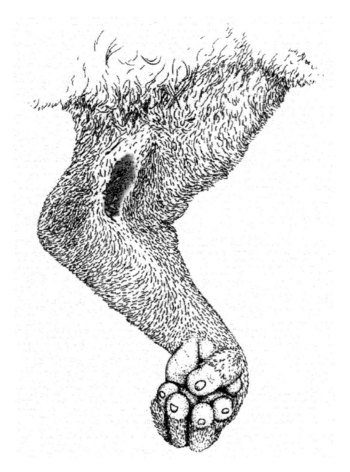

FIGURE 12.1. A drawing of the brachial gland by Helga Schulze

While lorises are predatory as well as frugivorous, their prey consists of insects and vertebrates so small relative to lorises that no venom should be needed to kill or immobilize these prey prior to being ingested by lorises. Little is known about predation of lorises and whether their bite is used as a defensive weapon against predators. They are unable to fend off predators with their bites, and they have been captured and killed by orangutans (Utami & van Hooff, 1997), snakes (Wiens & Zitzmann, 1999), and hawk-eagles (*Spizaetus cirrhatus*) (C. van Schaik, 2005). In contrast, the bites they inflict on social enclosure mates are common and severe. These wounds show a large affected area with a loss of fur, prolonged edema, are slow-healing, and often are life-threatening (Rasmussen, 1986).

The toxin hypothesis has lead to a number of attempts to characterize the protein contents of the gland, beginning with Alterman & Hale (1991), Alterman (1995), and, most recently, Krane et al. (2003). Recognizing that lorises are a prosimian species whose biology is not fully understood, we re-examined different components of exudate from both the pygmy and slow lorises (including both the volatile low molecular weight metabolite and non-volatile higher molecular weight protein fractions) and present two hypotheses for the dual use of the gland. In addition to being a defensive toxin reservoir, the strong-smelling glandular secretion displays all the components necessary for it to play an important functional role in olfactory communication.

Methods

Brachial gland samples were obtained from a colony of male and female slow lorises (*Nycticebus bengalensis*) (Groves, 1998) and pygmy lorises (*Nycticebus pygmaeus*) housed at the San Diego Zoo, and *N. bengalensis*, *N. coucang coucang*, and *N. pygmaeus* housed at the Singapore Zoo. Samples (from four individuals of each species) were obtained from fluid collected on the surface of the gland using capillary action in 2-μl micropipettes, and stored in glass vials at $-20°C$ until analysis.

Volatile and semi-volatile compounds were extracted (35 min) from the vapor surrounding the micropipette using a solvent-free solid phase matrix extractor (SPME) containing a 65–μm polydimethylsiloxane fiber (Supelco, Bellefonte, PA). Fiber contents were analyzed by capillary GC-MS, with a Hewlett-Packard 5890 Gas Chromatograph-5970 MSD, controlled by HP/UX Chem Station software. The GC used a Supelco 60- m 0.25- mm ID, low-polarity SPB-octyl column (Supelco, Bellefonte, PA) operated isothermically at 75°C for 9 min followed by a 1.6°/min ramp gradient up to 210°C. A splitless injection was used with an injection temperature of 250°C. Helium was used as the carrier gas with a 7-psi column head pressure. Relative retention times and fragmentation spectra of peaks obtained by GC-MS were matched with those of known standards for identification. The remaining, scientifically unnamed compounds were used for analysis because they possessed a reproducible retention time and mass fragmentation pattern and could be classified according to chemical function. These compounds were assigned a number based on their column retention time.

Samples (5 μl) of exudate oil were diluted in methanol, centrifuged, and examined for non-volatile polar compounds with a Perkin-Elmer Sciex API-III instrument (Alberta, Canada) modified with a nanoESI source from Protana A/S. Medium-sized palladium-coated, borosilicate glass capillaries from Protana A/S were used for sample delivery. The instrument was operated in the negative (−) mode with ISV voltage set to 600 V, IN voltage set to 110 V and the ORI voltage set to 90 V. A curtain gas of ultra pure nitrogen was pumped into the interface at a rate of 0.6 L/min to aid evaporation of solvent droplets and to prevent particulate matter from entering the analyzer region.

On-line LC-MS of brachial gland samples dissolved in 0.1% TFA to a concentration of approximately 1 mg/ml were performed on a Phenomenex Jupiter

C-18 reverse phase column (150 × 1.0 mm, 5 μ, 300 Å); a gradient solvent A (0.1% TFA) and solvent B (90% acetonitrile in 0.1% TFA) were used at a flow rate of 50 μl/min. The gradient formation of a 1% gradient from 0–90% acetonitrile/0.1% TFA over 90 min was achieved using an Applied Biosystems 140B solvent delivery system. Electrospray mass spectra were acquired on a PE-SCIEX API 300 LC/MS/MS system with an Ionspray atmospheric pressure ionization source. Samples (25 μl) were injected manually into the LC-MS system and analyzed in the positive ion mode. Full scan data were acquired at an ionspray voltage of 4600 V and the orifice potential was set at 30 V over an ion range of 600–3000 m/z with a step size of 0.2 amu. Data processing was performed with the aid of the software package Biomultiview (PE-SCIEX, Canada).

Brachial gland fluid in the capillary was reconstituted with 10 μl of water by leaving the capillary side down in water. A portion of the reconstituted sample was added to 10 μl loading buffer, boiled for 5 min, and loaded onto an 18% tris-glycine gel (Novex cat. no EC6505). The gel was run for 60 min (running buffer tris 1.52 g, glycine 7.2 g, SDS 0.5 g, water 500 ml), stained (trichloroacetic acid 10 g, sulfosalicylic acid 10 g, coomassie blue 0.2 g, water 80 ml) for 120 min, and destained (methanol 10%, acetic acid 7.5%) overnight. A second portion was reduced prior to loading on the 18% tris-glycine gel.

After separation on the gel the α- (lower band) and β- (upper band) peptides were eluted from the gel and the amino terminal ends sequenced. Edman sequencing was perfomed on an Applied Biosystems PROCISE 494HT Protein Sequenator, using the Division of Biological Sciences Protein Sequencing Facility on the University of California, San Diego campus, Matthew Williamson, facility operator.

We analyzed 76 protein sequences obtained from the NCBI database (http://www.ncbi.nlm.nih.gov/entrez/query.fcgi?db=Protein&itool=toolbar). We used the program CLUSTAL-X (Thompson et al., 1997) to align the sequences, followed by visual inspection of the resultant alignment for errors. The final alignment consisted of 159 amino acid sites. A copy of the full sequence alignment can be obtained by emailing Dr. Bryan Grieg Fry at bgf@unimelb.edu.au. Phylogenetic trees were reconstructed using the maximum parsimony (MP) and neighbor-joining (NJ) (Saitou & Nei, 1987) methods. MP heuristic searches were conducted by implementing random stepwise taxon addition with TBR branch swapping and the PROTPARS weighting scheme (Felsenstein, 2001), which takes into account the number of changes required at the nucleotide level to substitute one amino acid for another. NJ searches were conducted using amino acid p distances, as the simple p distance generally gives better results in phylogenetic inference than more complicated distance measures for minimum evolution methods such as NJ (Takahashi & Nei, 2000). Statistical reliability was assessed using 100 and 1000 bootstrap replications for MP and NJ searches, respectively. In order to simplify sequence nomenclature and to minimize confusion, we refer to proteins by their NCBI accession number in the text and figures.

Results

When examined by GC-MS, brachial gland exudates contained a complex mixture of volatile and semi-volatile compounds. A representative profile for the *N. pygmaeus* is shown in Figure 12.2. A total of 212 different compounds were observed in two individual pygmy loris samples. The results from these were pooled and are listed in Table 12.1. Also shown in Table 12.1 is the proportion (averaged) of the compound present. SPME matrix peaks and other artifacts are not listed in Table 12.1. Not all compounds were definitively identified by name, but based on fragmentation pattern and GC column retention time could still be recognized as unique compounds. In Table 12.1, these compounds are referred to as numbers, the numerical value reflecting each individual compound's relative retention time on the SPB-35 GC column. Minor amounts of a wide variety of aromatic compounds were identified, consistent with dietary absorption from a frugivorous species, and a concurrent difficulty in complete metabolism of this chemical class of compounds. The remaining identified compounds were a series of C_4–C_7 aldehydes, ketones, and acetates.

The compounds present in the brachial gland exudate from a *N. bengalensis* are listed in Table 12.2. There were 68 different compounds in the profile, about half of which were found in the pygmy loris using an identical method of analysis. A comparison of Tables 12.1 and 12.2 indicate that 33 of the 68 *N. bengalensis* compounds (48%) were unique to this species and not detected in the larger table of data from the pygmy loris. A disproportionately large signal of m-cresol is characteristic of slow loris scent, which, when mixed with other compounds, may contribute to the reported strong odor from this gland. It does not appear as if the dominant component of the exudate oil itself is a low molecular weight

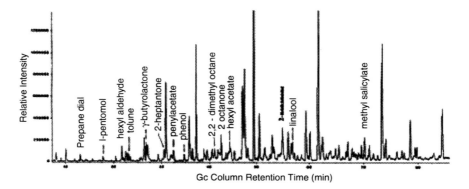

FIGURE 12.2. GC-MS profile of the volatile and semi-volatile components of pygmy loris (*N. pygmaeus*) brachial gland exudate. Peaks are expressed according to relative abundance. The x-axis is the time of elution from the GC column and the y-axis is the relative intensity of each peak

TABLE 12.1. Volatile and semi-volatile compounds found in pygmy loris (*N.pygmaeus*) brachial gland exudate as determined by GC-MS. Compounds are identified by relative area, followed by either the name (if known) or the relative retention time (if the identity has not yet been confirmed).

Name or Identification No.	Relative Amount
methanol	50
3,4-dimethyl heptane	1
0.952	5
methyl salicylate	103
ethanol	22
m-xylene	2
0.960	7
1.248	13
acetic acid	23
0.643	2
0.963	1
1.256	26
n-butane	21
phenol	75
0.966	36
1.261	96
isoprene	15
p-xylene	22
p-cymene	5
1.272	26
chloroform	3
0.678	35
0.973	8
1-dodecene	2
1-butanol	10
0.684	250
0.974	9
1.286	26
0.255	1
0.687	11
acetophenone	3
1.303	60
0.260	110
0.691	65
0.983	1
1.307	36
2-methylbutyral-dehyde	3
styrene	36
0.986	9
1.317	20
1,2-propanediol	32
o-xylene	2
0.989	7
1.325	6

TABLE 12.1. (Continued).

Name or Identification No.	Relative Amount
n-valeraldehyde	27
0.724	12
0.992	7
1.339	49
butyric acid	5
0.727	10
0.993	3
1.348	53
0.292	8
0.737	20
3-nonanone	25
1.354	1
methyl butyrate	124
0.742	7
2-nonanone	44
1.358	9
0.312	14
1-heptanol	35
limonene	12
1.361	1
0.318	12
0.747	37
1.002	20
1.363	24
1-heptene	4
0.755	22
1.013	13
1.365	24
0.342	5
0.761	29
1.016	12
1.378	247
0.346	6
benzaldehyde	21
benzoic acid	10
1.389	9
1-pentanol	23
0.762	40
1.025	25
1.400	75
0.372	1
0.767	15
1.030	2
1.411	12
0.375	14
6-Me-5-hepten-2-one	6
linalool	13
1.424	17
isovaleric acid	1
0.771	4
1.027	14

TABLE 12.1. (Continued).

Name or Identification No.	Relative Amount
1.432	9
0.392	2
2,2-dimethyl octane	62
nonyl aldehyde	35
1.437	10
0.397	32
0.774	13
1.047	8
1.447	43
3-hexanone	26
0.777	6
1.056	7
1.465	36
ethyl butyric acid	2
0.782	66
1.060	2
1.485	2
hexyl aldehyde	72
0.785	43
1.071	220
1.499	36
n-valeric acid	45
0.790	15
1.078	21
1.501	20
toluene	580
2-octanone	103
1.082	148
1.504	32
"anti" 2-methyl butyraldehyde oxime	49
3-octanone	34
1.107	117
1.509	2
0.806	10
resorcinol	3
1.511	24
"syn" 2-methyl butyraldehyde oxime	27
0.808	14
n-undecane	91
1.519	2
0.810	31
1.122	146
1.528	52
0.472	1
0.817	31
1.126	19
1.537	26
1-octene	25
n-propyl benzene	18
1.127	4
dodecyl aldehyde	5

TABLE 12.1. (Continued).

Name or Identification No.	Relative Amount
0.499	14
hexyl acetate	78
1.133	30
0.502	20
0.829	35
1.139	8
0.510	17
octyl aldehyde	23
1.140	35
butyrolactone	98
0.828	1
1.146	51
0.525	42
0.838	43
1.153	12
1-hexanol	81
0.858	3
1.158	45
isopentyl acetate	15
p-cresol	5
1.162	63
4-heptanone	23
0.871	50
1.170	93
0.561	4
0.879	36
1.171	1
0.580	17
m-cresol	22
1.183	1
2-heptanone	26
0.887	16
1.190	1
0.592	26
ethyl hexanoate	13
1.191	53
0.593a	445
0.918	1
1.197	17
0.593b	600
0.926	71
2-decanone	123
0.605	7
0.928	14
1.210	74
2-heptanol	26
0.932	4
1.216	53
0.615	31
0.936	1
1.221	41

TABLE 12.1. (Continued).

Name or Identification No.	Relative Amount
pentyl acetate	32
0.941	8
1.226	38
0.618	11
1-octanol	3
n-decyl aldehyde	23
0.622	25
0.945	8
1.233	22
hexanoic acid	39
0.946	3
1.237	94

TABLE 12.2. Volatile and semi-volatile compounds found in slow loris (*N. bengalensis*) brachial gland exudate as determined by GC-MS. Compounds are identified by relative area, followed by either the name (if known) or the relative retention time (if the identity has not yet been confirmed).

Name or Identification No.	Relative Amount
methanol	21
0.704	62
1.107	52
ethanol	18
1-heptanol	10
1.141	24
n-butane	4
0.760	6
1.158	9
2-butanol	22
benzaldehyde	16
1.191	3
acetic acid	102
6-Me hepten-2-one	16
1.230	15
0.228	7
0.777	3
1.237	19
2-methyl butyraldehyde	4
3-octanone	3
1.278	13
2-methyl butyronitrile	99
0.787	69
1.297	88
0.314	20
0.822	4

TABLE 12.2. (Continued).

Name or Identification No.	Relative Amount
1.378	34
0.329	944
2-octanol	6
1.395	15
0.357	92
0.844	12
1.409	36
0.496	4
0.856	5
1.465	19
"anti" 2-methyl butyraldehyde oxime	109
0.871	865
1.472	46
0.885	3
1.496	4
"syn" 2-methyl butyraldehyde oxime	30
m-cresol	1503
1.504	13
0.906	26
1.532	4
0.514	187
0.922	9
1.538	2
1-hexanol	23
0.929	29
1.556	5
0.583a	24
0.986	14
dodecyl aldehyde	4
0.583b	12
1.002	5
1.591	3
2-heptanol	9
1.027	97
0.617	58
1.039	30
phenol	86
1.052	117
0.691	25
1.056	4
0.693	29
1.081	24

hydrocarbon, for such a compound should have dominated the volatile profiles in Tables 12.1 and 12.2.

To examine the exudate oil contents by a different approach, samples from both loris species were examined by nanoESI-MS. Although this instrument will not detect neutral molecules, it is sensitive to charge-bearing compounds of a higher

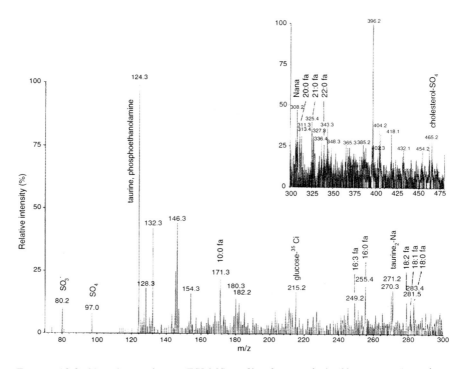

FIGURE 12.3. Negative mode nanoESI-MS profile of pygmy loris (*N. pygmaeus*) exudate. Tentative identifications are listed above the mass per charge value of the peaks. Abbreviations: fa: fatty acids, NANA: N-acetylneuraminic acid

molecular weight than that detectable by GC-MS. The negative (–) mode profile of pygmy loris exudate is shown in Figure 12.3 and tentative identifications of the peaks are labelled. Although the sugars glucose, neuraminic acid, and a variety of fatty acids (fa) were detected, none were present in amounts sufficient to constitute the exudate oil itself. Easily seen by this instrument, but notably absent from the profile were phospholipids.

To detect the presence of higher molecular weight compounds (proteins), SDS-PAGE gels were run on exudates from both species. The gels for both the *N. pygmaeus* and *N. bengalensis* lorises showed single large bands of approximate molecular weight 18,000 daltons. After reduction, the 18 kDa bands, run in identical gel systems, revealed a pair of bands of approximate molecular weight 7 kDa and 8 kDa, as shown in Figure 12.4 (pygmy loris right panel, slow loris left panel). Since the reduction was not 100% complete, some of the original 18 kDa protein can be seen in both gels. The 7 kDa and 8 kDa bands were isolated and used for amino acid sequencing.

LC-MS analysis of the brachial gland secretion from both species also revealed that each contained a single dominant protein component with a molecular weight 17.6 kDa, as shown in Figure 12.5. The pygmy loris contained two isoforms

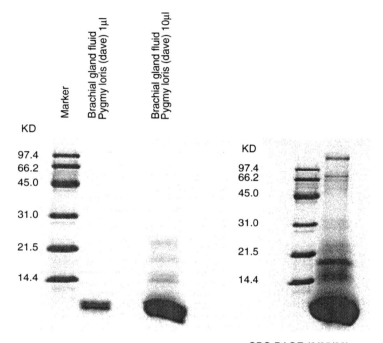

FIGURE 12.4. SDS-PAGE gels of the protein component of *N. pygmaeus* (left panel) and *N. bengalensis* (right panel) loris brachial gland exudate. After reduction with TCEP, a pair of bands of approximate molecular weight α 7.9 kDa and β 9.8 kDa

(17671 and 17601 Da), as did both species (*N. bengalensis, N. coucang*) of slow loris (17649 and 17610 Da). Reduction of the disulfide bonds in the 17.6kDa peptide revealed that it was a heterodimer of two smaller peptides, molecular weights 7.8 kDa (α-chain) and 9.8 kDa (β-chain).

The amino terminal sequences of the two chains are shown in Figure 12.6. The initial 35 amino acids of the α-chain sequence reported by Krane et al. (2003) is essentially identical to our sequence, with the single exception in position six, where they found Leu, but we observed Val. At position Val-25 we see some chains with Ala-25 substituted for Val. Krane et al. (2003) also reported the sequence of the first 31 amino acids of the β-chain, which was identical to ours. They reported the presence of Cys in position-1 but we were unable to confirm this. Using the N-terminal sequences of the loris toxin, database searching (http://www.ncbi.nlm.nih.gov/BLAST/) revealed a number of proteins with a high degree of similarity (Figure 12.6). Further, genome mining

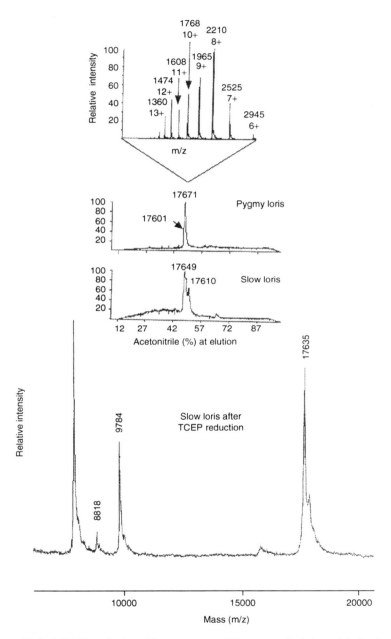

FIGURE 12.5. LC-MS analysis of *N. pygmaeus* and *N. coucang* loris brachial gland secretions. Reconstructed masses are given above each peak

(A)

```
1.  SGCKLLEDMVEKTINSDISIPEYKELLQEF-IDSDAAAEAMGKFKQCFLNQSHRTLKNFGL
2.  GLCPALQRKVDLFLNGTT--EEYVQYLKQF-NENRDVLDNAENIKKCSDRTLTEEDKAQAT
3.  GICPAIKEDVCLFLNGTS--EEYVEYVKQY-KNDPEILENTEKIKQCVDSTLTEKDKAHAT
4.  AICPAVEKHANLFLKGTT--DEFLNNAKNF-VKSSEVLEN--------------------
5.  EICPAVKRDVDLFLTGTP--DEYVEQVAQY-NALPVVLENARILKNCVDAKMTEEDKENAL
6.  DICPVVTKDVDLFLVGTP--DEYVDHVAQY*TSS-LILSNARKLKNCFNGKLADEDKRHVL
7.  DICPGFLQVLEALLLGSE--SNYEAALKPF-NPASDLQNAGTQLKRLVD-TLPQETRINIV
8.  EFCPALVSELLDFFFISE--PLFKLSLAKF-DAPPEAVAAKLGVKRCTD-QMSLQKR----
```

```
1.  MMHTVYDSIWCN--MKSNNQSHRTLKNFGLMMHTVYDSIWCNMKSN-
2.  SLIR--------------TLTEEDKAQATSLI----------------
3.  AFILSLYMSG-------STLTEKDKAHATAFILSLYMSG--------
4.  ----------------------------------------------
5.  SVLDKIYTSPLC-----AKMTEEDKENALSVLDKIYTSPLC------
6.  S---------------GKLADEDKRHVLS-----------------
7.  KLTEKILTSPLCEQDLRVTLPQETRINIVKLTEKILTSPLCEQDLRV
8.  ------------------SLIAEVLVKILKKCSV------------
```

(B)

```
1.  VKMAETCPI-FYD--VFFAVANGNELLLDLSLTKVNATEPERTAMKKIQDCYVENGLISRVLD
2.  -------WPLPFSGPSLIG-LIT---NLLNSFSDK-SGISSWFGSITGELA*
3.  -------AP-FVG--AYVKILGGNRLALNAYLSMFQATAAERVAFEKIQDCFNEEPLTTKLKS
```

```
1.  GLVMIAIN------EYCMGEAVQNTVEDLKLNTLGR
2.
3.  PQIMMSILFSSECKAYYPEDSVNKMADMFKLDSIN-
```

FIGURE 12.6. NH$_2$-terminal amino acid sequence comparison of the *N.pygmaeus* loris α- and β-chains that make up the 18 kDa major peptide of brachial gland exudate. (a) Comparison between the *N. pygmaeus* loris α-chain sequence and members from each clade of the α-chain superfamily: (1) Secretoglobin AAC79996; (2) Mouse salivary androgen binding protein AAM08259; (3) Mouse putative protein XP_142918; (4) Loris brachial gland secretion; (5) Domestic cat allergen Fel d1-A CAA44344; (6) Human genome putative protein (ensemble, http://www.ensembl.org, accession # AC020910.7.1.203201, location 112772-112948); (7) Uteroglobin NP_037183; and (8) Lipophilin NP_006542. (b) NH$_2$-terminal amino acid sequence comparison between the *N. pygmaeus* loris β-chain sequence and two members with similar β-chains. (1) Domestic cat allergen Fel d 1-B P30440; (2) Loris brachial gland secretion β-chain; and (3) Mouse salivary androgen binding protein beta AAH24677

(http://www.ensembl.org) revealed a match with an α-chainlike human gene, located on chromosome-19 (accession # AC020910.7.1.203201, location 112772-112948) and a highly similar chimpanzee match (database location AADA0-1344989, genomic location: chromosome 20, 36466963 to 36467142). Both the human and chimpanzee putative proteins have a stop codon in the middle of the sequence, and as a result may be degenerative pseudogenes. Phylogenetic reconstruction of the molecular evolutionary history revealed a strong association between the alpha-chain of the loris protein and the cat allergen alpha-chain as well as the human and chimpanzee genome sequence (Figure 12.7).

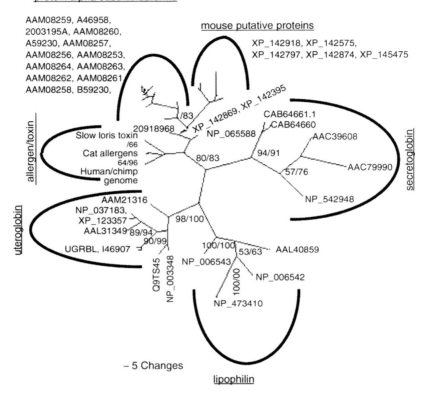

FIGURE 12.7. Phylogenetic analysis of proteins showing similarity to the *N. bengalensis* toxin α-chain. Shown is the maximum parsimony tree. Bootstrap values are the result of 100 replicates (maximum parsimony) or 1000 replicates (neighbor-joining). Accession numbers for representative sequences of the major clades are given (http://www. ncbi.nlm.nih.gov/entrez/query.fcgi?db=Protein&itool=toolbar). Domestic cat allergen is Fel dl-A CAA44344, human genome is from ensemble, http://www.ensembl.org, accession # AC020910.7.1.203201, location 112772-112948 and chimpanzee genome is from ensemble, http://www.ensembl.org, accession # AADA01344989, genomic location: chromosome 20, 36466963 to 36467142

Discussion

Like other nocturnal prosimians occupying individual home ranges or territories with limited social contact, lorises are specialized in olfactory signalling. It has been described that lorises scent mark with copious quantities of urine and that the introduction of lorises to new environments is accompanied by vigorous urine marking and sniffing behavior (Tenaza & Fitch, 1984). A second site of olfactory-mediated chemical communication is the brachial gland exudate. Once expressed, this secretion dries to amorphous, amber-colored crystals, and the hair on the loris head and neck can become encrusted with the solid. Results from the mass spectrometer (Tables 12.1 and 12.2) show that oil from each species is unique and complex, with more than 68 (*N. bengalensis*) and 200 (*N. pygmaeus*) volatile and semi-volatile components. The qualitative and quantitative differences seen in the scent profiles of the two loris species can code for extensive information, including sex, age, nutritional status, health, and dominance of the sender (Hagey & MacDonald, 2003). The likely recipient (and interpreter) of such a coded message is another loris, as social contact is a clearly recognized source of agitation in these species. Since 48% of the compounds found in slow lorises were not found in pygmy lorises, chemical analysis of the brachial scent may be a useful technique to resolve taxonomy issues among *Nycticebus*.

Brachial gland secretion is not an immediate response to stress or being chased. The characteristic odor is not noticeable to the researcher trying to catch the animal, and lorises only stop to lick this gland after the encounter is completely over (Fitch-Snyder, 1996). It thus seems unlikely that the loris would have time to mount a chemical defense to a stealth attacker. This behavior implies that the BGE functions either to deter a predator or to warn other lorises of the danger, or perhaps both.

Many animal scents show a strong component of bacterial commensalism dominated by bacterial metabolites, particularly low molecular weight short-chain acids (Albone, 1984). Both loris exudates had little or none of these bacterial metabolites, reflecting their origin as a fresh glandular secretion. When dried, metabolic acids form salts, and when embedded in a lipid matrix have fairly long-lasting odors. The short aldehydes, ketones, and aromatics seen in loris exudates (and lack of short-chain acids) are consistent with reports that fresh exudate emits a very distinctive volatile short-range odor that is rapidly lost to evaporation. To the olfaction of a recipient loris, the brachial scent must represent a powerful, but rapidly decaying, burst of detailed information.

The exudate oil in both species contained isoforms of a 17.6 k protein, which was composed of α/β-heterodimeric subunits (α-chain MW 7880, β-chain MW 9784) linked together by two disulfide bridges. Sequencing of the α/β-chains showed that the loris brachial gland peptide is a new member of the secretoglobin (uteroglobin/Clara cell 10 k) family. For a list of 40 peptide members in 11 species, see Klug, 2000). Based on the close sequence homology with domestic cat Fe-dI chain I peptide, loris peptide could be assigned to subfamily 4, using the nomenclature of Klug (2000). The sequences of seven of these family members are listed in Figure 12.6, including the human genome match. The secretoglobin family is characterized by small lipid-loving peptides found as major constituents in a variety of mammalian secretions. These proteins are all α/β-homo- and heterodimers stabilized by two or three intramolecular cystine disulfide bonds (Lehrer et al., 2000). In what is termed the uteroglobin-fold, the α- and β-monomers are formed from grouping four α-helices, and (for the two monomers) the combined eight α-helix bundle folds to form a pocket for the binding of different hydrophobic molecules (Callebaut et al., 2000). This simple structural motif of the uteroglobin fold stands in sharp contrast to the wide array of biological activities assigned to this group of proteins.

In the loris 17.6 k protein, the smaller β-chain forms a slightly pyramidal shaped lid that is hinged along one edge by the two disulfide bridges to the larger α-chain, forming a unit roughly in the shape of a cigar box. The β-subunit has a shallow hydrophobic center in the lid, which sits over a similar but deeper pocket in the α-chain box, which may act as a molecular snare for a small hydrophobic molecule. A potential for molecular docking of hydrophobic molecules like progesterone, polychlorinated biphenyls, and retinol has been shown using the crystal structure of human uteroglobin (Pattabiraman et al., 2000). Other than the disulfide bridges located together in the "hinge," only these interacting hydrophobic regions hold the lid to the box. The behavior of the loris may add to an understanding of how the molecular snare functions. When the snare is in the hydrophobic environment of the secreted oil on the arm, the lid is free to open, and the box can accept a signaling molecule. One function for the box would be to hold a species-specific message, and the varying compositions of the α/β-chains in different species support this idea. In the case of the mouse salivary binding protein, the signaling molecule is an androgen (Karn & Russell, 1993).

In the loris, the message molecule enters the gland oil through an equilibrium state with blood serum. When the loris licks the gland, the external environment of the box is mixed with water and the lid closes, ensnaring the message molecule. These are not swallowed, but are deposited on the external fur to be conveniently dropped or rubbed off. For related molecules like lipophilins secreted in tears, it has been speculated that the protein could function as a snare to capture pheromones or other lipophilic molecules from the atmosphere, as the tear drains into the nasal cavity (via the nasolacrimal ducts) and conveys them to receptors for further sensory processing (Lehrer et al., 2000). Although the function of domestic cat allergen peptide is not known, cats heavily contaminate their environments with Fe-dl (Morgenstern et al., 1991), using the protein not as a toxic defense, but as a species recognition molecule. Rather than possessing a brachial gland like the loris, small cats directly add salivary allergen proteins to inanimate objects during cheek rubbing, as well as transferring them to their body fur with washing behaviors (Mellen, 1993). The idea that species recognition systems share a close relationship with immune recognition has been investigated by Palumbi & Metz (1991).

The key question here is why invest so much to put a single molecule in a snare. From a communications viewpoint there is not much information in a single molecule. It could easily report on the sex of the originator, but urine could serve the same purpose. Licking of the box will deliver the message to vomeronasal glands inside the oral cavity, and in this hydrophobic environment the box will reopen. Many members of the secretoglobin family are excreted in saliva and urine. The need for the loris to use a brachial gland as its vehicle for secretion may reflect either the need for added olfactory components to accompany the molecular snare (the complex list of compounds in Tables 12.1 and 12.2) or a uniquely hydrophobic message for the box.

Although the secretoglobin family has been associated with many functions (progesterone binding, transglutaminase substrate, protease inhibitor, phospholipase A_2 inhibitor, calcium binding, anti-inflammatory activity, immunomodulatory activity, prevention of renal disease) (Singh & Katyal, 2000), none of these is assumed to be particularly toxic. However, humans suffer severe effects (and have even died) from loris bites (Wilde, 1972). The cause is not certain, but is likely due to an anaphylactic shock, as loris researchers readily develop allergies to the glandular secretions (Fry, unpubl. obs.). Similar to the dual functionality of cat allergen, loris glandular secretion likely evolved as a communication molecule, and it is a toxin only for certain (incidentally) susceptible species, like humans. Further work will be necessary to elucidate the toxic actions on humans of the secretions from these fascinating animals.

Acknowledgments. H. Fitch-Snyder would like to dedicate this chapter to Mike Minch, who did the initial chemical analysis of loris brachial gland extract and co-authored a manuscript on this topic in 1989. He, unfortunately, died before his contributions were published. We wish to thank Matthew Williams and his staff at the Division of Biological Sciences Protein Sequencing Facility, University of California, San Diego campus, for their assistance in performing the peptide sequence analysis, and Dr. Paolo Martelli at the Singapore Zoo for all of his help. The authors also thank Richard Tenaza and three anonymous reviewers for their insightful comments and helpful suggestions.

References

Albone, E.S. (1984). *Mammalian semiochemistry: The investigation of chemical signals between mammals*. New York: Wiley.

Alterman, L., and Hale, M.E. (1991). Comparison of toxins from brachial gland exudates of *Nycticebus coucang* and *N. pygmaeus*. *Amer. Jour. Phys. Anthrop.*, 12: 43A (abstract).

Alterman, L. (1995). Toxins and toothcombs: Potential allopecific chemical defenses in *Nycticebus* and *Perodicticus*. In L. Alterman, G.A. Doyle, and M.K. Izard (Eds.), *Creatures of the dark: The nocturnal prosimians* (pp. 413–424). New York: Plenum Press.

Callebaut, I., Poupon, A., Bally, R., Demaret, J.-P., Housset, D., Delettré, J., Hossenlopp, P., and Mornon, J.-P. (2000). The uteroglobin fold. In A.B. Mukherjee and B.S. Chilton (Eds.), *The uteroglobin/clara cell protein family* (pp. 90–110). New York: Academy of Sciences.

Groves, C.P. (1998). Systematics of tarsiers and lorises. *Primates*, 39(1): 13–27.

Hagey, L.R., and MacDonald, E. (2003). Chemical cues identify gender and individuality in Giant pandas (*Ailuropoda melanoleuca*). *Jour. Chem. Ecol.*, 29: 1479–1488.

Karn, R.C., and Russell, R. (1993). The amino acid sequence of the alpha subunit of mouse salivary androgen-binding protein (ABP) with a comparison to the partial sequence of the beta subunit and to other ligand-binding proteins. *Biochem. Genetics*, 31: 307–319.

Klug, J. (2000). Uteroglobin/clara cell 10-kDa family of proteins: Nomenclature committee report. The uteroglobin fold. In A.B. Mukherjee and B.S. Chilton (Eds.), *The uteroglobin/clara cell protein family* (pp. 348–353). New York: Academy of Sciences.

Krane, S., Itagaki, Y., Nakanishi, K., and Weldon, P.J. (2003). "Venom" of the slow loris: Sequence similarity of prosimian skin gland protein and Fel dl cat allergen. *Naturwissenschaften*, 90: 60–62.

Lehrer, R.I., Nguyen, T., Zhao, C., Ha, C.X., and Glasgow, B.J. (2000). Secretory lipophilins: A tale of two species. In A.B. Mukherjee and B.S. Chilton (Eds.), *The uteroglobin/clara cell protein family* (pp. 59–67). New York: Academy of Sciences.

Mellen, J. (1993). A comparative analysis of scent-marking, social and reproductive behavior in 20 species of small cats (*Felis*). *Amer. Zool.*, 33: 151–166.

Morgenstern, J.P., Griffith, I.J., Brauer, A.W., Rogers, B.L., Bond, J.F., Chapman, M.D., and Kuo, M.-C., (1991). Amino acid sequence of Fel-dl, the major allergen of the domestic cat: protein sequence analysis and cDNA cloning. *Proc. Natl. Acad. Sci.*, 88: 9690–9694.

Palumbi, S.R., and Metz, E.D. (1991). Strong reproductive isolation between closely related tropical sea urchins (genus *Echinometra*). *Mol. Biol. Evol.*, 8: 227–239.

Pattabiraman, N., Matthews, J.H., Ward, K.B., Mantile-Selvaggi, G., Miele, L., and Mukherjee, A.B. (2000). Crystal structure analysis of recombinant human uteroglobin and molecular modeling of ligand binding. In A.B. Mukherjee and B.S. Chilton (Eds.), *The uteroglobin/clara cell protein family* (pp. 113–127). New York: Academy of Sciences.

Rasmussen, D.T. (1986). A review of bite wounds in lorises. *Zoo Zen*, 2(2): 6–8. Zoo Outreach Organization/CBSG, India.

Saitou, N., and Nei, M. (1987). The neighbor-joining method: A new method for reconstructing phylogenetic trees. *Mol.Biol.Evol.* 4(4): 406–25.

Singh, G., and Katyal, S.L. (2000). Clara cell proteins. In A.B. Mukherjee and B.S. Chilton (Eds.), *The uteroglobin/clara cell protein family* (pp. 43–58). New York: Academy of Sciences.

Takahashi, K., and Nei, M. (2000). Efficiencies of fast algorithms of phylogenetic inference under the criteria of maximum parsimony, minimum evolution, and maximum likelihood when a large number of sequences are used. *Molecular Biol. and Evol.*, 17: 1251–1258.

Tenaza, R. and Fitch, H. (1984). The slow loris. *Zoonooz*, Vol LVIII, No. 4. Zoological Society of San Diego.

Thompson, J.D., Gibson, T.J., Plewniak, F., Jeanmougin, F., Higgins, D.G. (1997). The CLUSTAL_X windows interface: Flexible strategies for multiple sequence alignment aided by quality analysis tools. *Nucleic Acids Res.*, 25(24): 4876–82.

Utami, S., and van Hooff, J.A.R.A.M. (1997). Meat-eating by adult female Sumatran orangutans (*Pongo pygmaeus abelii*). *Amer. Jour. Primatol.*, 43: 159–165.

Wiens, F., and Zitzmann, A. (1999). Predation on a wild slow loris (*Nycticebus coucang*) by a reticulated python (*Python reticulatus*). *Folia Primatol.*, 70: 362–364.

Wilde, H. (1972). Anaphylactic shock following bite by a slow loris, *Nycticebus coucang*. *Amer. Jour. Tropical Medicine Hygiene*, 21: 592–594.

Part 3
Anti-Predator Strategies of Non-Nocturnal Primates

13
Anti-Predator Strategies in a Diurnal Prosimian, the Ring-Tailed Lemur (*Lemur catta*), at the Beza Mahafaly Special Reserve, Madagascar

Lisa Gould and Michelle L. Sauther

Introduction

With the dramatic increase in research on Madagascar's lemurs during the past few decades, it is now feasible to both document anti-predator behavior and to test predictions regarding the effect of predation pressure on the behavioral ecology of lemurs. In 1994 Goodman raised much interest by his suggestion that, in the absence of large, extant predators on Madagascar, anti-predator behaviors and strategies in lemurs were an artifact of a behavioral repertoire that existed before the extinction of a very large eagle, *Stephanoaetus mahery*. However, both before and subsequent to Goodman's argument, numerous studies of both diurnal and nocturnal lemurs revealed that both extant avian and mammalian predators pose a real predation threat. (Sauther, 1989, 2002; Overdorff & Strait, 1995; Wright & Martin, 1995; Gould, 1996; Gould et al., 1997; Wright, 1998; Schwab, 1999; Karpanty & Grella, 2001; Karpanty 2003). In this chapter, we first present information on predation risk, group size, and foraging in the ring-tailed lemur (*Lemur catta*), and then we examine sex differences in predator vigilance, canopy level differences in vigilance, and how alpha females contribute to anti-predator strategies in this species.

Ring-tailed lemurs inhabit a wide range of habitat in south and southwestern Madagascar, ranging from gallery (riverine) forest to xerophytic, spiny thorn scrub, and limestone forest, and a even a sub-alpine habitat in the central southeastern part of the island (Jolly, 1966; Budnitz & Dainis, 1975; Sussman, 1977; Tattersall, 1982; Mittermeier et al., 1994; Goodman & Langrand, 1996). Ring-tailed lemurs spend anywhere from 3–75% of their time on the ground, depending upon the month and season (Sauther, 2002), and they are therefore frequently exposed to both ground and aerial predators (Jolly, 1966; Sussman, 1972; Sauther, 1989, 2002; Gould, 1996).

L. catta is a medium-sized diurnal lemur with a mean weight of 2.2 kilograms (Sussman, 1991). Therefore, it is not as vulnerable to predation as much smaller, nocturnal lemur species; however, it is still under considerable predator pressure from a number of endemic and introduced species in their geographic range. These include the Madagascar harrier hawk (*Polyboroides radiatus*), the Madagascar buzzard (*Buteo brachypterus*), boa constrictor (*Boa mandtria*), hognose snake (*Leioheterondon madagascariensis*), and Indian civet (*Viverricla indica*) (Sauther, 1989, 2002; Goodman et al., 1993; Gould, 1996). Also, in some areas, the fossa (*Cryptoprocta ferox*), a medium-sized (7–12 kilogram) viverrid carnivore is a predator of *L. catta* (Goodman et al., 1993). In addition to these endemic predators, village dogs, feral domestic cats and a hybrid wild cat can be a serious predation threat to ring-tailed lemurs living in forests near human habitation (Sauther, 1989; Gould, 1996).

At the Beza Mahafaly Special Reserve in southwestern Madagascar actual predation has been observed and recorded. Infants have been taken by the Madagascar harrier hawk, and bones of both adults and juveniles have been found under the nests of this raptor (Ratsirarson, 1985; Goodman et al., 1993). Infants have also been preyed upon by Indian civets, feral cats, and dogs in the reserve (Sauther, 1989; Goodman et al., 1993). Numerous predation attempts by raptors have also been observed. For example, a migrating male sitting alone in an exposed spot near the top of a tree was nearly taken by a harrier hawk (Gould, 1994, 1996), and harrier hawks have been observed flying into the canopy where ring-tailed lemur groups were feeding or resting. An infant exploring in a tree away from the mother was followed by a small raptor until an adult lemur retrieved it (Gould, 1994, 1996). Sauther (2002) notes that the size of the harrier hawk (60–62 cm in body length) prevents it from moving easily in a closed canopy; however, harrier hawks have been observed in more open environments at Beza Mahafaly, flying above ring-tailed lemur groups or perched in dead trees watching groups of lemurs.

Dog and cat predation may also be especially important. On one occasion a feral cat was observed stalking a group of ring-tailed lemurs feeding on the ground, and on numerous occasions single ring-tailed lemurs have been "treed" by roaming dogs that wait at the base of the tree. We have also found both ring-tailed lemur and sifaka hair within dog scat as well as the predated remains of ring-tailed lemurs (Figure 13.1).

Types of Anti-Predator Behaviors

Ring-tailed lemurs exhibit a large repertoire of anti-predator behaviors, described below, which include both vocal signals and behavioral responses such as mobbing and vigilance.

Vigilance

Predator vigilance occurs when a ring-tailed lemur ceases the activity in which it is engaged, sits, or stands upright with ears facing forward and visually scans

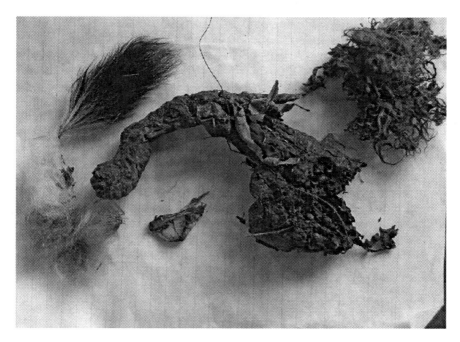

FIGURE 13.1. Remains of a female ring-tailed lemur after predation, likely by a feral dog. Only portion of the stomach, fur, tail and vertebra remain (Photo: M.L. Sauther)

the surrounding environment (Gould, 1996). Animals engage in this "vigilance sequence" when sightings or vocalizations of predators occurred, and/or when they hear other groups of nearby ring-tailed lemurs or Verreaux's sifakas emitting anti-predator calls.

Vocalizations

Ring-tailed lemurs engage in "representational signaling," that is, they emit particular vocalizations for particular kinds of predators (Jolly, 1966; Sussman, 1972; Sauther, 1989; Macedonia, 1990). A "click" vocalization occurs when an animal is agitated or startled by the presence of a potential predator (Jolly, 1966). Sauther (1989) noted that if one or more animals engage in an initial click series they are immediately joined by much of the rest of the troop, which then move into the canopy or bushes and scan the ground. With terrestrial predators the clicking sequence can change to a series of sharp "yaps" while the group keeps watch on the potential predator until it has moved away (Jolly, 1966; Sussman, 1972; Sauther, 1989; Gould, 1996). If the predator is a low-flying or swooping raptor, a few individuals will click, the group will look up, scream, and quickly drop from high to lower canopy (Jolly, 1966; Sauther, 1989; Gould, 1996).

Sauther (1989, 2002) described anti-predator behavior in *L. catta* and examined the relationship between group size, anti-predator behaviors, habitat use, and predator sensitive foraging in two groups of wild ring-tailed lemurs at the Beza Mahafaly Special Reserve. She found that (1) when predation risk is high, ring-tailed lemurs avoided risky foraging areas, particularly terrestrial foraging; (2) that smaller groups of ring-tailed lemurs avoided open areas more frequently than animals in larger groups; and (3) that groups that avoided more open areas had reduced foraging/feeding measures.

Gould (1996, 1997) investigated sex differences in predator vigilance in two groups of ring-tailed lemurs, during the birth and lactation season at Beza Mahafaly Special Reserve, when risk of predation on infants is high. The goal of that study was to test whether males were benefiting females during this period, and to test costs and benefits to females of tolerating male residence in social groups, since female are dominant in this species and serve as the primary resource defenders (Jolly, 1966; Sussman, 1977; Sauther, 1992) The contribution of the alpha female to anti-predator vigilance was also examined.

Methods

Sauther collected predator sensitive foraging data on two groups of ring-tailed lemurs in the eastern part of the Beza Mahafaly Special Reserve as part of a larger study of feeding ecology on this species in 1988–1989 (1992, 1993, 1998, 2002). A total of 1,800 hours of focal animal data were collected. One group contained between 14 and 16 individuals, the other contained 6–8 animals. To examine predator sensitive foraging specifically, focal animal data (Altmann, 1974) were collected at 5-min intervals on all adult and sub-adult animals in the two groups and all behaviors were recorded. Nearest neighbor data were also collected every 15 min in order to examine whether animals were either more or less cohesive while foraging, and whether predator pressure affects cohesion. If an individual's nearest neighbor was <3 m, "group feeding" was recorded, as animals at this proximity were usually feeding in the same patch. If the nearest neighbor was 8 m or more away, "solitary feeding" was recorded. This was done because at a distance of 8 m the animal was usually outside of any other animal's food patch and thus it was truly feeding solitarily (Sauther, 1992). If nearest neighbors were between 3 and 7 m apart, intermediate foraging was recorded. Foraging location was recorded by a tree quadrat method, wherein a tree is divided into thirds along the vertical and horizontal axes and quadrats are numbered 1–9. Ground foraging was assigned the number 10. Each group's location was mapped every 15 min to determine ranging patterns and foraging effort. If the animals entered new areas of their home range each month, or if they foraged outside of their home range, a "new hectare" designation was recorded.

Predator pressure was determined by recording all encounters with potential diurnal predators throughout an annual cycle. Monthly predator encounters were assessed as "low" if 0–3 encounters occurred in a one-month period, and "high" if the encounters numbered greater than 3. In addition, all instances of predator vigilance were recorded as defined earlier in this chapter.

In Gould's study, 424 continuous time focal animal data sessions (Altmann, 1974) of 15-min duration were collected on 15 adult ring-tailed lemurs in two groups (Red group = 4 adult females, 2 adult males; Green group = 5 adult females, 4 adult males) at the Beza Mahafaly reserve between early October and mid-December, 1994. Females give birth from late September to late October at this site, and by four weeks of age, nursing ring-tailed infants begin to explore the environment on their own (Gould, 1990), becoming vulnerable to predation as the weeks progress.

Anti-predator vigilance was scored both when actual predators were seen or heard in the area and when the animals engaged in vigilance behavior toward anything in the environment that could have been a predator (e.g., a sudden unidentified sound occurring on the ground or in the trees, the spotting of an animal moving through the trees that was not a lemur, hearing a raptor in the distance, dogs barking). An instance of vigilance was scored when the animals engaged in the behaviors described as "vigilant" earlier in this chapter. Each focal animal's total frequency of anti-predator vigilance was divided by the number of focal animal sessions collected on that animal (which ranged from 27 to 30 sessions) to obtain a rate per focal animal session.

Sauther's interval data were analyzed using randomization tests (Edgington, 1980) in which a t-test with systematic data permutation was used to determine statistically significant differences between the two study groups. Gould's continuous-time data were analyzed using non-parametric analysis of variance tests for small samples to test sex differences in vigilance and the chi-square test to determine vigilance differences in canopy level.

Results

Predator Sensitive Foraging

When predator pressure was high, the smaller group foraged more often in the middle level of trees compared to periods of low predator pressure ($t = 2.76$, $p = 0.02$), and avoided terrestrial foraging. Predator pressure did not seem to affect foraging behavior or foraging level for the larger group. The larger group foraged in the low level of trees when predation risk was high, but they also continued to forage on the ground.

The smaller group, through avoiding terrestrial foraging when predation risk was high, had a significantly lower intake of leaves compared to when predation risk was low ($t = 2.40$, $p = 0.04$), as many of their leaf-food resources are found on the ground in the form of herbaceous vegetation. The smaller group also had a lower intake of fruit during periods of high predation risk ($t = 2.08$, $p = 0.04$). Food intake of the larger group was not significantly affected by predation risk.

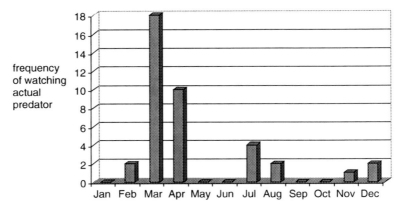

FIGURE 13.2. Frequency of vigilance towards an actual predator in the vicinity (predator watching) over a one-year period (1988–1989)

When entering new hectares of the home range or foraging outside of the home range, spatial cohesion, measured as "group nearest neighbor pattern" was positively correlated with number of new hectares entered for the smaller group ($r = 0.51$, $p = 0.049$). In other words, when smaller groups entered new areas they maintained close proximity to one another. In the larger group individuals actually spread out and fed some distance from one another.

In fact, there was a strong correlation between foraging without a nearest neighbor and new hectares entered ($r = 0.72$, $p = 0.008$). The larger group entered more new hectares when foraging than did the smaller group ($t = 2.60$, $p = 0.03$), and even though such behavior was correlated with a higher frequency of predator encounters ($t = 2.53$, $p = 0.03$), there was a positive correlation between foraging in these new areas and fruit feeding ($r = 0.75$, $p = 0.005$).

Comparing the total frequency of male versus female vigilance toward an actual predator by month ("predator watching") revealed no sex differences ($t = 0.633$, $p = 0.533$). However, total monthly vigilance behavior toward a predator did vary, peaking during March and April when infants are being weaned (Figure 13.2).

Anti-Predator Vigilance Towards Actual or Potential Predators

No sex difference was found in rates of vigilance behavior towards a real or potential predator in Gould's two study groups; however, females spent more time in anti-predator vigilance ($U = 11$, $p = 0.051$) than did males. Pooling the females and males from each group together, higher-ranking females were vigilant significantly more often than were lower-ranking females (U-test, $U = 0$, $p = 0.04$), but no relationship was found between rank and rates of anti-predator vigilance in males. The alpha female in each group was vigilant towards potential predators significantly more often than all other adults in her group (single sample against the mean test, Green group, $t_s = 3.54$, df = 7, $p < 0.01$; Red group $t_s = 3.49$, df = 4, $p < 0.05$, Figure 13.3).

The study animals were significantly more vigilant when on the ground, compared with low, middle, or high canopy (chi-square goodness of fit, $\chi^2 = 128.59$, df = 3, $p < 0.001$, Figure 13.4). Red group contained three fewer adults than did Green group, but there was no between-group difference in overall rates of anti-predator vigilance, nor were there between-group differences in proportion of vigilance on the ground, or in low, medium or high canopy.

Discussion

Primates vulnerable to predation must be able to balance alertness toward potential predators with getting enough food to meet their nutritional needs and with conducting other daily activities. Different types of anti-predator strategies have evolved depending upon variables such as habitat, body size, density and types of predators in the area, group size, and degree of arboreality or terrestriality.

As *L. catta is* a medium-sized Malagasy primate, and the most terrestrial of all lemurs, it is vulnerable to both the large avian predators and the terrestrial predators described in the introduction. As with most primates, arboreal food resources are very important to ring-tailed lemurs; but this species also depends greatly upon terrestrial vegetation (Sauther, 2002) and thus, ring-tailed lemurs must be watchful at all levels of the forest.

Canopy Level and Vigilance

We found that ring-tailed lemurs at Beza Mahafaly reserve were markedly more vigilant while on the ground, compared to how vigilant they were at any canopy level. This pattern has been found in a number of primate species of similar or slightly larger body size which spend some time foraging terrestrially (e.g., wedge-capped capuchins (de Ruiter 1986; Miller, 2002); red-fronted brown lemurs in a dry forest (Rasolofoson, 2002); brown capuchins (Hirsch, 2002); red colobus and red-tailed monkeys (Treves, 2002). And, predictably, smaller-bodied saddleback and moustached tamarins exhibit heightened vigilance when in the lower canopy, even when found in mixed-species associations (Smith et al., 2004). Smith et al. (2004) and Peres (1993) suggest that in the Neotropics, increased vigilance on the ground or in low canopy relates to both numerous terrestrial predators as well as the fact that many raptors are ambush predators, which take their prey by swooping down from a stationary perch. The velocity with which these raptors swoop into the lower canopy could greatly increase the probability of a successful predatory attempt, which could explain why the tamarins in their studies were more vigilant at lower levels. Some of the smaller raptor species at Beza Mahafaly have been observed watching infant lemurs while on mid to low-canopy stationary perches (Gould, pers. obs). The large raptors at Beza commonly use the ambush method, swooping down on their prey (including chickens) from above

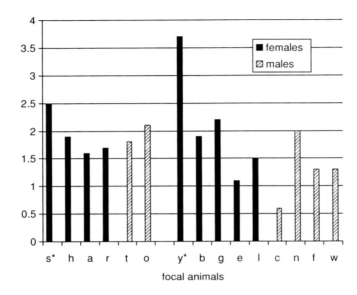

FIGURE 13.3. Rates of vigilance (rate per 15-min. focal session) towards an actual or potential predator by each focal animal in the two study groups over the birth/laction study period, 1994. The first animal presented for each group (s* and y*) were the alpha females in their respective groups

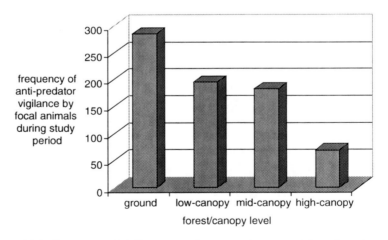

FIGURE 13.4. Frequency of vigilance towards an actual or potential predator (e.g. unfamiliar/unidentified sound or sighting) at different forest levels by all focal animals during the birth/lactation season in 1994

and especially within open areas (Sauther, 1989). Thus, when spending time in exposed terrestrial areas, of which there are many in this habitat, ring-tailed lemurs need to be vigilant toward these "ambush" avian predators.

Group-Size and Anti-Predator Strategies in Relation to Feeding and Foraging

Group size certainly plays a role in anti-predator strategies in ring-tailed lemurs. Larger groups might be able to take more foraging chances, particularly with respect to feeding on ground vegetation. Sauther (2002) found that 48% of the leaf diet of ring-tailed lemurs at Beza Mahafaly came from plants located on the ground. By not foraging terrestrially while predation pressure was high, the smaller group of ring-tailed lemurs in the study incurred a cost in terms of reducing leaf intake, thus, the benefits of engaging in predator avoidance foraging behaviors must be balanced against the cost of obtaining important protein resources (e.g., fewer leaves in the diet) at certain times of the year.

For the larger group, while fruit foraging efficiency increased in the new areas, the behavior of spreading out and therefore being less cohesive could potentially be costly in terms of predator spotting. It is also possible that by having more "eyes and ears" predator detection is enhanced in larger groups making the trade-offs more feasible. The smaller group, by being more spatially cohesive while foraging in new zones, was exhibiting predator sensitive foraging. Overdorff et al. (2002) found similar group-size effects on anti-predator strategies in rainforest lemurs. In three sympatric species (*Eulemur fulvus rufus* (rufous brown lemur), *Propithecus edwardsi* (Milne-Edwards' sifaka), and *Eulemur rubriventer* (red-bellied lemur)) at the Ranomafana site in the southeastern rainforest, the two species living in the larger, multi-male–multi-female groups (*E. f. rufus* and *P. edwardsi*) used all levels of the canopy and exploited a much wider range of food items than did the monogamous *E. rubriventer*. *E. f. rufus* and *P. edwardsi* also fed more on ground vegetation and soil, and were less likely to have a nearest neighbor when feeding, whereas the red-bellied lemurs primarily fed in the upper canopy were highly cohesive. Miller (2002) also found that in wedge-capped capuchins, members of larger groups utilize resources from risky areas and thus increase their foraging opportunities compared to animals residing in smaller groups.

Another strategy related to group size is that the smaller ring-tailed lemur group in Sauther's study formed a mixed-species association with groups of Verreaux's sifaka at the time of year when infants of both species had just been weaned, and thus were more independent but extremely vulnerable to predation.

Infant Vulnerability and Vigilance Behavior

While no sex differences were found in vigilance behavior during both studies, total vigilance behaviors toward actual predators did vary by month during the 1987–88 study. The highest frequency of vigilance behavior occurred during the months of March and April. Peak weaning of ring-tailed lemurs occur during March, and by April most infants are weaned (Sauther, 1989). Thus, ring-tailed lemurs exhibit the most vigilance toward actual predators during a time period when infants are still not adult body size but are engaging in solitary feeding and presumably more vulnerable to predation.

Sex and Rank Differences in Anti-Predator Behavior

One prediction in relation to sex differences in vigilance in female philopatric species is that males should offer females enhanced predator protection through vigilance in exchange for tolerance in a social group (Baldellou & Henzi, 1992). In similar-sized primates such as vervet monkeys, white-faced capuchins, and rufous brown lemurs (in a dry forest), males are often more vigilant than females (Baldellou & Henzi, 1992; Rose, 1994; Rose & Fedigan, 1995; Rasolofoson, 2002), although in brown capuchins, sex differences in vigilance were not detected (Hirsch, 2002). Since female *L. catta* are dominant to males and are the primary resource defenders (Jolly, 1966; 1984), theoretically, males should benefit females by being more vigilant against potential predators, but they were not more vigilant, at least not during the birth and lactation season when this study was conducted.

The lack of a sex difference in vigilance in this case may relate to a phylogenetic trend in dominance patterns—white capuchin males are dominant to females and rufous brown lemurs are co-dominant (Fedigan, 1990; Pereira & McGlynn, 1997; Overdorff, 1998), and perhaps the dominant sex in a species is the more vigilant. Lewis (2005) found that female Verreaux's sifaka, also a female dominant species, were more likely to alarm call in the presence of a predator than were males. Nonetheless, male ring-tailed lemurs in this study did devote a similar percentage of time to anti-predator vigilance as did females, and it can be argued that they still contribute to group protection through their vigilance (Gould, 1996). We also suggest that males may serve as low-cost sentinels in the group, since females have priority of access to all resources (Sauther, 1993; Gould, 1996). Also, females may be more vigilant toward actual or potential predators in the birth and infant-rearing season and may relax their vigilance when offspring are not as vulnerable. Information on total vigilance by adult females and males at other times of the year would be useful toward determining if this is the case.

During this study the alpha females from both groups exhibited significantly more anti-predator vigilance than did other group members. In white-faced capuchins, alpha males are significantly more vigilant (Rose, 1994; Rose & Fedigan, 1995). Such heightened vigilance by these individuals has prompted Gould et al. (1997) to suggest that there may be certain behaviors that alpha animals engage in either more often or uniquely, and that these are characteristically found in any animal occupying the top rank in a group. For example, on one occasion when one of the study groups was drinking from standing water on the road just outside of the reserve, the alpha female stood guard, sometimes bipedally, while each group member that was on the road at the time drank. Thus, although the alpha female has priority of access to all resources in her group, which may incur a cost to lower-ranking group members, she is providing a benefit to those animals through a high degree of vigilance.

The "Startle Response": An Innate Anti-Predator Strategy?

Researchers who have studied ring-tailed lemurs in both wild and semi-free-ranging captive situations have noted that, no matter how habituated, when groups are on the ground, any sudden or unfamiliar sound in the environment causes the animals scatter immediately. Usually the individuals jump into the nearest tree, where they will often remain for some time. Taylor (1986) has termed such behavior the "startle response," and it can occur even at the slightest sound, such as that made by a researcher stepping on a stick in the forest. Since this response is also found in animals in captivity, it could be innate and makes sense as an anti-predator strategy that has evolved in a relatively small primate that spends considerable time on the ground.

In summary, we can suggest several anti-predator strategies in ring-tailed lemurs:
(1) Ring-tailed lemurs exhibit heightened vigilance while foraging or engaging in other activities on the ground. (2) Larger groups may take more chances while foraging on both the ground and in new areas, as more animals are available to watch for predators in a larger group. Smaller groups may be more cohesive and exhibit heightened predator sensitive foraging, such as avoiding terrestrial food patches, a response that is beneficial with respect to avoiding predation, but can also incur a nutritional cost. (3) Smaller groups may form a mixed-species association with Verreaux's sifaka in geographic areas where they are sympatric and respond to each other's alarm calls. (4) Group members appear to be most vigilant toward actual predators during periods of high infant vulnerability, e.g., during and just after weaning. (5) Males provide a low-cost sentinel service, particularly during the lactation and infant-rearing season when offspring are most vulnerable, even though males are not significantly more vigilant than females during this period. (6) Alpha females may provide a high degree of vigilance in their female-philopatric groups, thereby enhancing survival of their own and their female relatives' offspring. (7) When ring-tailed lemurs are terrestrial they exhibit the "startle response" at the first sign of potential danger. This response is possibly an innate anti-predator behavior, which arose during the evolution of this relatively small, semi-terrestrial primate species.

Acknowledgments. Our research at the Beza Mahafaly Special Reserve in 1988–89 and 1994 was made possible through the kind assistance of Mme. Berthe Rakotosamimanana, M. Benjamin Andriamihaja, Dr. Pothin Rakotomanga, Dr. Andrianansolo Ranaivoson, Joseph Andrianmampianina, Mme. Celestine Ravaoarinoromanga, and research permission from the School of Agronomy at the University of Antananarivo, and Direction des Eaux et Fôret, Madagascar. L. Gould's research described in this chapter was funded by an I.W. Killam Post-doctoral Research Fellowship and the Boise fund of Oxford, and that of Michelle

Sauther was funded by grants from the National Science Foundation, National Geographic Society, and L.S.B. Leakey Foundation. We also thank Sharon Gursky and Anna Nekaris for inviting us to participate in this volume.

References

Altmann, J. (1974). Observational study of behaviour: Sampling methods. *Behaviour*, 49: 227–265.

Baldellou, M., and Henzi, S.P. (1992). Vigilance, predator detection and the presence of supernumerary males in vervet monkey troops. *Anim. Behav.*, 43: 451–461.

Budnitz, N., and Dainis, K. (1975). *Lemur catta*: Ecology and behavior. In I. Tattersall and R.W. Sussman (Eds.), *Lemur biology* (pp. 219–235). New York: Plenum.

Edgington, E.S. (1980). *Randomization tests*. New York: Marcel-Dekker, Inc.

Fedigan, L.M. (1990). Vertebrate predation in *Cebus capucinus*: Meat eating in a Neotropical monkey. *Folia Primatol.*, 54: 196–205.

Goodman, S.M. (1994). The enigma of antipredator behavior in lemurs: Evidence of a large extinct eagle on Madagascar. *Inter. Jour. Primatol.*, 15: 129–134.

Goodman, S.M, O'Connor, S., and Langrand, O. (1993). A review of predation on lemurs: Implications for the evolution of social behavior in small, nocturnal primates. In P.M. Kappeler and J.U. Ganzhorn, (Eds.), *Lemur social systems and their ecological basis* (pp. 51–66). New York: Plenum Press.

Goodman, S.M., and Langrand, O. (1996). A high mountain population population of the ringtailed lemur *Lemur catta* on the Andringitra Massif, Madagascar.*Oryx*, 30: 259–268.

Gould, L. (1994). Patterns of affiliative behavior in adult male ringtailed lemurs (*Lemur catta*) at the Beza-Mahafaly Reserve, Madagascar. Doctoral dissertation. Washington University, St. Louis, MO.

Gould, L. (1996). Vigilance behavior in naturally occurring ringtailed lemurs (*Lemur catta*) during birth and lactation season. *Inter. Jour. Primatol.*, 17: 331–347.

Gould, L., Fedigan, L.M., and Rose, L.M. (1997). Why be vigilant? The case of the alpha animal. *Inter. Jour. Primatol.*, 18: 401–414.

Hirsch, B.T. (2002) Social monitoring and vigilance behavior in brown capuchin monkeys (*Cebus apella*). *Behav. Eco. Sociobiol.*, 52: 458–464.

Jolly, A. (1966). *Lemur behavior*. Chicago: Univ. of Chicago Press.

Jolly, A. (1984). The puzzle of female feeding priority. In M. Small (Ed.), *Female primates: Studies by women primatologists* (pp. 197–215). New York: Alan R. Liss.

Karpanty, S.M. (2003). Rates of predation by diurnal raptors on the lemur community of Ranomafana National Park, Madagascar. *Amer. Jour. Phys. Anthropol. (Suppl.* 36): 126–127.

Karpanty, S.M., and Grella R. (2001). Lemur responses to diurnal raptor calls in Ranomafana National Park, Madagascar. *Folia Primatol.*, 72: 100–103.

Lewis, R.J. (2005). Sex differences in vigilance in Verreaux's sifaka: Are males providing a predator detection service? *Am. J. Phys. Anthropol. (Suppl.* 40): 138.

Macedonia, J.M. (1990). What is communicated in the antipredator calls of lemurs: Evidence from playback experiments with ring-tailed and ruffed lemurs. *Ethology*, 86: 177–190.

Miller, L.E. (2002). The role of group size in predator sensitive foraging decisions for wedge-capped capuchin monkeys (*Cebus olivaceus*). In L.E. Miller (Ed.), *Eat or be eaten: Predator sensitive foraging among primates* (pp. 195–106). Cambridge: Cambridge Univ. Press.

Mittermeier, R.A., Tattersall, I., Konstant, W.R., Meyers, D.M., and Mast, R.B. (1994). *Lemurs of Madagascar.* Conservation International Tropical Field Guide Series. Washington, DC: Conservation International.

Overdorff, D.J. (1998). Are *Eulemur* species pair-bonded? Social organization and mating strategies in *Eulemur fulvus rufus* from 1988–1995 in southeast Madagascar. *Amer. Jour. Phys. Anthropol.*, 105: 153–166.

Overdorff, D.J., and Strait, S.G. (1995). Life history and predation in *Eulemur rubriventer* in Madagascar. *Amer. Jour. Phys. Anthropol.*, 100: 487–506.

Overdorff, D.J., Strait, S.G., and Seltzer, R.G. (2002). Species differences in feeding in Milne-Edwards' sifakas (*Propithecus diadema edwardsi*), Rufus lemur (*Eulemur fulvus rufus*), and red-bellied lemurs (*Eulemur rubriventer*) in southeastern Madagascar: Implications for predator avoidance. In L.E. Miller (Ed.), *Eat or be eaten: Predator sensitive foraging among primates* (pp. 126–137). Cambridge: Cambridge Univ. Press.

Pereira, M.E., and McGlynn, C. (1997). Special relationships instead of female dominance for redfronted lemurs, *Eulemur fulvus rufus. Amer. Jour. Phys. Anthropol.*, 43: 239–258.

Peres, C.A. (1993). Anti-predation benefits in a mixed-species group of Amazonian tamarins. *Folia Primatol.*, 61: 61–76.

Rasolofoson, R.D.W. (2002). Stratégies anti-prédatrices *d'Eulemur fulvus rufus* (Audebert, 1800) dans la forêt dense sèche de Kirindy, Morondava, Madagascar. *Lemur News*, 8: 29.

Ratsirarson, J. (1985). *Contribution à l'Étude comparative de l'eco-ethologie de Lemur catta dans deux habitats différents de la Réserve Spéciale de Beza Mahafaly.* Université de Antananarivo: Memoire de le fin d'etudes.

Rose, L.M. (1994). Benefits and costs of resident males to females in white-faced capuchins, *Cebus capucinus. Am. J. Phys. Anthropol.* 32: 235–248.

Rose, L.M., and Fedigan L.M. (1995). Vigilance in white-faced capuchins, *Cebus capucinus*, in Costa Rica. *Anim. Behav.*, 49: 63–70.

de Ruiter, J.R. (1986). The influence of group size on predator scanning and foraging behaviour of wedge capped capuchin monkeys (*Cebus olivaceus*). *Behaviour*, 98: 240–258.

Sauther, M.L. (1989). Antipredator behavior in troops of free-ranging *Lemur catta* at Beza Mahafaly Special Reserve, Madagascar. *Inter. Jour. Primatol.*, 10: 595–606.

Sauther, M.L. (1992). The effect of reproductive state, social rank and group size on resource use among free-ranging ringtailed lemurs (*Lemur catta*) of Madagascar. Doctoral dissertation. Washington University, St. Louis, MO.

Sauther, M.L. (1993). Resource competition in wild populations of ringtailed lemurs (*Lemur catta*): Implications for female dominance. In P.M. Kappeler and J.U. Ganzhorn (Eds.), *Lemur social systems and their ecological basis* (pp. 135–152). New York: Plenum Press.

Sauther, M.L. (1998). The interplay of phenology and reproduction in ringtailed lemurs: Implications for ringtailed lemur conservation *Folia primatol.*, 69(Suppl. 1): 309–320.

Sauther, M.L. (2002). Group size effects on predation sensitive foraging in wild ring-tailed lemurs (Lemur catta). In L.E. Miller (Ed.), *Eat or be eaten: Predator sensitive foraging among primates* (pp. 107–125). Cambridge: Cambridge Univ. Press.

Schwab, D. (1999). Predation on *Eulemur fulvus* by *Accipiter henstii* (Henst's goshawk). *Lemur News*, 4: 34.

Smith, A.C., Kelez, S., and Buchanan-Smith, H. (2004). Factors affecting vigilance within wild mixed-species troops of saddleback (*Saguinus fuscicollis*) and moustached tamarins (*S. mystax*). *Behav. Ecol. Sociobiol.*, 56: 18–25.

Sussman, R.W. (1972). An ecological study of two Madagascan primates: *Lemur fulvus rufus* Audebert and *Lemur catta* Linnaeus. Doctoral dissertation. Duke University, North Carolina.

Sussman, R.W. (1974). Ecological distinctions between two species of Lemur. In R.D. Martin, D.A. Doyle, and C. Walker (Eds.), *Prosimian biology* (pp. 75–108). London: Duckworth.

Sussman, R.W. (1977). Socialization, social structure and ecology of two sympatric species of lemur. In S. Chevalier-Skolnikoff and F.E. Poirier (Eds.), *Primate bio-social development: Biological, social, and ecological determinants* (pp. 515–528). New York: Garland Publishing, Inc.

Sussman, R.W. (1991). Demography and social organization of free-ranging *Lemur catta* in the Beza Mahafaly Reserve, Madagascar. *Amer. Jour. Phys. Anthropol.*, 84: 43–58.

Tattersall, I. (1982). *The primates of Madagascar*. New York: Columbia Univ. Press.

Taylor, L.L. (1986). Kinship, dominance, and social organization in a semi-free ranging group of ringtailed lemurs (*Lemur catta*). Doctoral dissertation, Washington University, St. Louis, MO

Treves, A. (2002). Predicting predation risks for foraging arboreal monkeys. In L.E. Miller (Ed.), *Eat or be eaten: Predator sensitive foraging among primates* (pp. 222–241). Cambridge: Cambridge Univ. Press.

Wright, P.C. (1998). Impact of predation risk on the behaviour of *Propithecus diadema edwardsi* in Ranomafana National Park, Madagascar. *Inter. Jour. Primatol.*, 16: 835–854.

Wright, P.C., and Martin, L.B. (1995). Predation, pollination and torpor in two nocturnal prosimians: *Cheirogaleus major* and Mic*rocebus rufus* in the rain forest of Madagascar. In L. Alterman, G.A. Doyle, and M.K. Izard (Eds.), *Creatures of the dark: The nocturnal prosimians* (pp. 45–60). New York: Plenum Press.

14
Howler Monkeys and Harpy Eagles: A Communication Arms Race

Ricardo Gil-da-Costa

Introduction

Predation is considered by many researchers to be a selective pressure and strong evolutionary driving force in natural ecosystems. Predation phenomena are dynamic interactions that by definition need more than one agent: at least one predator and one prey. The interaction gets exponentially more complicated when we consider multiple agents and different strategies. These predator-prey interactions can be viewed as evolutionary arms races. There have been numerous studies on prey adaptations (Blumstein et al., 2000; Hauser & Caffrey, 1994; Marler et al., 1992; Endler, 1991; Cheney & Seyfarth, 1990; Hauser & Wrangham, 1990; Ryan et al., 1982), but few report both detailed adaptive responses to predation and ways predators can improve their killing efficiency (Berger et al., 2001). This lack of knowledge is even more striking for predation upon primates (Shultz et al., 2004; Gil-da-Costa et al., 2003; Zuberbühler, 2000a; Zuberbühler et al., 1999).

The adaptation of each agent can take many forms, such as anatomical, physiological, and/or behavioral modifications. In this report I will focus on behavioral modifications that seem to be elicited by communication. More specifically, I will describe a case study of harpy eagles (*Harpia harpyja*) and howler monkeys (*Alouatta palliata*) in the Barro Colorado Island, Panama. Here, the behavioral adaptations of both predator and prey will be reviewed, followed by a brief discussion on data that might provide insights into the neural basis of these adaptations. This chapter will conclude with final considerations and potential applications.

Many field studies have expanded our knowledge on primate communication. Amongst them are the African vervet monkey studies. Vervets possess a specific alarm call system, one that discriminates between leopards, snakes, and eagles. This system includes both call production, with acoustically distinct alarm calls, and perception, with appropriate behavioral escape strategies (Cheney & Seyfarth, 1990; Seyfarth et al., 1980; Struhsaker, 1967). Also, more recently Zuberbühler's (1999) work on Diana monkeys showed again the use of species-specific alarm calls, but now functionally expanding it to *predator-deterrent* calls. In this case call production by the prey contributes to a decrease of the attack rate of stealth predators, since these predators become aware of being spotted. All these findings have strengthened the case for non-human primates as being capable of creating acoustic labels for mental representations and using them as part of their surviving strategies (Zuberbühler, 2000b). Although the core of these studies is commonly associated with conspecific (i.e., intra-species) communication, several cases have been reported on the use of het-erospecific (i.e., inter-species) communication, including communication between different primate species (Zuberbühler, 2000b), primates and birds (Hauser & Wrangham, 1990), and primates and mammal stealth predators (Zuberbühler et al., 1999). In these cases one species seems to be exploiting the other species' acoustic signals.

Seyfarth & Cheney (2003) argued: "In animal communication natural selection favors callers who vocalize to affect the behavior of listeners, and listeners who acquire information from vocalizations, using this information to represent their environment" (p. 250–291). I will argue that this is precisely what happens in the reported interaction between harpy eagles and howler monkeys, and therefore I approach it as a *communication arms race*. It should be noted that, unlike what is offered in previous reports, here I discuss the vocal signals produced by the predator, not by the prey.

My colleagues and I took advantage of a unique situation where two radio-collared harpy eagles were introduced to Barro Colorado Island (BCI), Panama. BCI is home to several primate species, including the howler monkey and has been a biological reserve since 1923, continuously monitored by resident biologists. Harpies have not existed in this region of Panama for at least the last 50 to 100 years (Willis & Eisenmann, 1979).

Methods

Predator: Harpy Eagles (Harpia harpyja)

Harpy eagles are powerful raptor predators. Their name comes from the Greek word *harpe*, referred by Aristotle and others as probably mythological, winged creatures with a vulture's body, strong claws, and a woman's face. This eagle, although in some areas close to extinction, can be found in Neotropical low-land forests of Central and South America. It is the largest raptor species in America, and certainly one of the largest worldwide (Brown & Amadon, 1968). This species is sexual dimorphic with females reaching twice the size of males. A female can grow to an impressive 35 to 41 inches in body length with a 6-foot wingspan and weight between 10 to 20 pounds (The Peregrine Fund, unpublished data). As part of their amazing hunting traits, they have thick tarsus with large hind talons and sharp claws, which together can exert tremendous pressure upon a selected prey. Although they are specialized in hunting arboreal mammals, their diet includes monkeys, sloths, iguanas, large birds, and even the occasional ter-restrial prey as big as a deer. The harpy predation strategy can vary depending on prey and environment type (Gil-da-Costa et al., 2003; Palleroni, 2003; Touchton et al., 2002; Rettig, 1978; Fowler & Cope, 1964). Harpy eagles have exceptional aerobatic skills that allow rapid attacks through the trees within the forest canopy.

In this study two adult harpies (one male, one female) were used. The Peregrine Fund, an international organization for the conservation of birds of prey in nature, had previously bred the two eagles and ran radio-tracked releases on the Panamanian mainland. Both eagles, the male (J) and female (MV), had experienced over 9 months of freedom in a natural habitat, including active primate killing, before they were released in BCI during 1999 at 19 and 20 months of age, respectively. At the time of their release into the BCI ecosystem, the eagles were already radio-tagged so that, as in the mainland, it was possible to locate and follow them at all times. This provided us detailed records of locations, general behavior, hunting strategies, predation attempts, and kill rates for both eagles. During a period of over 15 months harpy eagles were present in BCI. The pair flew over almost the entire island (1564 ha). However, they preferred hunting grounds on the west side of BCI, spending about 75% of their time within 100-ha home ranges on the western extreme of the island. Altogether, this allowed for a period of almost continuous observation of their natural behavior in the wild for 450 days—240 days in the mainland and 210 days in BCI.

Prey: Howler Monkeys (Alouatta palliata)

The mantled howler monkeys (*Alouatta palliata*) present in Barro Colorado Island are anthropoid primates from the Family: Cebidae. Adult males are estimated to weigh from 16 to 20 pounds with an average length of 45 inches and adult females from 12 to 18 pounds with an average length of 43 inches (Carpenter, 1965). The mantled howler diet consists mostly of leaves, fruits, and flowers from canopy trees. According to a census of the BCI population done in 1977, 65 troops exist in the island, with an average of 19 individuals *per* troop (Wong & Ventocilla, 1995).

Previous studies made on the BCI howler population focused on foraging, physiology, population density, population growth and social behavior (Froehlich et al., 1981; Gaulin et al., 1980; Milton, 1980). Although extensive research has gone into these issues, only a few studies address this genus' vocal repertoire, where their vocalizations were functionally associated with inter-troop spacing, conspecific competition, and sexual selection (Sekulic & Chivers, 1986; Baldwin & Baldwin, 1976; Chivers, 1969). The harpy eagles seemed to use the loud howler roars, a characteristic vocalization from this species, as their localizer of the monkey troops (Palleroni, Touchton & Gil-da-Costa, unpublished data). Before the harpies' introduction the primate populations in BCI (Geoffrey's marmosets, capuchins, spider, and howler monkeys) had no significant mammal, bird, or snake predation (Carpenter, 1965). With the lack of relevant predators, pathogens seem to have been the controlling factor of the howler monkeys population in the island (Milton, 1996). Howlers are parasitized by larvae of the cuterebrid (*Alouattamyia baeri*), resulting in relevant host mortality. Until 1999 the lack of growth of the howler population on BCI was probably mostly a consequence of the primary and secondary effects of this parasitism.

Behavioral Observations

Barro Colorado Island has great logistic conditions offered by the Smithsonian Tropical Research Institute. It houses several laboratory and dormitory amenities while an organized network of trails enables access and orientation. Starting in July 2000, two extra observers joined the Peregrine Fund team and divided their daily time (approximately six hours of observation) between following the harpy eagles and recording both their behavior as well as the encountered howler troops; and scouted, localized, and recorded the behavior of howler troops without the eagles' presence. The observers positioned themselves between approximately 10 to 30 m of the howler groups and at varied distances from the eagle. Data collection from this part of the study was mostly done in a qualitative manner. Observed descriptions of harpy and howler behaviors, both when the animals were on their own and when they interacted, were recorded. Observation time and duration for each event were varied and determined opportunistically in the field.

Playback Experiments

Using howler troops previously localized during the observational part, the playback study included two groups—the "exposed group" and the "control group." The exposed group consisted of 10 troops of howler monkeys dispersed within BCI (3 troops off Stanley trail, 3 troops off Armour trail, 2 troops off Zetek trail, 1 troop off Snyder Molino trail, and 1 troop off Pena Blanca trail). The control group consisted of 5 troops in the Gigante Peninsula. Each group of howlers was only tested a single time.

The following protocol was used: First we located both the male and female harpy eagles using radio telemetry, and then we moved a minimum of 1000 m away from the eagles, but within their home range (only in BCI); next, we located a group of howler monkeys and established a position within visual and auditory proximity to the group; two experimenters remained between 5 and 15 m from the group while a third one, carrying a speaker, moved to an occluded position approximately 30 m away from the howler group. We waited between 15 to 30 min to allow the animals to habituate to our presence. Once the animals were calm, a sample started as soon as an appropriate subject was selected. Throughout the trial one experimenter recorded a 60-sec focal sample on an adult male approximately every 2 min. During a sample, the recorded information included foliage density, spatial position, vigilance rate (percentage of time spent scanning per 60-sec sample), direction of scanning and "other behavioral activities" ("resting," "moving," "foraging," and "socializing"). The resting, moving, foraging, and socializing activities were scored as either "present" or "absent" in each 60-sec sample.

We never conducted focal samples on animals that were in a position scored over 1 on Cords (1990) scale for foliage density, which classifies foliage as "sparse" (0), "medium" (1), or "dense" (2). We used Treves' (1997) definition of "vigilance" as scanning beyond arm's reach. Resting was scored when the focal animal had its eyes closed; socializing included social and allogrooming, as well as play. Since videotaping was not possible due to the density of the forest, one experimenter recorded behavioral data onto a handheld computer, while a second experimenter scored the spatial position (absolute and relative) of each visible animal within the group, once every 6min, meaning one sample at the beginning and two after playback for a total of 3 samples. Behavioral data were collected for 5 min prior to playback, during playback and 10 min post-playback, although sometimes the pre-playback recording period was extended to 6 or 7 min due to factors unrelated to our experiment objective (e.g., subjects moved briefly out of view, equipment problems, etc.). This sampling was based on the established times for assessing baseline behaviour and stimulus effect and decay from previous reported studies, as well as pilot data collected prior to the playback study.

The third experimenter, who occluded the speaker, controlled the stimuli playback. In the 2000 season, the playbacks consisted of different exemplars of the call produced by the female harpy during prey pursuit; a playback presentation included two bouts of calls separated by a 120-sec silence interval. During the 2001 study period, the BCI howlers' response to 8 different calls (4 from harpy eagles, including the male and female introduced on BCI, plus another 4 from a male and female unfamiliar to the howler population) was tested. Different exemplars were used to avoid problems of pseudoreplication and test for discrimination of harpy calls by individual, sex, and familiarity. Other colleagues and I also broadcast other acoustic stimuli as an acoustic control condition. The control stimuli were tinamou (*Tinamus major*) and bald eagle (*Haliaeetus leucocephalus*) calls. The tinamou is native to BCI, and thus, its calls are likely to be familiar to howlers, but non-threatening. Like harpies, the bald eagle is a raptor, but is non-native to Panama; its calls are therefore unfamiliar to the study population of howlers. During the 2000 study season we broadcast the harpy calls using a Sony DAT TCD-D8 recorder and a portable Sony SRS speaker (frequency response = 70 Hz to 20 kHz; mean amplitude: 67.2 dB SPL; range: 58.3 to 79.8 dB SPL at 10 m from the speaker); during the 2001 study season, we used a Sony D-191 CD player and a portable Cambridge Soundworks customized speaker (frequency response = 60 Hz to 18 kHz with broadcast values at 10 m as follows: harpy—mean amplitude: 66.66 dB SPL; range: 58.3 to 78.4 dB SPL; tinamou—mean amplitude: 67.8 dB SPL; range: 57.65 to 79.2 dB SPL; bald eagle—mean amplitude: 67.2 dB SPL; range: 57.98 to 78.8 dB SPL. The speaker was occluded at a mean distance of 30(+/–5) m from the closest animal within the howler group. The mean duration of playback stimuli was 159 sec (range: 148–168 sec). We did not initiate playbacks in groups that remained agitated after 30 min of our arrival and aborted every trial in which the focal animal was out of the observer's visual range for more than 2 min. Following these aborted attempts, we searched for another group of howlers, and moved to a distance of approximately 1000 m if the playback stimulus had been broadcast. Approximately 35% of the trials were aborted.

Results

The Predator's Perspective: Harpy Eagle Assessment Calls

Attacks with successful kills covered a period of 294 days, during which the female captured prey every 4.39 days and the male every 3.71 days. Taking a closer look at the type of prey captured, it was verified that 34.78% of female and 12% of the male prey were primates (Palleroni & Hauser, 2003; Touchton et al., 2002).

The surprising finding, however, was the strategy these eagles seemed to use to hunt their primate prey. Instead of rapid stealth attacks, the eagle would perch in a tree nearby, in plain view of the monkey troop, observe the animals and then utter a series of calls (Gil-da-Costa et al., 2003) (Figure 14.1). These calls always presented the same acoustic structure, being composed by two different elements. A first element uttered only once at the beginning, followed by multiple repetitions of the second element. Observations on and off BCI indicated that this hunting behavior occurred before prey pursuit and only when engaging primate prey (Palleroni & Gil-da-Costa, unpublished data). In the 35 observed cases where the howler monkeys displayed a coordinated defensive response (i.e., increased vigilance, group repositioning, etc.) after the harpy calls, the eagles either moved to a different troop or delayed the attack, hunting later by stealth. When the howlers' response was minimal or chaotic the eagles' approached and in most cases performed the attack. Therefore, the question of whether the eagle delayed the attack, moved to search for another troop, or struck seemed to be contingent on the preys'

FIGURE 14.1. Female adult harpy eagle (MV) uttering a call during the study. The tagging and the radio antenna used for tracking are noticeable in this photograph of the eagle. Waveform and spectrogram of harpy eagle vocalizations used in the study

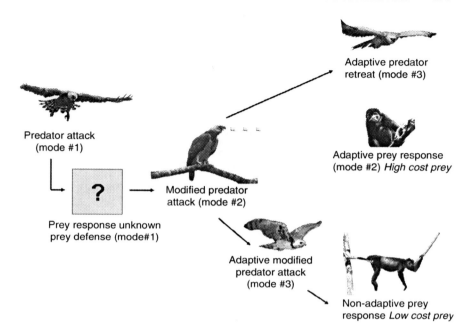

FIGURE 14.2. Communication arms race. In "Predator attack (mode #1)" the eagle spots its prey from the distance and attacks by surprise; In "Prey response unknown—Prey defense (mode #1)" the predator does not know a priori the prey's defense strategy, which makes it vulnerable to rapid prey protective responses. The prey is unaware of the predator's potential attack, therefore it is only possible to produce a *last minute* response; in "Modified predator attack (mode #2)—Predator-assessment call" The predator's attack is modified to a probing strategy, vocalizing in full view and observing the prey's response *before* attacking; when the prey's response is non-adaptive, either disregarding the eagle or displaying a panic chaotic response, then "Adaptive modified predator attack (mode #3)" is elicited and the eagle moves to a rapid striking attack.; in contrast, when howlers show an anti-predator response "Adaptive prey response (mode #2)", including increased vigilance, upward scanning, and coordinated group protective measures (mothers and infants move closer to trunk, males to more distal positions in branches), then "Adaptive predator retreat (mode #3)" occurs, with the eagle flying away in search of another troop

behavioral response (Figure 14.2). While this behavior was observed with both howler and capuchin monkeys, we focus here on the howler monkey predation since it accounted for 81% of the female and 100% of the male primate captures (Palleroni, in prep.; Touchton et al., 2002).

The Prey's Perspective: Howler Monkeys "Live and Learn"

Based on the data collected in the first part of our study in which we documented naturally occurring encounters between the harpy eagles and howler troops, we believe that the reintroduction of these raptors may have triggered an adaptive

(a)   (b)

FIGURE 14.3. a. Vigilant howler monkey. b. Adult and young howler monkeys at rest

anti-predator response by the primate populations. When the monkey troop seemed to react with a vigilance increase (Figure 14.3a) and group protective measures (such as the adult males moving to distal branch positions, placing themselves between the eagle and rest of the troop, and females carrying their infants to more occluded positions near the trunk, where it is harder for the eagle to maneuver) the eagle would typically leave and search for another group to prey, or return later to the first troop but then hunt by stealth. In the cases where monkeys either remained calm (Figure 14.3b) or reacted with random panic agitation, the eagle initiated pursuit in a rapid flight, through the trees, directed at its chosen animal target (Figure 14.4), in most events making a kill. Considering the strong association between this harpy call and the predatory attacks, one would expect that an adaptive learning mechanism could arise from it. Since these calls are given in a hunting context and the prey's response to it influences the predator's attack strategy, we gave the call a probe function, naming it a "predator-assessment" call. Also, to further explore this possibility we ran a playback study using the howlers as our primate subjects.

Experimental Playback

Using the harpies' hunting strategy of calling prior to attack to our advantage, we broadcasted their species-specific call to various howler troops. The assumption was that this call would not only be associated with the eagles' presence but also with the recognition of the eagle as a predator. Building from previous playback studies where it was shown that both primate and non-primate animals were able to recognize and react to predators based on acoustic cues alone (Zuberbühler et al., 1999; Hauser & Caffrey, 1994; Cheney & Seyfarth, 1990; Hauser & Wrangham, 1990), here, myself and colleagues explored how fast this adaptive behavior emerges in howler monkeys and the specificity of their responses.

The program of localization and observation of howler monkey troops in BCI was extended to Gigante, an adjacent peninsula. The population in Gigante is estimated to be smaller than in BCI, but in both cases there have not been any

FIGURE 14.4. Harpy eagle (MV) in rapid flight through the canopy

significant predators for the last 50 to 100 years. Therefore, the Gigante peninsula, out of reach of the newly introduced harpy eagles, presented the ideal "control" population for our study.

The observational data collection started approximately one year after the raptors' introduction to the island. Once the necessary number of monkey troops was located and their natural occurring interactions with the eagles (in BCI) and their baseline behavior (in both BCI and Gigante) were recorded, it was possible to initiate the experimental playback part of the study.

Throughout two research seasons, 2000 and 2001, harpy eagle calls (from both familiar and unfamiliar harpies) were broadcasted from a hidden speaker to the exposed troops in BCI and to the control groups in Gigante. I refer to these as "BCI harpy present" and "Gigante control," respectively. During the 2001 season two acoustic controls, tinamou (*Tinamus major*) and bald eagle (*Haliaeetus leucocephalus*) calls were broadcasted to the BCI howlers. The former, a common bird in BCI that, although known to the monkeys, does not represent a threat, and the latter a raptor, like the harpy eagle, but one that does not exist in Panama, therefore unfamiliar to the monkeys. These playbacks were labeled "BCI control." Finally, also during the 2001 research season, approximately seven months after the eagles were withdrawn from the island, the howler troops in BCI were tested again with harpy eagle calls, assessed for potential maintenance or extinction of their specific anti-predator response. For analysis purposes these data were labeled "BCI harpy absent." As stated in the methods section, several parameters were tested and recorded during each playback condition. The behavioral parameters studied included vigilance rate, direction of scanning, display of other behavioral activities (resting, moving, foraging and socializing), and utterance of alarm calls.

Vigilance Rate

Focal behavioral samples were recorded during three different periods: pre-playback, playback, and post-playback. Within BCI it was possible to observe some variation in the level of predation exposure for the several tested groups. Nevertheless, there were no statistically significant differences within the studied troops in BCI, or in Gigante, regarding vigilance rate (BCI: $n = 60$, $H = 9.0$, $p = 0.4373$; Gigante: $n = 30$, $H = 4.0$, $p = 0.4060$) (Gil-da-Costa et al., 2003). However, when we compare vigilance rates between the exposed BCI howlers (BCI harpy present) and the Gigante troops (Gigante control) (Figure 14.5a), we find there is a highly significant difference across the two conditions ($n = 90$, $F = 16.553$, $p < 0.0001$) (Gil-da-Costa et al., 2003). This finding indicates a learned adaptive vigilance response in the BCI howlers that can be elicited by the harpy call alone. Moreover, this result was replicated one year later, seven months after harpy absence. There was no significant difference in vigilance rate between the conditions "BCI harpy present" and "BCI harpy absent" ($n = 96$, $F = 0.108$, $p = 0.8977$) (Gil-da-Costa et al., 2003). The vigilance rate was also shown to be specialized for the harpy eagle call, since it was not elicited by the other playback stimuli tested during the condition "BCI control" (Figure 14.5b). The vigilance rates for "BCI control" significantly differed from those during "BCI harpy absent" ($n = 60$, $F = 16.591$, $p = 0.0001$) and did not differ from "Gigante control" ($n = 54$, $F = 0.501$, $p = 0.0823$) (Gil-da-Costa et al., 2003).

Harpy calls from different individuals were used, namely from MV, J, and other unfamiliar harpies. There were no significant differences in the howlers' vigilance response when we compared responses to harpy calls from the different individuals (male vs. female harpies: $n = 36$, $F = 0.925$, $p = 0.6325$; MV vs. other harpies: $n = 36$, $F = 0.841$, $p = 0.4661$) (Gil-da-Costa et al., 2003). There was also no differential response between the two control stimuli (tinamou vs. bald eagle: $n = 24$, $F = 3.002$, $p = 0.4676$) (Gil-da-Costa et al., 2003). This indicates a recognition and behavioral response selective for the harpy eagle species call, independent of individual familiarity.

Further analysis of the temporal patterning of playback responses revealed significant differences between tested conditions. In the BCI howlers, both with harpy presence and absence, the vigilance increase seen during playback was maintained during the post-playback period, as one would expect, considering the danger level that this predator imposes. In the Gigante populations, however, the response significantly decreased between playback and post-playback periods, leading to the conclusion that the initial increased vigilance was novelty response rather than predator recognition (Figure 14.6). As stated before, the acoustic control stimuli never elicited a protective response.

(a)   Gigante Control vs. BCI-Harpy Present

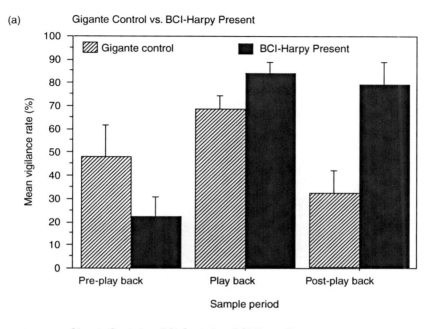

(b)   Gigante Control vs. BCI-Control vs. BCI-Harpy Absent

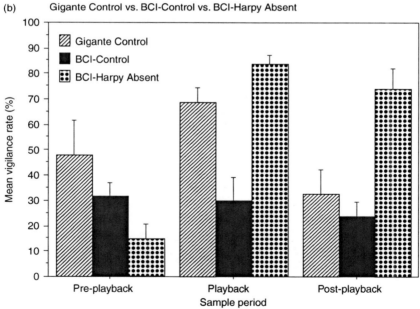

FIGURE 14.5. a. Mean vigilance rates of the howler monkey groups on BCI and in Gigante, during the period of harpy presence. b. Vigilance rates of the howler groups for the various experimental conditions. (Adapted from Gil-da-Costa et al., 2003)

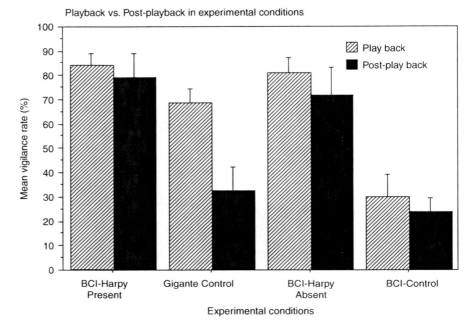

FIGURE 14.6. Mean vigilance rates of the howler groups during the playback and post-playback periods in the various experimental conditions. (Adapted from Gil-da-Costa et al., 2003)

Direction of Scanning

Further behavioral responses contributed to complement the picture of complex anti-predator behavior elicited by the harpy call playback. The BCI howler monkeys scanned upward significantly more than the other groups (BCI harpy present vs. Gigante control: $n = 60$, $\chi^2 = 13.33$, $p = 0.0025$; BCI harpy absent vs. Gigante control: $n = 44$, $\chi^2 = 9.6$, $p = 0.0165$; BCI harpy absent vs. BCI control: $n = 40$, $\chi^2 = 10.91$, $p = 0.0086$) (Gil-da-Costa et al., 2003). Even though the speaker was hidden far away and beneath the monkeys' position in the trees, in BCI the animals scanned upward to areas above them where typically a harpy eagle would perch. The Gigante howlers would orient toward the real sound source location, i.e., the speaker.

Display of Other Behavioral Activities

Quantitative study of the display level of other behavioral activities was carried out using a comparison of the percentage of time the animals would spend resting, moving, foraging, and/or socializing, as described in the methods section. When we compared the display level of these behaviors between the four experimental conditions, we found only one significant difference, namely, howler monkeys in BCI display resting behavior significantly less frequently than the troops in Gigante ($n = 128$, $\chi^2 = 6.196$, $p < 0.05$).

Also, descriptive reports show sex differences as part of a complex protective behavior. After hearing the eagle's call, adult males would often move to more distal positions in the branches, sometimes clustering in the direction of the potential eagle presence and assuming mobbing positions, but the females would pick up the younger infants and move closer to the trunk, as though trying to occlude themselves in harder to maneuver, more dense areas of the canopy. In some cases, different members of the troop would produce alarm calls, but this did not occur during every playback trial.

Altogether, these results show that howler monkeys acquired a selective anti-predator response in a period of one year or less of exposure to harpy eagles. This adaptive response suggests recognition and association of an acoustic cue to a mental representation of a specific predator threat. Also, it was shown that the howlers' sensitivity to the harpy call and appropriate protective behavior was maintained for at least seven months after the predators were removed. How long this protective behavior will be maintained remains an interesting open question.

Discussion

Another Piece of the Story

The study reported here refers to behavioral adaptations within the harpy eagle–howler monkey predator-prey interaction. However, another interesting piece of this dynamic puzzle is the neurophysiological adaptations underlying the behavioral changes. Previous work has presented convincing cases of prey adaptation to predators' acoustic signals at both the behavioral and neurophysiological levels. The bat-moth interaction is a well-studied and quite illustrative case. Noctuoid moths are under severe predation from bats as the moths conduct their night flights (Hoy et al., 1989). During a night time scenario sound is the main communication and cueing channel between predator and prey. The auditory system of noctuoid moths has been intensively studied as a model for anti-predator adaptations (Fullard, 2003; Roeder, 1975). Bats perform prey location by using a biosonar system, and moths have adapted hearing to be sensitive to the range of ultrasonic frequencies present in the bats' biosonar signals (Hoy et al., 1989; Roeder, 1975). When flying, moths can react to detection of the bats' ultrasounds by rapidly altering their behavioral output, in this case their flight pattern. Amongst different groups moths have evolved anti-predator abilities that range from ultrasonic hearing detection, to evasion strategies and bat sonar jamming techniques (by producing loud clicks) (Fullard, 2003; Rydell, 1998; Hoy et al., 1989; Roeder, 1975). Similar cases can be found in other phylogenetic groups and different ecosystems. In the ocean, for example, the American shad (*Alosa sapidissima*), a bony fish, developed ultrasonic hearing to match the frequencies of the echolocation system of one of its stronger predators, the dolphin (*Tursiops truncatus*) (Mann et al., 1998, 1997). This fish species seem to have developed behavioral and neurophysiological adaptations to counteract selective pressures from echolocating odontocete cetaceans. These systems demonstrate clear cases of an evolutionary arms race based on coupled behavioral and neurophysiological adaptations. The understanding of the underlying mechanisms of brain physiology relating to these rapid adaptations brings a valuable insight. Although there are no studies exploring the neural basis of acoustic perception in the howler monkey, there are interesting, and, I believe, relevant, findings in the harpy eagle.

The same head-orienting technique utilized to explore perceptual asymmetries during auditory processing in primates (Hauser & Andersson, 1994) and sea lions (Boye, 2005) was used to test harpy eagles. The technique consists of the playback of sounds from an occluded speaker centered behind the subject's back. The subjects' response is video-recorded and later blindly scored as the percentage of head-turns to each side for every stimulus category. Because an orienting bias increases the strength of the signal entering the leading ear, it is likely to create a processing bias with respect to the contralateral hemisphere. Therefore, a systematic head-turning to the right, for instance, would imply a left hemispheric bias.

Two groups of harpy eagles were tested. One group experienced in hunting howlers and another naïve to hunting this primate. Both groups were tested with sets of harpy eagle calls (conspecifics) and howler vocalizations. Results showed that both naïve and hunting expert eagles presented a left hemispheric bias for their own vocalizations. However, for the howler monkey calls the naïve harpies show a right hemisphere bias and the experienced harpies show a left hemisphere bias (Palleroni & Hauser, 2003). This finding provides evidence of an orienting response and, by implication, a brain hemispheric bias that can be altered by explicit hunting experience. This supports the idea of adaptive plasticity for the processing of acoustic cues, which can be molded within the animals' life by predation experience. It provides a rapid physiological adaptation that can accompany, and probably support, the adaptive behavioral strategy. Building from previous work in other primate species (Gifford et al., 2005; Gil-da-Costa et al., 2004; Poremba et al., 2004; Ghazanfar, 2003), it would be very valuable to explore the neural basis, and perhaps plasticity, of auditory processing in the howler monkey, expanding the understanding of auditory-driven interactions. Only with further knowledge regarding the occurrence of this type of brain adaptation across different taxonomic groups and in which contexts the adaptation appears, can we better understand the implications and functionality of this finding.

Conclusions

The findings reported here, besides providing additional support for previous general findings (Berger et al., 2001; Bshary, 2001; Caro, 1995; Woodland et al., 1980), go further by demonstrating: (i) how rapid primate prey adaptation can be within one generation; (ii) the maintenance of the response following predator absence; and (iii) the importance and use of a predators' call in prey assessment.

To minimize risk and increase efficiency a predator needs specific critical information about its target prey. Namely, the level of prey alertness and relative escape ability, its defense mechanisms, and its nutritional value (Hasson, 1991). My colleagues and I argue that by using this *predator-assessment* call, harpy eagles extract more accurate and extensive information regarding the first three items than they would by using mere visual observation alone, prior to an attack. By probing the potential defense strategies of the prey before engaging in confrontation, the predator can assess the attack risk and make an a priori decision, minimizing risks from surprise defenses during attack. The prey can also benefit from this advanced warning by developing defense/escape strategies and gaining more time to prepare to counteract the attack. One may then ask, if the predator is seemingly back at step one, what is the point of it all? Indeed, the predator loses the ambush/surprise advantage but gains a better control of the situation by learning about the preys' abilities and therefore choosing more vulnerable prey. This risk avoidance may provide an altogether better fitness. From the preys' perspective there is a clear advantage in developing specific anti-predator behaviors that can be elicited by this auditory cue prior to attack. We can speculate that this trend will gradually occur in all exposed prey, developing in them anti-predator behaviors and increasing their fitness. If this happens in the whole population it will ultimately cancel any advantage for the predator to call. At which point in time the predator would have to create another modified behavior, exploring alternative strategies.

And so the arms race continues. It is important to keep in mind, at all times, that predator-prey interactions are dynamic, and more than finding definitive solutions each agent tries to have at least a temporary advantage.

In an attempt to model predator-prey adaptive behaviors Jim & Giles (2000) used a genetic algorithm to evolve multi-agent communication systems for the predators in an artificial version of the predator-prey interaction. Their simulations show that predators' performance in prey pursuit increases with the evolution of a communication system. There are numerous studies reporting cooperative hunting between predators, from wolves to lions to chimpanzees, some of them presenting cases where communication plays a crucial role (McGregor, 2005; Stanford, 1998; Heinsohn, 1995; Boesch & Boesch, 1989; Schaller, 1972). However, these studies refer to communication between conspecifics. That is, multiple individuals from the same species communicate amongst themselves in order to perform elaborate group hunting strategies resulting in coordinated attacks. This is different from the harpy eagle–howler monkey case, where communication seems to occur between the different species, between predator and prey. One can hypothesize that, in a case like this, communication can ultimately lead to temporary beneficial adaptations in both species if the two of them can explore the use of the acoustic signal in a way that increases each one's fitness.

The inexistence of a protective response in the Gigante howlers that were only exposed to harpy predation over 75 years ago leads to the conclusion that those monkey populations lost their ability to recognize this predator's call and elicit an anti-predator response. The call of a predator that has long been extinct in an area will not re-elicit response in a prey who first hears it; hence, this prey is highly vulnerable to first encounters with such a predator. The situation in Gigante contrasts with the situation of the newly re-exposed population in BCI, which in a short period of time (less than one generation) developed recognition and specific anti-predator mechanisms.

As a final note I would like to stress the applicable conservation implications beyond the theoretical study of predator-prey interactions and animal communication. The extinction of large predators worldwide has provoked tremendous ecological imbalances, leaving mammal (including primate) prey without predation pressures. This loss of anti-predator response makes them extremely vulnerable to new predators and, as such, re-population attempts of extinct predators can have catastrophic consequences over prey populations (Gittleman & Gompper, 2001). The use of potential predator-assessment calls, as well as other relevant training cues in prey population assessments, prior to a full scale predator re-introduction, should help prepare populations for an adequate and balanced interaction. This kind of methodology could prevent, or at least significantly reduce, some of the negative impact of predator re-population over prey species.

Acknowledgments. I would like to acknowledge Marc Hauser, Alberto Palleroni, Janeene Touchton, J. P. Kelley, Angel Muela, Nicola Morton, Eduardo Santa-Maria, and Katia Herrera for their participation in the Barro Colorado Island study; the Smithsonian Tropical Research Institute for allowing us to conduct research in the Barro Colorado Island Reserve, the BCI-Smithsonian and The Peregrine Fund and staff, as well as Stanley Rand, for all their support; and Christian Ziegler, Alberto Palleroni and Angel Muela for their howler monkeys and harpy eagle photos. For their comments on the manuscript I would like to thank Jonathan Fritz, Carl Senior, Marla Zinni, and three anonymous reviewers. The original research was funded by grants from the Fundacao para a Ciencia e Tecnologia, The Peregrine Fund, and Harvard University.

References

Baldwin, J.D. and Baldwin, J.I. (1976). Vocalizations of howler monkeys (*A. palliata*) in southwestern Panama. *Folia primatol.*, 26: 81–108.

Berger, J., Swenson, J.E., and Persson, I. (2001). Recolonizing carnivores and naïve prey: Conservation lessons from Pleistocene extinctions. *Science*, 291: 1036–1039.

Blumstein, D.T., Daniel, J.C., Griffin, A.S., and Evans, C.S. (2000). Insular tammar wallabies (*Macropus eugenii*) respond to visual but not to acoustic cues from predators. *Behav. Ecol.*, 11(5): 528–535.

Boesch, C., and Boesch, H. (1989). Hunting behavior of wild chimpanzees in the Taï National Park. *Amer. Jour. of Phys. Anthropol.*, 78: 547–573.

Boye, M., Gunturkun, O., and Vauclair, J. (2005). Right ear advantage for conspecific calls in adults and subadults, but not infants, California sea lions (*Zalophus californianus*): Hemispheric specialization for communication? *European Jour. of Neuroscience*, 21: 1727–1732.

Brown, L., and Amadon, D. (1968). *Eagles, hawks and falcons of the world*. Vol. 2. New York: McGraw-Hill Book Co.

Bshary, R. (2001). Diana monkeys, *Cercopithecus diana*, adjust their anti-predator response behaviour to human hunting strategies. *Behav. Ecol. Sociobiol.*, 50: 251–256.

Caro, T.M. (1995). Pursuit-deterrence revisited. *Trends Ecol. Evol.*, 10: 500–503.

Carpenter, C.R. (1965) The howlers of Barro Colorado Island. In I. DeVore (Ed.). *Primate behavior. Field studies of monkeys and apes* (pp. 250–291). New York: Holt, Rinehart and Winston Inc.

Cheney, D.L., and Seyfarth, R.M. (1990). *How monkeys see the world*. Chicago: Chicago Univ. Press.

Chivers, D.J. (1969). On the daily behaviour and spacing of howling monkey groups. *Folia Primatol. (Basel)*, 10: 48–102.

Cords, M. (1990) Vigilance and mixed-species association of some East African forest monkeys. *Behav. Ecol. Sociobiol.*, 26: 297–300.

Endler, J.A. (1991). Interactions between predators and prey. In J.R. Krebs and N.B. Davies (Eds.), *Behavioral ecology* (pp. 169–202). Oxford: Blackwell Scientific Publications.

Fowler, J.M., and Cope, J.B. (1964). Notes on the harpy eagle in British Guiana. *The Auk*, 81: 257–273.

Froehlich, J.W., Thorington, R.W., and Otis, J.S. (1981). The demography of howler monkeys (*Alouatta palliata*) on Barro Colorado Island, Panama.*Inter. Jour. Primatol.*, 2: 207–236.

Fullard, J.H., Dawson, J.W., and Jacobs, D.S. (2003). Auditory encoding during the last moment of a moth's life. *Jour. Exp. Biol.*, 206: 281–294.

Ghazanfar, A.A. (Ed.). (2003). *Primate audition: Ethology and neurobiology*. Boca Raton: Florida: CRC Press.

Gaulin, S.J.C., Knight, D.H., and Gaulin, C.K. (1980). Local variance in *Alouatta* group size and food availability on Barro Colorado Island. *Biotropica*, 12: 137–143.

Gifford, G.W., MacLean, K.A., Hauser, M.D., and Cohen, Y.E. (2005). The neurophysiology of functionally meaningful categories: Macaque ventrolateral prefrontal cortex plays a critical role in spontaneous categorization of species-specific vocalizations. *Jour. Cog. Neurosci.*, 17: 1471–1482.

Gil-da-Costa, R., Palleroni, A., Hauser, M.D., Touchton, J., and Kelley, J.P. (2003). Rapid acquisition of an alarm response by a Neotropical primate to a newly introduced avian predator. *Proc. Roy. Soc. Lond. B*, 270: 605–610.

Gil-da-Costa, R., Braun, A., Lopes, M., Hauser, M.D., Carson, R.E., Herscovitch, P., and Martin, A. (2004). Toward an evolutionary perspective on conceptual representation: Species-specific calls activate visual and affective processing systems in the macaque. *Proc. Natl. Acad. Sci. USA*, 101: 17516–17521.

Gittleman, J.L., and Gompper, M.E. (2001) The risk of extinction—What you don't know will hurt you. *Science*, 291: 997–999.

Hasson, O. (1991). Pursuit-deterrent signals: Communication between prey and predators. *Trends Ecol Evol.*, 6: 325–329.

Hauser, M.D., and Wrangham, R.W. (1990). Recognition of predator and competitor calls in non-human primates and birds: A preliminary report.*Ethology*, 86: 116–130.

Hauser, M.D. and Andersson, K. (1994). Left hemisphere dominance for processing vocalizations in adult, but not infant, rhesus monkeys: Field experiments. *Proc. Natl. Acad. Sci.*, 91: 3946–3948.

Hauser, M.D., and Caffrey, C. (1994). Anti-predator response to raptor calls in wild crows, *Corvus brachyrhynchos hesperis*. *Anim. Behav.*, 48: 1469–1471.

Heinsohn, R.P., and Craig, P. (1995). Complex cooperative strategies in group-territorial African lions. *Science*, 269: 1260–1262.

Hoy, R., Nolen, T., and Brodfuehrer, P. (1989). The neuroethology of acoustic startle and escape in flying insects. *Jour. Exp. Biol.*, 146: 287–306.

Jim, K.C., and Giles, C.L. (2000). Talking helps: Evolving communicating agents for the predator-prey pursuit problem. *Artif. Life*, 6: 237–254.

Mann, D.A., Lu, Z., Hastings, M.C., and Popper, A.N. (1998). Detection of ultrasonic tones and simulated dolphin echolocation clicks by a teleost fish, the American shad (*Alosa sapidissima*). *Jour. Acoust. Soc. Am.*, 104: 562–568.

Mann, D.A., Lu, Z., and Popper, A.N. (1997). A clupeid fish can detect ultrasound. *Nature*, 389: 341.

Marler, P., Evans, C.S., and Hauser, M.D. (1992). Animal signals: Reference, motivation or both? In H. Papoucek, U. Jurgens, and M. Papoucek (Eds.), *Nonverbal vocal communication: Comparative and developmental approaches* (pp. 66–86). Cambridge: Cambridge Univ. Press.

McGregor, P.K. (Ed.). (2005). *Animal communication networks*. Cambridge: Cambridge Univ. Press.

Milton, K. (1980). *The foraging strategy of howler monkeys*. New York: Columbia Univ. Press.

Milton, K. (1996) Effects of bot fly (*Allouattamya baeri*) parasitism on a free-ranging howler monkey (*Alouatta palliata*) population in Panama. *Jour. Zool. Lond.*, 239: 39–63.

Palleroni, A., and Hauser, M.D. (2003). Experience-dependent plasticity for auditory processing in a raptor. *Science*, 299: 1195.

Palleroni, A. in prep. Harpy eagle development and behaviour.

Poremba, A., Malloy, M., Saunders, R.C., Carson, R.E., Herscovitch, P., and Mishkin, M. (2004). Species-specific calls evoke asymmetric activity in the monkey's temporal poles. *Nature*, 427: 448–451.

Rettig, N. (1978) Breeding behavior of the harpy eagle (*Harpia harpyja*). *The Auk*, 95: 629–643.

Roeder, K.D. (1975). Neural factors and evitability in insect behavior. *Jour. Exp. Zool.*, 194: 75–88.

Ryan, M.J., Tuttle, M.D., and Rand, A.S. (1982). Bat predation and sexual advertisement in a Neotropical frog. *American Naturalist*, 119: 136–139.

Rydell, J. (1998). Bat defence in lekking ghost swifts (*Hepialus humuli*), a moth without ultrasonic hearing. *Proc. Biol Sci.*, 265: 1373–1376.

Schaller, G.B. (1972). The Serengeti lion: A study of predator-prey relationships. Chicago: Univ. Chicago Press.

Sekulic, R., and Chivers, D.J. (1986). The significance of call duration in howler monkeys. *Inter. Jour. Primatol.*, 7(2): 183–190.

Seyfarth, R.M., Cheney, D.L., and Marler, P. (1980). Vervet monkey alarm calls: Semantic communication in a free-ranging primate. *Animal Behaviour*, 28: 1070–1094.

Seyfarth, R.M., and Cheney, D.L. (2003). Signalers and receivers in animal communication. *Annu. Rev. Psychol.*, 54: 145–173.

Shultz, S., Noë, R., McGraw, W.S., and Dunbar, R.I. (2004). A community-level evaluation of the impact of prey behavioural and ecological characteristics on predator diet composition. *Proc. R. Soc. Lond. B.*, 271: 725–732.

Stanford, C.B. (1998). *Chimpanzee and red colobus: The ecology of predator and prey*. Cambridge, MA: Harvard Univ. Press.

Struhsaker, T.T. (1967). Auditory communication among vervet monkeys. In S.A. Altman (Ed.), *Social communication among primates* (pp. 281–324). Chicago: Univ. Chicago Press.

Touchton, J.M., Hsu, Y. and Palleroni, A. (2002). Foraging ecology of captive-bred subadult harpy eagles (*Harpia harpyja*) on Barro Colorado Island, Panama. *Ornitologia neotropical*, 13: 365–379.

Treves, A. (1997). Self-protection in primates. Doctoral. Harvard University.

Willis, E.O., and Eisenmann, E. (1979). A revised list of birds of Barro Colorado Island, Panama. *Smithsonian Contributions to Zoology*, 291: 1–31.

Wong, M., and Ventocilla, J. (1995). *A day in Barro Colorado Island*. Panama: Smithsonian Tropical Research Institute.

Woodland, D.J., Jaafar, Z., and Knight, M.L. (1980). The 'pursuit-deterrent' function of alarm signals. *American Naturalist*, 115: 748–753.

Zuberbühler, K., Jenny, D., and Bshary, R. (1999). The predator deterrence function of primate alarm calls. *Ethology*, 105: 477–490.

Zuberbühler, K. (2000a). Causal knowledge of predators' behaviour in wild Diana monkeys. *Animal Behaviour*, 59: 209–220.

Zuberbühler, K. (2000b). Interspecies semantic communication in two forest primates. *Proc. R. Soc. Lond. B*, 267: 713–718.

15
Effects of Habitat Structure on Perceived Risk of Predation and Anti-Predator Behavior of Vervet (*Cercopithecus aethiops*) and Patas (*Erythrocebus patas*) Monkeys

Karin L. Enstam

Introduction

This chapter summarizes the ways in which habitat structure affects perceived risk of predation and responses to predators (i.e., anti-predator behavior) by cercopithecines (Superfamily: Cercopithecoidea), with specific reference to vervet (*Cercopithecus aethiops*) and patas (*Erythrocebus patas*) monkeys. Predation has long been thought to be an important selective pressure on primate behavior and sociality (e.g., Altmann, 1974; Busse, 1977; Struhsaker, 1981; van Schaik, 1983; Cheney & Wrangham, 1987; Cords, 1987; Isbell, 1991, 1994; Miller, 2002). Among Old World monkeys, predation has been argued to have favored traits such as large group size (e.g., van Schaik, 1983), multi-male groups (e.g., Henzi, 1988; van Schaik & Hörstermann, 1994), sexual dimorphism in canine size (e.g., Harvey & Kavanagh, 1978; Plavcan & van Schaik, 1994), and polyspecific associations (e.g., Gautier-Hion et al., 1983; Cords, 1987; Struhsaker, 1981, 2000), although others maintain that these traits have been selected for by feeding competition (Wrangham, 1980, 1983; Janson & Goldsmith, 1995), sexual selection (Andelman, 1986; Ridley, 1986; Altmann, 1990; Mitani et al., 1996), or, most recently, infanticide (van Schaik & Kappeler, 1997; Isbell et al., 2002).

Cercopithecines have a wide array of known and potential predators (Table 15.1) that differ greatly in hunting style (e.g., Kruuk & Turner, 1967; Boesch, 1994; Shultz, 2001). Thus, it is not surprising that they display a variety of behaviors in response to the threat of predation, including alarm calls (Seyfarth et al., 1980; Cheney & Wrangham, 1987; Cheney & Seyfarth, 1990; Isbell, 1994; Zuberbühler et al., 1999; Zuberbühler, 2001), cryptic behavior (i.e., silence: Hall, 1965; Tilson, 1977; Chism et al., 1983; Chism & Rowell, 1988; Wahome et al., 1993; Boesch, 1994; Isbell, 1994), and the formation of

TABLE 15.1. Confirmed and potential predators of cercopithecines[a] (data on guenons adapted from Enstam & Isbell, 2007 in *Primates in Perspective* by Bearder et al., copyright Oxford Univ. Press, Ltd. Reprinted with permission of Oxford Univ. Press, Inc.). Absence of data (blank spaces) indicates that data are unavailable; [a]"Confirmed" predators include species that have been observed preying on a particular cercopithecoid species, whether the attack was successful or not. Confirmed predators also include species that have left remains of monkeys in their nests or dung; [b]"Potential" predators are species researchers listed as possible predators, but have not been observed attacking, or attempting to attack the species in question. In general, potential predators are those that co-occur with monkeys and are known to take prey of equal or greater size than that species, even if they have not been observed preying on that species.

Species	Confirmed[a]	Potential[b]	Sources
GUENONS			
Cercopithecus aethiops	leopard, martial eagle, python, yellow baboon, black eagles	lion, spotted hyena, African wild cat, serval, black-backed jackal, cheetah, caracal	Haufstater, 1975; Haltnorth & Diller 1980; Isbell Seyfarth et al., 1980; Boshoff et al., 1991;Gevaerts, 1992; Enstam, 2002
C. ascanius	crowned eagle, chimpanzee		Cords, 1987, 1990, 2002a; Skorupa, 1989; Struhsaker & Leakey, 1990; Wrangham & Riss, 1990; Gevaerts, 1992; Colyn, 1994; Mitani & Watts, 1999; Mitani et al., 2001;Sanders et al., 2003
C. campbelli	leopard, crowned eagle, chimpanzee		Hoppe-Dominik, 1984; Oates et al., 1990; Shultz, 2001; Zuberbühler, 2001; Zuberbühler & Jenny, 2002; Shultz et al., 2004
C. cephus	crowned eagle, human	python, golden cat, leopard	Gautier-Hion & Gautier, 1976; Gautier-Hion et al., 1983; Gautier-Hion & Tutin, 1988
C. diana	chimpanzee, human, leopard, crowned eagle		Hoppe-Dominik, 1984; Boesch & Boesch, 1989; Oates et al., 1990; Zuberbühler et al., 1997; Zuberbühler et al., 1999; Shultz, 2001; Zuberbühler & Jenny, 2002; Shultz et al., 2004
C. lhoesti	crowned eagle	leopard, golden cat, python	Haltenorth & Diller, 1980; Skorupa, 1989; Struhsaker & Leakey, 1990; Gevaerts, 1992; Colyn, 1994; Mitani et al., 2001 Sanders et al., 2003
C. mitis	crowned eagle chimpanzee, human		Napier, 1981; Skorupa, 1989; Struhsaker & Leakey, 1990; Wrangham & Riss, 1990; Gevaerts, 1992; Mitani & Watts, 1999; Struhsaker, 2000
C. neglectus	crowned eagle	leopard, golden cat, python	Gautier-Hion & Gautier, 1976 Haltenorth and Diller, 1980; Gevaerts, 1992; Wahome et al., 1993; Colyn,1994

TABLE 15.1. (Continued).

Species	Confirmed[a]	Potential[b]	Sources
GUENONS			
C. nictitans	crowned eagle, human	python, golden cat, leopard	Gautier-Hion & Gautier, 1976; Gautier-Hion et al., 1983 Gautier-Hion & Tutin, 1988; Gevaerts; 1992; Zuberbühler & Jenny, 2002
C. petaurista	leopard, crowned eagle chimpanzee		Hoppe-Dominik, 1984; Zuberbühler & Jenny, 2002; Shultz et al., 2004
C. pogonias	crowned eagle, human	python, golden cat, leopard	Gautier-Hion & Gautier, 1976; Gautier-Hion et al., 1983; Gautier-Hion & Tutin, 1988
C. wolfi	crowned eagle	leopard	Zeeve, 1991; Gevaerts, 1992; Colyn, 1994
Erythrocebus patas	black-backed jackal, domestic dog	leopard, serval, caracal, African wild cat, lion, spotted hyena, martial eagle, chimpanzee, wild dog, baboon	Haltnorth & Diller, 1980; Chism et al., 1983; Chism & Rowell, 1988; Isbell & Enstam, 2002; Isbell, in prep; Isbell, unpubl. data
Miopithecus talapoin		leopard, golden cat, genet Nile monitor, crowned eagle	Gautier-Hion, 1971; Gautier-Hion & Gautier, 1976; Haltenorth & Diller, 1980
MANGABEYS			
Cercocebus torquatus	leopard, crowned eagle, chimpanzee		Hoppe-Dominik, 1984; Zuberbühler & Jenny, 2002
Lophocebus albigena	crowned eagle		Gautier-Hion & Tutin, 1988; Colyn, 1994; Skorupa, 1989; Struhsaker & Leakey, 1990; Mitani et al., 2001; Olupot & Waser, 2001; Sanders et al., 2003 Horn, 1987 Colyn, 1994
L. aterrimus	crowned eagle, human		
MACAQUES			
Macaca fascicularis		monitor lizard, reticulated python, clouded leopard, golden cat, tiger	Napier & Napier, 1967; Fittinghoff & Lindburg, 1980; van Schaik & van Noordwijk, 1985
M. mulatta	tiger, unidentified raptor	jackal, leopard, tiger	Lindburg, 1977; Edgaonkar & Chellam, 2002
M. radiata	leopard	tiger, domestic dog, hyena, wild dog, python	Fa, 1989; Ramakrishnan et al., 1999; Ramakrishnan & Coss, 2000; Edgaonkar & Chellam, 2002

TABLE 15.1. (Continued).

Species	Confirmed[a]	Potential[b]	Sources
BABOONS			
Papio anubis	crowned eagle, chimpanzee, leopard		Kruuk & Turner, 1967; Goodall, 1986; Mitani et al., 2001; Sanders et al., 2003
P. cynocephalus	lion, leopard, hyena		Altmann, 1980; Rasmussen, 1983; Condit & Smith, 1994
P. ursinus	lion, leopard, crocodile, black eagle	python, hyena, wild dog	Busse, 1980; Boshoff et al., 1991; Cheney et al., 2004
P. hamadryas	dog, leopard	Verreaux's eagle, lion, leopard, cheetah, wolf, hyena, jackal, crocodile	Kummer, 1968; Nagel, 1973; Haltnorth & Diller, 1980; Sigg, 1980; Biquand et al., 1992; Zinner & Pelaez, 1999; Swedell, 2006

polyspecific associations (Struhsaker, 1981; Gautier-Hion et al., 1983; Cheney & Wrangham, 1987; Cords, 1987; Struhsaker & Leakey, 1990; Isbell, 1994; Bshary & Noë, 1997; Noë & Bshary, 1997; Chapman & Chapman, 2000; Enstam & Isbell, 2007). Although cercopithecines sometimes harass, mob, attack and drive off, or kill predators (e.g., Altmann & Altmann, 1970; Lindburg, 1977; Gautier-Hion & Tutin, 1988; Boesch, 1994; Cowlishaw, 1994; Stanford, 1995, 1998; Iwamoto et al., 1996; Boesch & Boesch–Achermann, 2000), the majority of recorded responses involve fleeing from predators (e.g., Seyfarth et al., 1980; van Schaik et al., 1983; Cheney & Seyfarth, 1990; Isbell, 1994; Iwamoto, 1993; Boesch, 1994; Bshary & Noë, 1997; Boesch & Boesch–Achermann, 2000; Ramakrishnan & Coss, 2000; Bshary, 2001; Enstam & Isbell, 2002), often after an alarm call has been given.

Early warning of predator presence is apparently so vital for effective escape that a number of cercopithecine species respond to alarm calls given by sympatric (primate and non-primate) prey species (e.g., Gautier-Hion et al., 1983; Seyfarth & Cheney, 1990; Ramakrishnan & Coss, 2000; McGraw & Bshary, 2002). Furthermore, research on the responses of cercopithecines to alarm calls indicates that the "correct" response depends on both predator hunting style (e.g., Seyfarth et al., 1980; Cheney & Seyfarth, 1990; Zuberbühler, 2001; Bshary, 2001; Shultz et al., 2004) *and* habitat structure (e.g., Boesch, 1994; Stanford, 1995; Noë & Bshary, 1997; Enstam & Isbell, 2002).

Difficulties in Documenting Predation in Cercopithecines

Challenges to the importance of predation in shaping primate traits and behaviors come from several fronts. First, while some anti-predator behaviors, such as alarm calls, mobbing, and evasive maneuvers, are relatively easy for observers to document, it is considerably more difficult to document predator-directed vigilance (Janson, 2000). Among both cercopithecines and colobines, vigilance has been shown to increase with increasing predation risk, regardless of whether this increased risk is due to social factors (e.g., group size: Isbell & Young 1993; Hill & Cowlishaw, 2002; Shultz et al., 2004; position within the group: Steenbeek et al., 1999; nearest neighbor distances: Cowlishaw, 1998; Treves, 1998; Hill & Cowlishaw, 2002; Stanford, 2002; absence of neighbors: Steenbeek et al., 1999, but see Cords, 1990) or ecological factors (e.g., exposure to predators: Cords, 1990; Steenbeek et al., 1999; Sterk, 2002; Shultz et al., 2004; visibility: Cowlishaw, 1998; proximity to refuges: Cowlishaw, 1997a,b, 1998; Hill and Cowlishaw, 2002; unfamiliar habitat: Isbell et al., 1991, 1993). Further complications with documenting predator-directed vigilance arise because in some cercopithecines and colobines a large proportion of time dedicated to scanning is in fact directed at detecting potential competitors, infanticidal males, or mates (e.g., Keverne et al., 1978; Baldellou & Henzi, 1992; Cowlishaw, 1998; Steenbeek et al., 1999; Treves, 1999), rather than predators.

Second, traits (such a large group size; Hill & Weingrill, 2006; this volume) that may increase predator avoidance capability may also increase a group's ability to compete with other groups for food, while also increasing intragroup competition for food. This means that it can be difficult to separate the relative influences of predation and feeding competition on primate traits and behaviors (but see Cowlishaw, 1997a).

Finally, predation on cercopithecines is often difficult to observe (Cheney & Wrangham, 1987; Isbell, 1990; but see Busse, 1980; Gautier-Hion et al., 1983; Struhsaker & Leakey, 1990; Baldellou & Henzi, 1992; Condit & Smith, 1994; Stanford, 1998; Mitani et al., 2001, for observations of predation on specific cercopithecine species), so accurate predation rates are difficult to obtain. The result is that arguments that maintain predation has favored traits such as large group size and sexual dimorphism are largely based on the finding that these traits tend to vary with gross habitat type. Among cercopithecines, terrestrial monkeys tend to be larger, with larger group sizes, multiple males, and greater sexual dimorphism in canine and body size (Crook & Gartlan, 1966; Clutton-Brock & Harvey, 1977; Dunbar, 1988; but see Cheney & Wrangham, 1987; Isbell, 1994, for an alternative view), and terrestrial monkeys are often assumed to be (e.g., Dunbar, 1988; Plavcan & van Schaik, 1994), and in some habitats are (e.g., Shultz, 2004; but see Olupot & Waser, 2001; Zuberbühler & Jenny, 2002, for an alternative view) at greater risk of predation than their arboreal counterparts.

Predation Risk vs. Predation Rate

Given all of this, gaining an accurate picture of predation pressure on cercopithecines is difficult. There are two ways to measure the level of predation pressure on a population: predation rate and predation risk. Predation *rate* refers to the annual mortality rate within a population that is due to predation and represents the level of successful attacks after the prey have employed their anti-predator strategies (Hill & Dunbar, 1998). Predation *risk*, on the other hand, refers to the animals' perceptions of the likelihood of attack (regardless of whether it is successful or not) or their perceived danger in a habitat or area, based on the animals' behavior, as inferred by researchers (Hill & Dunbar, 1998; Hill & Lee, 1998; Stanford, 1998). Predation risk may be thought of as the probability of an individual or group encountering a predator, and it is this risk of predation on which animals base their anti-predator strategies (Hill & Dunbar, 1998; but see Vermeij, 1982, for an alternative view).

For primates in particular, both predation rate and predation risk can be difficult variables to measure. Accurate predation rates for many primate populations are difficult to obtain (Janson, 2000) because predators are rarely habituated to observers, and predation tends to occur when observers are absent (Isbell & Young, 1993; Stanford, 1995). Observers may return to their study groups to find animals missing, but with little or no direct evidence of their fate. Thus, researchers must estimate predation rates and do so using a variety of methods (see "How estimated" in Table 15.2), which may account for much of the variation in estimated predation rates presented in Table 15.2. The few studies in which predation by mammals on cercopithecines and colobines have been accurately estimated are those in which predators are habituated to human presence (e.g., Wrangham & Riss, 1990; Stanford et al., 1994; Stanford, 1995, 1998). These studies indicate that predation rates on Old World monkeys can be as high as 35% per year (Wrangham & Riss, 1990; Stanford et al., 1994).

When predators are not habituated and predation events are not directly observed, estimates of predation rates are determined by indirect methods including: (1) counting animals as victims of predation if they disappeared in apparently healthy condition within a short time (e.g., days) of the observer's last observation (Cheney et al., 1988; Isbell, 1990); (2) counting primate remains found under raptor nests (e.g., Struhsaker & Leakey, 1990; Shultz et al., 2004) or in the dung of mammalian predators (e.g., Karanth & Sunquist, 1995; Boesch & Boesch–Achermann, 2000; Bagchi et al., 2003); (3) counting unexplained disappearances as deaths based on the known mortality rate (i.e., known number of deaths per population size; Alberts & Altmann, 1995); and (4) locating collars from radio-collared individuals in the presence or absence of primate remains (Olupot & Waser, 2001). Estimates of predation rates of cercopithecines based on circumstantial evidence vary widely, from less than 1% to as much as 35% per year (Cheney & Wrangham, 1987; Table 15.2). Under certain circumstances, predation rates can be extremely high. For example, Isbell (1990) estimated the predation rate on Amboseli vervets (*C. aethiops*) in 1987 was at least 45% due to increased leopard (*Panthera pardus*) predation. Indeed, predation can greatly impact the size and structure of cercopithecoid populations (Stanford, 1998) and may lead, at least temporarily, to reduced group sizes (Isbell, 1990; Stanford, 1995, 1996; Isbell & Enstam, unpubl. data), lower population densities (Stanford, 1996, 1998), skewed adult sex ratios (Struhsaker & Leakey, 1990), more males per group (Stanford, 1998), or the elimination of groups altogether (Isbell, 1990; Stanford, 1998).

TABLE 15.2. Estimated predation rates of cercopithecines. [a] Unless noted otherwise under "How estimated"; [b] The authors of the original study did not distinguish between species; [c] multiple percentages are given in parentheses indicate the respective predation rates of the specified predators in the "Predators Responsible" column.

Species	Estimated Annual Predation Rate[a]	Predator(s) Responsible	Site	Reference	How Estimated
GUENONS					
Cercopithecus aethiops	45% (1987)	leopard(primarily)	Amboseli NP, Kenya	Isbell, 1990	Overall percent of all adult females that died/disappeared
	11% (1977–1986)	leopard, pythons, martial eagles	Amboseli NP, Kenya	Isbell, 1990	Overall percent of all adult females that died/disappeared
	56% (1992–2001)	leopard, martial eagle	Segera Ranch, Kenya	Isbell & Enstam, 2002	Overall percent of all adult Female that died/disappeared
C. diana	7% (1.5%, 5.5%[c])	leopard, crowned eagle	Taï forest, Cote d'Ivoire	Shultz et al., 2004	# individuals removed per km^2 by each predator species
	0.5–1%	chimp	Taï forest, Cote d'Ivoire	Boesch & Boesch–Achermann, 2000	
C. campbelli/ petaurista[b]	9.50% (1.5%, 7.75%, 0.25%[c])	leopard, crowned eagle, chimp	Taï forest, Cote d'Ivoire	Shultz et al., 2004	# individuals removed per km^2 by each predator species
C. petaurista	0.30%		Forest, Cote d'Ivoire	Boesch & Boesch–Achermann.2000	
C. mitis	0.58% (1/86–1/88)	crowned eagle	Kibale NP, Uganda	Struhsaker & Leakey. 1990	minimum % killed/year (separated by male & female. I added the numbers of each sex & divided by the minimum # of kills for each sex)

Species	% killed	Predators	Location	Reference	Notes
C. mona	0.30%				
C. ascanius	0.16% (1/86–1/88)	chimp crowned eagle	Taï Forest, Cote d'Ivoire Kibale NP, Uganda	Boesch & Boesch–Achermann,2000 Struhsaker & Leakey, 1990	minimum % killed/year (separated by male & female. I added the numbers of each sex & divided by the minimum # of kills for each sex)
Erythrocebus patas	34% (1992–2001)	leopard, cheetah	Segera Ranch, Kenya	Isbell & Enstam, 2002	Overall percent of all adult females that died/disappeared
MANGABEYS					
Cercocebus atys	0.40%	chimp crowned eagle	Taï Forest, Coted'Ivoire Kibale NP, Uganda	Boesch & Boesch–Achermann,2000 Struhsaker & Leakey, 1990	Minimum % killed/year (separated by male & female. I added the numbers of each sex & divided by the minimum # of kills for each sex)
Lophocebus albigena	2.78% (1/86–1/88)				
BABOONS					
Papio ursinus	8% (9/77–2/80)	leopard, lion	Moremi, Botswana	Busse, 1980	

Predation risk may be even more difficult to measure. Frequency of attempted predation, both successful and unsuccessful, can provide a reasonable estimate (Hill & Lee, 1998), but observers rarely witness predation attempts, successful or not. Estimating the risk of predation for cercopithecines and colobines is complicated because a number of factors may influence predation risk, including predator species (Isbell, 1990) and density (Stanford, 1995), prey preferences of individual predators (Kruuk, 1986; Kruuk & Turner, 1967; Isbell, 1990; Boesch, 1994; Cowlishaw, 1994; Stanford et al., 1994; Stanford, 1996), prey body weight or age (Struhsaker & Leakey, 1990; Boesch, 1994; Isbell, 1994; Stanford et al., 1994; Hill & Dunbar, 1998; Mitani & Watts, 1999), prey group size (Crook & Gartlan, 1966; Clutton-Brock & Harvey, 1977; van Schaik & van Noordwijk, 1985; but see also Isbell, 1994), proximity to humans (Isbell & Young, 1993), and habitat structure (Crook & Gartlan, 1966).

Importance of Habitat Structure

Aspects of habitat structure that may affect predation risk of cercopithecines include access to refuges (Stacey, 1986; Cowlishaw, 1997b; Hill & Weingrill, 2006; this volume), tree height (Boesch, 1994), and degree of obstructive cover (Altmann & Altmann, 1970; Rasmussen, 1983; Cowlishaw, 1994, 1997a; Hill & Weingrill, 2006; this volume). It has been assumed that predation risk is greater in savannahs than in rainforests because savannahs provide fewer refuges (e.g., trees) from predators (Crook & Gartlan 1966; Clutton-Brock & Harvey 1977; Dunbar 1988). Although this assumption is now being challenged (Isbell, 1994; Olupot & Waser, 2001), even within the same broad type of habitat (e.g., "savannah," "woodland," "rainforest") more subtle differences in structure may also influence primates' perceived risk of predation. For example, in savannahs few trees and short grass may actually lower predation risk because such habitats provide terrestrial stealth predators with little cover from which to hunt (Isbell, 1994; FitzGibbon & Lazarus, 1995). In contrast, savannah areas with more trees and tall grass may be riskier because terrestrial predators are provided with more cover for ambushes (Kruuk & Turner, 1967; Altmann & Altmann, 1970; Rasmussen, 1983; Isbell, 1994). This means that terrestrial cercopithecines may sometimes be at greater risk of predation from terrestrial predators, at least during the day, when they are on the ground nearer trees than when they are farther away. Subtle differences in habitat structure of forests can have similar effects on predation risk. For example, reduced canopy cover or tree height can increase risk of predation on red colobus monkeys (*Procolobus badius*) by chimpanzees (*Pan troglodytes*) (Boesch, 1994; Stanford, 1995).

Such subtleties suggest that it is no longer useful to identify predation risk simply by ecosystem type. Rather, looking more carefully at habitat structure within ecosystems may reveal more meaningful patterns. To illustrate the importance of habitat structure on the perceived risk of predation and anti-predator behavior of cercopithecines I focus now on the relationship between the structure of open *Acacia* woodland habitat and the behavior of two cercopithecine primates, vervet and patas (*Erythrocebus patas*) monkeys. Although the data presented below have been presented separately elsewhere (Enstam, 2002; Enstam & Isbell, 2002, 2004), I combine them here to illustrate the importance of examining *multiple* aspects of habitat structure (e.g., tree height, canopy cover, *and* ground cover) for their potential effects on anti-predator behavior and perceived risk of predation.

Methods

The Study Species and the Study Site

I studied one group of patas monkeys and two groups of vervet monkeys between October 1997 and September 1999 on Segera Ranch in central Kenya. During the study period the patas group declined in size from 51 to 20 individuals. Much the decline was associated with illness following unusually heavy El Niño rains (Isbell & Young, in preparation). The two vervet groups also declined during the study, from 30 to 9 and 10 to 5 individuals, respectively, and in June 1999 these two groups fused into one. The decline in the vervet group sizes was largely the result of suspected and confirmed predation (Isbell & Enstam, 2002). A detailed description of the data collection methods and statistical results are provided in Enstam (2002) and Enstam & Isbell (2002, 2004).

Patas monkeys are highly terrestrial primates that live in grassland and open woodland habitat below the Sahara Desert from northwest Senegal through Sudan to eastern Ethiopia, northern Uganda, central Kenya, and northern Tanzania (Isbell, submitted) and they possess a number of anatomical adaptations for cursorial locomotion, including long limbs (Hurov, 1987; Strasser, 1992; Gebo & Sargis, 1994) and digitigrade feet (Meldrum, 1991; Gebo & Sargis, 1994). The home range of the patas study group was about 4 km from the home range of the vervet study groups. Vervets are also highly terrestrial, although they spend more time in trees than do patas monkeys (Chism & Rowell, 1988) and do not possess the extreme cursorial adaptations of patas monkeys (Strasser, 1992; Gebo & Sargis, 1994). Vervet monkeys occupy savannah woodland habitats and are patchily distributed along waterways throughout the woodlands of sub-Saharan Africa (Wolfheim, 1983; Isbell & Enstam, submitted).

Vervet and patas monkeys are ideal subjects upon which to pursue a study of the effects of habitat structure on perceived risk of predation. First, they are closely related, thereby minimizing confounding factors resulting from different phylogenetic histories. This provides a clearer picture of the effects of ecology on their behavior. Indeed, recent studies suggest that vervets and patas are more closely related to one another than either is to any other cercopithecine (Groves, 1989, 2000; Disotell, 1996, 2000). Second, except for adult males, patas and vervets overlap in body size (adult female vervets—weight: 2.5–5.3 kg; length, excluding tail: 40—61 cm; adult female patas—weight: 4–7.5 kg; length,

TABLE 15.2. Signs of potential predators from November 1997– August 1999 in the home ranges of the study groups of vervet and patas monkeys (from Isbell & Enstam, 2002, reprinted with permission of Cambridge Univ. Press).[a] "Direct observations" indicate sightings made by observers. "Indirect observations" indicate sightings based on tracks, dung, and reliable cattle herders; [b] confirmed predator of vervets at Segera Ranch (martial eagle) or another site (baboon: Struhsaker, 1967c; Altmann & Altmann, 1970; Hausfater, 1976; Seyfarth et al., 1980b; Cheney & Sayfarth, 1981; leopard: Struhsaker, 1967c; Seyfarth et al., 1980b, martial eagle: Struhsaker, 1967c; Seyfarth et al., 1980b); [c] confirmed predator of patas at Segera Ranch (black-backed jackal) or another site (domestic dogs: Chism & Rowell, 1988); [d] numbers indicate the number of individual direct and indirect observations of predators. A "zero" indicates no observations during the study period.

Predator Species	Vervet Home Ranges		Patas Home Range	
	Direct obs.[a]	Indirect obs.[a]	Direct obs.	Indirect obs.
African wildcat (*Felis libyca*)	1[d]	0	10	0
Baboons (*Papio anubis*)[b]	8	0	28	0
Black-backed jackal (*Canis mesomelas*)[c]	3	0	93	1
Caracal (*F. caracal*)	0	0	2	0
Cheetah (*Acinonyx jubatus*)	4	1	3	0
Domestic dog (*C. familaris*)[c]	2	0	27	0
Leopard (*Panthera pardus*)[b]	3	5	0	0
Lion (*P. leo*)	1	3	4	18
Martial eagle (*Polemaetus bellicosus*)[b]	2	0	2	0
Serval (*F. serval*)	2	0	0	0
Spotted hyena (*Crocuta crocuta*)	0	4	0	3
Total	**26**	**13**	**169**	**22**

excluding tail: 50–60 cm; Haltenorth & Diller, 1980), and are thus (theoretically) vulnerable to predation from the same predators, reducing the likelihood that differences in anti-predator behavior are related to inherent differences in vulnerability. Third, at the Segera Ranch study site in Laikipia, Kenya, they share the same ecosystem (*Acacia* woodland) and, therefore, the same community of predators (Enstam & Isbell, 2002; Isbell & Enstam, 2002; Table 15.2), again reducing the chances that observed differences in behavior are due to differences in the predators that each study group encounters.

Habitat Structure of the Study Site

Although vervet and patas monkeys occupy the same ecosystem (i.e., open *Acacia* woodland), there are two habitat types at the Segera Ranch study site. While patas are found only in non-riverine habitat, vervets use both riverine and non-riverine habitats, sleeping in riverine habitat at night but foraging in both riverine and non-riverine habitats during the day. These two habitats differ in several aspects of habitat structure, including tree height and canopy cover, which appear to affect the animals' perceived risk of predation as well as their responses to both

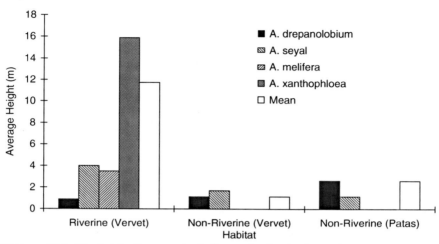

FIGURE 15.1. The height (in meters) of all trees in the riverine and non-riverine habitats. *Acacia melifera* did not occur in any transects in the patas home range, and *Acacia xanthophloea* did not occur in any transects in the non-riverine habitat (from Enstam & Isbell, 2002, reprinted with permission of Wiley-Liss)

nocturnal and diurnal predators. Areas along the river (i.e., riverine habitat) are dominated by *A. xanthophloea* (fever trees), while areas away from rivers (i.e., non-riverine habitat) are dominated by *A. drepanolobium* (whistling thorn acacias) (Enstam & Isbell, 2002). In addition to differing in species composition, the riverine and non-riverine habitats differ in structure since *A. xanthophloea* and *A. drepanolobium* differ in size and structure. Specifically, the average height (Figure 15.1) and degree of canopy cover of trees in the riverine habitat are significantly greater than those in the non-riverine habitat (Enstam & Isbell, 2002).

In addition, within the non-riverine habitat, there are two microhabitats that differ most obviously in height. I use the term "microhabitat" to refer to areas within the same general habitat type (i.e., non-riverine) that differ in key structural features, such as tree height (see Enstam & Isbell, 2004). Average tree height in the tall microhabitat was significantly taller than average tree height in the short microhabitat (Enstam & Isbell, 2004).

Results

Sleeping Site Choice

For cercopithecines and colobines at risk of predation by nocturnal predators, sleeping site selection may be an important anti-predator strategy and sleeping sites that afford as much security from terrestrial predators as possible should be preferred (Anderson, 1984). Some research on sleeping site selection has shown that distribution and availability of food (e.g., chacma baboons (*P. ursinus*), cf.

Hamilton, 1982; bonnet macaques (*M. radiata*), cf. Rahaman & Parthasarathy, 1969; black and white colobus (*Colobus guereza*), cf. von Hippel, 1998) presence of water (e.g., rhesus macaques (*M. mulatta*), cf. Lindburg, 1971), or conspecific groups (black and white colobus, cf. von Hippel, 1998) can influence the location of sleeping sites; but other research has indicated that selection of sleeping sites may also afford anti-predator benefits. For example, bonnet macaques and black and white colobus both prefer to sleep high up in tall trees with few or no low branches, apparently because such trees reduce access by terrestrial predators (von Hippel, 1998; Ramakrishnan & Coss, 2001). Safe, elevated sleeping sites are so important for some cercopithecines that, in some cases, their social systems have adapted to take advantage of the best possible sleeping sites. The fission-fusion social system of hamadryas baboons (*Papio hamadryas*), for example, which occupy the highlands of Ethiopia, Somalia, and Saudi Arabia, is apparently designed for life in an environment with few trees and low food abundance. In this species, multiple one-male units converge together at night in the form of a troop at one of the few safe sleeping sites in their habitat: sheer cliff faces (Kummer, 1968, 1995; Swedell, 2006).

Although closely related, patas and vervet monkeys display strikingly different sleeping site preferences and behaviors, which appear to be due in large part to differences in habitat structure, and studies of patas monkey sleeping site habits suggest that their dispersed sleeping patterns in both time and space may be an adaptation to avoid predation by nocturnal predators where only small trees exist (Chism et al., 1983). Vervets do not require such adaptations because they rely on much taller sleeping trees, which may reduce the risk of predation from terrestrial predators (see Anderson, 1984). For example, at night, patas scatter over a wide area and sleep singly in trees (Hall, 1965; Chism et al., 1983; Enstam, pers. obs.), whereas entire groups of vervets often sleep in the same tree (Struhsaker, 1967; Enstam, pers. obs.). Moreover, patas rarely sleep in the same trees on consecutive nights (Chism et al., 1983; Enstam, pers. obs.), whereas vervets frequently sleep in the same trees on multiple consecutive nights (Struhsaker, 1967, Enstam, pers. obs.). Also, unlike vervet monkeys (and other cercopithecoids), who typically give birth at night, patas monkeys typically give birth during the day, which may further reduce predation at night (Chism et al., 1983).

Diurnal Anti-Predator Behavior

Like sleeping site behavior, the diurnal behavior of vervet and patas monkeys in the presence of predators can be strikingly different and, I argue, is attributable to differences in habitat structure. In open woodland habitat trees are valuable refuges from terrestrial predators once primates get into them (Stacey, 1986; Cowlishaw, 1997b), but they vary in structure and height. Thus, some trees may be more effective refuges than others. Taller trees with overlapping canopies are expected to be more effective as refuges than shorter trees with little overlapping canopies because the former enable primates to get both up and away from terrestrial predators. Likewise, taller trees with thinner, more dense, and more vertical

TABLE 15.3. Responses of patas and vervet monkeys, excluding infants, to mammalian predator alarm calls (from Enstam & Isbell, 2002, reprinted with permission of Wiley-Liss). Each response was counted only once in analyses, regardless of the number of animals displaying that response; [a]includes arboreal scanning only and alarm calling while scanning arboreally; [b]numbers indicate the number of observations of the different types of anti-predator responses. A "zero" indicates the behavior was not observed; "N/A" indicates that the behavioral response is not applicable for the specific substrate.

| Response | Vervets | | | | Patas | |
| | Riverine habitat | | Non-riverine habitat | | Non-riverine habitat | |
	In tree	On ground	In tree	On ground	In tree	On ground
Arboreal scan[a]	36[b]	N/A	3	N/A	3	N/A
AC only	0	0	0	0	0	0
Climb tree	N/A	0	N/A	2	N/A	3
None	4	0	0	0	2	5
Descend, run	0	N/A	2	N/A	7	N/A
Run away	N/A	0	N/A	1	N/A	7
Bipedal scan	N/A	0	N/A	1	N/A	10
Climb & scan	N/A	3	N/A	1	N/A	6
Active defense	0	0	0	0	0	1
Total	**40**	**0**	**5**	**5**	**22**	**32**

branches might be expected to be more effective as refuges than smaller trees with thicker, less dense, and more horizontal branches because the latter may be more accessible to mammalian predators that can climb trees. In Amboseli National Park, Kenya, for example, leopards were found more often in umbrella trees (*Acacia tortilis*), which are shorter with thicker and less angled branches than fever trees (*A. xanthophloea*), the other available tree species (Isbell, pers. comm.).

At the Segera Ranch site, the same vervet study group used both the riverine and non-riverine habitats. Since the habitats differ in key structural features, I was able to examine how differences in tree height and canopy cover between the two *Acacia* woodland habitats affect the responses of vervet monkeys to alarm calls at mammalian predators and to compare their responses to those of the patas study group. Earlier studies of vervet responses to "leopard" alarm calls at Amboseli showed that vervets respond by climbing into tall trees, or remaining in them and not descending (Seyfarth et al., 1980). My research at the Segera Ranch study site supported these results, as the vervets responded as "typical vervets" to mammalian predator alarm calls (Enstam & Isbell, 2002; Table 15.3). Such behavior is a good strategy to avoid attack by terrestrial predators in riverine habitat, where trees are quite tall (11.8 m, on average at the Segera Ranch study site) (Enstam & Isbell, 2002).

Patas monkeys, on the other hand, respond quite differently to mammalian predators. While their primary response is to scan the environment (apparently in an attempt to locate the stimulus of the alarm call), their secondary response differs depending on the substrate they are occupying at the time of the alarm call. If they are in trees at the time of a mammalian predator alarm call, patas monkeys will descend the tree they are in and run away. Although this behavior is not part of the "typical" vervet repertoire in riverine habitat, when they were in the non-riverine habitat arboreal vervet monkeys were observed to descend and run away during a mammalian predator alarm call (Table 15.3). Similarly, while patas monkeys on the ground at the time of a mammalian predator alarm call will climb trees and scan the environment, just as vervets do (Table 15.3), they will also engage in behavior not seen in the typical vervet anti-predator repertoire. Specifically, patas monkeys will scan bipedally from the ground, run away (past the nearest trees without climbing them), or engage in "active defense" by attacking the predator if it is close to the group (Enstam & Isbell, 2002; Table 15.3).

Even though some behaviors exhibited by patas monkeys are not displayed by vervets in their "typical" (riverine) habitat does not mean that the vervets' anti-predator response repertoire is inflexible. Just as vervets (and patas) respond differently to different *types* of alarm calls (denoting different predators with different hunting techniques: Seyfarth et al., 1980; Enstam, pers. obs.), vervets also alter their repertoire of responses depending on habitat type. In the non-riverine habitat, vervets respond with patas-like behaviors, including descending trees, bipedally scanning, and running away from predators, rather than climbing the nearest tree (Enstam & Isbell, 2002; Table 15.3).

Height is not the only aspect of tree structure that may play a role in affecting perceived risk of predation and responses to predators. Tree density is significantly higher in the non-riverine habitat of both the vervet home ranges (Figure 15.2), but degree of canopy cover is lower. Degree of canopy cover is often an important measure for determining abundance of food resources (e.g., Chapman et al., 1994; Pruetz & Isbell, 2000; Wieczkowski, 2004), but it may also be important in terms of providing viable escape routes for monkeys under

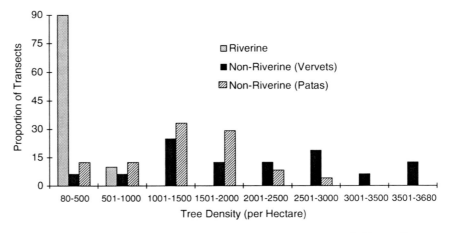

FIGURE 15.2. Tree density (in hectares) of riverine and non-riverine habitats. The non-riverine habitat has greater variation in tree density and greater average tree density (from Enstam & Isbell, 2002, reprinted with permission of Wiley-Liss)

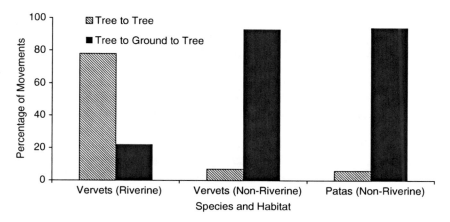

FIGURE 15.3. Percentage of movements between trees in which the focal animal remained arboreal (tree to tree movements) or descended one tree before climbing the next tree (tree to ground to tree movements) (from Enstam & Isbell, 2002, reprinted with permission of Wiley-Liss)

threat of predation. Because I was interested in discovering which habitat afforded greater opportunities to remain arboreal during a predator attack, I used a behavioral measure of canopy cover, namely, movements between trees by focal animals (Enstam & Isbell, 2002). Analysis of movements between trees indicates that the riverine habitat has more continuous canopy cover than the non-riverine habitat, because vervets in the former habitat were able to move between trees without descending significantly more often than the same group of vervets in the non-riverine habitat (Enstam & Isbell, 2002; Figure 15.3). This change in vervet behavior indicates that, at least to the monkeys, the non-riverine habitat has a more discontinuous canopy, and the results agree with data obtained from ecological measurements of average maximum crown diameter (Pruetz, 1999), which show that the canopy of the riverine habitat overlaps more extensively than the non-riverine habitat. When not under the threat of predation, vervets in the riverine habitat were able to remain arboreal significantly more often than vervets or patas in the non-riverine habitat. In the presence of mammalian predators this, in combination with the presence of appreciably taller trees, is especially significant because it means that vervets can increase their distance from predators both *vertically* (by climbing or remaining in tall trees) and *horizontally* (by moving between trees without descending) (see also Enstam & Isbell, 2002).

Such a strategy is unavailable in the non-riverine habitat because there the trees are short with discontinuous canopy cover, which makes them ineffective means by which an animal can increase vertical and horizontal distance from predators while remaining arboreal, particularly if those predators can climb trees (see also Enstam & Isbell, 2002). Monkeys may be able to avoid predation by mammalian predators that cannot effectively climb tall trees, such as cheetah (*Acinonyx jubatus*), if they can simply get high into tall trees before being attacked. "Tall" is the operative word, however, since lions (*P. leo*) are large enough to presumably push a small tree over or swat a monkey out of a short *Acacia drepanolobium* by standing bipedally. But vervets and patas do fall prey to mammalian predators, such as leopard, which can climb trees, and when under threat of predation by leopard, climbing a tree (even a tall tree) may not be sufficient. Under such circumstances, horizontal arboreal flight would be the best option, and this option is only available in the riverine habitat where canopy cover is relatively continuous.

Given the differences in tree structure between the two habitats it is not surprising that vervets adopt different anti-predator strategies in the non-riverine habitat; strategies that are comparable to those of patas monkeys. In a habitat filled with relatively short trees with little continuous canopy cover, the best anti-predator strategy appears to be to increase horizontal distance between oneself and one's predator as quickly as possible by running in the opposite direction of the predator, even if that means descending a tree one is occupying during the alarm call. This is exactly what both vervet and patas monkeys do in the non-riverine habitat.

These data on the same group of vervet monkeys indicate a number of important aspects about the effects of habitat structure on vervet anti-predator behavior that may also apply to other cercopithecine species. First, as with many other cercopithecine (and, indeed, primate) behaviors, the response of vervets to the threat of predation from terrestrial predators does not appear to be hard-wired, but rather is flexible and sensitive, both to the hunting strategy of the specific type of predator (Seyfarth et al., 1980), and the height and canopy cover of trees in the immediate habitat (Enstam & Isbell, 2002). While vervets in riverine habitat responded to mammalian predator alarm calls as "typical" vervets, when they were in the non-riverine habitat their responses were more similar to the responses of patas monkeys in the same habitat. In fact, in the non-riverine habitat, vervets responded to mammalian predator alarm calls with behaviors that were observed in patas in that habitat, but not in the same group of vervets in the riverine habitat. The change in the anti-predator behavior of the same group of vervet monkeys is apparently related to the limited number of refuges (i.e., tall trees with overlapping canopy) from large mammalian predators that exist in the *A. drepanolobium* habitat.

Even within the same habitat type, small differences in habitat structure can lead animals to prefer one microhabitat instead of another. The patas study group rarely entered the riverine habitat, using the non-riverine habitat almost exclusively (Enstam, pers. obs.; see also Chism & Rowell, 1988). But the non-riverine habitat is not uniform in structure. Rather, there is a very distinct and abrupt difference within the non-riverine habitat in tree height. During my two-year study the patas study group used 2,851 ha of their approximately 4,000-ha home range. Within these 2,851 ha the microhabitat with tall *A. drepanolobium* trees (hereafter called "tall microhabitat") comprised approximately 80% (2284 ha), while the microhabitat with apparently perennially short *A. drepanolobium* trees (hereafter called "short microhabitat") comprised approximately 20% (567 ha). A comparison of the number of observation days the patas group spent in each microhabitat relative to its size indicated that the patas group preferred the tall microhabitat, spending more days there than expected (Enstam & Isbell, 2004).

But what led the patas group to prefer the tall microhabitat? Among cercopithecines, habitat preference may be related to differences in resource availability (Clutton-Brock, 1975; Gautier-Hion et al., 1981; Harrison, 1983; Olupot et al., 1997) or predation risk and avoidance (Treves, 1997; Hill & Weingrill, 2006; this volume), or monkeys may attempt to trade off these variables by preferring microhabitats that provide either more resources but with greater risk of predation or greater safety with fewer resources (Cowlishaw, 1997a).

The patas monkeys appeared to prefer the tall microhabitat for the greater number of taller than average trees found there. In the tall microhabitat, focal animals were found in trees that were, on average 4.6 ± 0.16 m (range: 3.1–5.7 m) in height, and they climbed into trees that were significantly taller than the average height of trees in the non-riverine habitat (Enstam & Isbell, 2004; Figure 15.4). Moreover, when focal animals were in trees taller than average tree height (>3 m), they were found high up in the trees, at higher-than-average tree height (Enstam & Isbell, 2004; Figure 15.4). Finally, height of focal animals was correlated with tree height in the tall microhabitat (Figure 15.5), suggesting that the animals climbed as high into trees as the trees would allow (Enstam & Isbell, 2004).

Patas monkeys at the Segera study site obtain the majority of their food (83%) from *A. drepanolobium* trees (Isbell, 1998), with swollen thorns making up the main part of their diet (Isbell, 1998; Pruetz & Isbell, 2000). However, swollen thorns do not appear to be less available to patas monkeys in the short

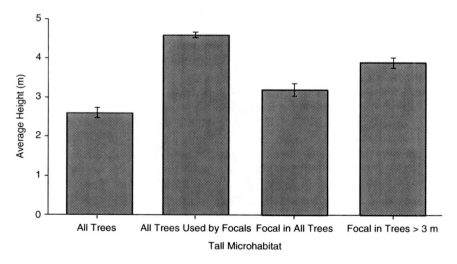

FIGURE 15.4. Average height of (a) trees (see Enstam & Isbell, 2002), (b) trees into which focal animals climbed, (c) focal animals in all trees they climbed, and (d) focal animals in trees >3.0 m tall in the microhabitat. Bars represent one standard error (from Enstam & Isbell, 2004, reprinted with permission of Karger)

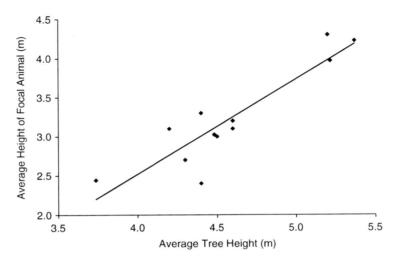

FIGURE 15.5. Correlation between average tree height and average focal animal height in the tall microhabitat (from Enstam & Isbell, 2004, reprinted with permission of Karger)

microhabitat. First, the density of *A. drepanolobium* trees did not differ between the short and tall microhabitats (Enstam & Isbell, 2004). Second, swollen thorns are found on all *A. drepanolobium* trees, regardless of height (Isbell, 1998). Third, patas monkeys typically feed on only 1–2 swollen thorns per tree due to the fact that the ants (*Crematogaster spp.*) that live on the *A. drepanolobium* trees defend the trees by biting intruders (Isbell, 1998; Isbell et al., 1998; Madden and Young, 1992; Young et al., 1997). Thus, both short and tall trees provide patas monkeys with as many swollen thorns as the monkeys can tolerate taking. Finally, the monkeys seem to prefer feeding in short trees (average feeding height for swollen thorns: 75 cm) (Pruetz, 1999), perhaps because feeding from the ground reduces the harassment by the ants that live on the trees.

Instead, it appears that patas preferred the tall microhabitat for its predation avoidance benefits. Focal animals spent more time scanning from taller than average trees (>3 m), and spent less time feeding and foraging there. Scanning from tall trees was also useful in detecting predators. In six focal samples for which the height of the focal animal was recorded while it gave an alarm call, patas monkeys were in trees that were significantly taller than average tree height, and they were significantly higher in these trees than the average tree height would have allowed (Enstam & Isbell, 2004). In five of the six cases, the focal animals were within a half-meter from the top of the tree while giving an alarm call (Enstam & Isbell, 2004).

Although access to resources is important, these results suggest that the microhabitat preference of patas monkeys at Segera Ranch is based more on predator detection, and the group minimized their used of a microhabitat within their home range despite the potential opportunity to feed longer or more efficiently on short trees that predominate there. As stated earlier in this chapter, the majority of responses by cercopithecoids to predator presence involves fleeing (e.g., Seyfarth et al., 1980; van Schaik et al., 1983; Cheney & Seyfarth, 1990; Isbell, 1994; Iwamoto, 1993; Boesch, 1994; Bshary & Noë, 1997; Boesch & Boesch–Achermann, 2000; Ramakrishnan and Coss, 2000; Bshary, 2001; Enstam and Isbell, 2002), and the effectiveness of that flight may be greatly enhanced by early detection. It is in the tall, rather than the short, microhabitat that patas monkeys gain a predator detection advantage based on their ability to scan from taller than average trees (Enstam & Isbell, 2004).

Trees are only one aspect of habitat structure that affects predation risk, however. Ground cover is also important to consider because ground cover affects the visibility of both predators and prey. Prey species that rely heavily on concealment to avoid being detected by predators often use protective cover to avoid detection (e.g., Lloyd et al., 2000; Fisher & Goldizen, 2001). On the other hand, species like diurnal cercopithecines that are unable to hide from predators rely on detecting the predator before being detected themselves. Cercopithecines may use a number of different strategies to increase their ability to detect predators, including living in (larger) groups, forming mixed-species associations, and using areas with reduced ground cover (e.g., Struhsaker, 1981; Rasmussen, 1983). Areas with reduced ground cover also provide the additional benefit of reducing hunting success in many mammalian predators that rely on cover to get as close to their prey as possible before attacking (e.g., Schaller, 1972; Eaton, 1974; van Orsdol, 1984; Bothma et al., 1994; Caro, 1994; Cowlishaw, 1994).

Because both vervet and patas monkeys are highly terrestrial (Chism & Rowell, 1998; Enstam & Isbell, 2002; Isbell, submitted), effects of ground cover may be a significant aspect of habitat structure affecting their perceived risk of predation and anti-predator behavior. After a wildfire burned the ground cover of a significant portion of the home range of a vervet study group, the vervets ranged significantly farther from the core of their home range along the river, moving into the burned area, where they had never been observed to go before the fire occurred (Enstam, 2002). Three lines of evidence indicate that the vervets entered the burned area because they perceived a lower risk of predation there. First, the wildfire significantly reduced grass cover and enabled the vervets to see significantly farther while on the ground in the burned area, potentially increasing their ability to detect predators at a greater distance (Enstam, 2002). Second, the burned area was safer because it contained fewer mammalian predators and poisonous reptiles for vervet monkeys to encounter (Enstam, 2002). Finally, female vervets scanned bipedally less often in the burned area than in the unburned areas (Enstam, 2002). Among cercopithecines, rates of scanning have been correlated with predation risk (e.g., Cowlishaw, 1998), so the reduced rate of scanning in vervets suggests that they felt less threatened in the burned area (Enstam, 2002).

Conclusions

Predation has been argued to have exerted strong selection pressures on primates, favoring a number of behavioral and morphological traits (e.g., Busse, 1977; Harvey & Kavanagh, 1978; Struhsaker, 1981; Gautier-Hion et al., 1983; Terborgh & Janson, 1986; Cords, 1987; Cheney & Wrangham, 1987; Henzi, 1988; van Schaik, 1983; Isbell et al., 1990; Isbell, 1994; van Schaik & Hörstermann, 1994; Plavcan & van Schaik, 1994; Struhsaker, 2000). Cercopithecines, like other primates, display a variety of behaviors in response to the threat of predation, including alarm calls and the formation of polyspecific associations. These responses are not always successful, however, and many cercopithecines, including vervet and patas monkeys, can suffer relatively high mortality due to predation (Isbell, 1990;

FIGURE 15.6. Evidence of suspected predation attempt on an adult female patas monkey. Note the healed wounds that resemble scratches on her right hip (photo: K.L. Enstam)

Chism et al., 1984; Isbell & Enstam, 2002; Table 15.2). Despite these sometimes high mortality rates, not all predation attempts are successful (Figure 15.6). It is possible that primates can extract themselves from death by responding appropriately based on both the hunting strategy of the predator and the structure of habitat they are in when the predator attacks. Vervet and patas monkeys exemplify the ability of cercopithecines to respond to predation risk with flexibility, altering their behavior when changes in risk occur. Predation risk and predation rate can be difficult variables to quantify. Since predation risk depends greatly on habitat structure, studies of the effects of *multiple aspects* of habitat structure on primate behavior (e.g., Boesch, 1994; Stanford, 1995; Cowlishaw, 1997a,b; Enstam & Isbell, 2002, 2004) promise to help primate behavioral ecologists gain a better understanding of the variables that affect predation risk, and ultimately, predation rate.

Acknowledgments. I would like to thank Sharon Gursky and Anna Nekaris for inviting me to contribute to *Primate Anti-Predator Strategies*. I thank the Office of the President, Republic of Kenya, for permission to conduct field research in Kenya; J. Mwenda, Acting Director of the Institute of Primate Research, for local sponsorship; and J. Ruggieri and J. Gleason, the owners of Segera Ranch, and P. Valentine, the manager, for logistical support and permission to work on Segera Ranch. I thank L. A. Isbell and three anonymous reviewers for providing valuable comments on an earlier draft of this chapter. This chapter was written in part while the author was supported by a Sonoma State University School of Social Sciences Summer Research Grant.

References

Alberts, S.C., and Altmann, J. (1995). Balancing costs and opportunities: Dispersal in male baboons. *Amer. Nat.,* 145: 279–306.

Altman, J. (1980). *Baboon mothers and infants.* Cambridge, MA: Harvard Univ. Press.

Altmann, J. (1990). Primate males go where the females are. *Animal Behav.,* 34: 193–194.

Altmann, S.A. (1974). Baboons, space, time, and energy. *Amer. Zool.,* 14: 221–248.

Altmann, S.A. and Altmann, J. (1970). *Baboon ecology.* Chicago: Univ. of Chicago Press.

Andelman, S.J. (1986). Ecological and social determinants of Cercopithecine mating patterns. In D.I. Rubestein and R.W. Wrangham (Eds.), *Ecological aspects of social evolution* (pp. 201–216). Princeton, NJ: Princeton Univ. Press.

Anderson, J.R. (1984). Ethology and ecology of sleep in monkeys and apes. *Adv. Stud. Behav.,* 14: 165–229.

Bagchi, S., Goyal, S.P., Sankar, K. (2003). Prey abundance and prey selection by tigers (*Panthera tigris*) in a semi-arid, dry deciduous forest in western India. *Jour. Zool., Lond.,* 260: 285–290.

Baldellou, M., and Henzi, S.P. (1992). Vigilance, predator detection and the presence of supernumary males in vervet monkey troops. *Animal Behav.,* 43: 451–461.

Biquand, S., Biquand-Guyot, V., Boug, A., and Gautier, J.-P. (1992). The distribution of *Papio hamadryas* in Saudi Arabia: Ecological correlates and human influence. *Inter. Jour. Primatol.,* 13: 223–243.

Boesch, C. (1994). Chimpanzee-red colobus monkeys: A predator-prey system. *Animal Behav.*, 47: 1135–1148.

Boesch, C., and Boesch, H. (1989). Hunting behavior of wild chimpanzees in the Taï National Park. *Amer. Jour. Phys. Anthropol.*, 78: 547–573.

Boesch, C., and Boesch-Achermann, H. (2000). *The chimpanzees of the Taï forest: Behavioural ecology and evolution.* Oxford: Oxford Univ. Press.

Boshoff, A.F., Palmer, N.G., Avery, G., Davies, R.A.G., and Jarvis, M.J.F. (1991). Biogeographical and topographical variation in the prey of the black eagle in the Cape Province, South Africa. *Ostrich*, 62: 59–72.

Bothma, J. du P., van Rooyen, N., Theron, G.K., and leRiche, E.A.N. (1994). Quantifying woody plants as hunting cover for southern Kalahari leopards. *Jour. Arid Environ.*, 26: 273–280.

Bshary, R. (2001). Diana monkeys, *Cercopithecus diana*, adjust their anti-predator response behaviour to human hunting strategies. *Behav. Ecol. Sociobiol.*, 50: 251–256.

Bshary R, and Noë R. (1997). Red colobus and Diana monkeys provide mutual protection against predators. *Animal Behav.*, 54: 1461–1474.

Busse, C.D. (1977). Chimpanzee predation as a possible factor in the evolution of red colobus monkey social organization. *Evolution*, 31: 907–111.

Busse, C.D. (1980). Leopard and lion predation upon chacma baboons living in the Moremi Wildlife Reserve. *Bots. Notes Rec.*, 12: 15–21.

Caro, T.M. (1994). *Cheetahs of the Serengeti plains: Group living in an asocial species.* Chicago: Univ. of Chicago Press.

Chapman, C.A., and Chapman, L.J. (2000). Interdemic variation in mixed–species association patterns: Common diurnal primates of Kibale National Park, Uganda. *Behav. Ecol. Sociobiol.*, 47: 129–139.

Chapman, C.A., Wrangham, R., and Chapman, L.J. (1994). Indices of habitat-wide fruit abundance in tropical forests. *Biotropica*, 26: 160–171.

Cheney, D.L., and Seyfarth, R.M. (1981). Selective forces affecting the predator alarm calls of vervet monkeys. *Behaviour*, 76: 25–61.

Cheney, D.L., and Seyfarth, R.M. (1990). *How monkeys see the world: Inside the mind of another species.* Chicago: Univ. of Chicago Press.

Cheney, D.L., and Wrangham, R.W. (1987). Predation. In B.B. Smuts, D.L. Cheney, R.M. Seyfarth, R.W. Wrangham, and T.T. Struhsaker (Eds.), *Primate societies* (pp. 227–239). Chicago: Univ. of Chicago Press.

Cheney, D.L., Seyfarth, R.M., Andelman, S.J., and Lee, P.C. (1988). Reproductive success in vervet monkeys. In T.H. Clutton–Brock (Ed.), *Reproductive success* (pp. 384–402). Chicago: Univ. of Chicago Press.

Cheney, D.L., Seyfarth, R.M., Fischer, J., Beehner, J., Bergman, T., Johnson, S.E., Kitchen, D. M., Palobmit, R.A., Rendall, D., and Silk, J.B. (2004). Factors affecting reproduction and mortality among baboons in the Okavango Delta, Botswana. *Inter. Jour. Primatol.*, 25: 401–428.

Chism, J. and Rowell, T.E. (1988). The natural history of patas monkeys. In A. Gautier-Hion, F. Bourliere, J.-P. Gautier, and J. Kingdon (Eds.), *A primate radiation: Evolutionary biology of the African guenons* (pp. 412–438). Cambridge: Cambridge Univ. Press

Chism, J., Olson, D.K., and Rowell, T.E. (1983). Diurnal births and perinatal behavior among wild patas monkeys: Evidence of an adaptive pattern. *Inter. Jour. Primatol.*, 4: 167–184.

Chism J., Rowell T.E., and Olson D. (1984). Life history patterns of female patas monkeys. In M.F. Small (ed.), *Female primates: Studies by women Primatologists* (pp. 175–190). New York: Alan R. Liss, Inc.

Clutton-Brock, T.H. (1975). Ranging behaviour of red colobus (*Colobus badius tephrosceles*) in the Gombe National Park. *Animal Behav.*, 23: 706–722.

Clutton-Brock, T.H. and Harvey, P.H. (1977). Primate ecology and social organization. *Jour. Zool. Soc. London*, 183: 1–39.

Condit, V.K., and Smith, E.O. (1994). Predation on a yellow baboon (*Papio cynocephalus cynocephalus*) by a lioness in the Tana River National Primate Reserve, Kenya. *Amer. Jour. Primatol.*, 33: 57–64.

Colyn, M. (1994). Données pondérales sur les primates Cercopithecidae d'Afrique Central (Bassin du Zaire/Congo). *Mammalia*, 58: 483–487.

Cords, M. (1987). *Mixed-species association of Cercopithecus monkeys in the Kakamega forest, Kenya.* Berkeley: Univ. of California Publications in Zoology.

Cords, M. (1990). Vigilance and mixed-species association of some East African forest monkeys. *Behav. Ecol. Sociobiol.*, 26: 297–300.

Cords, M. (2002). Foraging and safety in adult female blue monkeys in the Kakamega forest, Kenya. In L.E. Miller (Ed.), *Eat or be eaten: Predator sensitive foraging among primates* (pp. 205–221). Cambridge: Cambridge Univ. Press.

Cowlishaw, G. (1994). Vulnerability to predation in baboon populations. *Behaviour*, 131: 293–304.

Cowlishaw, G. (1997a). Trade-offs between foraging and predation risks determine habitat use in a desert baboon population. *Animal Behav.*, 53: 667–686.

Cowlishaw, G. (1997b). Refuge use and predation risk in a desert baboon population. *Animal Behav.*, 54: 241–253.

Cowlishaw, G. (1998). The role of vigilance in the survival and reproductive strategies of desert baboons. *Behaviour.* 135: 431–452.

Crook, J.H., and Gartlan, J.S. (1966). Evolution of primate societies. *Nature*, 210: 1200–1203.

Disotell, T.R. (1996). The phylogeny of Old World monkeys. *Evol. Anth.*, 5: 18–24.

Disotell, T.R. (2000). The molecular systematics of the Cercopithecidae. In P.F. Whitehead and C.J. Jolly (Eds.), *Old World monkeys* (pp. 29–56). Cambridge: Cambridge Univ. Press.

Dunbar, R.I.M. (1988). *Primate social systems.* London: Croom Helm.

Eaton, R.L. (1974). *The cheetah: The biology, ecology and behavior of an endangered species.* Malabar, FL: R.E. Krieger Publishing Co.

Edgaonkar, A., and Chellam, R. (2002). Food habit of the leopard, *Panthera pardus*, in the Sanjay Gandhi National Park, Maharashtra, India. *Mammalia*, 66: 353–360.

Enstam, K.L. (2002). Behavioral ecology of perceived risk of predation in sympatric patas (*Erythrocebus patas*) and vervet (*Cercopithecus aethiops*) monkeys in Laikipia, Kenya. Doctoral dissertation. University of California, Davis.

Enstam, K.L.. and Isbell, L.A. (2002). Comparison of responses to alarm calls by patas (*Erythrocebus patas*) and vervet (*Cercopithecus aethiops*) monkeys in relation to habitat structure. *Am. Jour. Phys. Anth.*, 119: 3–14.

Enstam, K.L., and Isbell, L.A. (2004). Microhabitat preference and vertical use of space by patas monkeys (*Erythrocebus patas*) in relation to predation risk and habitat structure. *Folia Primatol.*, 75: 70–84.

Enstam, K.L., and Isbell, L.A. (2007). The guenons (Genus: *Cercopithecus*) and their allies: Behavioral ecology of polyspecific associations. In C.J. Campbell,

A. Fuentes, K.C. MacKinnon, M.A. Panger, and S.K. Bearder *Primates in perspective* (pp. 252–273). Oxford: Oxford Univ. Press.

Fa, J.E. (1989). The genus *Macaca*: A review of taxonomy and evolution. *Mammal. Rev.*, 19: 45–81.

Fisher, D.O., and Goldizen, A.W. (2001). Maternal care and infant behaviour of the bridled nailtail wallaby (*Onychogalea fraenata*). *Jour. Zool.*, 255: 321–330.

Fittinghoff, N.A., Jr., and Lindburg, D.G. (1980). Riverine refuging in East Bornean *Macaca fascicularis*. In D.G. Lindburg (Ed.), *The macaques: Studies in ecology, behavior and evolution* (pp.182–214). New York: van Nostrand Reinhold Company.

FitzGibbon, C.D., and Lazarus, J. (1995). Antipredator behavior of Serengeti ungulates: Individual differences and population consequences. In A.R.E. Sinclair and P. Arcese (Eds.), *Serengeti II: Dynamics, management, and conservation of an ecosystem* (pp. 274–296). Chicago: Univ. of Chicago Press.

Gautier-Hion, A. (1971). L'ecologie du talapoin du Gabon. *Terre et Vie*, 25: 427–490.

Gautier-Hion, A., and Gautier, J.-P. (1976). Croissance, maturité sociale et sexuelle, reproduction chez les Cercopithecinés forestiers arboricoles. *Folia Primatol.*, 4: 103–118.

Gautier-Hion, A., and Tutin, C.E.G. (1988). Simultaneous attack by adult males of a polyspecific troop of monkeys against a crowned hawk eagle. *Folia Primatol.*, 51: 149–151.

Gautier-Hion, A., Gautier, J.-P., and Quris, R. (1981). Forest structure and fruit availability as complementary factors influencing habitat use by a troop of monkeys (*Cercopithecus cephus*). *Reveu d'Ecol.*, 35: 511–536.

Gautier-Hion, A., Quris, R., and Gautier, J.-P. (1983). Monospecific vs. polyspecific life: A comparative study of foraging and antipredatory tactics in a community of *Cercopithecus* monkeys. *Behav. Ecol. Sociobiol.*, 12: 325–335.

Gebo, D.L., and Sargis, E.J. (1994). Terrestrial adaptations in the postcranial skeletons of guenons. *Am. Jour. Phys. Anth.*, 93: 341–371.

Gevaerts, H. (1992). Birth seasons of *Cercopithecus, Cercocebus* and *Colobus* in Zaire. *Folia Primatol.*, 59: 105–113.

Goodall, J. (1986). *The chimpanzees of Gombe: Patterns of behavior*. Cambridge, MA: Harvard Univ. Press.

Groves, C.P. (1989). *A theory of human and primate evolution*. Oxford: Clarendon Press.

Groves, C.P. (2000). The phylogeny of the Cercopithecoidea. In P.F. Whitehead and C.J. Jolly, (Eds.), *Old World monkeys* (pp. 77–98). Cambridge: Cambridge Univ. Press.

Hall, K.R.L. (1965). Behaviour and ecology of the wild patas monkey, *Erythrocebus patas*, in Uganda. *Jour. Zool. Soc. London*, 148: 15–87.

Haltenorth, T., and Diller, H. (1980). *A field guide to the mammals of Africa including Madagascar*. London: William Collins, Sons and Co.

Hamilton, W.J., III. (1982). Baboon sleeping site preferences and relationships to primate grouping patterns. *Amer. Jour. Primatol.*, 2: 149–158.

Harrison, M.J.S. (1983). Patterns of range use by the green monkey, *Cercopithecus sabaeus*, at Mt. Assirik, Senegal. *Folia Primatol.*, 41: 157–179.

Harvey, P., and Kavanagh, M. (1978). Sexual dimorphism in primate teeth. *Jour. Zool. Lond.*, 186: 475–485.

Hausfater, G. (1976). Predatory behavior of yellow baboons. *Behaviour*, 61: 44–68.

Henzi, S.P. (1988). Many males do not a multimale troop make. *Folia Primatol.*, 51: 165–168.

Hill, R.A., and Cowlishaw, G. (2002). Foraging female baboons exhibit similar patterns of antipredator vigilance across two populations. In L.E. Miller (Ed.), *Eat or be eaten:*

Predator sensitive foraging among primates (pp. 187–204). Cambridge: Cambridge Univ. Press.

Hill, R.A., and Dunbar, R.I.M. (1998). An evaluation of the roles of predation rate and predation risk as selective pressures on primate grouping behaviour. *Behaviour*, 135: 411–430.

Hill, R.A., and Lee, P.C. (1998). Predation risk as an influence on group size in cercopithecoid primates: Implications for social structure. *J. Zool. Soc. Lond.*, 245: 447–456.

Hill, R.A., and Weingrill, T. (2006). Predation risk and habitat-specific activity patterns in chacma baboons (*Papio hamadryas ursinus*). In S. Gursky and K.A.I. Nekaris (Eds.), *Primates and their predators* (pp. 337–352). New York: Kluwer Academic Press.

Hoppe-Dominik, B. (1984). Etude du spectre des proies de la panthèra, *Panthera pardus*, dans le Parc National de Taï en Côte d'Ivoire. *Mammalia*, 48: 477–487.

Horn, A.D. (1987). The socioecology of the black mangabey (*Cercocebus aterriums*) near Lake Tumba, Zaire. *Am. J. Primatol.* 12: 165–180.

Hurvo, J.R. (1987). Terrestrial locomotion and back anatomy in vervets (*Cercopithecus aethiops*) and patas monkeys (*Erythrocebus patas*). *Am. J. Primatol.*, 13: 297–311.

Isbell, L.A. (1990). Sudden short-term increase in mortality of vervet monkeys (*Cercopithecus aethiops*) due to leopard predation in Amboseli National Park, Kenya. *Am. Jour. Primatol.* 21: 41–52.

Isbell, L.A. (1991). Contest and scramble competition: Patterns of female aggression and ranging behavior among primates. *Behav. Ecol.*, 2: 143–155.

Isbell, L.A. (1994). Predation on primates: Ecological patterns and evolutionary consequences. *Evol. Anth.*, 3: 61–71.

Isbell, L.A. (1998). Diet for a small primate: Insectivory and gummivory in the (large) patas monkey (*Erythrocebus patas pyrrhonotus*). *Amer. Jour. Primatol.*, 45: 381–398.

Isbell, L.A. (submitted). Species profile: Patas monkey (*Erythrocebus patas*). To appear In J. Kingdon, D. Happold, and T. Butynski (Eds.), *The mammals of Africa*. New York: Academic Press.

Isbell, L.A., Cheney, D.L., and Seyfarth, R.M. (1990). Costs and benefits of home range shifts among vervet monkeys (*Cercopithecus aethiops*) in Amboseli National Park, Kenya. *Behav. Ecol. Sociobiol.*, 27: 351–358.

Isbell, L.A., Cheney, D.L., and Seyfarth, R.M. (1991). Group fusions and minimum group sizes in vervet monkeys (*Cercopithecus aethiops*). *Amer. Jour. Primatol.*, 25: 57–65.

Isbell, L.A., Cheney, D.L., and Seyfarth, R.M. (1993). Are immigrant vervet monkeys (*Cercopithecus aethiops*) at greater risk of mortality than residents? *Animal Behav.*, 45: 729–734.

Isbell, L.A., Cheney, D.L., and Seyfarth, R.M. (2002). Why vervets (*Cercopithecus aethiops*) live in multimale groups. In M. Glenn and M. Cords (Eds.), *The Guenons: Diversity and adaptation African monkeys* (pp. 173–187). New York: Kluwer Academic/Plenum Publishers.

Isbell, L.A., and Enstam, K.E. (2002). Predator-(in)sensitive foraging in sympatric female vervets (*Cercopithecus aethiops*) and patas monkeys (*Erythrocebus patas*): A test of ecological models of group dispersion. In L.E. Miller (Ed.), *Eat or be eaten: Predator sensitive foraging among primates* (pp. 154–168). Cambridge: Cambridge Univ. Press.

Isbell, L.A., and Enstam, K.L. (submitted). Species profile: Vervet monkey (*Cercopithecus pygerythrus*). To appear In J. Kingdon, D. Happold, and T. Butynski (Eds.), *The mammals of Africa*. New York: Academic Press.

Isbell, L.A., Pruetz, J., Lewis, M., and Young, T.P. (1998). Locomotor activity differences between sympatric vervet monkeys (*Cercopithecus aethiops*) and patas monkeys

(*Erythrocebus patas*): Implications for the evolution of long hindlimb length in *Homo. Am. Jour. Phys. Anthropol.*, 105: 199–207.

Isbell, L.A., and Young, T.P. (1993). Human presence reduces predation in a free-ranging vervet monkey population in Kenya. *Animal Behav.*, 45: 1233–1235.

Iwamoto, T. (1993). The ecology of *Theropithecus gelada*. In N.G. Jablonski (Ed.), *Theropithecus: The rise and fall of a primate genus* (pp. 441–452). Cambridge: Cambridge Univ. Press.

Iwamoto, T., Mori, A., Kawai, M., and Bekele, A. (1996). Anti-predator behavior of Gelada baboons. *Primates*, 37: 389–397.

Janson, C.H. (2000). Primate socio-ecology: The end of a golden age. *Evol. Anth.*, 9: 73–86.

Janson, C.H., and Goldsmith, M.L. 1995. Predicting group size in primates foraging costs and predation risks. *Behav. Ecol.*, 6: 326–336.

Karanth, K.U., and Sunquist, M.E. (1995). Prey selection by tiger, leopard and dhole in tropical forests. *Jour. Animal Ecol.*, 64: 439–450.

Keverne, E.B., Leonard, R.A., Scruton, D.M., and Young, S.K. (1978). Visual monitoring in social groups of talapoin monkeys (*Miopithecus talapoin*). *Animal Behav.*, 26: 933–944.

Kruuk, H. (1986). Interactions between Felidae and their prey species: A review. In S.D. Miller and D. D. Everett (Eds.), *Cats of the world: Biology, conservation, and management* (pp. 343–374). Washington, DC: National Wildlife Federation.

Kruuk, H., and Turner, M. (1967). Comparative notes on predation by lion, leopard, cheetah and wild dog in the Serrengeti area, East Africa. *Mammalia*, 31: 1–27.

Kummer, H. (1968). *Social organization of hamadryas baboons*. Chicago: Univ. of Chicago Press.

Kumer, H. (1995). *In quest of the sacred baboon: A scientist's journey*. Princeton, NJ: Princeton Univ. Press. N.J.

Leutenegger, W., and Cheverud, J. (1982). Correlates of sexual dimorphism in primates: Ecological and size variables. *Inter. Jour. Primatol.*, 3: 387–403.

Lindburg, D.G. (1971). The rhesus monkey in North India: An ecological and behavioral study. In L.A. Rosenblum (Ed.), *Primate behavior: Developments in field and laboratory research* (pp. 1–106). New York: Academic Press.

Lindburg, D.G. (1977). Feeding behaviour and diet of rhesus monkeys (*Macaca mulatta*) in a Siwalik forest in North India. In T.H. Clutton-Brock (Ed.), *Primate ecology: Studies of feeding and ranging behaviour in lemurs, monkeys, and apes* (pp. 223–249). London: Academic Press, Inc.

Lloyd, P., Plaganyi, E., Lepage, D., Little, R.M., and Crowe, T.M. (2000). Nest-site selection, egg pigmentation and clutch predation in the ground-nesting Namaqua Sandgrouse *Pterocles namaqua*. *Ibis*, 142: 123–131.

Madden, D., and Young, T.P. (1992). Symbiotic ants as an alternative defense against giraffe herbivory in spinescent *Acacia drepanolobium*. *Oecologia*, 91: 235–238.

McGraw, W.G., and Bshary, R. (2002). Association of terrestrial mangabeys (*Cercocebus atys*) with arboreal monkeys: Experimental evidence for the effects of reduced ground predator pressure on habitat use. *Inter. Jour. Primatol.*, 23: 311–325.

Meldrum, D.J. (1991). Kinematics of the cercopithecine foot on arboreal and terrestrial substrates with implications for the interpretation of hominid terrestrial adaptations. *Amer. Jour. Phys. Anth.*, 84: 273–290.

Miller, L.E. (2002). An introduction to predator sensitive foraging. In L.E. Miller (Ed.), *Eat or be eaten: Predator sensitive foraging among primates* (pp. 1–17). Cambridge: Cambridge Univ. Press.

Mitani, J.C., and Watts, D.P. (1999). Demographic influences on the hunting behavior of chimpanzees. *Amer. Jour. Phys. Anth.*, 109: 439–454.

Mitani, J.C., Gros-Louis, J., and Manson, J.H. (1996). Number of males in primate groups: comparative tests of competing hypotheses. *Amer. Jour. Primatol.*, 38: 315–332.

Mitani, J.C., Sanders, W.J., Lwanga, J.S., and Windfelder, T.L. (2001). Predatory behavior of crowned hawk–eagles (*Stephanoaetus coronatus*) in Kibale National Park, Uganda. *Behav. Ecol. Sociobiol.*, 49: 187–195.

Nagel, U. (1973). A comparison of anubis baboons, hamadryas baboons, and their hybrids at a species border in Ethiopia. *Folia Primatol.*, 19: 104–165.

Napier, P.H. (1981). *Catalogue of primates in the British Museum (Natural History), part 2: Family Cercopithecidae, Subfamily Cercopithecinae.* London: British Museum (Natural History).

Napier, P.H. (1985). *Catalogue of primates in the British Museum (Natural History), part 3: Family Cercopithecidae, Subfamily Colobinae.* London: British Museum (Natural History).

Napier, J.R., and Napier, P.H. (1967). *A handbook of living primates: Morphology, ecology and behaviour of nonhuman primates.* New York: Academic Press.

Noë R., and Bshary R. (1997). The formation of red colobus–Diana monkey associations under predation pressure from chimpanzees. *Proc. Roy. Soc. Lond. B.*, 246: 253–259.

Oates, J.F., Whitesides, G.H., Davies, A.G., Waterman, P.G., Green, S.M., Desilva, G., and Mole, S. (1990). Determinants of variation in tropical forest primate biomass: New evidence from West Africa. *Ecology*, 71: 328–343.

Olupot, W., and Waser, P.M. (2001). Activity patterns, habitat use and mortality risks of mangabeys males living outside social groups. *Animal Behav.*, 61: 1227–1235.

Olupot, W. Chapman, C.A., Waser, P.M., and Isabirye-Basuta, G. (1997). Mangabey (*Cercocebus albigena*) ranging patterns in relation to fruit availability and the risk of parasite infection in Kibale National Park, Uganda. *Amer. Jour. Primatol.*, 43: 65–78.

Plavcan J.M., and van Schaik, C.P. (1994). Canine dimorphism. *Evol. Anth.*, 2: 208–214.

Pruetz, J.D. (1999). Socioecology of adult female vervet (*Chlorocebus aethiops*) and patas monkeys (*Erythrocebus patas*) in Kenya: Food availability, feeding competition, and dominance relationships. Doctoral dissertation. University of Illinois at Urbana–Champaign.

Pruetz, J.D., and Isbell, L.A. (2000). Correlations of food distribution and patch size with agonistic interactions in female vervets (*Chlorocebus aethiops*) and patas monkeys (*Erythrocebus patas*) living in simple habitats. *Behav. Ecol. Sociobiol.*, 49: 38–47.

Rahaman, H., and Parthasarathy, M.D. (1969). Studies on the social behaviour of bonnet monkeys. *Primates*, 10: 149–162.

Ramakrishnan, U., and Coss, R.G. (2000). Recognition of heterospecific alarm vocalizations by bonnet macaques (*Macaca radiata*). *Jour. Comp. Psych.*, 114: 3–12.

Ramakrishnan, U., and Coss, R.G. (2001). Strategies used by bonnet macaques (*Macaca radiata*) to reduce predation risk while sleeping. *Primates*, 42: 193–206.

Rasmussen, D.R. (1983). Correlates of patterns of range use of a troop of yellow baboons (*Papio cynocephalus*). II. Spatial structure, cover density, food gathering, and individual behaviour patterns. *Animal Behav.*, 31: 834–856.

Ridley, M. (1986). The number of males in a primate troop. *Animal Behav.*, 34: 1848–1858.

Sanders, W.J., Trapani, J., and Mitanti, J.C. (2003). Taphonomic aspects of crowned hawk-eagle predation on monkeys. *J. Human Evol.*, 44: 87–105.

Schaller, G.B. (1972). *The Serengeti lion: A study of predator-prey relations*. Chicago: Univ. of Chicago Press.

Seyfarth, R.M., and Cheney, D.L. (1990). The assessment by vervet monkeys of their own and another species' alarm calls. *Animal Behav.*, 40: 754–764.

Seyfarth, R.M., Cheney, D.L., and Marler, P. (1980). Vervet monkey alarm calls: Semantic communication in a free-ranging primate. *Animal Behav.*, 28: 1070–1094.

Shultz, S. (2001). Notes on interactions between monkeys and African crowned eagles in Taï National Park, Ivory Coast. *Folia Primatol.*, 72: 248–250.

Shultz, S., Noë, R., McGraw, W.S., and Dunbar, R.I.M. (2004). A community-level evaluation of the impact of prey behavioural and ecological characteristics on predator diet composition. *Proc. Roy. Soc. Lond. B.*, 271: 725–732.

Sigg, H. (1980). Differentiation of female positions in hamadryas one-male units. *Zeit. für Tierpsychol.*, 53: 265–302.

Skorupa, J.P. (1989). Crowned eagles *Stephanoaetus coronatus* in rainforest: Observations on breeding chronology and diet at a nest in Uganda. *Ibis*, 131: 294–298.

Stacey, P.B. (1986). Group size and foraging efficiency in yellow baboons. *Behav. Ecol. Sociobiol.*, 18: 175–187.

Stanford, C.B. (1995). The influence of chimpanzee predation on group size and anti-predator behaviour in red colobus monkeys. *Animal Behav.*, 49: 577–587.

Stanford, C.B. (1996). The hunting ecology of wild chimpanzees: Implications for the behavioral ecology of Pliocene hominids. *Am. Anth.*, 98: 96–113

Stanford, C.B. (1998). *Chimpanzee and red colobus: The ecology of predator and prey*. Cambridge, MA: Harvard Univ. Press.

Stanford, C.B. (2002). Avoiding predators: Expectations and evidence in primate antipredator behavior. *Inter. Jour. Primatol.*, 23: 741–757.

Stanford, C.B., Wallis, J., Matama, H., and Goodall, J. (1994). Patterns of predation by chimpanzees on red colobus monkeys in Gombe National Park, Tanzania, 1982–1991. *Am. Jour. Phys. Anth.*, 94: 213–228.

Steenbeek, R., Piek, R.C., van Buul, M., and van Hooff, J.A.R.A.M. (1999). Vigilance in wild Thomas' langurs (*Presbytis thomasi*): The importance of infanticide risk. *Behav. Ecol. Sociobiol.*, 45: 137–150.

Sterk, E.H.M. (2002). Predator sensitive foraging in Thomas langurs. In L.E. Miller (Ed.), *Eat or be eaten: Predator sensitive foraging among primates* (pp. 74–91). Cambridge: Cambridge Univ. Press.

Strasser, E. (1992). Hindlimb proportions, allometry, and biomechanics in Old World monkeys (Primates, Cercopithecidae). *Amer. Jour. Phys. Anth.*, 87: 187–214.

Struhsaker, T.T. (1967). Behavior of vervet monkeys (*Cercopithecus aethiops*). *University of California Publications in Zoology*, 82: 1–74.

Struhsaker, T.T. (1981). Polyspecific associations among tropical rain–forest primates. *Zeit. für Tierpscych.*, 57: 268–304.

Struhsaker, T.T. (2000). The effects of predation and habitat quality on the socioecology of African monkeys: Lessons from the islands of Bioko and Zanzibar. In P.F. Whitehead and C.J. Jolly (Eds.), *Old World monkeys* (pp. 393–430). Cambridge: Cambridge Univ. Press.

Struhsaker, T.T., and Leakey, M. (1990). Prey selectivity by crowned hawk-eagles on monkeys in the Kibale forest, Uganda. *Behav. Ecol. Sociobiol.*, 26: 435–443.

Swedell, L. (2006). *Strategies of sex and survival in hamadryas baboons: Through a female lens.* Upper Saddle River, NJ: Pearson/Prentice Hall.

Terborgh, J., and Janson, C. H. (1986). The socioecology of primate groups. *Ann. Rev. Ecol. and System*, 17: 111–135.

Tilson, R.L. (1977). Social organization of Simakobu monkeys (*Nasalis concolor*) in Siberut Island, Indonesia. *Jour. Mammal.*, 58: 202–212.

Treves, A. (1997). Vigilance and use of micro-habitat in solitary rainforest animals. *Mammalia*, 61: 511–525.

Treves, A. (1998). The influence of group size and near neighbors on vigilance in two species of arboreal primates. *Behaviour*, 135: 1–29.

Treves, A. (1999). Within-group vigilance in red colobus and redtail monkeys. *Amer. Jour. Primatol.*, 48: 113–126.

van Orsdol, K.G. (1984). Foraging behaviour and hunting success of lions in Queen Elizabeth National Park, Uganda. *Afr. Jour. Ecol.*, 22: 79–99.

van Schaik, C.P. (1983). Why are diurnal primates living in groups? *Behaviour*, 87: 120–144.

van Schaik, C.P. and Hörstermann, M. (1994). Predation risk and the number of adult males in a primate group: a comparative test. *Behav. Ecol. Sociobiol.*, 35: 261–272.

van Schaik, C.P., and Kappeler, P.M. (1997). Infanticide risk and the evolution of male-female association in primates. *Proc. Soc. Lond.*, 264: 1687–1694.

van Schaik, C.P., and van Noordwijk, M.A. (1985). Evolutionary effects of the absence of felids on the social organization of macaques on the island of Simeulue (*Macaca fascicularis fusca*, Miller 1903). *Folia Primatol.*, 44: 138–147.

Vermeij, G.J. (1982). Unsuccessful predation and evolution. *Amer. Nat.*, 120: 701–720.

von Hippel, F.A. (1998). Use of sleeping trees by black and white colobus monkeys (*Colobus guereza*) in the Kakamega Forest, Kenya. *Amer. Jour. Primatol.* 45: 281–290.

Wahome, J.M., Rowell, T.E., and Tsingalia, H.M. (1993). The natural history of de Brazza's monkey in Kenya. *Int. Jour. Primatol.*, 14: 445–466.

Wieczkowski, J. (2004). Ecological correlates of abundance in the Tana mangabey (*Cercocebus galeritus*). *Amer. Jour. Primatol.*, 63: 125–138.

Wolfheim, J.H. (1983). *Primates of the world: Distribution, abundance, and conservation.* Seattle: Univ. of Washington Press.

Wrangham, R.W. (1980). An ecological model of female-bonded primate groups. *Behaviour*, 75: 262–300.

Wrangham, R.W. (1983). Ultimate factors determining social structure. In R.A. Hinde (Ed.), *Primate social relationships* (pp. 255–262). Oxford: Blackwell Scientific Publications.

Wrangham, R.W., and Riss, E. (1990). Rates of predation on mammals by Gombe chimpanzees, 1972–1975. *Primates*, 31: 157–170.

Young, T.P., Stubblefield, C., and Isbell L.A. (1997). Ants on swollen-thorn acacias: Species coexistence in a simple system. *Oecologia*, 109: 98–107.

Zeeve, S.R. (1991). *Behavior and ecology of primates in the Lomako forest, Zaire.* Doctoral thesis. State University of New York, Stony Brook.

Zinner, D., and Pelaez, F. (1999). Verreaux's eagles (*Aquila verreauxi*) as potential predators of hamadryas baboons (*Papio hamadryas hamadryas*) in Eritrea. *Amer. Jour. Primatol.*, 47: 61–66.

Zuberbühler, K. (2001). Predator-specific alarm calls in Campbell's monkeys, *Cercopithecus campbelli*. *Behav. Ecol. Sociobiol.*, 50: 414–422.

Zuberbühler, K., and Jenny, D. (2002). Leopard predation and primate evolution. *Jour. Human Evol.*, 43: 873–886.

Zuberbühler, K., Jenny, D., and Bshary, R. (1999). The predator deterrence function of primate alarm calls. *Ethology*, 105: 477–490.

Zuberbühler, K., Noë, R., and Seyfarth, R.M. (1997). Diana monkey long–distance calls: Messages for conspecifics and predators. *Animal Behav.*, 53: 589–604.

16
Predation Risk and Habitat Use in Chacma Baboons (*Papio hamadryas ursinus*)

Russell A. Hill and Tony Weingrill

Introduction

Research into the importance of predation is underrepresented in primatology. Furthermore, the literature that does exist has produced inconsistent results. In part this reflects the difficulty one encounters in estimating predation pressure in natural environments. Here we present an introduction into how predation risk might be estimated in terrestrial environments, and we employ this model to explore patterns of habitat use in chacma baboons. The results suggest that baboons respond behaviorally to habitat-specific levels of predation risk, even in a low predator-density environment. This idea suggests that researcher primate perceptions of predation are not to be simply equated with what is observed in the predator-prey interactions; it is the breaking down of the predation process that offers considerable scope for understanding the impact of predation on primate behavior.

Predation pressure has long been assumed to be a powerful selective force on primate sociality (Alexander, 1974; van Schaik, 1983; Dunbar, 1988; Hill & Dunbar, 1998; Janson, 1998, 2003), although attempts at establishing its importance have reached contradictory conclusions. While some studies have reported positive relationships between group size and predation (Anderson, 1986; Dunbar, 1988; van Schaik & Hörsterman, 1994; Hill & Lee, 1998; Zuberbühler & Jenny, 2002), others have reported negative relationships (Isbell, 1994; Shultz et al., 2004), while still others have reported the absence of any relationship at all (Cheney & Wrangham, 1987). Much of the confusion in this debate, however, stems from a conflation of the effects of *predation risk* with *predation rate* to the extent that many of these studies have addressed fundamentally different aspects of predation (Hill & Dunbar, 1998).

Hill & Dunbar (1998) argued that "predation risk" and "predation rate" were separate elements of predation that generated disparate predictions about primate behavioral responses to the threat of predation. Observed predation rates in natural population *reflect* net predation risk *after* animals have invested in risk-reducing behavior; predation risk (or "*intrinsic predation risk*") (see Janson 1998), in contrast, represents "the probability that an animal living on its own and exercising no behavioral anti-predator strategies will succumb to a predator within a given time period" (Hill & Dunbar, 1998, p. 413). In essence, predation risk is an animal's own perception of the likelihood of that it will be subject to an attack by a predator, and it is this that acts as both the proximate and ultimate constraint on primate behavior. Although in this sense the definition is somewhat abstract, it nevertheless suggests that an individual's predation risk (as in its perception of that risk) is likely to be closely linked to its local environment. The challenge, therefore, is to identify the parameters within a primate's environment that contribute to its perception of the level of intrinsic predation risk.

The Dynamics of Predation Risk in Terrestrial Environments

Predation events are complex sequential dynamic processes that comprise a number of constituent elements (Lima & Dill, 1989; Endler, 1991). For the purposes of this chapter it is convenient to consider "predation risk" as a sequence of four components (Figure 16.1): (i) predator encounter; (ii) predator attack; (iii) prey capture; and (iv) individual capture probability. These categories are not necessarily exclusive and may overlap to a considerable extent. The order in which they are presented is also unlikely to be an accurate reflection of biological reality; predators are expected to select their target prey prior to attack and capture. Nevertheless, since here we are not interested in a specific individual per se, but rather in how predation risk averages out over individuals in a group, the current order at least provides a useful starting point. Thus, while it is recognized that the subdivision of predation risk into constituent components is not without its limitations, it is probably the most appropriate basis for a study of this type.

Probability of Encounter

Few studies are able to obtain accurate encounter rates between primates and their major predators, particularly since the presence of observers may in fact reduce the

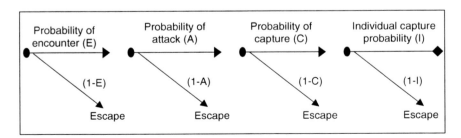

FIGURE 16.1. Schematic representation of individual predation risk in terrestrial primates. Predation risk is a sequence of four components: (i) predator encounter; (ii) predator attack; (iii) prey capture; and (iv) individual capture probability. Prey animals should attempt to interrupt the process as early as possible

frequency of predator-prey interactions (Isbell, 1991). Across populations, however, predator density is likely to provide a good working estimate of the frequency of interactions (Abrams, 1994; van Schaik & Hörsterman, 1994). While data on the density of predators are also limited, previous studies across populations and species have found positive relationships between primate group size and either estimated predator densities or categorical indices of the frequency of predator-prey relationships (Anderson, 1986; Hill & Lee, 1998; Hill, 1999). Within populations, predator habitat preferences are likely to provide a proxy for encounter rates. Leopards, for example, show preferences for dense vegetation and avoidance of open habitats (Bailey, 1993). Since leopards are the primary predators of baboons (Cowlishaw, 1994), it is likely that enclosed habitats present a high risk in relation to rates of encounters with predators.

Probability of Attack

Once a predator has encountered and detected potential prey, an attack decision is made based on the likelihood of attack success (although other factors such as hunger levels and the availability of alternative prey are also important) (Schoener, 1971; Elliot et al., 1977; van Orsdol, 1984). Attack success is determined by the probability that a predator is able to approach within a critical striking distance of the prey (Underwood, 1982; van Orsdol, 1984). Primates attempt to detect predators before they approach this critical distance through monitoring their local environment, although the effectiveness of such vigilance is restricted by habitat visibility. In general, attacks by leopards and lions, which are both ambush predators, are more successful as the degree of cover increases (van Orsdol, 1984; Bothma & Le Riche, 1986). These observations suggest that, for terrestrial primates at least, predation risk and the probability of predator attack should together be a positive function of the level of ground cover and thus be negatively related to habitat visibility.

Probability of Prey Capture

Once a predator attacks from ambush, prey will attempt to evade capture by trying to reach cover or a refuge (Dill & Houtman, 1990). Probability of escape is likely to be strongly influenced by the distance to a suitable refuge; this fact has been reflected in vigilance rates in baboons (Cowlishaw, 1998). Selective use of habitat may also be advantageous if the prey animal has greater agility over a substrate than its predator (Dunbar, 1986). In general, however, local refuge availability is likely to be the key parameter underlying the probability of prey capture for terrestrial primates.

Individual Capture Probability

The "individual capture probability" for an animal simply relates to the probability that, given a successful attack upon a group, a particular individual is the predated individual. This is a simple function of group size: the probability of an individual being the prey is a function of the number of individuals present, i.e., 1/N, where N is the prey group size; this is the *dilution effect* (Hamilton, 1971). Although it is counterintuitive to consider this the final stage of the predation process, since a predator will select its likely prey prior to attacking, it is useful to separate it from the other risk components since the level of risk is socially rather than environmentally determined. Across populations, therefore, group size (and thus individual capture probability) can be taken to indicate the degree to which individuals are responding to the perceived threat of predation. Observed variation in group size across populations should reflect the degree of risk resulting from the three habitat-related components (probability of encounter, probability of attack and probability of escape). Similarly, within populations, this approach allows us to examine the behavior of groups of different sizes in response to the habitat-specfic levels of predation risk in that environment.

Predation and Habitat Use in Baboons

Traditionally, studies of home range use in primates have tended to examine the relationship between intensity of habitat use and the spatial and temporal variability in food distribution. Several lines of evidence suggest, however, that predation risk may be an important influence on the ranging patterns of baboons. Stacey (1986) found that a small group at Amboseli remained in closer proximity to trees (refuges) than larger groups, and at Mkuzi the baboons tended to forage in areas of high tree density and avoid areas of tall grass where visibility was poor (Gaynor, 1994). In a more formal examination, Cowlishaw (1997a) found that at Tsaobis habitat use was influenced by group size, with smaller groups spending proportionally less time in high-risk habitats. There was also evidence that smaller troops spent more time on or close to refuges (Cowlishaw, 1997b).

Habitat choice may not be the only factor influenced by predation risk, and baboons appear to modify their behavior within habitats on the basis of the habitat-specific level of risk. At Tsaobis grooming and resting activity were almost exclusively restricted to the safest habitat, and moving was also predominantly confined to the safer areas (Cowlishaw, 1997a). Similarly, Stacey (1986) found that groups selected trees or other elevated areas while resting at Amboseli. In contrast, time spent in the high predation risk but high food availability habitat at Tsaobis was almost exclusively for feeding (Cowlishaw, 1997a). This suggests that where activities have no specific habitat requirements (such as sufficient food availability) these activities are conducted preferentially in low predation risk habitats.

The above studies provide general support that habitat characteristics—most notably visibility and refuge availability—influence perceived predation risk in baboons. The remainder of this chapter presents an examination of the role of predation risk in determining habitat use in a population of chacma baboons at De Hoop Nature Reserve, South Africa. In doing so, it further assesses the validity of the predation schema depicted in Figure 16.1 for determining habitat-specific predation risk.

TABLE 16.1. Home range composition, vegetation structure, habitat visibility, predation risk, and food availability of the major habitat types within the baboon home ranges at De Hoop.

Habitat Type	VT Home Range (%)	ST Home Range (%)	Bush Cover (%)	Tree Cover (%)	Visibility (m)	Visibility <10 m (%)	Predation Risk	Food Availability
Acacia woodland	16.0	11.2	55.8	34.4	4.6	90	high	high
Burnt acacia woodland	0.2	0	3.2	0.4	20.8	16.7	intermediate	low
Burnt fynbos	27.8	43.2	3.6	0.0	35.8	0	intermediate	low
Fynbos	25.9	32.0	54.0	3.4	13.7	72.5	high	low
Grassland	11.2	0.2	1.6	1.2	129.7	2.5	low	intermediate
Vlei	18.9	13.4	0.0	0.0	251.7	0	low	high
Cliffs	–	–	–	–	–	–	very low	very low

Methods

Study Population

De Hoop Nature Reserve, South Africa, is a coastal reserve situated close to Cape Agulhas, the southern tip of Africa. Vegetation on the reserve is classified as coastal fynbos, a unique and diverse vegetation type comprising Proteaceae, Ericaceae, Restionaceae, and geophyte species. Six distinct habitat types were classified on the basis of vegetation structure within the home range of the baboons (Table 16.1: see Hill (1999) for detailed descriptions and further information on the ecology of the reserve).

The data presented here were collected over a 10-month period (March to December 1997) from two groups of chacma baboons (*Papio hamadryas ursinus*): VT, which ranged in size from 40 to 44 individuals, and ST, which numbered 17 to 21 animals over the course of the study. Data were collected by means of instantaneous scan samples (Altmann, 1974) at 30-min intervals, with further 20-min focal samples with point samples at 2-min intervals collected for the all adults in VT. During 5 full day follows each month the position of the center of mass of the group (see Altmann & Altmann, 1970) was determined using a Magellan 4000 XL GPS for each scan sample. Cumulative home range areas and patterns of habitat use for the two troops were then established on the basis of the number and habitat composition of 4-ha quadrats entered by the groups during these day follows.

Results

Habitat-Specific Predation Risk at De Hoop

Although the *probability of encounter* with a predator is likely to be closely related to predator density across populations, within populations it is unlikely to be a constant factor since predator preferences for specific habitat types will undoubtedly be important. The distribution of leopard (*Panthera pardus*) extends well into the Western Cape and the Cape Agulhas region (Stuart et al., 1985), although leopard numbers have declined significantly over the last century (Norton, 1986). As a consequence the region is characterized by low leopard density. But while no leopards were known to be resident on De Hoop during the study, they were recorded soon after (Henzi et al., 2000). Although this suggests the local leopard population is migratory, passing through the study area on an intermittent basis, these transient individuals should nevertheless be associated with the densest vegetation when on the reserve. Probability of encounter should thus be greatest in habitats with highest bush level cover (see Table 16.1).

The *probability of attack* also increases in enclosed habitats because the decreased visibility increases the probability that a predator can stalk to within a critical attack distance. Attack distances for leopards are recorded to range from 5 m to 10 m (Kruuk & Turner, 1967; Bertram, 1982). As a consequence, the proportion of visibility below 10 m rather than visibility per se may be the important factor, since this is the critical distance that delimits susceptibility to leopard attacks. The visibility distances and the proportion of visibility below 10 m for each De Hoop habitat are given in Table 16.1.

We expect the *probability of capture* to related to the availability of refuges within each habitat. De Hoop is a relatively treeless environment (a characteristic of fynbos vegetation; Campbell et al., 1979; Moll et al., 1980), however, and thus none of the habitat types contained a significant number of trees of a sufficient height to operate as refuges. As a consequence, refuge density is not a habitat-specific parameter but instead relates to the availability of cliffs that are topographic features of the landscape and independent of habitat type. Refuge density is not habitat-specific, therefore, and the probability of prey capture can thus be considered relatively constant between habitat types.

Estimates of habitat-specific predation risk for six habitats with the baboon home ranges at the De Hoop were thus quantified on the basis of two parameters: probability of encounter and probability of attack (Table 16.1). In both cases, risk increased in more enclosed habitats, and vegetation structure (bush cover) and habitat visibility were closely linked (visibility: $r^2 = 0.816$, $F_{(1,5)} = 22.18$, $p = 0.005$; visibility < 10 m: $r^2 = 0.879$, $F_{(1,5)} = 36.16$, $p = 0.002$). In addition A seventh habitat, regenerating fynbos, is included in this analysis (visibility: 14.7 m; visibility < 10 m: 47.5%; bush cover: 30.6%; tree cover: 0.2 %). For comparison, information on habitat-specific food availability is also provided since this is likely to be another important factor in habitat choice, and it is clear that both predation risk and food availability vary considerably between habitats. It is important to remember, however, that for any habitat-specific predation risk, the risk per individual in ST will be absolutely much greater relative to individuals in VT. As a consequence, we would expect the behavior of individuals in ST to show elevated responses to predation risk relative to individuals in VT.

Predation Risk and Habitat Choice

In order to assess the relative preference for the different habitats by the two study troops, and thus determine the factors involved in habitat selection, it is important to control for the availability of the different habitat types. Monthly habitat preferences were therefore computed on the basis of Krebs' (1989) electivity index. The electivity index varies between +1 (strongly selected) and −1 (strongly avoided); it was calculated on the basis of the following formula:

$$EI = \frac{(h_i - p_i)}{(h_i + p_i)}$$

where EI is the electivity index, h_i is the observed proportion of time spent in habitat i, and p_i is the relative availability of habitat i in the home range of the troop. The mean electivity indices for the six habitat types are presented in Figure 16.2.

The relationships for three habitats are worth noting. As one would predict, both troops showed a clear preference for the high food availability, low predation risk vlei habitat. Similarly, both troops show avoidance of the low food availability, high predation risk climax fynbos. Interestingly, however, differences emerge when we consider acacia woodland, the high food availability but high predation risk habitat. While VT, the larger study group, showed a general preference for this habitat, ST avoided it. That individual capture probabilities were higher for the baboons in ST might suggest this group traded off food availability with predation risk in its patterns of habitat choice.

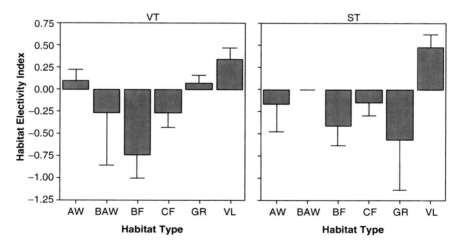

FIGURE 16.2. Mean monthly habitat electivity indices for each habitat type based upon the proportion of each habitat type within the home ranges of VT and ST. Data from quadrats containing sleeping sites are excluded to reduce artificial preferences for habitat types in close proximity to sleeping cliffs. AW: acacia woodland; BAW: burnt acacia woodland; BF: burnt fynbos; CF: climax fynbos; GR: grassland; VL: vlei

TABLE 16.2. Stepwise linear regression equations relating monthly habitat electivity indices to monthly food availability and habitat visibility. EI: electivity index; HF: habitat food availability (defined as the proportion of total home range food available in each habitat type); V: habitat visibility.

Troop	Equation	r^2	F	p
VT	EI $= 0.160 + 0.0795 \ln$ (HF)	0.463	(1, 40) 34.489	<0.001
ST	EI $= -0.275 + 0.132 \ln$ (HF) $+ 0.143 \ln$ (V)	0.388	(2, 27) 8.572	0.001

In order to determine whether this was the case, stepwise regression analysis was used to determine which factors best accounted for the observed monthly variation in habitat electivity indices. The best-fit models are given in Table 16.2. Both troops clearly show strong relationships between habitat electivity indices and relative food availability in that habitat. Furthermore, the coefficient is positive in both cases, indicating that high food availability habitats are more strongly selected relative to low food availability habitats. The best-fit model for ST also includes habitat visibility, suggesting that the baboons in this troop indeed traded food availability against predation and thus modified their habitat choice to minimize exposure to predation risk. Interestingly, if visibility is entered into the model for VT, a significant regression remains ($r^2 = 0.481$; $F_{(2,39)} = 18.103$; $p < 0.001$), but although the coefficient for visibility is in the predicted positive direction, it is not itself a significant component of the model.

Two possible explanations could explain these troop-specific patterns of ranging behavior and habitat choice. Firstly, because VT was approximately twice the size of ST, individuals in this troop did not experience the same level of individual capture probability and thus may not have perceived acacia woodland as high predation risk habitat in the same way members of ST did. Alternatively, it may be that while individuals in VT may have attempted to avoid high predation risk habitats, their higher foraging requirements may have constrained them to foraging in the high food availability habitats. As a consequence, VT may have had little latitude with respect to predation risk in terms of its patterns of habitat choice. Although the data are not available to explicitly test these hypotheses, the available evidence does lend some support to the latter explanation.

An anecdotal feature of the foraging strategy of VT while in acacia woodland was an apparent preference for feeding in trees on the edge of this habitat. Since high visibility grassland and vlei primarily surrounded acacia woodland, such a strategy would have increased visibility in some directions, thus reducing overall predation risk. Figure 16.3 displays the mean intensity of use of acacia woodland against the distance of that quadrat to the nearest quadrat containing either grassland or vlei habitat. It is clear that acacia woodland is used more intensively when in close proximity to high visibility habitats, and the differences between the distance categories are significant (ANOVA: $F_{(2,83)} = 6.028$, $p = 0.004$). Post hoc analysis reveals that it is the 0–200m category that differs significantly from the other two (Scheffé: 0–200 v 200–400, $p < 0.02$; 0–200 vs 400–600, $p < 0.025$). This relationship cannot be interpreted as merely a by-product of

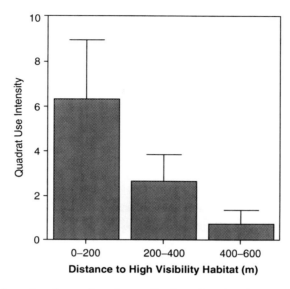

FIGURE 16.3. Intensity of use of acacia woodland quadrats in relation to the distance of this habitat from the nearest quadrat containing high visibility (low predation risk) habitat (either grassland or vlei)

declining intensity of habitat use with increasing distance away from sleeping sites and thus vlei habitat (which bordered many of the sleeping sites). A partial correlation controlling for distance to sleeping site maintains a strong negative correlation between intensity of acacia woodland use and distance to high visibility habitat ($r = -0.321$, df = 83, $p = 0.003$). This provides compelling support for the idea that while the baboons in VT may have been forced to feed in this high food availability habitat to satisfy their daily nutritional requirements, they were able to ameliorate their exposure to predation risk to a certain degree by preferentially using the habitat fringes. In doing so, the troop was able to maintain a higher level of visibility while feeding within the acacia woodland. The role of predation in shaping the habitat use of the larger study group, therefore, appears to operate at the microhabitat level.

Similar relationships appear to explain why the *climax fynbos* is not more strongly avoided by the baboons, despite the fact that it is of low food availability and high predation risk. Since a large proportion of sleeping sites were fringed by climax fynbos, the habitat could not be completely avoided. Nevertheless, it is clear that the intensity of climax fynbos use declines markedly with distance from sleeping site for both groups (Figure 16.4). Thus, while the baboons often needed to use this habitat in order to access the sleeping sites, they only did so when they were in close proximity to these refuges. For both troops, therefore, it is clear that predation shapes patterns of habitat use at both the habitat and microhabitat level.

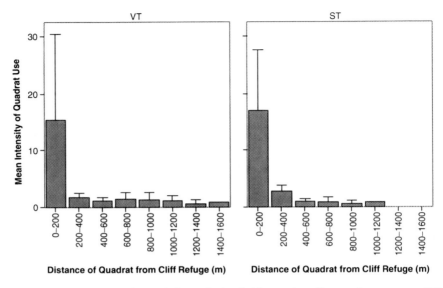

FIGURE 16.4. Intensity of use of climax fynbos habitat against distance from nearest cliff refuge for VT and ST

Predation Risk and Habitat Activity Patterns

Predation risk might not only shape how frequently certain habitats are used and the microhabitats selected within habitats types, but could also influence behavioral profiles within habitats. Cowlishaw (1997a) found that at Tsaobis, the baboons utilised certain habitats preferentially for certain activities. Grooming and resting activity were almost exclusively restricted to the safest habitat, while moving was also predominantly confined to the safer areas. On the other hand, time spent in the high predation risk but high food availability habitat was used almost exclusively for feeding. This suggests that where activities have no specific habitat requirements, such as sufficient food availability, then they are conducted preferentially in low predation risk habitats. Due to the wide variation in habitat visibility and food availability at De Hoop, we might thus expect to find differences in activity budgets between the various habitat types.

One issue that we need to address before habitat-specific patterns of activity can be assessed, however, is the fact that a large proportion of resting and grooming occurs on sleeping cliffs, which are not associated with any specific habitat type. As Cowlishaw (1997a) concedes, the apparent use of the Namib Hills habitat at Tsaobis for non-foraging activity may arise from the fact that these activities are conducted preferentially at dawn and dusk while the animals are on their sleeping cliffs. It is certainly true that the De Hoop baboons use cliffs almost exclusively for resting and grooming (Figure 16.5). As a consequence, it is important that activity profiles be assessed in a way that allows behavior on cliffs to be removed.

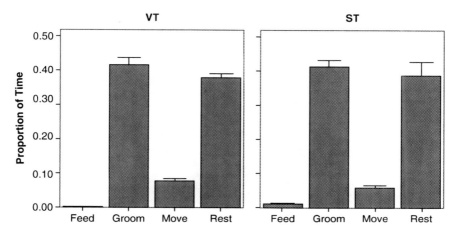

FIGURE 16.5. Mean time budgets while on cliffs for individuals in VT and ST

The importance of close cliff proximity on activity budgets can be seen in Figure 16.6, which displays the mean time budgets in the climax fynbos and vlei habitats for three distance to cliff categories: 0–10 m, 10–50 m, and 50 m+. Significant differences exist between distance categories for most of the activities in both habitats, although it is the non-foraging activities (grooming and resting) that show most pronounced effects. The relationships are most striking for grooming activity, particularly in climax fynbos where grooming represents almost sixty percent of activity within 10 m of a cliff. Such relationships do not appear to result from local resource depletion around sleeping sites causing animals to feed at greater distances from the refuge (and thus reducing feeding close to sleeping sites) since for fynbos in particular there is no relationship between proportion of time feeding and distance from sleeping site. It does appear, therefore, that where activities have no specific habitat requirements, such as sufficient food availability, then they are conducted preferentially in low predation risk habitats or in close proximity to refuges.

More complex analyses of habitat-specific activity patterns are complicated by the fact that, if percentage time budgets are considered, an increase in one activity must inevitably lead to an apparent decrease in other activities. It is thus difficult to determine whether high levels of a given behavior in a habitat reflect a preference for that activity or an avoidance of a different behavior. Furthermore, preferences for two activities are difficult to detect. It is also difficult to gauge what the expected level of activity should be in any habitat. One possible way to overcome this hurtle and assess preferences for conducting certain behaviors in different habitats is to restrict analyses to a single behavior. Feeding is a useful activity in this respect since we would expect the proportion of total feeding conducted in each habitat to correlate with the proportion of total home range food available within that habitat type. Figure 16.7 displays the mean preference for feeding

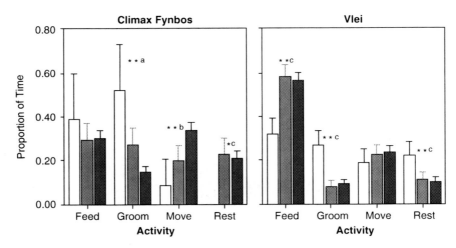

FIGURE 16.6. Comparison of the mean time budgets from focal samples for individuals in VT at various distances from the nearest cliff in the climax fynbos and vlei habitats. Open bars: 0–10 m; light bars: 10–50 m; dark bars: 50 m+. Asterisks indicate significant relationships (ANOVA: * $p < 0.05$; ** $p < 0.001$; Scheffé, $p < 0.05$: a: 0–10 v 10–50; 0–10 v 50+, 10–50 v 50+; b: 0–10 v 50+; 10–50 v 50+; c: 0–10 v 10–50; 0–10 v 50+)

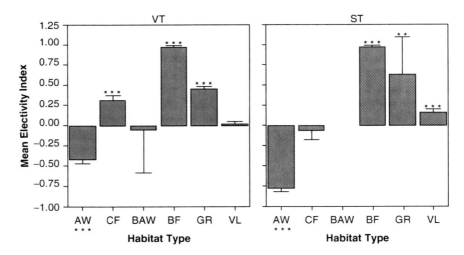

FIGURE 16.7. Mean individual electivity indices for feeding in each habitat type relative to food availability in these habitats. Asterisks indicate significant preferences or avoidances for feeding in that habitat (one-sample t-test: ** $p < 0.025$; *** $p < 0.001$). AW: acacia woodland; BAW: burnt acacia woodland; BF: burnt fynbos; CF: climax fynbos; GR: grassland; VL: vlei

in the different De Hoop habitats, where the electivity indices for each individual reflect the relative preference for feeding in each habitat against the baseline expected on the basis of food availability. The patterns are strikingly similar for both groups. Both troops show avoidance of feeding in high risk acacia woodland despite its high food availability, with preferences for feeding in the intermediate risk burnt fynbos and low risk grassland despite the lower food availabilities. ST also shows a significant preference for feeding in the high food availability, low risk vlei, although this preference is not significant for VT. While this again confirms a stronger predation response for ST due to higher individual capture probabilities, the fact that the preferences are not stronger for both groups probably reflects the fact that much of the food in this habitat is subterranean, with higher processing times and energetic costs of excavation. Finally, the apparent preference of individuals in VT for feeding in the low food availability, high risk climax fynbos almost certainly reflects the close proximity of this habitat to most of the sleeping sites within their home range. ST had fewer sleeping sites fringed by this habitat.

Discussion

Predation risk is clearly an important constraint on primate behavior, although it is essential we break down the predation process into its constituent parts in order to fully understand its effects. Interestingly, the results presented here suggest that the frequency of predator-prey interactions is just one element of an individual's perception of predation risk since the strong responses to predation risk at De Hoop occur in an environment of low predator density. This confirms that predation risk is not a simple function of the frequency of predator-prey interactions, and that, at the very least, evolved responses may persist in populations following local reduction or extinction of primary predators (Bouskila & Blumstein, 1992). As a consequence it is essential that other elements of predation risk be adequately quantified in order to fully understand the importance of behavior on current patterns of primate behavior.

For terrestrial primates it is clear that habitat visibility and the distribution of refuges are important elements of predation risk, since these factors determine the probability of encounter, attack, and successful prey capture in any habitat or population. The baboons at De Hoop show a general avoidance of high-risk habitats, with this pattern generally stronger for the smaller study group. Furthermore, even when higher risk habitats cannot be avoided, due to the presence of sleeping sites or the need to forage in areas of high food availability, predation shapes habitat choice at the microhabitat level. Use of climax fynbos declines markedly with distance from sleeping sites, suggesting that this habitat was used due only to its proximity to refuges, while VT used the habitat fringes when feeding in acacia woodland in order to maximize visibility. These results suggest that baboons are responding to subtle elements of their environment in gauging their current levels of predation risk.

Behavioral profiles within habitats are also modified on the basis of predation risk. Both resting and grooming are preferentially conducted on cliff refuges. In both vlei and climax fynbos, levels of resting and, in particular, grooming decline as distance to refuge increases. These relationships are unlikely to be an artifact of local resource depletion on foraging activity, but instead suggest an active preference for conducting these behaviors close to refuges. Similarly, the proportion of feeding activity conducted in different habitats is also influenced by predation. After controlling for availability of food in different habitats, the baboons show strong preferences for feeding in the low and intermediate risk habitats and avoidance of areas of high risk, even though high-risk habitats must be used to satisfy the troops' nutritional requirement. Overall, these results provide robust support for the idea that predation shapes the behavioral decisions of baboons, with clear responses evident in activity choice and habitat selection at both the habitat and microhabitat levels.

Predation risk is clearly a complex parameter, although through breaking down the predation process into a number of components significant progress in understanding the impact of predation on primate behavior. Since many of these components are features of the environment that can be readily quantified, this approach also allows us not only the opportunity to examine behavioral differences of groups between habitats within populations, but also to see how differences between populations may arise on the basis of local variation in predation risk. Hill & Cowlishaw (2002) illustrated that differences in vigilance levels between the De Hoop and Tsaobis baboon populations disappear once ecological differences between the populations are controlled for, suggesting that baseline anti-predator responses to predation risk are relatively consistent across populations. Breaking down the predation process thus affords us considerable scope for understanding the importance of predation risk in shaping patterns of primate behavior.

Acknowledgments. We thank Anna Nekaris and Sharon Gursky for the invitation to contribute to this volume and three anonymous referees for their helpful and constructive comments on an earlier version of the chapter. We are grateful to Louise Barrett, Peter Henzi, and Cape Nature Conservation for assistance and permission to work at De Hoop Nature Reserve.

References

Abrams, P.A. (1994). Should prey over estimate the risk of predation? *Am. Nat.*, 144: 317–328.
Alexander, R.D. (1974). The evolution of social behavior. *Ann. Rev. Ecol. Syst.* 5: 325–383.
Altmann, J. (1974). Observational study of behavior: sampling methods. *Behavior*, 49: 227–267.
Altmann, S.A., and Altmann, J. (1970). *Baboon ecology.* Chicago: Chicago Univ. Press.
Anderson, C.M. (1986). Predation and primate evolution. *Primates*, 27: 15–39.

Bailey, T.N. (1993). *The African leopard: A study of the ecology and behavior of a solitary felid*. New York: Columbia Univ. Press.

Bertram, B. (1982). Leopard ecology as studied by radio-tracking. *Symp. Zool. Soc. Lond.*, 49: 341–352.

du Bothma, J., and Le Riche, E.A.N. (1986). Prey preference and hunting efficiency of the Kalahari leopard. In S.D. Miller and D.D. Everett (Eds.), *Cats of the world: Biology, conservation and management* (pp. 389–414). Washington, DC: National Wildlife Federation.

Bouskila, A., and Blumstein, D.T. (1992). Rules of thumb for predation hazard assessment: Predictions from a dynamic model assessment. *Am. Nat.*, 139: 161–176.

Campbell, B.M., McKenzie, B., and Moll, E.J. (1979). Should there be more tree vegetation in the Mediterranean climate region of South Africa? *J. S. Afr. Bot.*, 45: 453–457.

Cheney, D.L., and Wrangham, R.W. (1987). Predation. In B.B. Smuts, D.L. Cheney, R. Seyfarth, R.W. Wrangham, and T.T. Struhsaker (Eds.), *Primate societies* (pp. 227–239). Chicago: Univ. of Chicago Press.

Cowlishaw, G.C. (1994). Vulnerability to predation in baboon populations. *Behavior*, 131: 293–304.

Cowlishaw, G.C. (1997a). Trade-offs between foraging and predation risk determine habitat use in a desert baboon population. *Anim. Behav.*, 53: 667–686.

Cowlishaw, G.C. (1997b). Refuge use and predation risk in a desert baboon population. *Anim. Behav.*, 54: 241–253.

Cowlishaw, G.C. (1998). The role of vigilance in the survival and reproductive strategies of desert baboons. *Behavior*, 135: 431–452.

Dill, L.M., and Houtman, R. (1989). The influence of distance to refuge on flight-initiation distances in the grey squirrel (*Sciurus carolinensis*). *Can. J. Zool.*, 67: 232–235.

Dunbar, R.I.M. (1986). The social ecology of gelada baboons. In D. Rubenstein and R.W. Wrangham (Eds.), *Ecological determinants of social evolution* (pp. 332–351). Princeton, NJ: Princeton Univ. Press.

Dunbar, R.I.M. (1988). *Primate social systems*. London: Chapman & Hall.

Elliot, J.P., McTaggart Cowan, I., and Holling, C.S. (1977). Prey capture by the African lion. *Can. J. Zool.*, 55: 1811–1828.

Endler, J.A. (1991). Interactions between predators and prey. In J.R. Krebs and N.B. Davies (Eds.), *Behavioral ecology: An evolutionary approach* (3rd ed.), (pp. 169–196). Oxford: Blackwell Scientific Publications.

Gaynor, D. (1994). Foraging and feeding behavior of chacma baboons in a woodland habitat. Doctoral thesis. University of Natal.

Hamilton, W.D. (1971). Geometry for the selfish herd. *J. Theor. Biol.*, 31: 295–311.

Henzi, S.P., Barrett, L., Weingrill, A., Dixon, P., and Hill, R.A. (2000). Ruths amid the alien corn: Males and the translocation of female chacma baboons. *S. Afr. J. Sci.*, 96: 61–62.

Hill, R.A. (1999). The ecological and demographic determinants of time budgets in baboons: Implications for the understanding of cross-populational models of baboon socioecology. Doctoral thesis. University of Liverpool.

Hill, R.A., and Cowlishaw, G. (2002). Foraging female baboons exhibit similar patterns of antipredator vigilance across two populations. In L.E. Miller (Ed.), *Eat or be eaten: Predator sensitive foraging among primates* (pp. 187–204). Cambridge: Cambridge Univ. Press.

Hill, R.A., and Dunbar, R.I.M. (1998). An evaluation of the roles of predation risk and predation rate as selective pressures on primate grouping behavior. *Behavior*, 135: 411–430.

Hill, R.A., and Lee, P.C. (1998). Predation risk as an influence on group size in Cercopithecoid primates. *J. Zool., Lond.*, 245: 447–456.

Isbell, L.A. (1991). Group fusions and minimum group sizes in vervet monkeys (*Cercopithecus aethiops*). *Am. J. Primatol.*, 25: 57–65.

Isbell, L.A. (1994). Predation on primates: Ecological patterns and evolutionary consequences. *Evol. Anthropol.*, 3: 61–71.

Janson, C.H. (1998). Testing the predation hypothesis for vertebrate sociality: Prospects and pitfalls. *Behavior*, 135: 289–410.

Janson, C.H. (2003). Puzzles, predation, and primates: Using life history to understand selection pressures. In P.M. Kappeler and M.E. Pereira (Eds.), *Primate life histories and socioecology* (pp. 103–131). Chicago: Univ. of Chicago Press.

Krebs, C. (1989). *Ecological methodology*. New York: Harper & Row.

Kruuk, H., and Turner, M. (1967). Comparative notes on predation by lion, leopard, cheetah and hunting dog in the Serengeti area, East Africa. *Mammalia*, 31: 1–27.

Lima, S. L., and Dill, L.M. (1989). Behavioral decisions under the risk of predation: A review and prospectus. *Can. J. Zool.*, 68: 619–640.

Moll, E.J., McKenzie, B., and McLachlan, D. (1980). A possible explanation for the lack of trees in the fynbos, Cape Province, South Africa. *Biol. Conserv.*, 17: 221–228.

Norton, P.M. (1986). Historical changes in the distribution of leopards in the Cape Province, South Africa. *Bontebok*, 5: 1–9.

van Orsdol, K.G. (1984). Foraging behavior and hunting success of lions in Queen Elizabeth National Park, Uganda. *Afr. J. Ecol.*, 22: 79–99.

van Schaik, C.P. (1983). Why are diurnal primates living in groups? *Behavior*, 87: 120–144.

van Schaik, C.P., and van Hörsterman, M. (1994). Predation risk and the number of males in a primate group: A comparative test. *Behav. Ecol. Sociobiol.*, 35: 261–272.

Schoener, T.W. (1971). Theory of feeding strategies. *Ann. Rev. Ecol. Syst.*, 2: 369–404.

Shultz, S., Noë, R., McGraw, W.S., and Dunbar, R.I.M. (2004). A community-level evaluation of the impact of prey behavioral and ecological characteristics on predator diet composition. *Proc. Roy. Soc. Lond. B*, 271: 725–732.

Stacey, P.B. (1986). Group size and foraging efficiency in yellow baboons. *Behav. Ecol. Sociobiol.*, 18: 175–187.

Stuart, C.T., Macdonald, I.A.W., and Mills, M.G.L. (1985). History, current status and conservation of large mammalian predators in Cape Province, Republic of South Africa. *Biol. Conserv.*, 31: 7–19.

Underwood, R. (1982). Vigilance behavior in grazing African antelopes. *Behavior*, 79: 81–107.

Zuberbühler, K., and Jenny, D. (2002). Leopard predation and primate evolution. *J. Human Evol.*, 43: 873–886

17
Reconstructing Hominin Interactions with Mammalian Carnivores (6.0–1.8 Ma)

Adrian Treves and Paul Palmqvist

Introduction

Several hominin genera evolved to use savanna and woodland habitats across Pliocene Africa. This radiation into novel niches for apes occurred despite a daunting array of carnivores (Mammalia, Carnivora) between 6.0 and 1.8 Ma (Figure 17.1). Many of these carnivores would have preyed on hominins if given the opportunity. In this paper we ask what the behavioral adaptations were that permitted hominins to survive and spread, despite this potentially higher risk of predation in ancient Africa.

When considering hominin anti-predator behavior, many scholars looked first to material culture, such as fire or weaponry (Kortlandt, 1980; Brain, 1981). However, the idea that deterrent fire or weaponry freed early hominins from threats posed by predators is unsatisfying for several reasons. First, the modern carnivores now roaming Africa are survivors of humanity's repeated and systematic campaigns to eradicate problem animals, trade in skins, and so on. (McDougal, 1987; Treves & Naughton-Treves, 1999), whereas Pliocene carnivores would not have had a history of conflict with armed hominins. Second, thousands of modern humans fell prey to leopards (*Panthera pardus*), lions (*P. leo*) and tigers (*P. tigris*) in the twentieth century despite their sophisticated weapons and fire (Turnbull-Kemp, 1967; McDougal, 1987; Treves & Naughton-Treves, 1999; Peterhans & Gnoske, 2001). Although, thorn branches, stone tools, fire brands, pointed sticks, or bones could potentially help to repel carnivores from their kills (Kortlandt, 1980; Bunn & Ezzo, 1993; Treves & Naughton-Treves, 1999), such weaponry seems wholly inadequate for personal defense when large carnivores achieve surprise, attack in a pack, or are accustomed to overcoming heavier prey defended by horns, hooves, or canines. Therefore, we assert that weaponry by itself does not nullify the risk posed by predators. Moreover, controlled use of fire and stone tool technology appear late in the archaeological record relative to the evolution of

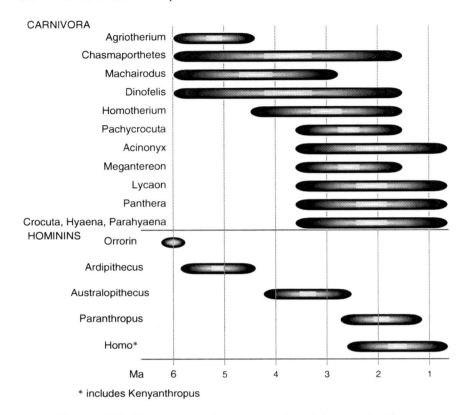

FIGURE 17.1. Time spans of paleopredator and hominin genera in Africa.

semi-terrestrial hominins in Pliocene Africa (Bellomo, 1994; Brain, 1994; Wolde-Gabriel et al., 1994; Brunet et al., 1997; Leakey et al., 1998; Haile-Selassie, 2001). Hominin anti-predator behavior remains a key puzzle of our human ancestry.

In the next section of this chapter we review African large carnivore ecology and hunting behavior in extant taxa and that reconstructed for Plio-Pleistocene forms ("paleopredators" hereafter). Following this, we review the anti-predator behavior of hominins by analogy with monkeys and apes; this analogy is parsimonious because of the observed cross-taxonomic consistency of their behavioral responses to predators. Vigilance behavior in relation to social organization is particularly informative. Finally, we integrate the two reviews to reconstruct the range of anti-predator behaviors open to hominins.

African Large Carnivores, Past and Present

Africa has long contained diverse carnivore communities (Figure 17.1). Carnivores have repeatedly radiated into various niches, including specializations for predation, active or passive scavenging, open-country or forested habitats, and small or large ungulate prey (Table 17.1).

Following Sunquist & Sunquist (1989) we define a "large carnivore" as any species with average individual or group body mass > 34 kg (e.g., *Hyaena hyaena* or *Lycaon pictus*, respectively). Subsequent reference to large/small prey relate to the carnivore under discussion.

Large carnivore diversity was greater in Africa's past than it is today (Figure 17.1). Between 6 and 3.6 Ma there were five genera of large carnivores without extant analogues (the long-legged ursid *Agriotherium*, the large coursing hyaenid *Chasmaporthetes*, and the saber-toothed felids Homotherium, *Machairodus* and *Dinofelis*). Then, from the mid-Pliocene (3.6 Ma), the archaic genera were joined by one large canid (*Lycaon lycaonoides*) (Martínez-Navarro & Rook, 2003), three new large felid genera (*Acinonyx, Megantereon* and *Panthera*), and four new genera of hyaenids (*Crocuta, Pachycrocuta, Hyaena*, and *Parahyaena*). At some sites, 8–10 species appear to have been coeval and broadly sympatric (Barry, 1987; Turner & Anton, 1997)(Figure 17.1). Niche separation under such conditions is not yet clear.

As the Pleistocene wore on (1.8 Ma onward) the archaic carnivores went extinct in Africa, partly as a result of a global carnivore guild turnover and species replacement (Figure 17.1). The African faunal turnover coincided with a decrease in woodland relative to grassland, more herd-living grazing ungulates, and fewer solitary or small-group-living large herbivores like giraffids, rather than from competition between the modern carnivore guild and archaic forms (Hendey, 1980; Turner, 1990; Werdelin & Turner, 1996; Turner & Anton, 1998).

Coexistence of hominins and carnivores is insufficient by itself to conclude that hominins evolved effective anti-predator defenses against such paleopredators. Coexistence would have had little selective impact if (a) carnivores did not kill Pliocene hominins regularly, or (b) if such predation were random with respect to hominin traits. Thus, in the following sections we assess whether paleopredators killed hominins regularly, and if so, were there consistent patterns of hominin-carnivore interactions that might have produced directional selection among hominins.

Habitat Selection

Carnivores generally go where prey are most abundant, but many will establish and defend territories year-round. Except for the leopard, all the extant African large carnivores are most abundant in open savannas and savanna-woodlands (variable mixtures of trees, grassland, and bushland where visibility is less than 100 m on average), coincident with highest ungulate densities (Table 17.1). Nevertheless, several carnivores can breed successfully within very arid regions or dense forest (Leakey et al., 1999; Bailey, 1993). The leopard is the greatest habitat generalist today, breeding from rainforest to desert, albeit preferring habitat with vegetation cover.

TABLE 17.1. African large carnivores, present and reconstructed.

Genus	Mass, kg	Brief description or common name	Predominant Hunting Behavior			Habitat*
			Activity	Grouping	Attack	
Acinonyx	ca. 50	cheetah	diurnal	solitary	stalking-coursing	open>woodland
Agriotherium	600–700	faster, more carnivorous than extant ursines	both?	solitary	opportunistic-coursing	woodland>open
Chasmaporthetes	50–70	long-legged hyena, felinoid teeth	both?	pack	coursing	open>woodland
Crocuta	ca. 60	spotted hyena	both	pack	coursing	open>woodland
Dinofelis	70–100	more powerful, less arboreal than leopard	both?	solitary	stalking	woodland>open
Homotherium	150–230	scimitar-toothed machairodont	both?	pack?	stalking-coursing	open>woodland
Hyaena, Parahyaena, Pliocrocuta	35–40	striped, brown, and Pliocene hyena	both?	solitary	opportunistic	all
Lycaon	25–40	African wild dog (L. pictus) or Pliocene form (L. lycanoides)	both	pack	coursing	open>woodland
Machairodus	130–230	scimitar-toothed machairodont	both?	pack?	stalking	open>woodland
Megantereon	80–110	dirk-toothed machairodont	both?	solitary	stalking	woodland>open
Pachycrocuta	100–130	giant hyena, robust bone-cracking teeth	both?	pack	opportunistic-coursing	all
Panthera leo	ca. 170	lion	both	pack	stalking	open>woodland
P. pardus	ca. 55	leopard	both	solitary	stalking	all

*Predominant habitat associations, e.g., "open>woodland" indicates less tree cover than grass cover.

Sources: Anyonge, 1993, 1996; Arribas & Palmqvist, 1999; Bailey, 1993; Barry, 1987; Berta, 1981; Boaz et al., 1979; Brain, 1981, 1994; Busse, 1980; Caro, 1987, 1989a,b; Cooke, 1991; Fanshawe & FitzGibbon, 1993; Ferretti, 1999; FitzGibbon, 1990a,b; Gerraads, 1997; Hendey, 1974, 1980; Holekamp et al., 1997; Hunt, 1996; Keyser, 1991; Leakey, 1999; Lewis, 1997; Marean, 1989; Martin, 1989; Martínez-Navarro and Palmqvist, 1995, 1996; Mills, 1989; Palmqvist, 2002; Palmqvist & Arribas, 2001; Palmqvist et al., 1996, 1999; Petter et al., 1994; Rook, 1994; Schaller, 1972; Taylor, 1989; Turnbull-Kemp, 1967; Turner & Anton, 1996, 1997, 1998; Turner, 1990, 1997; Werdelin, 1994.

As far as micro-site selection for hunting, only the leopard is known to hunt arboreal prey within 10–15 m of the ground. Leopards also kill in caves, cliff sides and houses (Simons, 1966; Turnbull-Kemp, 1967).

Among extinct carnivores, habitat use varied (Table 17.1). *Agriotherium* and machairodonts *Dinofelis* and *Megantereon* are believed to have selected more forested habitats based on their postcranial morphology, typical of stalking, ambush hunters. The latter two genera show relatively more robust forelimbs than hindlimbs. A comparative study of the postcrania in modern and Plio-Pleistocene carnivores shows *Dinofelis* resembles pantherine felids craniodentally, and its postcrania resembling modern prey-grappling lions, tigers, and leopards (Marean, 1989; Anyonge, 1996; Lewis, 1997). The postcrania of *Megantereon* reveal tree-caching and long-distance dragging capabilities, as in modern leopards and jaguars (*Panthera onca*) (Lewis, 1997; de Ruiter & Berger, 2000). *Homotherium* and *Machairodus* postcrania suggest cursorial tendencies in more open habitats, given their comparatively higher values for both brachial and crural indexes (Table 17.1). *Chasmaporthetes* and *Lycaon* have been associated with open-country habitats as well—although it should be noted that *Lycaon* today can hunt quite successfully in dense shrub land (Creel & Creel, 1995). The giant hyena *Pachycrocuta* was associated with more open habitats, particularly where medium to large ungulate carcasses were left by machairodont felids (Arribas & Palmqvist, 1998) (Table 17.1).

Associations of fossil hominins with remains of *Chasmaporthetes*, *Dinofelis*, *Homotherium*, *Machairodus*, Megantereon, and *Pachycrocuta* indicate sympatry in the period 6.0–1.8 Ma in habitats reconstructed as a mixture of woodlands and open country (Cooke, 1991; Keyser, 1991; Brain, 1994; Brantingham, 1998; Dominguez-Rodrigo & Pickering, 2003; Palmqvist et al., 2005). At a finer level, felid and hyaenid activity was considerable in and around the same caves with hominin remains (Brain, 1981, 1994; Turner, 1990). Deep caves would therefore have been dangerous resting sites (for vivid examples, see Simons, 1966; Brain, 1981). However, there seems to be evidence that hominins went voluntarily to caves used by paleopredators. For example, the presence of Plio-Pleistocene stone tools in South African caves without evidence of their manufacture (Brain, 1981) suggests that hominins came to some of these sites voluntarily (carrying tools) most likely, or, less likely, that predators transported their carcasses without losing the tools (e.g., in a portable container that neither fossilized nor dropped off the carcass when dragged).

The extant carnivores hunt by day and night, but seem to do so most often or most successfully between 19.00 and 07.00 with the exception of the diurnal cheetah (Table 17.1). There is some indication that carnivores hunt less by day when humans pose a threat to them (Turnbull-Kemp, 1967; van Schaik & Griffiths, 1996), a benefit hominins would not have enjoyed in the Pliocene for the reasons mentioned before. Moreover, observations of predation reveal that carnivores kill primates in the day as well as at night (reviews in Treves, 1999a; Boinski et al., 2000). Thus, hominins could not have escaped predation simply by diurnality.

In addition to some level of diurnal risk, hominins may have faced nocturnal threat at sleeping trees and caves, as do large primates today (Simons, 1966; Busse, 1980; Brain, 1994). Hominins might have preferentially selected smaller trees over larger ones and narrow or fragile ledges in caves rather than solid supports for sleeping sites. These microsites would provide earlier warning of nocturnal intrusion and impede the rapid approach of a predator such as *Agriotherium* in the early Pliocene and leopards or *Dinofelis* thereafter. In sum, hominins could not have escaped predation by using different habitats than large carnivores nor could they have done so by using them at different times of day, although such tactics may well have lessened their exposure to paleopredators.

Hunting Tactics

Stalking (ambush) predators rely on surprise or stealthy approach, followed by brief, high-speed pursuit (Taylor, 1989; Fanshawe & Fitzgibbon, 1993; Fitzgibbon & Lazarus, 1995). Their attacks are often aborted or fail when prey detect the predator in ambush or early in its approach. Hence, for stalking predators, the most vulnerable prey are the unwary, whether they are healthy adults, the young, or the old and infirm.

All the felids use ambush (stalking) to pounce or sprint after prey. They can accelerate rapidly but tire quickly. Among the extinct forms, *Dinofelis*, *Machairodus*, and *Megantereon* probably conformed to the felid pattern of stalking their prey using ambush, while *Acinonyx* and *Homotherium* show a mix of ambush and pursuit (coursing) adaptations with elongated and slender distal limbs suited to longer chases at higher speeds (Table 17.1). Since the Pliocene, *Acinonyx* has been specialized for longer chases, albeit still under 1 km.

In contrast to stalkers, coursing (pursuit) predators such as *Lycaon* and *Crocuta* often approach prey with little or no stealth; rather, they openly survey moving prey for weaknesses and chase their targets for distances often > 1 km (Table 17.1). Because most prey detect the overt approach of coursing predators before they begin a chase, prey vigilance is reduced in importance relative to rapid, agile, sustained flight or escape into refuge. Therefore, the most vulnerable prey are those nearest the predators, those far from refuge, or those that flee slowly (e.g., the aged or infirm). Both extinct *Lycaon* and *Chasmaporthetes* were probably pack-hunting coursers as well, simply on the grounds of morphology and relatedness to their modern relatives, described above (Table 17.1).

Pachycrocuta was substantially larger and probably slower than the spotted hyena *Crocuta*, judging from its shorter distal limb segments and more robust postcrania. Although these features do not preclude coursing, taphonomic analyses suggest reliance on scavenging behavior for this extinct hyena (Palmqvist et al., 1996) (Table 17.1).

Third, opportunistic hunters such as extant ursine bears and *Hyaena* rarely pursue rapidly fleeing prey and typically attack prey opportunistically by random search using generalized locomotion (Table 17.1). Prey typically avoid such predators by detecting them first and seeking appropriate refuge, but data are scant on prey vulnerability. Some opportunistic predators can pursue into refuges.

We assume that *Agriotherium* was such a predator, although its long limbs raise the possibility that it might have sprinted for short distances (Table 17.1). There is no evidence currently available on the feeding behavior of *Agriotherium*, but it is worth mentioning that the similarly proportioned short-faced bear *Arctodus*, the largest Pleistocene carnivore of North America, included more flesh in its diet than brown bears according to biogeochemical ($^{13}\delta C$, $-^{15}\delta N$) analyses of bone collagen (Bocherens et al., 1995; Matheus, 1995).

Finally, there is a long-standing and fascinating debate over the benefits of group living in carnivores; it revolves around whether pack hunters have higher hunting success than solitary predators, can hunt larger prey, retain their kills for longer time against scavengers, avoid predation better themselves, or defend their territories more effectively. Of course, these are not mutually exclusive functions, but researchers have found one benefit accruing under one set of conditions and not another, only to be contradicted by studies from other sites (for selected examples, see Packer et al., 1990; Fanshawe & Fitzgibbon, 1993; Fuller & Kat, 1993; Creel & Creel 1995; Holekamp et al., 1997). The flexibility of large carnivore grouping—fission-fusion sociality—permits individuals to respond to short-term changes in prey abundance and ease of capture by joining or leaving aggregations. For this reason we echo Turner & Anton (1997), who warn that one consider as full a range of carnivore behaviors as possible. In short, hominins would on occasion have encountered both solitary paleopredators and aggregations of them whatever the taxon. But would some taxa regularly form groups that might have posed an added risk to hominins?

Modern *Crocuta* and *Panthera leo* sometimes hunt in packs, while *Lycaon* virtually always does so. Other extant carnivores hunt in pairs, trios, or larger groups more rarely (Table 17.1). The extinct canids and hyaenids probably hunted in groups a majority of the time, judging from their extant relatives. *Pachycrocuta* may be a borderline case: Although excavations of den sites suggested solitary foraging (Palmqvist & Arribas, 2001), the possibility remains that groups of *Pachycrocuta* foraged together while only individuals (mothers?) returned to the dens to provision young. The large brains with more developed optical lobes than olfactory lobes of *Homotherium* and *Machairodus* may reflect pack-hunting behavior, in contrast to the smaller-brained, more olfactory *Megantereon* (Martin, 1989; Palmqvist et al., 2003). *Agriotherium* and *Dinofelis* are both reconstructed as solitary given the behavior of ursine bears and leopards, respectively. *Chasmaporthetes* is still too poorly known, although its phylogeny and anatomy suggest open-country coursing, and therefore pack hunting (Table 17.1).

From the perspective of prey, pack hunting usually increases the risk for prey, the number of prey killed per hunt, and the size of prey taken (see reference to pack hunting above). Prey defenses seem the same whether animals are hunted by packs or by single predators, but further study would be valuable.

In sum, coursing paleopredators would have placed a premium on refuge use by hominins when in open country, while the more numerous opportunistic and stalking predator genera (Figure 17.1, Table 17.1) using more forested or bush habitats would demand vigilance by hominins.

Prey Selection

Carnivores sometimes select their prey before the start of the hunt, but more typically they hunt any prey they encounter (Kruuk, 1972; Holekamp et al., 1997). This opportunism tends to protect primates, which usually occur at lower densities than medium to large ungulates. Nevertheless, primates appear regularly in the scat of carnivores and in observed predation (Brain, 1981; Bailey, 1993; Treves, 1999a). Predation on primates varies with conditions. For example, when leopards face competition from larger carnivores they increase their exploitation of monkeys (Seidensticker, 1983). Individual prey preferences vary within the same species by individual, age-sex class, habitat, season, etc. Over short periods, individual carnivores or even packs are known to specialize on a single type of prey to the exclusion of others. Indeed, reports of leopards specializing on primates are not uncommon (Brain, 1981; Hoppe-Dominik, 1984; Boesch, 1991). There are thousands of records of wild carnivore attacks on modern humans (Corbett, 1954; Turnbull-Kemp, 1967; McDougal, 1987; Sanyal, 1987; Peterhans & Gnoske, 2001). For example, colonial archives reveal at least 393 Ugandan men, women, and children were killed or injured by lions, leopards, and spotted hyenas in the last century. This grim toll would surely have been elevated without modern weaponry and coordinated campaigns to extirpate leopards and lions (Treves & Naughton-Treves, 1999). It was once thought that primarily the infirm or inexperienced carnivores would approach human settlements or attack humans, but systematic study leads us to reject this idea (Turnbull-Kemp, 1967; Linnell et al., 1999; Treves and Naughton-Treves, 1999; Peterhans & Gnoske, 2001).

The size of potential prey is also a consideration for most carnivores. Body mass of hominins has been estimated repeatedly (see reviews in Mathers & Henneberg, 1996; Hens et al., 2000). We follow McHenry (1992), using estimates based on regressions of hindlimb joint proportions to identify a *range* of body sizes that describes adults of both sexes for all hominin species between 6.0 and 1.8 Ma—rather than a *mean* for a particular species at a particular time. His upper and lower bounds span 29–52 kg for *Australopithecus, Paranthropus,* and Pliocene *Homo* (McHenry, 1992, 1994). Leopard-sized and larger carnivores routinely kill prey weighing over 52 kg (Palmqvist et al., 1996). As noted above, leopards are capable of killing adult modern humans and transporting much larger prey to caches. Because adult baboons such as *Parapapio jonesi* (mass 30–40 kg: Brain, 1981; Delson et al., 2000) fell prey to paleopredators it would not be parsimonious to suggest that adult hominins were immune, resistant, or avoided by large carnivores thanks to their size.

Fossil Anatomy, Isotope Frequencies, and Composition of Bone Assemblages

Several lines of evidence hint at the prey preferences of archaic predators and therefore the likelihood they regularly hunted hominins: (a) craniodental and postcranial morphology of carnivores plus bone assemblages bearing traces of carnivore foraging yield insights into dietary and hunting behavior; and (b) bone and enamel isotope measurements from fossils provide insights into diet. Below we briefly summarize a general consensus emerging from recent reviews.

(a) The saber-tooths *Homotherium, Machairodus,* and *Megantereon* were hyper-carnivores (>70% of their diet was meat) that could deflesh a carcass quickly, but rarely broke bones to access marrow; they would probably not have commonly transported meat to protect kills (Brain, 1981; Brantingham, 1998). Despite specialization for large ungulate killing, the saber-tooth felids would rarely have hunted hominins, yet hominins would still face danger if they encountered machairodonts at close quarters; the opportunism of large carnivores must always be kept in mind.

Equally carnivorous but more likely to focus on smaller prey and transport meat were the leopards and *Dinofelis.* At half a dozen southern African cave sites dated between 3.0 and 1.0 Ma, paleontologists have found fossil *Dinofelis* or leopards alongside fossils of many ungulates, at least 140 australopithecine hominins, and at least 324 baboons (Brain, 1981, 1994). Many of the large primate remains show characteristic patterns of damage by large felids and hyaenids (Brain, 1981; Keyser, 1991; Berger & Tobias, 1994; Turner, 1997; de Ruiter & Berger, 2000). Taphonomic evidence suggests that carnivores brought hominin remains to their dens. Some have proposed that leopards or *Dinofelis* specialized on large baboons and australopithecines, judging from the unusually high proportions of large primate fossils (Brain, 1981; Cooke, 1991).

Pliocene *Acinonyx* is reconstructed as a larger form (100 kg) than modern cheetahs, but still a specialist on small to medium-sized ungulates (Table 17.1). There is no evidence that hominins regularly fell prey to cheetahs given the size, specialized hunting behavior and timidity of the latter (e.g., Baeninger et al., 1977). The wolf-sized (45 kg) *Lycaon* of Pliocene Africa are reconstructed as hyper-carnivores—like extant *Lycaon* and wolves (*Canis lupus*)—that defleshed carcasses and, when undisturbed, cracked bones (Table 17.1). Their prey preferences were presumably the same as their extant relatives, i.e., medium to large ungulates (Rook, 1994; Palmqvist et al., 1999, 2003). Because healthy (non-rabid) wolves have sporadically killed modern humans in Eurasia, particularly women and children, in spates of encounters (Rajpurohit, 1998; Linnell & Bjerke, 2002), the paleocanids may have posed a sporadic threat to hominins as well.

The habits of *Agriotherium* have not been reconstructed in detail, although its dentition suggests it was more carnivorous than ursids today (Hendey, 1980; Petter et al., 1994; Miller & Carranza, 1996; Geraads, 1997). Modern grizzlies (*Ursus arctos*) are confrontational scavengers that steal kills from hyper-carnivorous wolves (Smith et al., 2003). *Agriotherium* seems capable of the same behavior, therefore, opportunistic but not regular attacks on hominins seem most likely; the frequency of such encounters would rise when hominins use the same foraging areas as bears, as is sometimes seen today (Rajpurohit & Krausman, 2000).

Pliocene hyenas appear very similar to today's striped and brown hyenas, taxa that are predominantly passive, non-confrontational scavengers (Table 17.1). They would probably not have posed a common threat to hominins. *Crocuta* have killed modern humans (Treves & Naughton-Treves, 1999), so the large paleohyenas presumably posed some threat to hominins, especially around carcasses. However there are no fossil assemblages of hyaenid prey or isotope data to support the idea that hyenas routinely killed hominins.

(b) Tooth enamel and bone C and N isotope measurements shed light on fossil carnivore diets. For example, carnivores eating mainly grazing animals that fed on C4 plants (tropical grasses) will have higher ratios of ^{13}C to ^{12}C in the hydroxyapatite and collagen of their tooth enamel and bones, respectively, than did carnivores who are mainly browsing animals that fed on C3 plants (trees, shrubs, forbs and tubers—plants that discriminate strongly against the heavy isotope of C). Lee-Thorp and colleagues (2000) showed South African paleopredators could be distinguished by their bone and enamel isotope ratios ($^{13}\delta C$). In their Pleistocene sample, lions ate the highest proportion of grazers, *Crocuta* was intermediate, while leopards ate more browsers and omnivores like baboons and hominins (Lee-Thorp et al., 1994; Sillen & Lee-Thorp, 1994; Spoonheimer & Lee-Thorp, 1999). Palmqvist et al. (2003) found the long-legged, ambush/coursing saber-tooth *Homotherium* and the coursing *Lycaon lycaonoides* (formerly *Canis falconeri*) of Venta Micena, Spain, had elevated heavy-nitrogen levels indicating a diet dominated by grazers, such as adult *Equus* and juvenile *Mammuthus*. By contrast, the short-legged *Megantereon,* comparatively depleted in ^{15}N, ambushed browsing ungulates such as megacerine deer in Spain's ancient forested habitats. Finally, *Pachycrocuta* shows intermediate ^{15}N values at this site, suggest it scavenged the prey of all the paleopredators described above (Palmqvist et al., 2003).

Leopard predation on robust australopithecines has been ruled unlikely based on tooth enamel carbon isotopes, as the $^{13}\delta C$ values of *Paranthropus* and *P. pardus* are similar. However, the isotopic enrichment between *Paranthropus* and *Dinofelis* agrees with that expected of prey and predator (Lee-Thorp et al., 1994; 2000). These finds are strong indication that hominins were regularly hunted.

Competition Between Carnivores and Hominins

Hominins may have competed directly with some paleopredators over prey or carcasses. After defleshing by hyper-carnivores, carcasses retained long bone marrow, epiphyseal grease, brains, and other axial elements accessible to bone-cracking scavengers. But were hominins usually the primary predators? the confrontational scavengers? or passive, non-confrontational scavengers? Despite years of debate and re-analyses of bone assemblages, experts disagree about which Plio-Pleistocene hominin competed with which carnivore and in what manner (Bunn & Ezzo, 1993; Blumenschine, 1995; Capaldo, 1997; Brantingham, 1998; Selvaggio, 1998; Arribas & Palmqvist, 1999; Dominguez-Rodrigo & Pickering, 2003; Palmqvist et al., 2005).

In the earliest period, prior to 3.6 Ma, there were no *Crocuta* or *Pachycrocuta*—confrontational scavengers—but *Agriotherium* and perhaps *Chasmaporthetes* may have stolen kills from *Machairodus,* a primary predator. *Dinofelis* may have been a primary predator and occasional confrontational scavenger when an opponent was smaller. Without the availability of close analogues for the archaic genera and scant taphonomic evidence, it is premature for us to speculate what role the earliest hominins may have played in scavenging. We have more evidence for predation or scavenging by hominins during the carnivore-rich late Pliocene (3.6–1.8 Ma) of Africa.

Hominin attempts to defend or steal a carcass could have increased the risk of attack by several carnivores approaching from many sides and using stealth. Although twentieth century humans have chased single large carnivores from the latter's kills using little or no weaponry, these carnivores had ample reason to fear humans, as mentioned previously (Sunquist & Sunquist, 1989; Treves & Naughton-Treves, 1999). Pliocene hominins engaging in confrontational scavenging would have had to overcome several obstacles to success. First, stealing a kill from paleopredators would have demanded very frequent vigilance and repeated, effective threats: The displaced carnivores would probably have remained nearby while newcomers continued to arrive. A scavenging hominin would have had to process a carcass and defend it while maintaining high levels of vigilance— mutually incompatible activities for a single individual (Treves, 2000). Many of these obstacles suggest the need for cohesive, coordinated group activity. Second, effective deterrent threats might have required weapons. Whatever weaponry used would have had to deter large carnivores with a habit of killing horned ungulates or primates with canines. Yet, missiles would be difficult to retrieve after use and hand-held weaponry would reduce the efficiency of butchery. Third, the optimal scavenging party size depends on per capita meat yield, which will increase with carcass size (Creel & Creel, 1995), but large carcasses are usually better defended and more attractive to multiple scavengers for longer periods.

All the complicating factors mentioned would reduce the time available for butchery or increase the risk to individual butchers (Brantingham, 1998; Lupo, 1998; Monahan, 1998). Coordination of activity among hominins would help but would require trust and practice. Then, assuming success, meat transport would require safe refuges from scavengers pursuing the encumbered hominins. At the moment, confrontational scavenging of the sort envisioned above appears an unlikely route to regular meat acquisition, hence we side more with authors who envision hominins as primary predators or as scavengers of unattended fresh carcasses who fled when challenged, rather than confrontational scavengers (Brantingham, 1998; Dominguez-Rodrigo & Pickering, 2003).

Anti-Predator Behavior and Hominin Reconstruction

We divide anti-predator behavior into two discrete strategies that correspond to different stages in a predator encounter. In the first stage we place all behavioral tactics displayed in the absence of predators, behaviors aimed at reducing the likelihood of encounter. The primary tactics of stage one are inconspicuousness, avoidance of dangerous locations, and vigilance oriented to early detection of a predator. The second stage begins when predators are encountered. The corresponding anti-predator behaviors will reflect the immediacy of the threat, although the exact steps and sequence will vary with the type of predator, type of primate prey, cost-benefit ratio of prey responses, and with the physical context (Ydenberg & Dill, 1986; Lima, 1993; Treves, 2002). The primary tactics in stage two are monitoring of predators, escape, deterrence, and hiding among other targets (selfish herd). Each tactic has requirements that make the tactic useful in some situations but not in others. Because anti-predator behavior has been studied for decades we refer the reader to more general reviews (Edmunds, 1974; Klump & Shalter, 1984; Cheney & Seyfarth, 1990; Lima, 1990; Goodman et al., 1993; Treves, 1999a; Boinski et al., 2000; Miller & Treves, 2006), but we cite primary sources for anti-predator behavior of living hominoids.

In the Absence of Predators

Primates reduce the likelihood of encounter with predators by avoiding dangerous areas, behaving inconspicuously, or surveying their physical surroundings for danger. Avoidance of known dangerous areas is probably universal among primates, but the role of learned versus innate avoidance is unclear. As a result, we know little about how primates respond to changes in predator communities or changes in their encounters with carnivores—issues of importance when we consider hominin-carnivore interactions. Inconspicuousness depends on small group size or coordination of activities among associates. The larger a group, the more sounds, smells, and other signs that may be detectable to predators.

Apes often rely on inconspicuousness and avoidance of risky areas, especially after they encounter a predator. For example, chimpanzees (*Pan troglodytes*) in Senegal were more silent than usual when they were crossing broad grassland, ostensibly to avoid detection by the abundant large carnivores (Tutin et al., 1981). Lowland gorillas (*Gorilla gorilla*) moved quickly and quietly after encountering a leopard (Fay et al., 1995). Aché hunter-gatherers (*Homo sapiens*) moved camp to avoid a jaguar (Hill & Hurtado, 1995), and Indian villagers stayed in their settlements after tigers and leopards attacked some villagers who had gone into the forest (Corbett, 1954). Early hominins would likely have avoided areas such as dark caves, treeless habitat, high grass, and rocky outcrops, at least until these areas had been thoroughly surveyed for danger. It would seem conspicuous behaviors (tool-making, loud display, mating, play, etc.) would have been most safely performed high in trees or on rock ledges. However, hominin tool making appears to have occurred at lacustrine and riverine edges (Bunn & Ezzo 1993; Capaldo 1997; Dominguez-Rodrigo & Pickering, 2003; Palmqvist et al., 2005); the risk at such sites remains to be determined.

Surveillance of surroundings also seems universal among primates. Visual and auditory monitoring may forewarn primates of impending encounters with predators and help the primates respond appropriately. Vigilance reduces uncertainty about a given location but uncertainty resurfaces after individuals leave an area or otherwise interrupt monitoring, hence vigilance must be continuously renewed.

Vision is particularly useful in providing precise information about predator type, location, and movement. Auditory vigilance can complement visual monitoring, especially in visually obstructed microsites.

Non-primates who keep their heads down suffer higher predation rates than those who survey their environment (FitzGibbon, 1989). Equivalent data are not yet available for wild primates, but they do spend more time scanning their surroundings when risk is elevated (Treves, 2000). We have little quantitative data on vigilance in apes or humans, but the few data resemble those of monkeys (Wirtz & Wawra, 1986; Setiawan et al., 1996; Treves, 1997; Watts, 1998). Locational features, such as the density of foliage and associates, will modify the effectiveness of visual vigilance and, presumably, auditory vigilance as well (e.g., running water or noises produced by non-predators). Visual obstructions were associated with less time spent vigilant in two studies (Hill & Cowlishaw, 2002; Treves, 2002). Therefore, hominins using Pliocene African savanna-woodlands might have invested more in visual vigilance than those in closed, forested habitats. We discuss vigilance further below as it is intricately tied to social organization.

After Predator Encounter

Once potential prey animals have been detected by a predator, their particular anti-predator response will depend on their detecting the predator in turn and on its hunting tactics. At one extreme predators may remain undetected throughout the attack sequence. Nocturnal predation tends toward this extreme (Busse, 1980; Peetz et al., 1992; Wright, 1998), as does predation with complete surprise (Chapman, 1986; Peres, 1990). Attack by complete surprise followed by death leaves prey with only one recourse: to practice safety in numbers. We discuss aggregation further in "Trade-Offs Between Anti-Predator Aggregation and Vigilance" below.

If a predator is detected before it kills its prey, primates display several effective anti-predator tactics. Many individuals will produce alarm calls to warn associates some protect themselves without warning others. When primates have detected a predator they may produce predator warnings to deter further approach by that predator (Zuberbühler, 2000). Mobbing calls are used to attract attention to a predator or intimidate it. Chimpanzees and humans give alarm and mobbing calls (Corbett, 1954; Goodall, 1986; Hiraiwa-Hasegawa et al., 1986; Boesch, 1991; Tsukahara, 1993). Hominins would presumably have done the same.

In addition, all primates escape. We have found no convincing descriptions of primates using the "confusion effect" (i.e., escape not in a direct line to refuge, but in coordinated, evasive action confusing to the observer), to avoid predation, the kind of effect that is seen in some fish or open-country herds of ungulates (Edmunds, 1974). Moreover, primates virtually always flee to refuge rather than try to outdistance their attackers. Refuges for terrestrial primates include some trees and cliffs, while arboreal primate forms rapidly change levels. Humans and apes also commonly flee from predators and use refuges such as trees (Corbett, 1954; Boesch, 1991; Tsukahara, 1993; Hill & Hurtado, 1995). Presumably, early hominins would have minimized forays away from refuge and maintained proximity to trees and cliff sides to improve their chances of escape from speedy predators.

More rarely, primates stand their ground to counterattack or mob predators. Of the two forms, mobbing appears to be less dangerous for the predator and is more common among primates much smaller than the predator. Mobbing involves two or more prey animals making repeated advances on a predator, usually while vocalizing and displaying in a conspicuous fashion. The predator is often distracted or repelled by persistent approaches. Adult males, acting alone or in small parties, are more likely to attack predators than other classes of individuals (Gautier-Hion & Tutin, 1988; Cowlishaw, 1994). Baboon counterattacks have been described most often. Sometimes adult male baboons coordinate a counterattack on a leopard or cheetah and may deliver serious injuries (Brain, 1981; Bailey, 1993; Cowlishaw, 1994), but at other times the males flee the scene (Smuts, 1985). The likelihood of counterattack by primates appears to depend on the size difference between predator and prey.

Silverback gorillas sometimes defend their groups from predators and hostile conspecifics by using intimidation displays. Chimpanzees have pursued and even killed cornered leopards (Boesch & Boesch, 1981; Hiraiwa-Hasegawa et al., 1986). Chimpanzees have attacked stuffed leopard models with sticks and stones (Kortlandt, 1980, 1989). However, healthy lioness-sized or larger carnivores may be too formidable, even for male apes in groups (Tsukahara, 1993). Counterattack with hand weapons may be an especially effective anti-predator tactic in some situations, but we have very little systematic evidence of this. It is doubtful that simple projectiles can deter coursing predators that do not abort pursuit easily or packs of carnivores emboldened by their own numbers. Moreover, a weapon does not provide protection if its wielder is surprised. Therefore, we doubt that hominins counterattacked carnivores in packs or lion-sized carnivores in the Pliocene.

Trade-Offs Between Anti-Predator Aggregation and Vigilance

Aggregation of individuals of one or more species has complex effects on predator detection. On the one hand, groups may detect predators earlier thanks to having many eyes and ears (Galton, 1871). Also, if associates warn each other in time, unwary individuals may remain safe (Bednekoff & Lima, 1998a,b). On the other hand, large groups may be more conspicuous to predators leading to higher rates of attack (Foster & Treherne, 1981; Fitzgibbon & Lazarus, 1995; Wright, 1998; Treves, 2000). Moreover, individuals in large groups may detect predators less quickly or reliably if larger groups contain more rivals and unfamiliar animals that must be monitored. For example, chimpanzee vigilance frequency was higher in large parties (Treves, 1997). Three solitary chimpanzees (an adult female and two juvenile females) averaged less of their time spent vigilant ($19.2 \pm 13.3\%$) than did nine chimpanzees observed in parties of 2–13 individuals who spent $46.5 \pm 26.3\%$ of their time vigilant. Most of this extra time was spent watching associates (50% of samples contained at least one glance at an associate); excluding this subset of samples the average time spent vigilant was $20.9 \pm 18.2\%$ in chimpanzee parties. Time spent monitoring associates is a cost additional to competition between individuals in large chimpanzee parties (Goodall, 1986; Chapman et al., 1995). We doubt hominins organized themselves as did the forest chimpanzees described above because the risk of conspicuousness and the added costs of vigilance in large, competitive parties would have been prohibitive in the carnivore-rich terrestrial habits under consideration here.

Our conjecture leaves open the possibility that hominins formed quiet, cohesive groups with less distracting competition. If vigilance were coordinated in some fashion (i.e., outwardly directed mainly, or asynchronous: cf. Horrocks & Hunte, 1986; Koenig, 1995), having many eyes would be advantageous.

Many studies of birds and ungulates have shown decreases in individual vigilance in larger groups. This has been interpreted most often as the animals in groups relaxing their individual efforts at vigilance because wary associates will give warnings (Lima, 1995; Bednekoff & Lima, 1998a). Increasing vigilance with larger group size as described above for forest chimpanzees is rare among animals in general (Elgar, 1989; Treves, 2000). In two cases it has been associated with the attraction of multiple, solitary scavengers to a single carcass (Knight & Knight, 1986; Jones, 1998). Indeed, individual primates rarely if ever reduce vigilance with the absolute number of conspecific associates (Treves, 2000). More often, primates relax their vigilance when associates are positioned nearby, regardless of total group size (but see Hill & Cowlishaw, 2002; Cowlishaw et al., 2004 for recent refinements). Moreover, one sees the highest vigilance among dominant animals and mothers of neonates (Gould et al., 1997; Treves et al., 2001, 2003). In short, individual vigilance in primates is shaped strongly by inter-individual proximity and social relationships. Therefore, safety may depend on establishing familiarity, trust, and reciprocity with a few individuals who will warn others when a predator is detected.

Leaving aside predator detection and aggregation, prey in a group usually enjoy dilution of risk—the inverse relationship between group size (N) and per capita risk (Foster & Treherne, 1981). However, dilution of risk rarely follows a simple inverse relationship (1/N) for primate prey because individuals within groups vary in their vulnerability to predation (Treves, 2000). This would have held for hominin groups with mothers and young. Grouping may also generate predator confusion or enable more effective counterattack against predators.

In sum, the net protection afforded by large groups fluctuates in a delicate balance of costs and benefits that are contingent on many local factors. This makes it difficult to conclude that hominins would have formed large groups in response to the putative higher risk of predation in the Pliocene.

Social Organization Reconstructed for Hominins

We have virtually no evidence for the foraging group size of the earliest hominins. Reconstructions of social organization that try to account for phylogeny suggest hominins lived in societies similar to those of chimpanzees or bonobos (*Pan paniscus*) (Foley, 1987; Ghiglieri, 1989). Today, chimpanzees rarely form parties exceeding 10 individuals (Chapman et al., 1994; Doran, 1997). Current evidence suggests that larger groups of apes are constrained by food availability (Chapman et al., 1995; Doran & McNeilage, 1998). However, the radiation of hominin taxa and extinction of several lineages cautions us against extrapolating uncritically from the social organization of living apes.

Consideration of the habitat rather than the phylogeny of the earliest hominins suggests the larger groups of savanna-woodland baboons (averaging 30–50 animals) may help us understand early hominin foraging groups in an open environment. Although habitat by itself does not ordain a certain group size, the foods available to the earliest hominins could have determined upper limits on aggregations. Bone and enamel isotope values from southern African fossils suggest that Plio-Pleistocene *Homo* and *Paranthropus* ate a higher-quality, more varied diet than either of two penecontemporaneous baboon species that focused on C3 plants (Lee-Thorp et al., 1994; Spoonheimer & Lee-Thorp, 1999). Given the larger body mass of hominins (see above) and this higher-quality diet, which presumably included variable amounts of animal proteins, average foraging group sizes exceeding 20 seem highly unlikely. However, we cannot reject this possibility yet.

Conclusions

Hominin ancestors of 6.0–1.8 Ma shared habitats with diverse genera of large carnivores that were opportunistic or with generalized predators that had no reason to fear hominins. In all likelihood, the hominins could not have avoided all encounters with these carnivores by virtue of diurnality, habitat selection, or body size. Nor could the hominins have deterred all attacks with weapons in this period. Given the existence of numerous ambush predators between 3.6–1.8 Ma, hominins would have experienced strong selection for efficient vigilance. Large parties of apes organized like those of chimpanzees are conspicuous and costly in terms of individual vigilance, competition for food and agonistic social interactions, hence we propose early hominin foraging parties would have adopted more cohesive and calmer social organization to maintain efficient vigilance and reduce conspicuousness to carnivores during diurnal foraging. Groups formed of trusted and familiar individuals often forage and travel with high levels of inter-individual proximity, experience minimal conflict, and coordinate vigilance more easily (Rasa, 1986, 1989; Koenig, 1994). For these reasons we rule out large (>20) hominin groups and particularly large, multi-male groups—like those of many baboon populations—as probable ancestral, anti-predator tactics.

Considering the range of anti-predator behaviors among monkeys and apes has helped researchers define the most likely adaptations of early hominins. Hominins would generally have avoided predator encounter through vigilance, minimizing time spent in dangerous areas, and behavioral inconspicuousness. Forays into open country would have been limited in extent and duration by access to refuges, whereas hominins foraging within woodlands would have been constrained by the demands of efficient, unobstructed vigilance because of the numerous ambush predators in Pliocene Africa. When close encounters occurred, the hominins would flee to refuge or counterattack. Counterattack would have been more likely by larger hominins against leopard-sized or smaller carnivores but virtually unknown against lioness-sized opponents in the period considered here. Hominins armed with weapons may have counterattacked more often, but we find no compelling evidence that material culture sheltered hominins from ambush and stalking predators before the advent of controlled fire. Frequent formation of foraging parties larger than 15–20 individuals seems unlikely given the dietary evidence presently available, although avoidance of nocturnal predation may have involved the formation of larger sleeping groups. Nevertheless, the essential nocturnal anti-predator adaptation was the use of trees or cliffs inaccessible to most large carnivores; this adaptation was important until the advent of controlled fire in the Pleistocene.

Modern humans may retain traces of some of the anti-predator adaptations of our ancestors. In particular, predictable behavioral responses and aversion to areas with dense vegetation or areas without suitable refuge (e.g., wide, open areas) should both be deeply embedded in human cognitive and perceptual abilities. These predictions are not trivial given that taxa differ based on selective pressures imposed by ancestral environments (Byers, 1997). Some animals perceive holes as refuges, while others perceive dense vegetation or open areas as avenues for escape (Lima, 1993). Experiments with sleeping sites, vigilance and group formation could test these ideas about ancestral human anti-predator adaptations; these would be analogous to the fruitful studies of brain and behavioral responses to strangers (reviewed in Treves & Pizzagalli, 2002).

In the following section, we consider some terrestrial mammalian taxa that live in environments with high predation pressure and display social organizations that share one or more of the following characteristics: inconspicuous, minimal internal conflict, or coordinated vigilance. For each we make predictions about the fossil record if one or more lineages of hominins had displayed such a social organization, and we make predictions about modern human behavior assuming we retain ancestral anti-predator adaptations.

Medium-Sized, Inconspicuous Groups

Individuals in groups of 10–15 animals can detect threats early and warn associates efficiently if distractions due to associates are few. For example, the Asian Hanuman langur (*Semnopithecus entellus*) forms large groups (averaging 29 members in 22 populations: Treves & Chapman, 1996), yet noisy, costly competition over resources seems to be muted by a combination of kinship bonds and even distribution of resources (Borries, 1993; Borries et al., 1994; Koenig, 1998). Male-male fighting is infrequent within groups because one male often monopolizes mates and evicts rivals. However, this calm evaporates when multiple males compete (Boggess, 1980; Borries, 2000). If modern humans retain traces of such a social organization, one should see higher vigilance among males watching for non-group rivals, and a significant increase in distractions and within-group vigilance when male rivals co-reside in a group. Hominins displaying such a social organization between 6.0–1.8 Ma would show marked sexual dimorphism associated with polygynous mating. Their dentition might also reflect the use of evenly distributed, low-quality foods, such as foliage or grasses.

Small Groups with Male Protector

Small, inconspicuous groups with a protective individual occur among terrestrial primates (e.g., gorillas: Doran & McNeilage, 1998). One version would include females attracted to watchful males, where female-female rivalry would be strong because the male's protective sphere would not be infinitely divisible among many females. If modern humans retain traces of this social organization, one should see higher vigilance among males than females and the greatest increase in within-group vigilance when multiple females are present in a group. Among early hominins, one would expect strong sexual dimorphism with polygynous mating, but dentition would reflect a high-quality diet due to low group size.

Small, Cooperative Groups

Small groups within which individuals cooperate in anti-predator behavior can survive under heavy predation pressure. The use of coordinated vigilance or sentinel systems is particularly important in such conditions because one or two individuals survey the surroundings while the remainder of the group forages uninterrupted. Upon detection of a predator, the sentinel gives a visual or acoustic signal as an alarm and the group takes defensive action. Modern humans use sentinels, of course. Sentinel systems are also seen today in many cooperatively breeding species (Wickler, 1985; Savage et al., 1996), but also among less cooperative groups that must forage silently (Horrocks & Hunte, 1986). Of particular relevance may be the social mongooses Herpestidae found in African woodland-savannas. High levels of cooperation and reciprocity appear critical under heavy predation pressure (Rasa, 1986, 1989); pressure that leads to the retention of juveniles and sub-adults in their natal groups (NB: also a modern human trait). If modern humans show traces of this social organization, the sexes will be equally vigilant, and familiar associates may readily coordinate defensive behavior. Hominins using this system would show little sexual dimorphism and delayed maturation, as in modern humans. Dentition would reflect a high-quality diet due to low group size.

Solitary Foragers

This form of inconspicuous social organization is seen in orangutans among the living apes and has been interpreted as a response to food scarcity (Sugardjito et al., 1987), and perhaps to avoidance of threats posed by conspecifics rather than predators (Setiawan et al., 1996; Treves, 1998). Nevertheless, early hominins might have foraged alone and aggregated only at superabundant resources or at sleeping sites. If modern humans retain traces of such a social organization, one should expect no coordination of vigilance within their groups and increases in vigilance with party size, particularly when reproductive females encounter non-father, adult males. Fossil hominins displaying such a system would presumably show extreme sexual size dimorphism (Rodman & Mitani, 1987) and evidence of high-quality diets.

Speculation about the behavior and social organization of ancient hominins is often dissatisfying because we will never be confident about the details. However, hominin anti-predator behavior demands further scrutiny. Enough data have accumulated to refine our hypotheses. We propose that the adaptive solution to the higher predation pressure of the end Miocene and Pliocene was a social adaptation that preceded any elaboration of material culture.

Acknowledgments G. Laden, L. Naughton, B. Richmond, and M. Wilson gave helpful comments. A. Treves' work on chimpanzees was made possible thanks to the Kibale Chimpanzee Project, R. Wrangham and the Ugandan Research Council. P. Palmqvist's work was financed by the Spanish Ministry of Science and Technology.

References

Anyonge, W. (1993). Body mass in large extant and extinct carnivores. *Jour. of Zool. Lond.*, 231: 339–350.

Anyonge, W. (1996). Locomotor behaviour in Plio-Pleistocene sabre-tooth cats: A biomechanical analysis. *Jour. of Zool. Lond.*, 238: 395–413.

Arribas, A., and Palmqvist, P. (1998). Taphonomy and palaeoecology of an assemblage of large mammals: hyaenid activity in the lower Pleistocene site at Venta Micena (Orce, Guadix-Baza Basin, Granada, Spain). *Geobios*, 31 (suppl.): 3–47.

Arribas, A., and Palmqvist, P. (1999). On the ecological connection between sabre-tooths and hominids: Faunal dispersal events in the lower Pleistocene and a review of the evidence for the first hominin arrival in Europe. *Jour. of Arch. Sci.*, 26: 571–585.

Baeninger, R., Estes, R., and Baldwin, S. (1977). Anti-predator behaviour of baboons and impalas toward a cheetah. *East African Wildlife Journal*, 15: 327–329.

Bailey, T.N. (1993). *The African leopard: Ecology and behavior of a solitary felid*. New York: Columbia Univ. Press.

Barry, J.C. (1987). Large carnivores (Canidae, Hyaenidae, Felidae) from Laetoli. In M.D. Leakey and J.M. Harris (Eds.), *Laetoli: A Pliocene site in Tanzania* (pp. 235–258). London: Clarendon Press.

Bednekoff, P.A., and Lima, S.L. (1998a). Randomness, chaos and confusion in the study of anti-predator vigilance. *Tree*, 13: 284–287.

Bednekoff, P.A., and Lima, S.L. (1998b). Re-examining safety in numbers: Interactions between risk dilution and collective detection depend upon predator targeting behaviour. *Proc. of the Roy. Soc. Lond. B*, 265: 2021–2026.

Bellomo, R.V. (1994). Methods of determining early hominin behavioral activities associated with the controlled use of fire at FxJj 20 Main, Koobi Fora Kenya. *Jour. of Human Evol.*, 27: 173–195.

Berger, L.R., and Tobias, P.V. (1994). New discoveries at the early hominid site of Gladysvale, South Africa. *South African Jour. of Sci.*, 90: 223–226.

Berta, A. (1981). The Plio-Pleistocene hyaena *Chasmaporthetes ossifragus* from Florida. *Jour. of Vertebr. Paleontol.*, 1: 341–356.

Blumenschine, R.J. (1995). Percussion marks, tooth marks, and experimental determinations of the timing of hominin and carnivore access to long bones at FLK Zinjanthropus, Olduvai Gorge, Tanzania. *Jour. of Human Evol.*, 29: 21–51.

Boaz, N.T., Gaziry, A.W., and El-Aurnati, A. (1979). New fossil finds from the Libyan Upper Neogene site of Sahabi. *Nature*, 280: 137–140.

Bocherens, H., Emslie, S., Billiou, D., and Mariotti, A. (1995). Stable isotopes (13C, 15N) and paleodiet of the giant short-faced bear (*Arctodus simus*). *Comptes Rendus de l'Academie des Sciences, Paris*, t. 320(série II a): 779–784.

Boesch, C. (1991). The effects of leopard predation on grouping patterns in forest chimpanzees. *Behaviour*, 117: 220–242.

Boesch, C., and Boesch, H. (1981). Sex differences in the use of natural hammers by wild chimpanzees: A preliminary report. *Jour. of Human Evol.*, 10: 585–593.

Boggess, J. (1980). Intermale relations and troop membership changes in langurs (*Presbytis entellus*) in Nepal. *Inter. Jour. of Primatol.*, 1: 233–263.

Boinski, S., Treves, A., and Chapman, C.A. (2000). A critical evaluation of the influence of predators on primates: Effects on group movement. In S. Boinski and P.A. Garber (Eds.), *On the move: How and why animals travel in groups* (pp. 43–72). Chicago: Univ. of Chicago Press.

Borries, C. (1993). Ecology of female social relationships: Hanuman langurs (*Presbytis entellus*) and the van Schaik model. *Folia Primatol.*, 61: 21–30.

Borries, C. (2000). Male dispersal and mating season influxes in Hanuman langurs living in multi-male groups. In P.M. Kappeler (Ed.), *Primate males.* (pp. 146–158). Cambridge: Cambridge Univ. Press.

Borries, C, Sommer, V., and Srivastava, A. (1994). Weaving a tight social net: Allogrooming in free-ranging female langurs (*Presbytis entellus*). *Inter. Jour. of Primatol.*, 15: 421–444.

Brain, C.K. (1981). *The hunters or the hunted? An introduction to African cave taphonomy.* Chicago: Univ. of Chicago Press.

Brain, C.K. (1994). The Swartkrans palaeontological research project in perspective: Results and conclusions. *South African Jour. of Sci.*, 91: 220–223.

Brantingham, P.J. (1998). Hominid–carnivore coevolution and invasion of the predatory guild. *Jour. of Anthropol. Arch.*, 17: 327–353.

Brunet, M., Beauvilain, A., Geraads, D., Guy, F., Kasser, M., Mackaye, H.T., Maclatchy, L.M., Mouchelin, G., Sudre, J., and Vignaud, P. (1997). Chad: A new Pliocene hominid site. *Comptes Rendus de l'Academie des Sciences Serie II–A: Sciences de la Terre et des Planetes*, 324: 341–345.

Bunn, H.T., and Ezzo, J.A. (1993). Hunting and scavenging by Plio-Pleistocene hominids: Nutritional constraints, archaeological patterns, and behavioural implications. *Jour. of Arch. Sci.*, 20: 365–398.

Busse, C. (1980). Leopard and lion predation upon chacma baboons living in the Moremi Wildlife Reserve. *Botswana Notes & Records*, 12: 15–20.

Byers, J.A. (1997). *American pronghorn: Social adaptations and the ghosts of predators past.* Chicago: Univ. of Chicago Press.

Capaldo, S.D. (1997). Experimental determinations of carcass processing by Plio-Pleistocene hominins and carnivores at FLK 22 (Zinjanthropus), Olduvai Gorge, Tanzania. *Jour. of Human Evol.*, 33: 555–597.

Caro, T.M. (1987). Cheetah mothers' vigilance: Looking out for prey or for predators? *Behav. Ecol. and Sociobiol.*, 20: 351–361.

Caro, T.M. (1989a). The brotherhood of cheetahs. *Natural History*, 6: 50–56.

Caro, T.M. (1989b). Determinants of asociality in felids. In V. Standen and R.A. Foley (Eds.), *Comparative socioecology: The behavioural ecology of humans and other mammals* (pp. 41–74). Oxford: Blackwell Scientific.

Chapman, C.A. (1986). Boa constrictor predation and group response in white-faced Cebus monkeys. *Biotropica*, 18: 171–172.

Chapman, C.A, Wrangham, R.W., and Chapman, L.J. (1995). Ecological constraints on group size: An analysis of spider monkey and chimpanzee subgroups. *Behav. Ecol. and Sociobiol.*, 36: 59–70.

Chapman, C.A, White, F.J., and Wrangham, R.W. (1994). Party size in chimpanzees and bonobos. In R.W. Wrangham, W.C. McGrew, F.B.M. de Waal, and P. Heltne (Eds.), *Chimpanzee cultures* (pp. 41–58). Cambridge: Harvard Univ. Press.

Chellam, R., and Johnsingh, A.J.T. (1993). Management of Asiatic lions in the Gir Forest, India. *Symp. of the Zool. Soc. of Lond.*, 65: 409–424.

Cheney, D.L., and Seyfarth, R.M. (1990). *How monkeys see the world.* Chicago: Univ. of Chicago Press.

Cooke, H.B.S. (1991). *Dinofelis barlowi* (Mammalia, Carnivore, Felidae) cranial material from Bolt's Farm collected by the University of California African expedition. *Paleontologia Africana*, 28: 9–22.

Corbett, J. (1954). The man-eating leopard of Rudrapayang. London: Oxford Univ. Press.

Cowlishaw, G. (1994). Vulnerability to predation in baboon populations. *Behaviour*, 131: 293–304.

Cowlishaw, G., Lawes, M.J., Lightbody, M., Martin, A., Pettifor, R., and Rowcliffe, J.M. (2004). A simple rule for the costs of vigilance: Empirical evidence from a social forager. *Proc. of Roy. Soc. Lond. B*, 271: 27–33.

Creel, S.R., and Creel, N.M. (1995). Communal hunting and pack size in African wild dogs, *Lycaon pictus. Animal Behaviour*, 50: 1325–1339.

de Ruiter, D.J., and Berger, L.R. (2000). Leopards as taphonomic agents in dolomitic caves: Implications for bone accumulations in the hominid-bearing deposits of South Africa. *Jour. of Arch. Sci.*, 27: 665–684.

Delson, E, Terranova, C.J, Jungers, W.L, Sargis, E.J, Jablonski, N.G., and Dechow, P.C. (2000). Body mass in Cercopithecidae (Primates, Mammalia): Estimation and scaling in extinct and extant taxa. *Anthropological Papers of the American Museum of Natural History*, 83: 1–159.

Dominguez-Rodrigo, M., and Pickering, T.R. (2003). Early hominid hunting and scavenging: A zooarcheological review. *Evol. Anthropol.*, 12: 275–282.

Doran, D. (1997). Influence of seasonality on activity patterns, feeding behavior, ranging and grouping patterns in Tao chimpanzees. *Inter. Jour. of Primatol.*, 18: 183–206.

Doran, D., and McNeilage, A. (1998). Gorilla ecology and behavior. *Evol. Anthropol.*, 6: 120–131.

Edmunds, M. (1974). *Defence in Animals: A Survey of Anti-Predator Defences.* London: Longman.

Elgar, M.A. (1989). Predator vigilance and group size in mammals and birds: A critical review of the empirical evidence. *Biol. Review*, 64: 13–33.

Fanshawe, J.H., and Fitzgibbon, C.D. (1993). Factors influencing the hunting success of an African wild dog pack. *Animal Behaviour*, 45: 479–490.

Fay, J.M., Carroll, R., Peterhans, J.C.K., and Harris, D. (1995). Leopard attack on and consumption of gorillas in the Central African Republic. *Jour. of Human Evol.*, 29: 93–99.

Ferretti, M.P. (1999). Tooth enamel structure in the hyaenid *Chasmaporthetes lunensis lunensis* from the Late Pliocene of Italy, with implications for feeding behavior. *Jour. of Vert. Paleontol.*, 19: 767–770.

FitzGibbon, C.D. (1989). A cost to individuals with reduced vigilance in groups of Thomson's gazelles hunted by cheetahs. *Animal Behaviour*, 37: 508–510.

FitzGibbon, C.D. (1990a). Mixed-species grouping in Thomson's gazelles: The anti-predator benefits. *Animal Behaviour*, 39: 1116–1126.

FitzGibbon, C.D. (1990b). Why do hunting cheetahs prefer male gazelles? *Animal Behaviour*, 40: 837–845.

Fitzgibbon, C.D., and Lazarus, J. (1995). Anti-predator behavior of Serengeti ungulates: Individual differences and population consequences. In A.R.E. Sinclair, P. Arcese (Eds.), *Serengeti II: Dynamics, management and conservation of an ecosystem* (pp. 274–296). Chicago: Univ. of Chicago Press.

Foley, R. (1987). *Another unique species*. Oxford: Oxford Univ. Press.

Foster, W.A., and Treherne, J.E. (1981). Evidence for the dilution effect in the selfish herd from fish predation on a marine insect. *Nature*, 293: 466–467.

Fuller, T.K., and Kat, P.W. (1993). Hunting success of African wild dogs in southwestern Kenya. *Jour. of Mammalogy*, 74: 464–467.

Galton, F. (1871). Gregariousness in cattle and men. *MacMillan's Magazine*, 23: 353–357.

Gautier-Hion, A., and Tutin, C.E.G. (1988). Simultaneous attack by adult males of a polyspecific troop of monkeys against a crowned hawk eagle. *Folia Primatol.*, 51: 149–151.

Geraads, D. (1997). Pliocene Carnivora from Ahl al Oughlam (Casablanca). *Geobios*, 30: 127–164.

Ghiglieri, M.P. (1989). Hominoid sociobiology and hominin social evolution. In P. Heltne and L. Marquardt (Eds.), *Understanding chimpanzees* (pp. 370–379). Cambridge, MA: Harvard Univ. Press.

Goodall, J. (1986). *The chimpanzees of Gombe: Patterns of behaviour*. Cambridge, MA: Harvard Univ. Press.

Goodman, S.M., O'Connor, S., and Langrand, O. (1993). A review of predation on lemurs: Implications for the evolution of social behavior in small, nocturnal primates. In P.M. Kappeler and J.U. Ganzhorn (Eds.), *Lemur social systems and their ecological basis* (pp. 51–66). New York: Plenum Press.

Gould, L., Fedigan, L.M., and Rose, L.M. (1997). Why be vigilant? The case of the alpha animal. *Inter. Jour. of Primatol.*, 18: 401–414.

Haile-Selassie, Y. (2001). Late Miocene hominins from the Middle Awash, Ethiopia. *Nature*, 412: 178–181.

Hendey, Q.B. (1974). The late Cenozoic Carnivora of the south-western Cape Province. *Annals of the South African Museum*, 63: 1–363.

Hendey, Q.B. (1980). Agriotherium (Mammalia, Carnivora, Ursidae) from LangeBannweg, South Africa, and relationships of the genus. *Annals of the South African Museum*, 81: 1–109.

Hens, S.M., Konigsberg, L.W., and Jungers, W.L. (2000). Estimating stature in fossil hominins: Which regression model and reference sample to use? *Journal of Human Evolution*, 38: 767–784.

Hill, K.R., and Hurtado, A.M. (1995). *Aché life history: The ecology and demography of a foraging people.* New York: Aldine-DeGruyter.

Hill, R.A., and Cowlishaw, G. (2002). Foraging female baboons exhibit similar patterns of anti-predator vigilance across two populations. In L. Miller (Ed.), *Eat or be eaten: Predator sensitive foraging in nonhuman primates* (pp. 187–204). Cambridge: Cambridge Univ. Press.

Hiraiwa-Hasegawa, M., Byrne, R.W., Takasake, H., and Byrne, J.M.E. (1986). Aggression toward large carnivores by wild chimpanzees of Mahale Mountains National Park, Tanzania. *Folia Primatol.*, 47: 8–13.

Holekamp, K.E., Smale, L., Berg, R., and Cooper, S.M. (1997). Hunting rates and hunting success in the spotted hyena (*Crocuat crocuta*). *Jour. of Zool. Lond.*, 242: 1–15.

Hoppe-Dominik, B. (1984). Etude du spectre des proies de la panther Panthera pardus, dans le Parc National de Tao en Cote d'Ivoire. *Mammalia*, 48: 477–490.

Horrocks, J.A., and Hunte, W. (1986). Sentinel behaviour in vervet monkeys: Who sees whom first? *Animal Behaviour*, 34: 1566–1567.

Hunt, R.M. (1996). Biogeography of the Order Carnivora. In J.L. Gittleman (Ed.), *Carnivore behavior, ecology and evolution*, Vol. 2. (pp. 485–541). Ithaca, NY: Cornell Univ. Press.

Jones, M. (1998). The function of vigilance in sympatric marsupial carnivores: The eastern quoll and the Tasmanian devil. *Animal Behaviour*, 56: 1279–1284.

Keyser, A.W. (1991). The palaeontology of Haasgat: A preliminary account. *Palaeontologia Africana*, 28: 29–33.

Klump, G.M., and Shalter, M.D. (1984). Acoustic behavior of birds and mammals in the predator context. *Zeitschrift fur Tierpsychologie*, 66: 189–226.

Knight, S.K., and Knight, R.L. (1986). Vigilance patterns of bald eagles feeding in groups. *The Auk*, 103: 263–272.

Koenig, A. (1994). Random scan, sentinels or sentinel system? A study in captive common marmosets (*Callithrix jacchus*). In J.J. Roeder, J.R. Anderson, N. Herrenschmidt (Eds.), *Current primatology, Vol. 11 Social development, learning and behaviour* (pp. 69–76). Strasbourg: University Louis Pasteur.

Koenig, A., Beise, J., Chalise, M.K., and Ganzhorn, J.U. (1998). When females should contest for food-testing hypotheses about resource density, distribution, size, and quality with Hanuman langurs (*Presbytis entellus*). *Behav. Ecol. and Sociobiol.*, 42: 225–237.

Kortlandt, A. (1980). How might early hominins have defended themselves against large predators and food competitors? *Jour. of Human Evol.*, 9: 79–112.

Kortlandt, A. (1989). The use of stone tools by wild-living chimpanzees. In P. Heltne and L. Marquardt (Eds.), *Understanding chimpanzees* (pp. 146–147). Cambridge, MA: Harvard Univ. Press.

Kruuk, H. (1972). *The spotted hyena.* Chicago: Univ. of Chicago Press.

Leakey, L.N., Milledge, S.A.H., Leakey, S.M., Edung, J., Haynes, P., Kiptoo, D.K., and McGeorge, A. (1999). Diet of striped hyaena in northern Kenya. *African Journal of Ecology*, 37: 314–326.

Leakey, M.G., Feibel, C.S., McDougall, I., Ward, C., and Walker, A. (1998). New specimens and confirmation of an early age for *Australopithecus anamensis*. *Nature*, 393: 62–66.

Lee-Thorp, J.A., Thackeray, J.F., and van der Merwe, N. (2000). The hunters and the hunted revisited. *Jour. of Human Evol.*, 39: 565–576.

Lee-Thorp, J.A., van der Merwe, N.J., and Brain, C.K. (1994). Diet of *Australopithecus robustus* at Swartkrans from stable carbon isotopic analysis. *Jour. of Human Evol.*, 27: 361–372.

Lewis, M.E. (1997). Carnivoran paleoguilds of Africa: Implications for hominin food procurement strategies. *Jour. of Human Evol.*, 32: 257–288.

Lima, S.L. (1990). The influence of models on the interpretation of vigilance. In M. Bekoff and D. Jamieson (Eds.) *Interpretation and explanation in the study of animal behavior* (pp. 246–267). Boulder, CO: Westview Press.

Lima, S.L. (1993). Ecological and evolutionary perspectives on escape from predatory attack: A survey of North American birds. *The Wilson Bulletin*, 105: 1–47.

Lima, S.L. (1995). Back to the basics of anti-predatory vigilance: The group-size effect. *Animal Behaviour,* 49: 11–20.

Linnell, J.D.C., Odden, J., Smith, M.E., Aanes, R., and Swenson, J.E. (1999). Large carnivores that kill livestock: Do problem individuals really exist? *Wildlife Society Bulletin*, 27: 698–705.

Linnell, J.D.C., Solberg, E.J., Brainerd, S., Liberg, O., Sand, H., Wabakken, P., and Kojola, I. (2003). Is the fear of wolves justified? A Fenniscandian perspective. *Acta Zoologica Lituanica*, 13: 1–160.

Lupo, K.D. (1998). Experimentally derived extraction rates for marrow: Implications for body part exploitation strategies of Plio-Pleistocene hominin scavengers. *Jour. of Arch. Sci.*, 25: 657–675.

Marean, C.W. (1989). Sabertooth cats and their relevance to early hominin diet and evolution. *Jour. of Human Evol.*, 18: 559–582.

Marean, C.W., and Ehrhardt, C.L. (1995). Paleoanthropological and paleoecological implications of the taphonomy of a sabertooth den. *Jour. of Human Evol.*, 29: 515–547.

Martin, L.D. (1989). Fossil history of the terrestrial Carnivora. In J.L. Gittleman (Ed.), *Carnivore behavior, ecology and evolution* (pp. 536–568). Ithaca: Comstock.

Martinez-Navarro, B., and Palmqvist, P. (1995). Presence of the African *machairodont Megantereon whitei* (Broom, 1937) (Felida, Carnivora, Mammalia) in the Lower Pleistocene site of Venta Micena (Orce, Granada, Spain), with some considerations of the origin, evolution and dispersal if the genus. *Jour. of Arch. Sci.*, 22: 569–582.

Martinez-Navarro, B., and Palmqvist, P. (1996). Presence of the African saber-toothed felid *Megantereon whitei* (Broom, 1937) (Mammalia, Carnivora, Machairodontinae) in Apollonia–1 (Mygdonia Basin, Macedonia, Greece). *Jour. of Arch. Sci.*, 23: 869–872.

Martínez-Navarro, B., and Rook, L. (2003). Gradual evolution in the African hunting dog lineage. Systematic implications. *Comptes Rendus Académie des Sciences Palevol.*, 2: 695–702.

Mathers, K., and Henneberg, M. (1996). Were we ever that big? Gradual increase in hominin body size over time. *Homo*, 46: 141–173.

Matheus, P.E. (1995). Diet and co-ecology of Pleistocene short-faced bears and brown bears in eastern Beringia. *Quaternary Research*, 44: 447–453.

McDougal, C. (1987). The man-eating tiger in geographical and historical perspective. In R.L. Tilson and U.S. Seal (Eds.), *Tigers of the world* (pp. 435–448). Park City, NJ: Noyes.

McHenry, H.M. (1992). Body size and proportions in early hominins. *Amer. Jour. of Phys. Anthropol.*, 87: 407–431.

McHenry, H.M. (1994). Behavioral ecological implications of early hominin body size. *Jour. of Human Evol.*, 27: 77–87.

Miller, W.E., and Carranza, O.C. (1996). *Agriotherium schneideri* from the hemphillian of Central Mexico. *Jour. of Mammalogy*, 77: 568–577.

Mills, M.G.L. (1989). The comparative behavioral ecology of hyenas: The importance of diet and food dispersion. In J.L. Gittleman (Ed.), *Carnivore behavior, ecology and evolution* (pp. 125–142). Ithaca: Comstock Assocs.

Monahan, C.M. (1998). The Hadza carcass transport debate revisited and its archaeological implications. *Jour. of Arch. Sci.*, 25: 405–424.

Packer, C., Scheel, D., and Pusey, A.E. (1990). Why lions form groups: Food is not enough. *The American Naturalist*, 136: 1–19.

Palmqvist, P. (2002). On the presence of *Megantereon whitei* at the south Turkwel hominin site, northern Kenya. *Jour. of Paleontol.*, 76: 923–930.

Palmqvist, P., and Arribas, A. (2001). Taphonomic decoding of the paleobiological information locked in a lower Pleistocene assemblage of large mammals. *Paleobiology*, 27: 512–530.

Palmqvist, P, Martinez-Navarro, B., and Arribas, A. (1996). Prey selection by terrestrial carnivores in a lower Pleistocene paleocommunity. *Paleobiology*, 22: 514–534.

Palmqvist, P., Arribas, A., and Martínez-Navarro, B. (1999). Ecomorphological study of large canids from the lower Pleistocene of southeastern Spain. *Lethaia*, 32: 75–88.

Palmqvist, P., Martínez-Navarro, B., Toro, I., Espigares, M.P., Ros-Montoya, S., Torregrosa V., and Pérez-Claros, J.A. (2005). A re-evaluation of the evidence of human presence during Early Pleistocene times in southeastern Spain. *L'Anthropologie*, 109: 411–450.

Palmqvist, P., Grocke, D.R., Arribas, A., and Farina, R.A. (2003). Paleoecological reconstruction of a lower Pleistocene large mammal community using biogeochemical (δ^{13}C, δ^{15}N, δ^{18}O, Sr: Zn) and ecomorphological approaches. *Paleobiology*, 29: 205–229.

Peetz, A.A., Norconk, M.A., and Kinzey, W.G. (1992). Predation by jaguar on howler monkeys (*Alouatta seniculus*) in Venezuela. *Amer. Jour. of Primatol.*, 28: 223–228.

Peres, A. (1990). A harpy eagle successfully captures an adult male red howler monkey. *The Wilson Bulletin*, 102: 560–561.

Peterhans, J.C.K., and Gnoske, T.P. (2001). The science of 'man-eating' among lions *Panthera leo* with a reconstruction of the natural history of the 'Maneaters of Tsavo.' *Jour. of East African Natural History* 90: 1–40.

Petter, G., Pickford, M., and Senut, B. (1994). Presence of the genus Agriotherium (Mammalia, Carnivora, Ursidae) in the late Miocene of the Nkondo Formation (Uganda, East Africa). *Paleontology*, 319: 713–717.

Rajpurohit, K.S. (1998). Child lifting wolves in Hazaribagh, India. *Ambio*, 28: 163–166.

Rajpurohit, R.S., and Krausman, P.R. (2000). Human–sloth–bear conflicts in Madhya Pradesh, India. *Wildlife Society Bulletin*, 28: 393–399.

Rasa, O.A.E. (1986). Coordinated vigilance in dwarf mongoose family groups: 'The watchman's song' hypothesis and the costs of guarding. *Ethology*, 71: 340–344.

Rasa, O.A.E. (1989). The cost and effectiveness of vigilance behaviour in the dwarf mongoose: Implications for fitness and optimal group size. *Ethology, Ecology and Evolution*, 1: 265–282.

Rodman, P.S., and Mitani, J.C. (1987). Orangutans: Sexual dimorphism in a solitary species. In B.B. Smuts, D.L. Cheney, R.M. Seyfarth, R.W. Wrangham, and T.T. Struhsaker (Eds.), *Primate societies* (pp. 146–154). Chicago: Univ. of Chicago Press.

Rook, L. (1994). The Plio-Pleistocene Old World canis (Xenocyon) ex gr falconeri. *Bollettino della Societa Paleontologica Italiana*, 33: 71–82.

Sanyal, P. (1987). Managing the man-eaters in the Sundarbans Tiger Reserve of India—A case study. In R.L. Tilson and U.S. Seal (Eds.), *Tigers of the world* (pp. 427–434). Park City, NJ: Noyes.

Savage, A., Snowdon, C.T., Giraldo, L.H., and Soto, L.H. (1996). Parental care patterns and vigilance in wild cotton-top tamarins (*Sanguinas oedipus*). In M. Norconk, A. Rosenberger, P. Garber (Eds.), *Adaptive radiation of Neotropical primates* (pp. 187–199) New York: Plenum Press.

Schaller, G.B. (1972). *The Serengeti Lion: A study of predator–prey relations.* Chicago: Univ. of Chicago Press.

Seidensticker, J. (1983). Predation by Panthera cats and measures of human influence in habitats of South Asian monkeys. *Inter. Jour. of Primatol.*, 4: 323–326.

Selvaggio, M.M. (1998). Evidence for a three-stage sequence of hominin and carnivore involvement with long bones at FLK Zinjanthropus, Olduvai Gorge, Tanzania. *Jour. of Arch. Sci.*, 25: 191–202.

Setiawan, E., Knott, C.D., and Budhi, S. (1996). Preliminary assessment of vigilance and predator avoidance behavior of orangutans in Gunung Palung National Park, West Kalimantan, Indonesia. *Tropical Biodiversity*, 3: 269–279.

Sillen, A., and Lee-Thorp, J.A. (1994). Trace element and isotopic aspects of predator–prey relationships in terrestrial foodwebs. *Palaeogeography, Palaeoclimatology, Palaeoecology*, 107: 243–255.

Simons, J.W. (1966). The presence of leopard and a study of the food debris in the leopard lairs of the Mount Suswa Caves, Kenya. *Bulletin of the Cave Exploration Group of East Africa*, 1: 51–61.

Smith, D.W., Peterson, R.O., and Houston, D.B. (2003). Yellowstone after wolves. *Bioscience*, 53: 330–340.

Smuts, B.B. (1985). *Sex and friendship in baboons.* Hawthorne, Aldine.

Spoonheimer, M., and Lee-Thorp, J.A. (1999). Isotopic evidence for the diet of an early hominin, *Australopithecus africanus. Science*, 283: 368–370.

Sugardjito, J., te Boekhorst, I.J.A., and van Hooff, J.A.R.A.M. (1987). Ecological constraints on the grouping of wild orang-utans (*Pongo pygmaeus*) in the Gunung Leuser National Park, Sumatra, Indonesia. *Inter. Jour. of Primatol.*, 8: 17–41.

Sunquist, M.E., and Sunquist, F.C. (1989). Ecological constraints on predation by large felids. In J.L. Gittleman (Ed.), *Carnivore behavior, ecology and evolution* (pp. 283–301). Ithaca: Comstock Assocs.

Taylor, M.E. (1989). Locomotor adaptations of carnivores. In J.L. Gittleman (Ed.), *Carnivore behavior, ecology and evolution* (pp. 382–409). Ithaca: Comstock Assocs.

Treves, A. (1997). Vigilance and use of micro-habitat in solitary rainforest mammals. *Mammalia*, 61: 511–525.

Treves, A. (1999a). Has predation shaped the social systems of arboreal primates? *Inter. Jour. of Primatol.*, 20: 35–53.

Treves, A. (2000). Theory and method in studies of vigilance and aggregation. *Animal Behaviour*, 60: 711–722.

Treves, A. (2002). Predicting predation risk for foraging, arboreal monkeys. In L. Miller (Ed.), *Eat or be eaten: Predator sensitive foraging in nonhuman primates* (pp. 222–241). Cambridge: Cambridge Univ. Press.

Treves, A., and Chapman, C.A. (1996). Conspecific threat, predation avoidance and resource defense: Implications for grouping in langurs. *Behav. Ecol. and Sociobiol.*, 39: 43–53.

Treves, A., Drescher, A., and Ingrisano, N. (2001). Vigilance and aggregation in black howler monkeys (*Alouatta pigra*). *Behav. Ecol. and Sociobiol.*, 50: 90–95.

Treves, A., Drescher, A., and Snowdon, C.T. (2003). Maternal watchfulness in black howler monkeys (*Alouatta pigra*). *Ethology*, 109: 135–146.

Treves, A., and Naughton-Treves, L. (1999). Risk and opportunity for humans coexisting with large carnivores. *Jour. of Human Evol.*, 36: 275–282.

Treves, A., and Pizzagalli, D. (2002). Vigilance and perception of social stimuli: Views from ethology, and social neuroscience. In M. Bekoff, C. Allen, G. Burghardt (Eds.), *The cognitive animal: Empirical and theoretical perspectives on animal cognition* (pp. 463–469). Cambridge, MA: MIT Press.

Tsukahara, T. (1993). Lions eat chimpanzees: The first evidence of predation by lions on wild chimpanzees. *Amer. Jour. of Primatol.* 29: 1–11.

Turnbull-Kemp, P. (1967). *The leopard.* Cape Town: Howard Timmins.

Turner, A. (1990). The evolution of the guild of larger terrestrial carnivores during the Plio-Pleistocene in Africa. *Geobios*, 23: 349–368.

Turner, A. (1997). Further remains of Carnivora (Mammalia) from the Sterkfontein hominin site. *Palaeontologica Africana*, 34: 115–126.

Turner, A., and Anton, M. (1996). The giant hyaena, *Pachycrocuta brevirostris* (Mammalia, Carnivora, Hyaenidae). *Geobios*, 29: 455–468.

Turner, A., and Anton, M. (1997). *The big cats and their fossil relatives.* New York: Columbia Univ. Press.

Turner, A., and Anton, M. (1998). Climate and evolution: Implications of some extinction patterns in African and European machairodontine cats of the Plio-Pleistocene. *Estudios Geologicos (Madrid)*, 54: 209–230.

Tutin, C.E.G., McGrew, W.C., and Baldwin, P.J. (1981). Responses of wild chimpanzees to potential predators. In A.B. Chiarelli and R.S. Corruccini (Eds.), *Primate behavior and sociobiology* (pp. 136–141). New York: Springer-Verlag.

van Schaik, C.P., and Griffiths, M. (1996). Activity patterns of Indonesian rain forest mammals. *Biotropica*, 28: 105–112.

Watts, D.P. (1998). A preliminary study of selective visual attention in female mountain gorillas (*Gorilla gorilla beringei*). *Primates*, 39: 71–78.

Werdelin, L., Turner, A., Solounias, N. (1994). Studies of fossil hyaenids: The genera Hyaenictis Gaudry and Chasmaporthetes Hay, with a reconsideration of the Hyaenidae of Langebaanwegm South Africa. *Zoological Journal of the Linnean Society* 11: 197–217.

Werdelin, L., and Turner, A. (1996). Turnover in the guild of larger carnivores in Eurasia across the Miocene–Pliocene boundary. *Acta Zoologica Cracovensia* 39: 585–592.

Wickler, W. (1985). Coordination of vigilance in bird groups. The 'watchman's song' hypothesis. *Zeitschrift fur Tierpsychologie*, 69: 250–253.

Wirtz, P., and Wawra, M. (1986). Vigilance and group size in Homo sapiens. *Ethology*, 71: 283–286.

WoldeGabriel, G., White, T.D., Suwa, G., Renne, P., de Heinzelin, J., Hart, W.K., and Helken, G. (1994). Ecological and temporal placement of early Pliocene hominins at Aramis, Ethiopia. *Nature*, 371: 330–333.

Wright, P.C. (1998). Impact of predation risk on the behaviour of Propithecus diadema edwardsi in the rain forest of Madagascar. *Behaviour*, 135: 483–512.

Ydenberg, R.C., and Dill, L.M. (1986). The economics of fleeing from predators. *Advances in the Study of Behavior*, 16: 229–251.

Zuberbühler, K. (2000). Interspecies semantic communication in two forest primates. *Proc. of Roy. Soc. Lond. B*, 267: 713–718.

Index

A

Acacia drepanolobium, 319, 324–326
Acacia tortilis, see Umbrella trees
Acacia xanthophloea, see Fever trees
Accipiter henstii, see Henst's goshawk
Accipiter madagascariensis, 86, 113
Acinonyx jubatus, see Cheetah
Acinonyx spp., 323–324, 357–358,
    360, 363
    Acrantophis madagascariensis, 156, 161
African civet *(Civetticus civetta),* 226
African crowned eagle *(Stephanoaetus
    coronatus),* 6, 8, 32
    effects of predation, 14
African hunting dogs *(Lycaon pictus),*
    51, 357
African large carnivores, 356
    after predator encounter, 367–368
    anti-predator behavior and hominin
        reconstruction, 365–366
    avoidance of dangerous areas, 365–366
    competition between carnivores and
        hominins, 364–365
    habitat selection of, 357, 359–360
    hunting tactics of, 360–361
    prey selection of, 360–362
    social organization reconstructed for
        Hominins, 369–370
    trade offs between anti-predator
        aggregation and vigilance, 367–369
African rainforests, 19
Agriotherium, 357, 359–361, 363–364
Alarm calls system, 105

of lemur species, 106–107
Alosa sapidissima, see American shad
Alouatta spp., 49
    Alouatta caraya, see Black and gold
        howler monkey
    Alouattamyia baeri, see Cuterebrid
    Alouatta palliata, see Mantled howler
        monkey
    Alouatta pigra, see Black howling
        monkeys
Amboseli National Park, 32, 45, 321
American shad *(Alosa sapidissima),* 301
Ampijoroa Forestry Station, 64–65
Animal nocturnal behavior to moonlight,
    study of
    activity level with moonlight level,
        183–186
    anti-predator behavior of primates,
        160–163, 179
    data analysis, 183–185
    influence of moonlight on highest levels
        in trees, 185–188
    interspecific comparisons, 192–193
    lunar neutrality, 174–179
    lunar philia, 174–179, 194
    lunar phobia, 174–179, 194
    moon effects on predators, 193
    moon effects on prey, 193–196
    observation procedures and behavior
        variables, 182–183
    predation rate and risk influence in
        primates sociality, 193
    study site and subjects, 181–182

vocalization and moonlight levels,
187–190
Animal vocalization as indication to
predation risk, study of
of *Galago moholi*, 212–218
of *Otolemur crassicaudatus*, 208–212,
215–218
study subjects and methods, 207–208
Ankarafantsika National Park, 64–65,
101, 110
Anti-predator behavior of vervet and patas,
study of
diurnal anti-predator behavior, 320–327
habitat structure effects and, 316–318,
320, 324
predation risk vs predation rate, 312–316
sleeping site choice of, 319–320
study site and subjects, 317–318
Anti-predator strategies, types of
aerial anti-predator strategy, 106–107
panic cry anti-predator strategy, 107
semantic predator recognition strategy,
107
terrestrial carnivore anti-predator
strategy, 106
terrestrial snake anti-predator strategy,
106
Aotus spp., 42, 149, 152–153, 159,
180–181, 194–195, 215
Aotus azarai, see Owl monkeys
Apostatic selection, 146–147, 159,
162–163
Applied Biosystems PROCISE 494HT
Protein Sequenator, 256
Aquila verreauxi, see Verreaux's eagles
Arboreal-diurnal primates, 33, 36, 39
Arboreal monkeys, 13
Arctocebus aureus, see Sympatric
perodicticine
Arctocebus calabarensis, 180
Arctodus, see Short-faced bear
Asian black bears *(Selenarctos thibetanus)*,
38
Asian Lorises, study of strategies by,
222–224, 234
chemical communication of pottos,
229–230
chemical communication of slender
lorises, 228–229

freezing and fleeing behavior of slender
lorises, 226
study sites and subjects, 224–225
vocalizations and display of pottos, 226
vocalizations and display of slender
lorises, 225–226
Asio madagascariensis, see Madagascar
long-eared owls
Ateles, 50
Attack, probability of, 341
Australopithecus, 136, 362
Australopithecus afarensis, 136
Avahi laniger, see Eastern woolly lemurs
Avahi occidentalis, 73, 136
Axis axis, 51
Axis porcinus, see Hog dear
Aye-aye *(Daubentonia madagascariensis)*,
37, 107

B
Bald eagle *(Haliaeetus leucocephalus)*,
293, 297–298
Bamboo lemur *(Hapalemur griseus)*, 81,
89, 92, 113–114, 138, 148, 161
Bangarus caeruleus, 226
Barking deer *(Muntiacus muntjak)*, 51
Barn owls, 193
Bdeogale nigripes, 180, 228
Beza Mahafaly Special Reserve, 113, 181,
197, 275–276, 278
Biomultiview, 256
Black and gold howler monkeys (Alouatta
caraya), 162
Black-and-white ruffed lemurs *(Varecia
variegata variegata)*, 116–117, 157
Black howling monkeys *(Alouatta pigra)*,
150–152, 157, 163
Black lemurs *(Eulemur macaco)*, 113, 156,
162
Black-mantled tamarins *(Saguinus
nigricollis)*, 159
Blue jays *(Cyanocitta cristata)*, 162
Blue monkeys *(Cercopithecus mitis)*, 29,
51
Boa constrictor *(Boa manditra)*, 77, 156,
161, 276
Boa manditra, see Boa constrictor
Bonnet macaques *(Macaca radiata)*,
241–242, 320

Brachial gland exudate, study of use of
    methodology, 255–256
    result analysis of, 257–268
    volatile and semi-volatile compounds,
        257–268
Brachyteles, 50
Brown mouse lemur *(Microcebus rufus),*
    84, 93, 110–111
Bubo africanus, see Eagle owls
Bubo viginianus, see Great-horned owl
Budongo forest, 13
Buteo brachypterus, 78, 113, 138, 154,
    180, 276
Buteo jamaicensis, 114, 118
Buteo lineatus, 116–117

C
Callithrix flaviceps, 159, 161
Campbell's monkeys, 17
Canids, 28–30, 38, 47, 49, 361
Canis aureus, see Golden jackal
Canis familaris, 115
Canis latrans, see Coyote
Canis lupus, see Wolves
Canis mesomelas, 208
Caprimulgus vociferous, see
    Whippoorwills
Cathemeral and diurnal lemurs, 113–118
Cebus spp., 29, 50, 153
    Cebus albifrons, 161
Cephalophus spp., 11
Cercocebus atys, see Sooty mangabeys
Cercopithecus aethiops, 32, 210, 308, *see
    also* Anti-predator behavior of vervet
    and patas, study of
Cercopithecus campbelli, 4
Cercopithecus diana, see Diana monkeys
Cercopithecus mitis, see Blue monkeys
Cercopithecus neglectus, see DeBrazza
    monkeys
Cercopithecus nictitans, 4
Cercopithecus petaurista, 4
Cervus duvanceli, see Swamp deer
Cervus unicolor, see Sambar
Chacma baboons, predation risk and habitat
    use of
    dynamics of predation risk, in terrestrial
        environments, 339

habitat-specific predation risk at
    De Hoop, 342–343
    individual capture probability, 341–342
    methodology, 343
    predation and habitat use in, 342–343
    predation risk and habitat activity
        patterns, 348–351
    predation risk and habitat choice,
        345–348
    probability of attack, 341
    probability of encounter, 340–341
    probability of prey capture, 341
Chacma baboons *(Papio hamadryasl),* 319,
    339–352
Chasmaporthetes, 357–361, 364
Cheetah *(Acinonyx jubatus),* 49, 323–324,
    359, 363, 368
Cheirogaleidae, 37
Cheirogaleus major, 93
Cheirogaleus medius, 73, 107, 110–111
Chimpanzees *(Pan troglodytes),* 6, 316,
    366
    pant hoots, 16–17
    sonso, 13
    of Taï forest, 6–7
    wild, 8
Chitawan park, 37
Chordeiles minor, see Nighthawks
Ciccaba woodfordii, see Wood owl
Civetticus civetta, see African civet
CLUSTAL-X, 256
Cognitive functions, in primates, 4, 17–18,
    21
Colobus guereza, 320
Colobus polykomos, 4
Common poorwill *(Phalaenoptilus
    nuttallii),* 193
Coquerel's dwarf lemur *(Mirza coquereli),*
    105, 107–111, 161, 174
Coracopsis vasa, see Vasa parrot
Coturnix coturnix japonica, 162
Coyote *(Canis latrans),* 50
Crematogaster spp., 326
Crocodylus palustris, 29
Crocuta spp., 51, 357, 360–361, 363–364
    Crocuta crocuta, 51
Crowned hawk-eagles, 32, 45, 50
Crowned lemurs *(Eulemur coronatus),* 161

Cryptic behavior, of primates, 18, 158,
      160–161
Cryptoprocta fossa, 180
Cryptoprocta spp., 35, 63–65, 70, 73–74,
      77, 110, 115, 138–139, 141 *see
      also Cryptoprocta fossa*, study of
      predatory behavior of
      Cryptoprocta ferox, 152, 276
Crytoprocta ferox, study of predatory
      behavior of
   as predator on primates in
      Ankarafantsika, 72–74
   scat samples and analysis, 65–66
   statistical analyses of, 66–67
   study site, 64–65
   temporal and geographic variation in
      body mass of prey items, 67–70
Cuon alpinus, 38
Cuterebrid *(Alouattamyia baeri)*, 291
Cyanocitta cristata, *see* Blue jays

D

Daubentonia madagascariensis,
      see Aye-aye
DeBrazza monkeys *(Cercopithecus
      neglectus)*, 5
De Hoop Nature Reserve, 342–343
Dendroapsis polylepis, 208
Diana monkeys *(Cercopithecus diana)*,
      4–6, 14, 17–19, 32
   cryptic behavior of, 18
   putty-nosed monkey association, 6, 19
Dicrurus forficatus, *see* Drongos
Dinofelis, 357–361, 363
Diurnal prosimian, study of anti-predator
      strategies of
   canopy level, 283
   group size and predator strategies, 278,
      281, 283
   infant vulnerability and vigilance
      behavior, 283–285
   innate anti-predator strategy, 285
   methodology, 276–277
   predator sensitive foraging behavior,
      278–279
   sex and rank differences in anti-predator
      behavior, 284–285
   types of anti-predator behaviors,
      276–278

vigilance behavior, 279–280, 283
Drongos *(Dicrurus forficatus)*, 156
Dwarf lemurs *(Microcebus coquereli)*, 29

E

Eagle owls *(Bubo africanus)*, 208, 214
Eastern gray squirrel *(Sciurus carolinensis)*,
      118
Eastern woolly lemurs *(Avahi laniger)*, 84,
      92, 111
Eliurus myoxinus, *see* Non-predatory
      rodent
Encounter, probability of, 340–342
Eravikulam National Park, 51
Erythrocebus patas, 308, 317, *see also*
      Anti-predator behavior of vervet and
      patas, study of
Estimated predation rate (EPR), 42–43, 51
Eulemur spp., 36–37, 73, 84, 89–92, 104
      130, 147–148, 152–153, 159–160,
      162–163
   Eulemur coronatus, *see* Crowned lemurs
   Eulemur fulvus, *see* Sanford's brown
      lemurs
   Eulemur fulvus rufus, 29, 81, 89,
      113–115, 155, 161, 283
   Eulemur macaco, *see* Black lemurs
   Eulemur macaco macaco, 156
   Eulemur mongoz, *see* Mongoose lemur
   Eulemur rubriventer, *see* Red-bellied
      lemur
   inter-species variability in cathemeral
      activity among, 147
   taxa of, 147–149
Eutriorchis astur, *see* Madagascar serpent
      eagle

F

Falco spp., 242
Felids, 28–30, 32, 36, 40, 46–49, 180, 226,
      357, 359–360, 363
Felis concolor, *see* Puma
Felis pardalis, *see* Ocelot
Felis tigrina, *see* Oncilla
Felis viverrinas, *see* Fishing cat
Felis wiedii, *see* Margay
Felis yagouroundi, *see* Jaguarundi
Fever trees *(Acacia xanthophloea)*, 319,
      321

Ficus trees, 243
Fishing cat *(Felis viverrinas)*, 225
Flexibility, notion of, 21
Forest falcon *(Micrastur semitorquatus)*, 41
Fork-marked lemur *(Phaner furcifer)*, 107, 110, 174, 223
Fossa fossana, see Malagasy civet
Fruit bat *(Pteropus rufus)*, 155, 157

G
Galago alleni, 128
Galagoides demidovii, 216
Galagoides rondoensis, 216
Galagoides thomasi, 216
Galagoides zanzibaricus, 173–174, 195
Galago moholi, 127–128, 130–139, 174, 180, 193, 195, 206–207, 211–218, 223
Galago senegalensis, 129, 207
Galidia elegans, 36, 78, 138, 154
Galidictis spp., 36
Gallus, 66
Genetta servilina, 226
Genetta tigrina, 139, 208
Giant hyena *(Pachycrocuta)*, 357–361, 364
Gliding effects, 130
Golden brown mouse lemur *(Microcebus ravelobensis)*, 107, 119
Golden cat *(Profelis aurata)*, 228
Golden jackal *(Canis aureus)*, 38
Golden palm civet *(Paradoxurus zeylonensis)*, 225
Gorilla, 29, 44, 368, 372
Gorilla gorilla, 366
Gray mouse lemur *(Microcebus murinus)*, 107, 110–112, 117, 119
Great-horned owl *(Bubo viginianus)*, 116–117, 152–153
Guiana crested eagle *(Morphnus guianensis)*, 49–50, 153
Guinea baboons *(Papio papio)*, 51
Guinea fowls, 18
Guttera pulcheri, see Guinea fowls

H
Habitat structure and anti-predator behavior, *see* Anti-predator behavior of vervet and patas, study of

Haliaeetus leucocephalus, see Bald eagle
Hanuman langurs *(Presbytis entellus)*, 37–38
Hapalemur spp., 36, 84, 86, 90–93, 104, 138, 161, 197
    Hapalemur griseus, see Bamboo lemur
    Hapalemur griseus alaotrensis, 148, 151
    Hapalemur griseus griseus, 81, 89, 113–114
    Hapalemur griseus occidentalis, 161
    Hapalemur simus, 84
Harpia harpyja, see Harpy eagle
Harpy eagle *(Harpia harpyja)*, 41, 153, 161, 289–290, 294, 297–304
Hawk-eagles *(Spizaetus cirrhatus)*, 254
Hemitragus hylocrius, see Nilgiri tahr
Henst's goshawks *(Accipiter henstii)*, 37, 49, 77, 111, 113, 155, 180
Herpestid, 28
Hewlett-Packard 5890 Gas Chromatograph-5970 MSD, 255
Hog dear *(Axis porcinus)*, 51
Hominin anti-predatory behavior, study of
    small cooperative groups, 372
    small groups with male protector, 372
    solitary foragers, 372–373
Homo sapiens, 366
Homotherium, 357–361, 363–364
Hyaena hyaena, 357
Hyaenids, 28–30, 38, 47, 50, 357, 361, 363
Hylocicla mustelina, see Wood thrush

I
Individual capture probability, 340–342
Indri spp., 36, 101, 119, 130, 133, 138–139
    Indri indri, 105

J
Jaguar *(Panthera onca)*, 40, 159, 359, 366
Jaguarundi *(Felis yagouroundi)*, 40
Japanese macaques *(Macaca fuscata)*, 38

K
Kanha Park, 37
Kanneliya Forest Reserve, 224
Kibale forest range, 14, 33, 51
Kirindy forest range, 80, 110, 155
Kiwengoma Forest Reserve, 33

L

L. l. lydekkerianus, 222–223, 231
L. l. nordicus, 224, 231
L. tardigradus tardigradus, see
    Southwestern Ceylon red
    slender loris
Lagothrix, 50
Large-eared bushbaby *(Otolemur*
    crassicaudatus), 135–139, 206–212,
    215–218
Leaping behavior, kinetics and
    kinematics of
    in *Avahi occidentalis,* 136
    in *Cryptoprocta,* 138
    and efficiency of transport, 130
    in *Galago moholi,* 127, 134–135
    and high take-off angles, 134–135
    leaping mechanics, 144–145
    leaping mechanisms, 128–129
    leaping style, 128–129
    in *Lepilemur ruficaudatus,* 137
    and musculoskeletal load, 130–131
    in *Otolemur crassicaudatus,* 134,
        136–137, 139
    as specialization, 129–131
    in *Tarsius bancanus,* 131, 134–140
    and transport speed, 131–134
Leioheterondon madagascariensis, 276
Lemur antipredator strategies,
        investigation of, *see also Lemur*
        catta, anti-predator strategies of
    of cathemeral and diurnal lemurs,
        113–118
    methods, 101
    of nocturnal lemurs, 107–112
    predation and relation to variation in life
        history and ecology, 101–105
    strategies and alarm call systems,
        105–107
    terrestrial carnivore anti-predator
        behaviors, 119
Lemur catta, 37, 113, 117–118, 130, 149,
        155, 163, 275, *see also Lemur catta,*
        anti-predator strategies of
Lemur catta, anti-predator strategies of
    canopy level, 281–283
    group size and predator strategies, 283
    infant vulnerability and vigilance
        behavior, 283–284

    innate anti-predator strategy, 285
    methodology, 278–279
    predator sensitive foraging behavior,
        279–280
    sex and rank differences in anti-predator
        behavior, 284–285
    types of antipredator behaviors, 276–278
    vigilance behavior, 279–283
Lemur predators, based on Goodman, 102
Lemur specialist, 64, 73–74
Leontopithecus rosalia, 159
Leopards *(Panthera pardus),* 6, 8–12,
        14–15, 17–18, 20, 32, 37, 40, 45, 51,
        77, 228, 289, 313, 321, 341, 344,
        355, 359–368
    daytime activities of, 9
    effect of predation, 14–17
    prey spectrum of, 11
    of savannah, 9, 12
    as scavengers, 12
Lepilemur edwardsi, 73, 107, 136
Lepilemur leucopus, 173–174, 179, 181
    see also Animal nocturnal behavior
        to moonlight, study of
    moonlight effects on activity budgets,
        191–194
    self-grooming behavior of, 185, 189, 191
    vocalizations of, during moonlight,
        174–179
Lepilemur mustelinus, see Weasel sportive
    lemur
Lepilemur mustelinus leucopus, 136
Lepilemur ruficaudatus, see Red-tailed
    sportive lemurm
Lepilemur seallii, 92
Life-dinner principle, 146
Lophotibis cristata, 86
Loris spp., 39–40, 128, 131, 180, 222–235,
        253, 257, 263, 267, 269–271, *see also*
        Asian Lorises, study of strategies by
    Loris tardigradus, 180
Lycaon spp., 359–361, 363
    Lycaon lycaonoides, 357, 364
    Lycaon pictus, see African hunting dogs

M

M. semitorquatus, 41
Macaca spp., 40
    Macaca fascicularis, 51

Macaca fuscata, see Japanese macaques
Macaca mulatta, see Rhesus macaques
Macaca radiata, see Bonnet macaques
Machairodontine *(Machairodus),* 357, 359–361, 363–364
Machairodont *(Megantereon),* 357–361, 364
Machairodus, see Machairodontine
Machiavellian intelligence hypothesis, 21
Madagascar boa *(Sanzinia madagascariensis),* 77105, 178
Madagascar buzzard, 76
Madagascar harrier hawk *(Polyboroides radiatus),* 36, 86, 108–109, 116, 136, 152, 178, 214
Madagascar long-eared owls *(Asio madagascariensis),* 36, 48, 136, 178
Madagascarophis colobrinus, 108
Madagascar serpent eagle *(Eutriorchis astur),* 79, 81, 107, 111, 113, 155
Madagascar tree boa *(Sanzinia madagascariensis),* 107, 113, 180
Mahale mountains, of Tanzania, 32
Malagasy brown-tailed mongoose *(Salanoia concolor),* 36
Malagasy civet *(Fossa fossana),* 36
Malagasy owls, 37
Malagasy predator diets, 47
Malagasy prosimians, 36, 43
Mangabeys, 4, 8, 14, 20, 33
Mantled howler monkey *(Alouatta palliata),* 137, 149, 157–159, 161, 163, 291
Margay *(Felis wiedii),* 40, 161
Megantereon, see Machairodont
Meru-Betiri Reserve, 37, 51
Meta-analysis, 28–29, 31, 40, 49
Micrastur, 40
Micrastur semitorquatus, see Forest falcon
Microcebus spp., 29, 43, 73–74, 104, 112, 127, 138, 179, 197
    Microcebus coquereli, see Dwarf lemurs
    Microcebus murinus, see Gray mouse lemur
    Microcebus ravelobensis, see Golden brown mouse lemur
    Microcebus rufus, see Brown mouse lemur
Mid-flight rotation, 130

Milne-Edward's sifakas *(Propithecus diadema edwardsi),* 36, 81, 84–86, 89, 91, 93, 114, 157, 283
Mimicry, 223, 233
Miopithecus talapoin, see Talapoin monkey
Mirza coquereli, see Coquerel's dwarf lemur
Mixed-species associations, 5–7, 281, 327
Mobbing behavior, of primates, 110, 113, 116, 119, 246
Mongoose lemur *(Eulemur mongoz),* 148, 152
Morphnus guianensis, see Guiana crested eagle
Moustached tamarins *(Saguinus mystax),* 159, 161, 281
Multi-male groups, as adaptation to predation, 3
Mungotictis decemlineata, see Narrow-striped mongoose
Muntiacus muntjak, see Barking deer
Mustelid families, 28, 159

N
Naja naja, 208
Naja nigricillis, 208
Nandinia binotata, see Palm civet
Narrow-striped mongoose *(Mungotictis decemlineata),* 36
Nasalis, 39
Neotropical raptor species, 40, 47, 49
Ngogo study site, in Kibale, 33
Nighthawks *(Chordeiles minor),* 193
Nilgiri langurs, 37, 51
Nilgiri tahr *(Hemitragus hylocrius),* 51
Niokolo-Koba National Park, 50
Nocturnal lemurs, 74, 100–101, 104, 107–113, 138, 153, 180–181, 275–276
Non-predatory rodent *(Eliurus myoxinus),* 111–112
Nyctereutes procyonoides, 38
Nycticebus spp., 39–40, 128, 194, 222, 233, 253, 269
    Nycticebus bengalensis, see Slow lorise
    Nycticebus coucang, see Slow lorise
    Nycticebus coucang coucang, 255
    Nycticebus pygmaeus, see Pygmy lorise

O

Ocelot *(Felis pardalis)*, 40
Olfactory behavior, of primates, 224–225, 229, 233
Olive colobus monkeys *(Procolobus verus)*, 4, 20
Oncilla *(Felis tigrina)*, 40
Otolemur spp., 134, 136–137, 139
    Otolemur crassicaudatu, see Large-eared bushbaby
    Otolemur garnetti, see Small-eared bushbaby
Otus rutilus, 37
Owl monkeys *(Aotus azarai), see* 42, 94, 148–149, 152–153

P

Pachycrocuta, see Giant hyena
Palm civet *(Nandinia binotata)*, 179, 226, 228
Panthera leo, 32, 324, 355, 361
Panthera onca, see Jaguar
Panthera pardus, see Leopards
Panthera tigris, see Tigers
Pan troglodytes, see Chimpanzees
Papio hamadryas, see Chacma baboons
Papio papio, see Guinea baboons
Paradoxurus zeylonensis, see Golden palm civet
Paranthropus, 362, 364, 370
Parapapio jonesi, 362
Perca fluviatilis, 162
Periyar Tiger Reserve, 37, 51
Perkin-Elmer Sciex API-III instrument, 255
Peromyscus leucopus, see White-footed mice
Phalaenoptilus nuttallii, see Common poorwill
Phaner furcifer, see Fork-marked lemur
Phenomenex Jupiter C-18 reverse phase column, 255–256
Philippine eagle *(Pithecophaga jeffery)*, 39
Pithecophaga jeffery, see Philippine eagle
Pleistocene climate, 19
Poaching activity, by humans, 6–7
    effects of, 13
Poiana leightoni, see West African linsang
Polemaetus bellicosus, 208

Polyboroides nests, 88,
Polyboroides radiatus, see Madagascar harrier hawk
Polyspecific associations, among primates, 5–6, 19–20, 50, 241, 338, 308–311, 328
Pongo, 39
Predation and primitive cognitive evolution, 3–22
    general assumptions in adaptation to predation, 3
Predation on lemurs, observations and experiments of
    activity cycle and risk of predation, 93–95
    body mass and risk of predation, 92
    experimental data analysis, 83–84
    experiments, 81–83
    intra-species response to experiments, 89–90
    lemur behavioral responses to multiple-predator community, 91
    lemur social aggregations and risk of predation, 91–92
    observations at raptor nest sites, 81
    observations of predation on lemurs from reported kills and scat, 84–86
    predation rates and anti-predator tactics, 90–91
    raptor nest site observations, 86–88
    review of reported kills, 80–81
    risk of predation and birth synchrony of lemurs, 93–94
    species rarity and hibernation, 93
    study site, 79–80
Predation on primates, biogeographical analysis of
    in Africa, 31, 32–34
    in Asia, 31, 37–40
    estimated predation rates, 42–44
    frequency of occurence in primates in predator diets, 44–49
    in Madagascar, 31, 34–37
    methodology, 28–29
    in Neotropics, 31, 40–42
Predation rate, 14, 27 , 42–43, 50, 86, 91–93, 127, 173, 196, 312–316, 329, 339

Predation risk, 27, 83, 100–101, 104–107,
    113–114, 118–119, 146, 158–159,
    162, 173–174, 179, 193, 195, 197,
    206–218, 206, 232, 275, 278–279,
    312–316
    dynamics of, in terrestrial environments,
        339–340
    and habitat activity patterns, 348–351
    and habitat choice, 342, 344–346, 351
Predator-assessment call, 158, 295–296,
    303–304
Predator fauna, 3, 4, 7
Predator-prey interactions, study of
    direction of scanning, 292, 298, 300
    display of other behavioral activities,
        300–301, 328
    experimental play back, 296–298
    methods, 290–294
    predator perspective, 294–295
    prey perspective, 295–296, 303
    vigilance rate, 298–300, 341
Predator-prey relationships, 28
Predator sensitive foraging, 278–280,
    283, 285
Presbytis, 37, 39
Presbytis entellus, see Hanuman langurs
Prey capture, probability of, 340–341,
    344, 351
Primate cathemerality and predation
    behavioral crypticity, 147, 152–153, 160
    escape behavior in relation to size,
        158–159
    mobbing behavior of primates, 110,
        113, 161
    predator confusion by polymorphism,
        159–160
    and sexual dichromatism, 147, 159,
        162–163
    social groups and predation, 147,
        152–153–158, 160–161
    in squirrel monkey, 41, 161
    taxonomic distribution of, 147–150
PrimateLit, 101
Procolobus badius, see Red colobus
    monkeys
Procolobus verus, see Olive colobus
    monkeys
Profelis aurata, see Golden cats
Propithecus diadema, 133

Propithecus diadema edwardsi, 36, 114,
    157, see also Milne-Edward's sifakas
Propithecus edwardsi, 81, 84–86, 89–91,
    93, 283
Propithecus pardus, 364
Propithecus potto edwardsi, 223–225
Propithecus ursinus, 319
Propithecus v. verreauxi, see Verreauxi
    sifakas
Prosmians, leaping kinematics and kinetics
    of, see Leaping behavior, kinetics
    and kinematics of
Protana A/S, 255
PROTPARS weighting scheme, 256
Pteropus rufus, see Fruit bat
Puma (Felis concolor), 36, 40
Putty-nosed monkeys, 4, 6, 14, 17–19
Pygmy lorise (Nycticebus pygmaeus), 234,
    253, 255, 257, 264, 269, 271
Python reticulatus, 39, 242
Python sebae, 208

R
Ranomafana National Park
    activity cycle of lemurs and predation
        evidence in, 93–94
    Lemur spp., in, 77–93
    primate prey of Accipiter henstii at, 88
    primate prey of Polyboroides radiatus
        at, 88
    Talatakely Trail System (TTS) in, 80, 85
Ranthambhore forest, 37
Raptors, 28–30, 32, 37, 39, 41–43, 46–47,
    49–50, 77–78, 80, 85–86, 91–93,
    101, 106–107, 110–111, 113–115,
    119, 138–139, 153–154, 156–157,
    159–161, 163, 180–181, 224, 249,
    276, 281, 295, 297
Rattus rattus, 86
Red-bellied lemur (Eulemur rubriventer),
    84, 86, 92, 114, 152, 155, 283
Red colobus monkeys (Procolobus badius),
    4, 13–14, 20, 29, 316
Red-fronted brown lemurs, 29, 114–116,
    281
Red ruffed lemur (Varecia variegata), 84,
    86, 93, 114, 149, 152, 157
Red-tailed sportive lemurm (Lepilemur
    ruficaudatus), 110, 137

Reptiles, 6, 28–30, 46–47, 66, 86, 88,
    101, 327
Rete mirabile, 128
Rhesus macaques, 37, 320
Ring-tailed lemurs, 113, 115–118, 149,
    275–285
Rudd *(Scardinius eryophthalmus)*, 162

S
Saguinus fuscicollis, 159, 161
Saguinus imperator, 161
Saguinus labiatus, 161
Saguinus mystax, see Moustached tamarins;
    Tamarins
Saguinus nigricollis, see Black-mantled
    tamarins
Saimiri spp., 50
    Saimiri sciureus, see Squirrel monkey
Salanoia concolor, see Malagasy
    brown-tailed mongoose
Sambar *(Cervus unicolor)*, 51
Sanford's brown lemurs *(Eulemur fulvus)*,
    153, 161–162
Sanzinia madagascariensis, see
    Madagascar boa; Madagascar tree
    boa
Scardinius eryophthalmus, see Rudd
Scent marking, 224–225, 228–229,
    233–234
Sciurus carolinensis, see Eastern gray
    squirrel
Segera Ranch site, 317–318, 321, 326, 329
Selenarctos thibetanus, see Asian black
    bears
Setifer setosus, 67
Short-faced bear *(Arctodus)*, 361
Short microhabitat, 319, 324, 327
Signaling, 105, 121, 157, 206, 217–218,
    225, 242, 270, 277
Slow lorise *(Nycticebus bengalensis)*,
    38, 216, 232, 251, 253, 255, 257,
    264–265, 269
Small-eared bushbaby *(Otolemur
    garnettii)*, 194
Sniffing, 225, 229, 269
Solid phase matrix extractor (SPME),
    255, 257
Sooty mangabeys, 4, 8, 14,
Sooty mangabeys *(Cercocebus atys)*, 4,

Southwestern Ceylon red slender loris
    (L. tardigradus tardigradus), 224
Spectral Tarsiers, behavior of, towards
    avian and terrestrial predators
    data collection, 243–245
    field site, 243
    result analysis, 245–248
Spizaetus cirrhatus, see Hawk-eagle
Squirrel monkey *(Saimiri sciureus)*, 161
Startle response behavior, 285–286
Stephanoaetus coronatus, see African
    crowned eagle
Stephanoaetus mahery, 275
Swamp deer *(Cervus duvanceli)*, 51
Sympatric perodicticine *(Arctocebus
    aureus)*, 223, 231
Sympatric pottos, 179

T
Taï monkeys, hunting behavior of
    predators on
    adaptations to chimpanzee predation, 8,
        13–14
    adaptations to eagle predation, 14
    adaptations to human predation, 13
    adaptations to leopard predation, 14–17
    chimpanzee hunting, 7–8
    by crowned eagles, 8–9
    habitats of, 4–5
    by human poachers, 6–7
    interaction effects, 20
    by leopards, 9–12
    polyspecific associations of, 5–6, 19–20
    predation and primate cognitive
        evolution, 17–18, 21
Taï National Park, 4
Talapoin monkey, 33
Talatakely study site (TTS), 80
Tamarindus indica, 181
Tamarins *(Saguinus mystax)*, 159, 161
Tangkoko Nature Reserve, 243
Tarsius spp., 40, 129, 135, 137, 195–196
    Tarsius bancanus, see Western tarsier
    Tarsius spectrum, see Wild spectral
        tarsier
Temporal crypticity strategy, 147, 152, 156
Terborgh's model, of evolutionary
    adaptation, 50
Tigers *(Panthera tigris)*, 37, 355

Tinamus major, 293, 297
Trachypithecus spp., 39
    Trachypithecus cristata, 51,
    Trachypithecus johnii, see Nilgiri langur
Tsaobis grooming, 342
Tursiops truncatus, 301
Tyto alba, 138, 180
Tyto alba affinis, 37
Tyto soumagnei, 37

U
Umbrella trees *(Acacia tortilis),* 321
UV properties, of primate urine, 234

V
Vanga curvirostris, 29
Varanus indicus, 242
Varanus spp., 208
Varecia spp., 36, 101, 119, 149–150, 157, 159, 163
    Varecia variegata, see Red ruffed lemur
    Varecia variegata variegata, see
        Black-and-white ruffed lemur
Vasa parrot *(Coracopsis vasa),* 81, 85
Vatoharanana study site (VATO), 80
Verreauxi sifakas *(Propithecus v.*
    verreauxi), 37, 73, 113 115, 130,

Verreaux's eagles *(Aquila verreauxi),* 46, 208
Visual systems, of predators and prey, 192–194
Viverra tangalunga, 242
Viverricula indica, 36, 67, 225 274
Viverricula indica majori, 225
Vivverid, 28
Vulpes vulpes, 38

W
Weasel sportive lemur *(Lepilemur mustelinus),* 111
West African linsang *(Poiana leightoni),* 226
Western tarsier *(Tarsius bancanus),* 131, 134–140
Whippoorwills *(Caprimulgus vociferous),* 193
White-footed mice *(Peromyscus leucopus),* 195
Wild spectral tarsier *(Tarsius spectrum),* 139, 174, 223
Wolves *(Canis lupus),* 38, 303, 363
Wood owl *(Ciccaba woodfordii),* 208
Wood thrush *(Hylocicla mustelina),* 118

Printed in the United States
66488LVS00001B/136-162

9 780387 348070